COMPUTER, COMMUNICATION AND ELECTRICAL TECHNOLOGY

PROCEEDINGS OF THE INTERNATIONAL CONFERENCE ON ADVANCEMENT OF COMPUTER COMMUNICATION AND ELECTRICAL TECHNOLOGY (ACCET-2016), WEST BENGAL, INDIA, 21–22 OCTOBER 2016

Computer, Communication and Electrical Technology

Editors

Debatosh Guha
University of Calcutta, India

Badal Chakraborty
Bidhan Chandra Krishi Viswavidyalya, India

Himadri Sekhar Dutta
Kalyani Government Engineering College, India

 CRC Press
Taylor & Francis Group
Boca Raton   London   New York   Leiden

CRC Press is an imprint of the
Taylor & Francis Group, an **informa** business

A BALKEMA BOOK

CRC Press/Balkema is an imprint of the Taylor & Francis Group, an informa business

© 2017 Taylor & Francis Group, London, UK

Typeset by V Publishing Solutions Pvt Ltd., Chennai, India
Printed and bound in the UK and the US

Published by: CRC Press/Balkema
        P.O. Box 11320, 2301 EH Leiden, The Netherlands
        e-mail: Pub.NL@taylorandfrancis.com
        www.crcpress.com – www.taylorandfrancis.com

ISBN: 978-1-138-03157-9 (Hbk)
ISBN: 978-1-315-40062-4 (eBook)

Table of contents

Preface   ix

Computer technology

A novel method for edge detection in a gray image based on human psychovisual
phenomenon and Bat algorithm   3
S. Dhar, S. Alam, M. Santra, P. Saha & S. Thakur

Implementation of salient region based secured digital image watermarking   9
A. Basu, R. Karmakar, A. Chatterjee, S. Datta, S. Sarkar & A. Mondal

Handwritten Devanagari numerals recognition using grid based Hausdroff distance   15
S. Bhowmik, S. Sen, N. Hori, R. Sarkar & M. Nasipuri

Image steganography using a histogram of discrete cosine transform   19
B.G. Banik, M.K. Poddar & S.K. Bandyopadhyay

A parallel cell based architecture for real time morphological edge detection   27
P. Roy & B.K. Das

A study on Bangla handwritten compound character recognition using LDP   33
M. Das, S. Halsana, A. Banerjee & A. Sahoo

Handwritten signature recognition and verification using artificial neural network   39
A. Roy, S. Dutta & M. Das

Breast cancer detection using feature selection and active learning   43
S. Begum, S.P. Bera, D. Chakraborty & R. Sarkar

Image contrast enhancement using a crossover-included hybrid artificial bee
colony optimization   49
S.K. Mondal, A. Chatterjee & B. Tudu

Supervised machine learning vs. Lexicon-based text classification for sentiment analysis:
A comparative study   55
S. Gupta & S. Mandal

Network anomaly detection using a fuzzy rule-based classifier   61
S. Ghosh, A. Pal, A. Nag, S. Sadhu & R. Pati

Hybrid VVRKFA-SLFN classifier for improved pattern classification   67
S. Ghorai

Investigations on user perspective evaluation of some reliability aspects of web services   75
S. Medhi, A. Bora & T. Bezboruah

Load balancing of the unbalanced cost matrix in a cloud computing network   81
R.K. Mondal, P. Ray, E. Nandi, P. Sen & D. Sarddar

Deadline-driven task scheduling in clustered heterogeneous distributed systems   85
L. Datta

Survey and analysis of mining requirements in spatio-temporal database   91
D. Dasgupta, S. Roy, S. Singha Roy & A. Chakraborty

A verifiable and cheating-resistant secret sharing scheme   97
A.K. Chattopadhyay, A. Nag & K. Majumder

Communication technology

A new handoff management scheme for reducing call dropping probability with efficient
bandwidth utilization   105
D. Verma, A. Agarwal & P.K. Guha Thakurta

Graph theory based optimum routing path selection method for wireless sensor network   111
B. Sharma, N. Brahma & H. Choudhury

Real time motion detection system using low power ZigBee based wireless sensor network   115
R. Tribedi, A. Sur & S. Bose

Optimal path selection for AODV routing protocol in MANET   121
S. Manna, A.K. Mondal & P. Sarcar

A study on outage-minimizing routing in a multi-hop wireless network   125
V.K. Karwa, S. Ghosh & S. Basak

Analyzing image transmission quality using filter and C-QAM   131
Md. Khaliluzzaman, D. Kumar Chy & K. Deb

Low-cost and secured biometric system-based EVM with instant counting by wireless
transmission of data to a central place   139
A. Ghosal, S. Chakraborty, S. Maity, A. Paul & P. Roy

Performance analysis of the CO-OFDM system in a CR network   143
S. Nandi, M. Sarkar, A. Nandi & N.N. Pathak

High temperature electrical properties of thin Zirconium di-oxide films on ZnO/Si layers   149
S.K. Nandi

Realization of all-optical frequency-encoded dibit-based OR and NOR logic gates
with simulated verification   153
P.P. Sarkar, S. Hazra, B. Ghosh, S.N. Patra & S. Mukhopadhyay

Analysis of optical gain in a Tin-incorporated group IV alloy-based transistor laser
for a mid-infrared application   159
R. Ranjan & M.K. Das

An efficient low-power 1-bit full adder using a multi-threshold voltage scheme   163
R. Kumar, S. Roy & A. Bhattacharyya

Power efficient, high frequency and low noise PLL design for wireless receiver applications   169
B. Supraja, N. Ravi & T.J. Prasad

Oscillating wave propagation inside DNG material based 1D photonic crystal   175
B. Das & A. Deyasi

Comparative analysis of filter performance in DNG material based photonic crystal structure   181
A. Deyasi, S. Ghosh, R. Dutta & V. Shaw

Sn-concentration dependent absorption in strain balanced GeSn/SiGeSn quantum well   185
P. Pareek & M.K. Das

Performance analysis of ZnO/c-Si heterojunction solar cell   189
S.S. Anwer Askari & M.K. Das

Electrical Technology

Line to line fault detection in a multi-bus power system by harmonic analysis   195
D.K. Ray, S. Chattopadhyay, K.D. Sharma & S. Sengupta

Loss minimization of power network by reconfiguration using GA and BPSO 199
S. Pal, J.N. Bera & S. Sengupta

A comparative study of the polarization–depolarization current measurements on different
polymeric materials 203
A. Kumar, N. Haque, R. Ghosh, B. Chatterjee & S. Dalai

DTCWT based approach for power quality disturbance recognition 209
S. Chakraborty, A. Chatterjee & S.K. Goswami

Optimised fractional order PID controller in automatic generation control 215
A. Mohanty, D. Mishra, K. Mohan, P.K. Ray & S.P. Mohanty

Fault detection in an IEEE 14-bus power system with DG penetration using wavelet transform 221
P.K. Ray, B.K. Panigrahi, P.K. Rout, A. Mohanty & H. Dubey

Detection of faults in a power system using wavelet transform and independent
component analysis 227
P.K. Ray, B.K. Panigrahi, P.K. Rout, A. Mohanty & H. Dubey

Effect of ultrasonic pretreatment on the osmotic drying of ash gourd during
Murabba processing 233
D. Mandal, N. Nath & P.K. Sahoo

Analytical model of MEMS-based piezoresistive pressure sensor using Si_3N_4 diaphragm 239
K. Das & H.S. Dutta

Alcohol sensor-based cost-effective, simple car ignition controller 245
P. Deb, A. Seth, S. Bhattacharya, M. Chakraborty & A. Roy

Study of a contactless capacitive-type linear displacement sensor 249
S. Nayak, S. Das & T.S. Sarkar

Study of the liquid-level-sensing method based on a capacitive rotary sensor 255
T.S. Sarkar, A. Hore, S. Das, T. Chowdhury, B. Chakraborty & H.S. Dutta

Comparison of two and four electrode methods for studying the impedance variation
during cucumber storage using Electrical Impedance Spectroscopy (EIS) 261
A. Chowdhury, D. Ghoshal, T.K. Bera, B. Chakraborty & M.L. Naresh Kumar

WHT-based tea quality prediction using electronic tongue signals 267
P. Saha, S. Ghorai, B. Tudu, R. Bandyopadhyay & N. Bhattacharyya

Multiple inhomogeneity phantom imaging with a LabVIEW-based Electrical
Impedance Tomography (LV-EIT) System 273
T.K. Bera, S. Bera, J. Nagaraju & B. Chakraborty

Electrical Impedance Spectroscopy (EIS) based fruit characterization: A technical review 279
T.K. Bera, S. Bera, A. Chowdhury, D. Ghoshal & B. Chakraborty

Congestion constraint corrective rescheduling in the competitive power market with
the integration of a wind farm 289
S. Gope, A.K. Goswami & P.K. Tiwari

Survey on solar photovoltaic system performance using various MPPT techniques
to improve efficiency 295
B. Pakkiraiah & G. Durga Sukumar

An improved strategy of energy conversion and management using PSO algorithm 307
MD.T. Hoque, A.K. Sinha & T. Halder

A study on variability of anthropometric data of various pinna patterns of human ear
for individualization of head related transfer function 313
S. Chaudhuri, D. Dey, S. Bandyopadhyay & A. Das

Preterm birth prediction using electrohysterography with local binary patterns 319
F. Francis, M. Bedeeuzzaman & T. Fathima

Arrhythmia classification by nonlinear kernel-based ECG signal modeling 325
S. Ghorai & D. Ghosh

Characterization of obstructive lung diseases from the respiration signal 331
S. Sarkar, S. Pal, S. Bhattacherjee & P. Bhattacharyya

Studies on a formidable dot and globule related feature extraction technique
for detection of melanoma from dermoscopic images 337
S. Chatterjee, D. Dey & S. Munshi

A method for automatic detection and classification of lobar ischaemic stroke
from brain CT images 343
A. Datta & A. Datta

A new modified adaptive neuro fuzzy inference system-based MPPT controller
for the enhanced performance of an asynchronous motor drive 349
B. Pakkiraiah & G. Durga Sukumar

Study of nonlinear phenomena in a free-running current controlled Ćuk converter 357
P. Chaudhuri & S. Parui

Study of the power fluctuation of the DC motor in chaotic and non-chaotic drive systems 363
M. Roy, P. Roy & S. Bhattacharya

Author index 369

Preface

This volume contains selected papers presented at the International Conference on Advancement of Computer Communication and Electrical Technology (ACCET 2016) held in Berhampore, West Bengal, India 21-22 October 2016. The meeting was organized by Murshidabad College of Engineering & Technology and sponsored by TEQIP-II, a world Bank Project maintained by the National Project Facilitation Unit (N PIU), India and technically sponsored by IET (UK), Kolkata chapter, IEEE young professionals and the International Frequency Sensor Association (IFSA).

113 papers from different countries were submitted and after an initial review of the papers by at least two reviewers 66 papers were presented at ACCET 2016 and selected for publication in the proceedings. The topics covered in the conference include Computer Science Engineering, Electronics & Communication Engineering and Electrical Engineering.

Reviewing papers of the ACCET 2016 was a challenging process that relies very much on the goodwill of scientists working in the field. We invited more than 40 researchers from relevant fields to review papers for presentation at the conference and publication in the Proceedings published by CRC Press. We would like to thank all the reviewers for their time and effort in reviewing the documents.

Finally, we would like to thank all of the members of the ACCET 2016 team who have given their constant support and countless time to prepare the papers for publication. We would like to express our special appreciation to the conference convener Er. Subir Das, for his efforts in helping to organize the conference and this proceeding. The ACCET 2016 proceeding is the collaborative work of a large group of people, and everyone should be proud of the outcome.

Prof. Debotosh Guha
Dr. Badal Chakraborty
Dr. Himadri Sekhar Dutta
Editors.

Computer technology

A novel method for edge detection in a gray image based on human psychovisual phenomenon and Bat algorithm

Soumyadip Dhar, Saif Alam, Mouli Santra, Priyanka Saha & Sukirti Thakur
RCC Institute of Information Technology, Kolkata, West Bengal, India

ABSTRACT: Edge detection is a primary and important task in image processing. In this a paper we propose a novel method for edge detection in a gray image based on Human Psycho Visual (HVS) phenomenon. The method automatically detects the De-Vries Rose, Weber and Saturation regions in an image as a HVS system, based on image statistics using Bat Algorithm (BA). This is followed by adaptive threshoding in each region to detect the edge points. The performance of the proposed method is found to be superior than that of the conventional and stare-of-the-art method for edge detection on standard data set.

1 INTRODUCTION

An Edge in an image is a noteworthy local change in the gray level intensity of an image. Edge detection is a basic tool for feature detection and extraction from an image by identifying the change in the pixel values in an image. But Due to some unavoidable reasons such as distortion, intensity variation and overlapped boundaries, it is usually impossible to extract complete object contours or to segment the whole objects. So detecting the proper edges in an image is a challenging task.

In the literature, several methods were proposed to detect the edges in an image. The popular conventional edge detection includes the edge detection by Canny, Robert, Sobel and Prewitt (Senthilkumaran & Rajesh, 2009). Canny's method finds the edges by looking for local maxima of the gradient, which is calculated using the derivative of a Gaussian filter. Robert and Sobel's technique perform 2D spatial gradient measurement on an image and emphasize the regions of high spatial frequency that correspond to edges. The Prewitt Edge filter is used to detect edges based applying a horizontal and vertical filter in sequence. Both filters are applied to the image and summed to form the final result.

Apart from the conventional gradient based techniques described above, some researchers used different techniques for detecting the edges. Hu and Tian (2006) integrated multi-directions characteristic of structure elements and image fuzzy characteristic to detect the edges using mathematical morphology. Junna and Jiang (2012), and Kaur and Garg (2011) proposed the edge detection technique based on mathematical morphology.

A bilateral filtering technique for edge detection was proposed by Seelamantula and Jose (2013), and wavelet based approach for edge detection was adopted by An (2013).

All the methods described above did not take into consideration the property of the Human Visual System (HVS) (Buchsbaum (1980)) to detect the edges in an image. As a result, the performances of the methods are still below the expectation. Kundu and Pal (1986) detected the edge points in an image by identifying the De Vries-Rose, Weber and saturated regions in HVS. But the proper identifications of the three regions were not fully adaptive and depended on human intervention. This is the major motivation for the current investigation for the better solution.

In this paper, we propose an edge detection technique in a gray image based on HVS and a relatively new evolutionary algorithm, the bat algorithm. The performance of the proposed method is compared with some popular methods of edge detection and a state-of-the-art method and found to be superior to the other methods.

In the proposed method the three regions (De Vries-Rose, Weber, and Saturated) in an image are estimated to adopt the efficiency of a human visual system and locate the edge points correctly. But the exact estimation of the three regions demands the correct determination of the parameters involved with the regions. The parameters depend on the image statistics. The incorrect estimation of the parameters may lead to poor performance. Thus, an efficient evolutionally algorithm is required to find the parameters and the regions correctly and adaptively. Hence in the proposed method the bio-inspired bat algorithm is used efficiently to find them correctly, regulated by the nature of a particular image. Thus, the novelty of the method is that it identifies the HVS regions adaptively depending on the image statistics and detects the edge points automatically.

The paper is organized as follows. Section 2 describes the HVS. The edge detection using HVS is reported in section 3. Section 4 and section 5 describes the Bat algorithm and its evaluation function. The proposed algorithm and the results are discussed in section 6 and section 7 respectively.

2 HUMAN VISUAL SYSTEM (HVS)

In HVS (Buchsbaum 1980) visual information is intense at points of large spatial variation of light intensity in an image. The human eye scans the image and compares intensity of each portion with its immediate background. It now divides the portions of the image into three different regions, De-Vries Rose, Weber, and Saturation regions. The basis of the division in these regions is the change in threshold value with the increase in background intensity. If B is the background intensity and ΔB_T is the change in threshold value, then the characteristics response of the three regions is shown in $\log \Delta B_T - \log B$ plane in the Figure 1.

So, the relations between $\log B$ and $\log \Delta B_T$ in the Weber, De-Vries rose and Saturation regions are given by Eq. (1), Eq. (2) and Eq. (3) respectively.

$$\log \Delta B_T = \log K_1 + \log B \tag{1}$$

$$\log \Delta B_T = \log K_2 + \frac{1}{2} \log B \tag{2}$$

$$\log \Delta B_T = K_3 + 2 \log B \tag{3}$$

Here K_1, K_2, K_3 are the constants of proportionality.

3 EDGE DETECTION USING HVS

The problem of edge detection is to find an appropriate threshold value which may be global, local or dynamic. The value is such that, if the spatial difference at a point exceeds that threshold, then only it will be considered as a valid edge. Here the threshold values for detecting edges at three different regions are given in the Eq. (1), Eq. (2) and Eq. (3). Kundu and Pal (1986) showed that for a particular point in an image having intensity B_p if we have

$$\frac{\Delta B}{\sqrt{B}} \geq K_2 \text{ when } \alpha_2' B_1 \geq B \geq \alpha_1' B_1 \tag{4a}$$

$$\frac{\Delta B}{B} \geq K_1 \text{ when } \alpha_3' B_1 \geq B \geq \alpha_2' B_1 \tag{4b}$$

$$\frac{\Delta B}{B} \geq K_3 \text{ when } B \geq \alpha_3' B_1 \tag{4c}$$

with $\Delta B = |B_p - B|$ then and only then it can be considered as a detectable edge point having edge intensity ΔB. The B is calculated for each pixel by taking the weighted average of its 8 neighborhood pixels. Here B_1 represents the maximum value of B. In the above equations $\alpha_2' B_1 \geq B \geq \alpha_1' B_1, \alpha_3' B_1 \geq B \geq \alpha_2' B_1$ and $B \geq \alpha_3' B_1$ represent the background intensity B in the De-Vries rose, Weber and Saturation region respectively for $0 < \alpha_1' < \alpha_2' < \alpha_3' < 1$. The parameters α_1', α_2' and α_3' are used for region demarcation and required human intervention.

Although the edges are detected using adaptive thresholding at the three different regions, accurate demarcations of the three regions are difficult. This is due to the difficulty in finding the proper combinations of α_1', α_2' and α_3'. So in the proposed method we use Bat algorithm as an evolutionary algorithm. The algorithm finds the optimal combinations of the parameters based on image statistics to identify the three regions and detects the edge points automatically.

4 BAT ALGORITHM (BA)

Yang (2011) developed a swarm intelligence optimization algorithm through simulating the interesting behavior of a bat, which combines major advantages of simulated annealing and particle swarm optimization.

Three generalized rules are used for implementing the bat algorithm:

1. All bats use echolocation to sense distance, and they also 'know' the difference between food/prey and background barriers in some magical way.

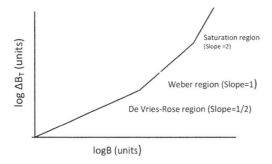

Figure 1. The incremental threshold ΔB_T vs background intensity B.

2. Bats fly randomly with velocity v_i at position x_i with a fixed frequency $f_{\min}$, varying wavelength λ and loudness A_0 to search for prey. They can automatically adjust the wavelength of their emitted pulses and adjust the rate of pulse emission $r \in [01]$ depending on the proximity of the target.

3. Although the loudness can vary in many ways, here it is assumed that the loudness varies from a large (positive) A_0 to a minimum constant value $A_{\min}$.

Main steps of the algorithm are given below:

1. Initialization; Repeat
2. Generation of new solutions;
3. Local searching;
4. Generation of a new solution by flying randomly;
5. Finding the current best solution; until (requirements are met).

In BA algorithm, initialization of the bat population is performed randomly. Generating new solutions is performed by moving virtual bats according to the following equations:

$$f_i = f_{\min} + (f_{\max} - f_{\min})\beta \qquad (5)$$

$$v_i^t = v_i^{t-1} + (x_i^t - x^*)f_i \qquad (6)$$

$$x_i^t = x_i^{t-1} + v_i^t \qquad (7)$$

where $\beta \in [01]$ is a random vector drawn from a uniform distribution. Here x^* is the current global best location (solution) which is located after comparing all the solutions among all the bats. Initially, each bat is randomly assigned a frequency which is drawn uniformly from $[f_{min}\ f_{max}]$. A random walk with direct exploitation is used for the local search that modifies the current best solution according to the equation:

$$x_{new} = x_{old} + \varepsilon A^t \qquad (8)$$

where $\varepsilon \in [01]$ a random number, while A^t is the average loudness of all the bats at this time step.

5   EVALUATION FUNCTION

We need an evaluation function to quantify the desired edge point extraction. The present algorithm considers the evaluation function as the fitness function of BA. The total variance across the edge points and non edge points are taken as the evaluation functions. That means the evaluation function V is given by

$$V = Var(eg) + Var(neg) \qquad (9)$$

where *eg* and *neg* represent the set of edge and non-edge pixels. *Var(.)* is the variance of the pixels. The objective of the BA is to construct the parameters for the three regions depending on image statistics with least variance. So, the reciprocal of evaluation function $E = 1/V$ is used as the fitness function, where BA will try to find out the maximum value of it.

6   PROPOSED METHOD AND IMPLEMENTATION

At first the background intensity B at each pixel position, maximum background intensity B_1 and ΔB is calculated. Now the following steps are executed. The initial bat population $P = 50$ and total number of iterations = 40.

Step 1:   Input image and, initialize the P parameter vectors of $(x_i = (\alpha_1', \alpha_2', \alpha_3')_i)$ and, A_i, v_i, r_i, f_i.

Step 2:   For each x_i Divide the image into three different regions using the ranges in the Eq. (4a), Eq. (4b) and Eq. (4c).

Step 3:   Threshold each pixel for edge point detection by Eq. (4a), Eq. (4b) and Eq. (4c) by checking their background intensity for each region.

Step 4:   Merge the edge pixel for making *eg* and non-edge pixel for making *neg* and calculate E.

Step 5:   Repeat step 6-step 10 until no change in evaluation function or the number of iterations completes.

Step 6:   Generate new solutions by the Eq. (5), Eq. (6) and Eq. (7).

Step 7:   if $rand > r_i$ then random walk around one of best solutions.

Step 8:   Generate a new solution by flying randomly and calculate E.

Step 9:   If $rand < A_i$ and $E(x_i) < E(x_i^*)$ then accept new solutions. Decrease A_i and increase r_i.

Step 10:   Find the current best x^*.

Step 11:   Display the edge points using the parameters in x^*.

7   RESULTS, COMPARISONS AND DISCUSSIONS

The proposed method was applied on 100 benched marked data taken from (http:\\wisdom.weizmann.ac.il) and 25 Non Destructive Testing images (NDT) from (http:\\mehmetsezgin.net). The size of the images varies from 200×200 to 300×300. Images with bimodal and multimodal histogram are taken for consideration. In non destructive

5

testing applications, one important task is to extract the defective regions of the materials. So detecting the proper edges in those images is an important task. Detecting edges in the NDT images are difficult due to the multimodal histogram of the images. Some of the results of edge detection by the proposed methods are shown in the Figure 2. From the visual inspection, it is clear that the proposed method is quite efficient in detecting the edge points. The reason is that the HVS system adaptively thresholded each of the pixels in the De-Vries Rose, Weber and Saturation regions the for edge point detection. Our method demarcated the three regions properly depending on the image statistics. We compared the performance of the proposed method with Canny, Sobel and An's (2013) method. The qualitative results are shown in the Figure 3.

For quantitative performance measure we used the figure of merit of Pratt (IMP) by Abdou and Pratt (1979). The IMP assesses the similarity between two contours. It is defined as

$$IMP = \frac{1}{\max(N_I, N_B)} \sum_1^{N_B} \frac{1}{1 + \nu \times d_i^2} \qquad (10)$$

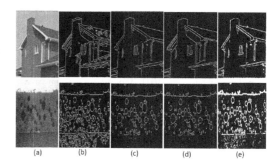

Figure 3. Edge detection by different methods (a) Test image (b) Canny's method (c) Sobel' method (d) An's method (e) Proposed method.

Table 1. Average performance of different methods on 25 NDT images.

Methods	Canny	Sobel	An	Proposed
Average IMP	0.75	0.72	0.80	0.82

where N_I and N_B are the edge points in the test image and its corresponding ground truth. d_i is the istance between an edge pixel and the nearest edge pixel of the ground truth and ν is an empirical calibration constant and we used $\nu = 1/9$, the optimal value established by Abdou and Pratt (1979). The value of IMP varies from [0 1] where 1 represents the optimal truth i.e. the result coincides with ground truth. The Table 1 shows the average IMP of different methods from 25 NDT images.

The table shows that the proposed method on average performs better the other methods compared here. The possible reason is that the three regions in HVS were detected correctly by the evolutionary BA algorithm. The correct estimation of the regions helped to identify the edge points correctly.

8 CONCLUSIONS

In this paper, we have proposed a novel method for edge detection in a gray image. The method used the human psycho visual phenomenon for edge points detection in an image. The De-Vries Rose, Weber and Saturation regions detected by the human visual system was determined here and the edge points were generated automatically from the regions based on the image statistics. For automatic and optimal detection of the three regions bat algorithm was used judiciously. The proposed method performed better both qualitatively and quantitatively than that of the other conventional and state-of-the-art method for edge detection.

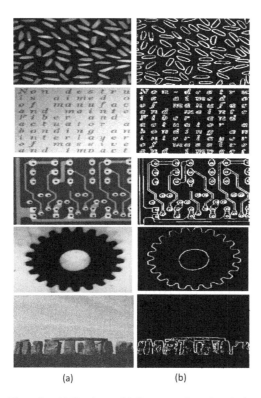

(a)   (b)

Figure 2. (a) Test image (b) Corresponding edges by the proposed method.

Current research is going on for the detection of the edge points from the color images.

REFERENCES

Abdou, I.A., & Pratt, W. (1979). Quantitative design and evaluation of enhancement/thresholding edge detectors. In proceedings of IEEE. 67, 753–766. doi: 10.1109/PROC.1979.11325.

An, V.N. (2013). Edge detection using wavelets. VNU Journal of Natural Sciences and Technology. 29, 1–9. Retrieved from http://js.vnu.edu.vn/index.php/NST/article/viewFile/1241/1207.

Buchsbaum, G. (1980). An analytical derivation of visual nonlinearity. IEEE transactions on Biomedical Engineering. 27, 237–242. doi: 10.1109/TBME.1980.326628.

Hu, D., & Tian, Z. (2006, November). A Multi-directions Algorithm for Edge Detection Based on Fuzzy Mathematical Morphology, Proceeding of the International Conference on Artificial Reality and Telexistence Workshops (ICAT'06'), 361–364, Hangzhou, China. doi:10.1109/ICAT.2006.15.

Junna, S., & F, Jiang. (2012, October). An Algorithm of Edge Detection Based on Soft Morphology. In proceeding of the International Conference on Signal Processing (ICSP), 66–169, Beijing, China. doi:10.1109/ICoSP.2012. 6491626.

Kaur, B., & Garg, A. (2011, April). Mathematical Morphological Edge Detection For Remote Sensing Images. In proceeding of the International Conference on Electronics Computer Technology (ICECT). 324–327. Kanyakumari, India. doi:10.1109/ICECTECH.2011.5942012.

Kundu, M. K., & Pal, S.K. (1986). Thresholding for edge detection using human psychovisual phenomenon. Pattern Recognition Letter. 4, 433–441. doi: 10.1016/0167-8655(86)90041-3.

Seelamantula, C. S., & Jose, A. (2013, May). Bilateral Edge Detectors. In proceeding of the International Conference on Acoustics, Speech and Signal Processing. 1449–1453. Vancouver, Canada. doi: 10.1109/ICASSP.2013.6637891.

Senthilkumaran, N., & Rajesh, R. (2009). Edge Detection Techniques for Image Segmentation and A Survey of Soft Computing Approaches", International Journal of Recent Trends in Engineering, 1, 250–255. Retrieved from http://academypublisher.com/ijrte/vol01/no02/ijrte0102250254.pdf.

Yang, X.S. (2011). A new metaheuristic bat-inspired algorithm. Nature Inspired Cooperative Strategies for Optimization. (NICSO). 284, 65–74. doi:10.1007/978-3-642-12538-6_6.

Computer, Communication and Electrical Technology – Guha, Chakraborty & Dutta (Eds)
© 2017 Taylor & Francis Group, ISBN 978-1-138-03157-9

Implementation of salient region based secured digital image watermarking

Abhishek Basu, Ranit Karmakar, Ankita Chatterjee, Soumita Datta & Subhodeep Sarkar
Department of Electronics and Communication Engineering, RCC Institute of Information Technology, Beliaghata, Kolkata, India

Ankur Mondal
Department of Computer Science and Engineering, GNIT, Sodepur, Kolkata, India

ABSTRACT: In this paper, a secured technique of digital watermarking by embedding a biometric watermark on a digital media efficiently has been proposed. Using biometrics, the digital media is solely characterized by an individual as his intellectual property. Using this technique, the authors can enhance the security of digital signals without the distortion of digital media. The biometric data, here iris, is embedded into the cover image pixels bit wise. Covariance saliency technique is used to identify the salient pixels of the watermarked image which is to replace the significant bits of the cover image. Various attacks and experimental results shows the improved imperceptibility and robustness of this method with less error percentage.

1 INTRODUCTION

In the modern era of advanced technologies, the sharing of digital data has been widely popular with the help of high speed internet. Due to this there has been immense growth on online multimedia businesses. As a consequence, there has been an essential need of digital copyright for security and authenticity (Bender W., 1996, Sarkar S., 2010, Basu A., 2013). However due to the increase in the internet revenues, customers and the company became likely more concerned about the security and the authenticity of this system. Henceforth, for the authentication and security of the digital media, information hiding techniques plays a vital role in probing a solution to the challenges faced. This technique is nothing but hiding information (Samanta D. 2008) in an encrypted way within a digital audio, images, and video files. The most authentic and secure technique of data hiding is digital watermark which completely characterizes any digital media as one's intellectual property. Digital watermarking (Peticolas F.A.P., 1999) can be said to be as the process by which a data, called a watermark, (Singh N., 2013) is to be embedded into another digital media so that the watermark could be extracted afterwards to complete the authentication process. In this paper we would discuss the secure way of hiding a biometric data into a digital media such as image by means of digital watermarking. This focuses on the covariance saliency technique. Biometric has been introduced for authenticity and ultimate security of a digital media. Biometric information here iris is being hidden in a cover image in an efficient way such that the quality of the crucial information is not corrupted (Hong & Chen, 2012). Several impairments were made to prove the robustness and resiliency of this technique. Experimental results validate the improved robustness of the technique with lesser perpetual error. Rest of the paper organized as Embedding and extraction, Result and discussion and conclusion.

2 EMBEDDING AND EXTRACTION

This technique is used to check the saliency regions of the image and the watermark is embedded on the darker regions which is less perciptable area of the cover image. Properties like robustness and imperceptibility are hard to perceive at the same instant. A Capacity Map is formed by saliency mapping which is been generated using covariance saliency technique. Within the non-salient regions of the cover image, the watermark pixels are to be hidden using the adaptive LSB (Least-Significant Bit) replacement method. The binary bits of the watermark image is hidden various number of times which achieves high capacity (Basu A., 2013, D. Soumyendu, 2008) and allows high perceptual transparency. Due to the higher sensitivity of the LSB technique various experiments can disrupt the embedded biometric watermark. But the proposed method had efficiently solved each problem.

2.1 Encoder

The biometric watermark is inserted into the cover image using this encoding algorithm. The watermarked image obtained after applying the watermark on the cover image cannot be distinguished from the original one.

The steps of the encoding process are the following:

Step 1: The watermarked image W(x) of size (IxJ) has been taken, which is subjected to edge detection and then converted to binary bits, which can be defined as,

$$W(x) = w(i, j) \mid 0 \le i \le I, 0 \le j \le J, w(i, j) \in [0,1] \tag{1}$$

Step 2: The cover Image C(X) of size (MxN) acts as the carrier of biometric information carrier. It is converted to grayscale and resized as per requirement can be defined as,

$$C(x) = \{c(m, n) \mid 0 \le m \le M, 0 \le n \le N, c(m, n) \in [0,1,2,3, \ldots, 255] \tag{2}$$

Step 3: Saliency mapping of the cover Image is performed using covariance saliency technique which is given by,

$$Sal(x) = \{sal(m, n) \mid 0 \le m \le M, 0 \le n \le N, sal(m, n) \in \mathbf{R} \wedge 0 \le sal(m, n) \le 1 \tag{3}$$

Step 4: The saliency values ranges from 0 to 1. Depending on these values of a pixel, the pixel is mapped into 6 different regions. This mapping forms the Hiding Capacity Map (HCM) which can be defined as:

$$\begin{aligned}
HCM(x) = \{&hcm(m, n) \mid 0 \le m \le M, 0 \le n \le N, \\
&hcm(m, n) = f(sal(m, n): hcm(m, n)\} \\
&= hcm1 \ for \ 0 \le sal(m, n) \le sal1 \\
&= hcm2 \ for \ sal1 \le sal(m, n) \le sal2 \\
&= hcm3 \ for \ sal2 \le sal(m, n) \le sal3 \\
&= hcm4 \ for \ sal3 \le sal(m, n) \le sal4 \\
&= hcm5 \ for \ sal4 \le sal(m, n) \le sal5 \\
&= hcm6 \ for \ sal5 \le sal(m, n) \le sal6 \\
&(hcm1, hcm2, hcm3, hcm4, hcm5, \\
&\quad hcm3) \in (0,1,2,\ldots,255) \\
&and \ (sal1, sal2, sal3, sal4, sal5) \\
&\in \mathbf{D}: \mathbf{D} \in \mathbf{R} \wedge 0 \le \mathbf{D} \le 1
\end{aligned} \tag{4}$$

Step 5: The adaptive LSB replacement is carried out depending on the HCM value. We have chosen Least Significant Bit (LSB) Replacement technique as we are implementing invisible watermarking (Basu A., 2015).

The watermarked image is generated in which the biometric data is embedded though these above mentioned steps.

2.2 Decoder

With the help of some similar kind of processes decoding of the watermarked is to be done where the embedded watermark is to be retrieved from the cover image. The extraction process is said to be successful when the output recovered image and the original image is compared and they match identically.

The block diagram and the steps of the decoding process is as follows:

Step 1: The watermarked image I(x) is produced at the receiving end which is given by

$$I(x) = \{i(m, n) \mid 0 \le m \le M, 0 \le n \le N, i(m, n) \in [0,1,2,\ldots,255]\} \tag{5}$$

Step 2: Saliency map of the watermarked image is formed by creating its Hiding Capacity Map. Now from the saliency map we perceive the hiding capacity map as

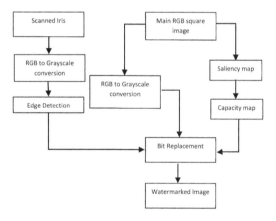

Figure 1. Block diagram of Encoder.

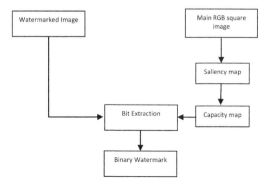

Figure 2. Block diagram of Decoder.

$SAL(x) = \{sal(m, n) \mid 0 \le m \le M, 0 \le n \le N,$
$\quad sal(m, n) \in {}^\wedge \mathbf{R}\ 0 \le sal(\mathrm{m, n}) \le 1]\}$ (6)

Step 3: The Encoded Image and the HCM is used to retrieve the biometric information by decoding process. The HCM is used to highlight the number of LSB replaced per pixel which can be represented by,

$HCM(x) = \{hcm(m, n) \mid 0 \le m \le M, 0 \le n \le N, hcm$
$\quad (m, n) = f(sal(m, n){:}hcm(m, n)\}$

Figure 3. (I) Main Image, (II) Binary Watermark Image, (III) Extracted Watermark.

$= hcm1\ for\ 0 \le sal(m, n) \le sal1$
$= hcm2\ for\ sal1 \le sal(m, n) \le sal2$
$= hcm3\ for\ sal2 \le sal(m, n) \le sal3$
$= hcm4\ for\ sal3 \le sal(m, n) \le sal4$
$= hcm5\ for\ sal4 \le sal(m, n) \le sal5$
$= hcm6\ for\ sal5 \le sal(m, n) \le sal6$
$(hcm1, hcm2, hcm3, hcm4, hcm5,$
$hcm6) \in (0,1,2,\ldots,255)\ and\ (sal1,$
$sal2, sal3, sal4, sal5) \in$
$\mathbf{D{:} D} \in \mathbf{R}^\wedge 0 \le \mathbf{D} \le 1$ (7)

3 RESULT AND DISCUSSION

The development of the proposed algorithm has yielded and the results as shown below. The implementation has been carried out on 1024 × 1024 pixel RGB mode cover images. The hiding watermark biometrics i.e. the Iris Scan is of 256 × 256 pixels. The covariance saliency (Edrem E.,

Table 1. Results of robustness.

Parameters → Images ↓	PSNR	Structural content	Mutual information	BER scaling	BER ratio	JC scaling	WPNRS	Payload capacity
BABOON	35.883	0.9675	4.3832	1528907	0.1823	0.9997	46.8409	2.9159
LENA	35.697	0.9658	4.4562	1574638	0.1877	1	Infinite	3
ABSTRACT	35.938	0.9638	4.4284	1540562	0.1836	0.996	36.0454	2.965
COLOTED CHIP	36.344	0.9766	4.4273	1401149	0.167	1	Infinite	2.6806
FOOTBALL	35.722	0.9513	3.8702	1569573	0.1871	1	Infinite	2.9954
GANTRYCRANE	35.681	0.9544	3.6431	1561326	0.1861	0.9993	43.3364	2.9632
GREENS	35.712	0.9959	4.4122	1572934	0.1875	0.9925	32.5608	3
LICHTENSTEIN	35.799	0.9645	4.4696	1560146	0.186	0.9997	48.3631	2.9807
TOYSNOFLASH	35.750	0.958	4.4395	1564241	0.1865	0.9929	33.3291	2.9937
YELLOWLILY	35.820	0.9648	4.1875	1559119	0.1859	0.9893	31.3005	2.9691
BEAUTIFUL NATURE	35.451	0.9705	4.992	1611330	0.1921	0.9523	20.7138	2.9989
SUPERMAN	35.874	0.9451	4.0272	1515142	0.1806	0.9823	24.8498	2.8553
LION MANE	35.649	0.9699	4.9311	1579752	0.1883	0.997	37.5219	3
MILKY WAY	35.673	0.9456	3.9921	1574930	0.1877	0.9891	29.8387	2.9799
WALPAPER TROPICAL	35.830	0.9717	4.8647	1545373	0.1842	0.9982	40.3358	2.9884
AVERAGE	35.788	0.9643	4.368286	1550608	0.18484	0.992553	28.33574	2.95240

Table 2. Results of imperceptibility.

	PSNR	Structural content	Mutual information	BER scaling	BER ratio	JC scaling	WPNRS
CROP	Infinite	1	0.6167	0	0	1	Infinite
COMPLEMENT	Infinite	1	0.6167	0	0	1	Infinite
ROTATE 90	Infinite	1	0.6167	0	0	1	Infinite
NOISE salt pepper	85.1562	0.9998	0.6143	13	1.98E-04	0.9998	52.1734
SCALING 1000	53.9386	0.7952	0.081	17207	0.2626	0.7056	17.3307
SCALING 0.999	56.5847	0.913	0.187	9356	0.1428	0.8381	21.8814
ERODE	56.135	0.9506	0.1207	10375	0.1583	0.8252	24.2187

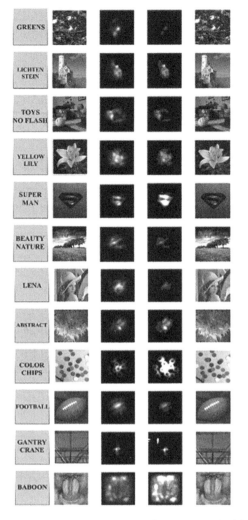

Figure 4. (I) Cover Image, (II) Saliency Map, (III) Hiding Capacity Map, (IV) Embedded Image.

2013) technique for information hiding provides the opportunity to hide the biometric information in different regions of saliency; as a consequence the information can be embedded several times. In Figure 3 the authors have publicized the images that have been used for hiding the information, along with it are the saliency maps and hiding capacity maps for each of those cover images and ultimately the encoded image. The simulation has been carried out in MATLAB R2013a and has produced a subtle quality of extracted watermarks even after exposure to different attacks. Imperceptibility results for the prosed technique have been shown in the given Table 2.

The images are subjected to attacks such as crop, compliment, rotate 90, noise salt pepper, scaling 1000, scaling 0.999, erode, dilate and scaling 2048.

The embedded image upon attacks and their respective extracted watermarks are shown in the Figure (4). The performance result after the attacks are recorded in the table for robustness in Table 1. The obtained average value of the PSNR for the scheme is 35.7887 and that of payload capacity is 2.95240667. Results in comparison to some similar techniques are depicted in Table 3.

Table 3. Proficiency comparison results.

Sl. no.	Method	PSNR	Payload capacity
Saliency based watermarking			
1.	Proposed method	35.788	2.952
2.	Saliency based (VAM) method (Sur A., 2009)	–	0.0017
Other watermarking techniques			
3.	Pairwise LSB matching (Yang C. H., 2008, Xu H., 2010)	35.05	2.25
4.	Mielikainen's Method (Mielikainen L., 2006)	33.05	2.2504
5.	Reversible Data Hiding Scheme (Gui X., 2014)	34.26	1

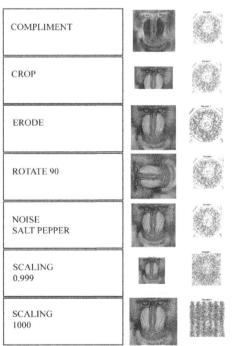

Figure 5. Attacks and extracted biometrics.

4 CONCLUSION

In this paper, the authors have proposed a digital watermarking algorithm based on saliency mapping technique that is covariance saliency method. Based on the non-salient regions, hiding capacity map is obtained and the watermark is embedded. Biometric watermark provides an additional security. A superior Capacity mapping with improved imperceptibility and robustness is obtained from the experimental results. An efficiency comparison with some other recent algorithms gives an insight of the effectiveness of the proposed algorithm. This paper will be a healthier foundation for the potential researchers in the digital image watermarking area to get acquainted with this proposed method and modify its algorithm to enhance the effectiveness of the projected technique in upcoming future.

REFERENCES

Basu A., Chatterjee A., Karmakar R., Datta S., Sarkar S. (2015), "On the implementation of secured digiatal watermarking framework", FRCCD 2015, pp. 41–49, ISBN 978-93-8592-600-6.

Basu A., Sarkar S. K., "On the Implementation of Robust Copyright Protection Scheme using Visual Attention Model," Information Security Journal: A Global Perspective, Taylor & Francis, vol. 22, no. 1, pp. 10–20, May. 2013.

Bender W. et al. (1996), "Techniques for data hiding", IBM systems Journal, vol. 35, Nos 3 & 4, pp. 313–335.

Erdem E., Erdem A. (2013), "Visual saliency estimation by nonlinearly integrating features using region covariances" Journal of Vision, 13(4): 11, pp. 1–20.

Gui X., Li X., Yang B. (2014), "A high capacity reversible data hiding scheme based on generalized prediction-error expansion and adaptive embedding," Signal Processing, vol. 98, pp. 370–380.

Hong, W. Chen, T.S. (Feb. 2012), "A Novel Data Embedding Method Using Adaptive Pixel Pair Matching," IEEE Transactions on Information Forensics And Security, vol. 7, no. 1, pp. 176–184.

Mielikainen J. (2006), "LSB Matching Revisited," IEEE Signal Processing Letters, vol. 13, no. 5, pp. 285–287.

Petitcolas F.A.P. et al. (1999), "Information Hiding—A Survey", Proceedings of the IEEE, Vol. 87, No. 7, pp. 1062–1078.

Samanta D. et al. (2008), "SET Based Logic Realization of a Robust Spatial Domain Image Watermarking", 5th International Conference on Electrical and Computer Engineering, ICECE 2008, 20–22, Dhaka, Bangladesh.

Sarkar S., Roy A. et al. (2010), "Real Time Implementation of QIM Image Watermarking", International Conference of Communication, Computer and Devices, Kharagpur, India, Paper Identification Number: 164.

Singh N., Jain M., Sharma S. (2013), "A Survey of Digital Watermarking Techniques", International Journal of Modern Communication Technologies & Research, vol. 1(6).

Soumyendu, D., D. Subhendu, B. Bijoy (2008), "Steganography and Steganalysis: Different Approaches", Information Security Consultant, International Journal of Computers, Information Technology and Engineering (IJCITAE), vol. 2, no. 1, Serial Publications.

Sur A., Sagar S. S., Pal R., Mitra P., Mukherjee J. (2009), "A New Image Watermarking Scheme Using Saliency Based Visual Attention Model, "India Conference, IEEE, pp. 1–4.

Xu H., Wanga J., Kim H. J. (2010), "Near-Optimal Solution to Pair Wise LSB Matching Via an Immune Programming Strategy," Information Sciences, vol. 180, no. 8, pp. 1201–1217.

Yang C. H. (2008), "Inverted Pattern Approach to Improve Image Quality of Information Hiding by LSB Substitution," Pattern Recognition, vol. 41, no. 8, pp. 2674–2683.

Computer, Communication and Electrical Technology – Guha, Chakraborty & Dutta (Eds)
© 2017 Taylor & Francis Group, ISBN 978-1-138-03157-9

Handwritten Devanagari numerals recognition using grid based Hausdroff distance

Showmik Bhowmik
Department of Computer Science and Engineering, Jadavpur University, Kolkata, West Bengal, India

Shibaprasad Sen
Department of Computer Science and Engineering, Future Institute of Engineering and Management, Kolkata, West Bengal, India

Naoto Hori
Department of Chemistry, University of Texas, Austin, USA

Ram Sarkar & Mita Nasipuri
Department of Computer Science and Engineering, Jadavpur University, Kolkata, India

ABSTRACT: Handwritten digit recognition plays a vital role in the automation of various aspect of our life, which starts from office automation to postal automation and many more. Handwritten digit recognition is basically a 10-class problem, but the writing styles of different individuals impose the challenge for the recognition process. In this paper, a method is devised for the recognition of handwritten Devanagari digits. For that purpose, a grid based Hausdroff distance feature descriptor is developed to represent the handwritten digits at the feature space. An appropriate classifier is then selected for the classification by comparing the performance of five well known classifiers. The proposed method is evaluated using a dataset of 6,000 images of Devanagari digits and on an average it has achieved 93.03% recognition accuracy.

1 INTRODUCTION

Development of an automated handwritten document recognition system always finds the attention to the researchers, due to its huge scope of utility in human society. Thus from the last few decades a large group of researchers from various organization situated at almost every corner of this world has indulged themselves in the development of the same. But to build such a system, varied set of problems need to be solved and handwritten numeral recognition is one of them. The recognition of handwritten numerals is itself a very hard nut to crack. The variety present in the writing styles of different individuals incorporates the randomness in the size, shape and orientation of the numerals, which impose the major obstacle to achieve the best recognition result.

Handwritten numeral recognition has a large area of applications, which starts from, handwritten pin code recognition in postal mails, recognition of handwritten bank cheque amount, numeric information processing in forms filled by hand to many others (Shrivastava & Gharde 2010). As a result, numerous methods have been reported by the researchers for the recognition of handwritten digits, written in different scripts (Singh et al. 2016).

Although a significant amount of research has been done but still researches in this field are going on with the goal of improving the accuracy. In this paper, a Hausdroff Distance (HD) based feature extraction method is developed to recognize the handwritten numerals, written in Devanagari script.

The rest of the paper is organized as follows: section 2 presents the state-of-the-art. In section 3, the proposed method is described and experimental results are discussed in section 4. Finally, section 5 concludes the paper.

2 PRESENT STATE-OF-THE-ART

In (Bajaj et al. 2002), authors have developed a recognition architecture based on connectionist network. In this work, the handwritten numerals are represented using three types of features *namely*, density features, moment features of the left, right, upper and lower profile curves and descriptive component features. After feature extraction, three neural classifiers are used for initial classification and finally the results of all classifiers are combined using connectionist schema. Method described in (Bhattacharya et al. 2006) has been built by following a two-stage approach for the recognition of handwritten Devanagari numerals. Initially, from each input image shape features, based on directional view based strokes, are extracted and then a Hidden Markov Model (HMM) and an Artificial Neural Network (ANN) based classifiers are used for the stage one classifications and finally the results generated by these classifiers are further combined using another ANN based classifier at the second stage.

In (Patil & Sontakke 2007), authors have used 16-dimentonal ring and Zernike features for the recognition of handwritten Devanagari numerals. The algorithm has used a general fuzzy hyperline segment neural network for the classification, which combines both supervised and unsupervised learning. Authors in (Pal et al. 2007) have initially segmented the input image into small blocks and from each block directional features are computed. After that, using Gaussian filter, these segmented blocks are downsampled and finally, the features from the downsampled blocks are fed to a modified quadratic classifier for the classification.

Method described in (Shrivastava & Gharde 2010) has extracted moment invariant and affine moment invariant features from each input image. The extracted features are then fed to a Support Vector Machine (SVM). In (Prabhanjan & Dinesh 2015), authors have presented a hybrid approach for the recognition of handwritten Devanagari numerals. In this work, Fourier descriptor is used to extract the global information and pixel density based features are extracted from different zones of the input image to get the local information. After that, the confidence score of four classifiers, *viz.,* Naïve Bayes (NB), Instance Based Learner (IBL), Random Forest (RF), and Sequential Minimal Optimization (SMO) are fused to perform the final classification. Authors in (Arya et al. 2015) have used Gabor filter based feature descriptor with three different filter sizes *viz.,* 7×7, 19×19 and 31×31. The final classification is performed using Nearest Neighbor and SVM classifiers.

3 PRESENT WORK

In the present work, initially the input numeral images are preprocessed and then a HD based feature descriptor is computed from each preprocessed image to represent them at the feature space. After that, the performances of five well known classifiers *viz.,* NB, SMO, RF, MLP and Bagging are compared in terms of their accuracy to select the appropriate one. Finally the classification is carried out using RF classifier. Figure 1 shows the flowchart of the entire process.

3.1 *Preprocessing*

Initially the input numeral images are smoothed with Gaussian filter to remove the noise and then

the resultant images are binarized using Otsu's method (Otsu 1975). To remove the isolated data pixels two morphological operators *viz.,* erosion and dilation are performed on all the binarized images. Finally, the bounding box for enclosing the numeral in each image is estimated.

3.2 *Feature extraction*

In the present work, Grid based Hausdroff distance (GHD) is used to design the feature descriptor for each of the handwritten Devanagari numerals. Any numeral recognition is basically a 10-class problem and each class of numerals follow a distinct shape characteristic. Thus it is assumed that due to the shape difference, the distance between the set of pixels of a particular region and the set of pixels of another particular region will also differ from one class of images to another. This is the main reason behind considering the distance based feature descriptor.

In the following subsection, the HD is explained first and then the implementation of GHD is described.

3.2.1 *Hausdroff Distance (HD)*
Consider X and Y are two non-empty subsets of a set M.

$$X = \left\{ x_1, x_2, \ldots \ldots \ldots x_n \right\} \text{and} Y = \left\{ y_1, y_2, \ldots \ldots \ldots y_n \right\}$$

The HD between X and Y can be defined as follows,

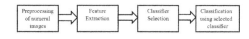

Figure 1.   Flowchart of the entire process.

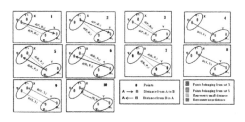

Figure 2.   Illustration of HD computation between two sets of points.

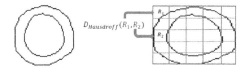

Figure 3.   Illustration of GHD for a sample image.

16

$$D_{Hausdroff}(X,Y) = \max\left(D(X,Y), D(Y,X)\right) \quad (1)$$

where,

$$D_{(X,Y)} = \max\left\{\forall_x \min\left\{\forall_y d(x,y)\right\}\right\} \quad (2)$$

$$D_{(Y,X)} = \max\left\{\forall_y \min\left\{\forall_x d(x,y)\right\}\right\} \quad (3)$$

Here, $d(x,y)$ represents the Euclidean distance between X and Y. Figure 2 demonstrates the HD computation.

3.2.2 Grid based Hausdroff Distance (GHD)

For estimating the GHD feature descriptor, first each sample image is decomposed into $N \times N$ grids. Then HDs from a particular grid to every other grids are computed. Before the decomposition, the contours of the input images are estimated. Here, focus is given on the shape of the numerals and as the contour pixels are sufficient to capture the shape nature, the non-contour pixels are eliminated. Hence, this not only preserves the shape information but also reduces the computational cost.

After the contour estimation, the sample digit with minimum bounding box is decomposed into $N \times N$ grids. The HD between these regions are computed as follows,

Consider a grid structure 'G' drawn on an input image 'I', which contains n grids and each grid 'R' contains m pixels.

$G = \{R_1, R_2, \ldots \ldots \ldots R_n\}$ and $R_i = \{P_1, P_2, \ldots \ldots \ldots P_m\}$ where, $i = 1,2,3,\ldots \ldots \ldots n$.

Then the GHD is computed as follows,

---
1. For all i=1 to n
2.   For all j=i+1 to n
3. $D_{hausdroff}(R_i, R_j)$
4.   End
5. End

For each image a feature vector of size $(n*(n-1))/2$ is extracted.

4 EXPERIMENTAL RESULTS AND ERROR ANALYSIS

For evaluating the present method, a dataset of 6,000 handwritten Devanagari numeral images (600 images per class) are used. During the experiment, initially, a suitable classifier is chosen by comparing the performance of five well known classifiers and finally the classification is carried out using the selected classifier.

In this subsection, first, the classifier selection is presented and then the experimental results are described. The error cases are analyzed at the end of this section.

4.1 Classifier selection

In this experiment the performance of five well known classifiers *namely*, NB, SMO, RF, MLP and Bagging are compared in terms of their recognition accuracies. For that purpose a dataset of 3000 images (300 images per numeral) are randomly selected from the main dataset and a data mining tool called WEKA is used. The comparison result, given in Table 1, shows that RF outperforms the other classifiers. Thus rest of the experiments is carried out using RF classifier.

During classifier selection, default parameters values are used for all the classifiers along with a 4×4 number of grids.

4.2 Experimental results

For the final experiment, entire dataset containing 6000 images of handwritten Devanagari numerals is used. The samples in the present dataset are collected from a large group of people having different educational and professional backgrounds. These images are collected in a preformatted A4 size datasheets, which are scanned with a resolution of 300 dpi. After that each image is cropped programmatically to prepare the isolated digits. Figure 4 shows some samples of the present dataset. For feature extraction, the optimal number of grid is also chosen experimentally. Features are extracted with 4×4, 5×5 and 6×6 number of grids. RF classifiers are trained with the feature set extracted with each of the above grid structures and their recognition performances on the test set are evaluated. The results are shown in Figure 5. Please note that in Figure 5, a minor improvement in recognition accuracy is observed while the grid number is increased from 5×5 to 6×6. But it is ignored in the present experimentation, as with increase of grid number, the number of features increases abruptly which in turn augments the computational time of the overall system (see Figure 5). Thus for the

Table 1. Performance comparison of the considered classifiers.

Sl#	Size of the dataset		Classifier	Accuracy (in %)
---	Train	Test		
1			SMO	81.33
2			NB	58.77
3	2100	900	Bagging	78.88
4			MLP	81.66
5			RF	85.66

experiment 5×5 number of grids are considered. A 5-fold cross validation method is followed during the final experiment. With the selected setup, the proposed system has achieved 93.03% recognition accuracy on an average. Table 2 presents some class-wise statistical measurements such as Recall, Precision and F-measure.

4.3 *Error case analysis*

Most of the samples present in the current dataset are correctly classified by the proposed system but some misclassifications are also observed. Two most probable reasons behind these misclassifications can be *unexpected elongation of stroke while writing the digit* (see Figure 6 (a)) and the *erroneous cropping* during database preparation (see Figure 6(b)). In both of these cases the expected shapes of the digits get hampered. As in the present feature computation, the entire focus is given on the shape information of the digits, these shape distortions lead to the misclassification.

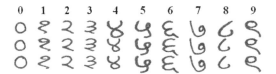

Figure 4.   Sample images from present dataset.

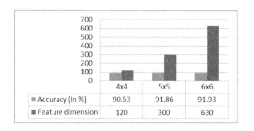

Figure 5.   Performance comparison using different grid size.

Table 2.   Class wise assessment obtained using 5-fold cross validation.

Class #	Precision	Recall	F-measure
0	0.96	0.98	0.97
1	0.85	0.91	0.88
2	0.93	0.91	0.92
3	0.90	0.91	0.91
4	0.94	0.91	0.93
5	0.95	0.91	0.93
6	0.91	0.93	0.92
7	0.96	0.92	0.94
8	0.95	0.96	0.96
9	0.86	0.91	0.88

Figure 6.   Example of some misclassified samples.

5   CONCLUSION

Handwritten digit recognition is one of the most researched areas in the domain of handwriting recognition. As said earlier, the research in this field is still continuing with the goal of improving the recognition module. In the present work, a handwritten Devanagari digit recognition method is developed. For that purpose, a GHD feature descriptor is designed. Finally the classification is carried out using RF classifier. The proposed system has achieved 93.03% on an average.

It is observed that misclassifications are found due to poor cropping of the images. So, in future, more care should be taken during data preparation. Also the proposed system is evaluated on a limited dataset. Hence, more number of samples would ascertain the effectiveness of the proposed system.

REFERENCES

Arya, S., Chhabra, I. & Lehal, G.S., (2015). Recognition of Devnagari Numerals using Gabor Filter. *Indian Journal of Science and Technology*, 8(27).

Bajaj, R., Dey, L. & Chaudhury, S., (2002). Devnagari numeral recognition by combining decision of multiple connectionist classifiers. *Sadhana*, 27(1), pp. 59–72.

Bhattacharya, U. et al., (2006). Neural combination of ANN and HMM for handwritten Devanagari numeral recognition. In *Tenth International Workshop on Frontiers in Handwriting Recognition*. Suvisoft.

Otsu, N., (1975). A threshold selection method from gray-level histograms. *Automatica*, 11(285–296), pp. 23–27.

Pal, U. et al., (2007). Handwritten numeral recognition of six popular Indian scripts. In *Document Analysis and Recognition, 2007. ICDAR 2007. Ninth International Conference on*. IEEE, pp. 749–753.

Patil, P.M. & Sontakke, T.R., (2007). Rotation, scale and translation invariant handwritten Devanagari numeral character recognition using general fuzzy neural network. *Pattern Recognition*, 40(7), pp. 2110–2117.

Prabhanjan, S. & Dinesh, R., (2015). Handwritten Devanagari Numeral Recognition by Fusion of Classifiers. *International Journal of Signal Processing, Image Processing and Pattern Recognition*, 8(7), pp. 41–50.

Shrivastava, S.K. & Gharde, S.S., (2010). Support vector machine for handwritten Devanagari numeral recognition. *International journal of computer applications*, 7(11), pp. 9–14.

Singh, P.K., Sarkar, R. & Nasipuri, M., (2016). A Study of Moment Based Features on Handwritten Digit Recognition. *Applied Computational Intelligence and Soft Computing*, 2016.

Computer, Communication and Electrical Technology – Guha, Chakraborty & Dutta (Eds)
© 2017 Taylor & Francis Group, ISBN 978-1-138-03157-9

Image steganography using a histogram of discrete cosine transform

Barnali Gupta Banik & Manish Kumar Poddar
Department of Computer Science and Engineering, St' Thomas College of Engineering and Technology, Kolkata, India

Samir Kumar Bandyopadhyay
Department of Computer Science and Engineering, University of Calcutta, Kolkata, India

ABSTRACT: Steganography is a technique to hide information in any digital media such as image, audio, video, or text and allow only the intended user to decode the hidden information. A technique has been developed and explained in this study to embed secret text message using histogram and Discrete Cosine Transformation (DCT) of the cover image. Here, four Least Significant Bits (LSB) of cover image are used to embed secret message. There are several methods of steganography using LSB, but the method proposed here is different in the approach of applying LSB technique and efficiency of the method according to PSNR and BER.

Keywords: Image Steganography, Discrete Cosine Transformation, Histogram, LSB Steganography

1 INTRODUCTION

Steganography is the ancient art of hiding secret information in cover object. With the expansion of digital media, steganography has become a secured data communication technique through it. The main objectives of steganography are:

- concealing secret message in a cover object without revealing its existence;
- recovery of secret information without tampering; and
- maximum bits to be embedded.

There are various types of steganography in digital media (Gupta Banik & Bandyopadhyay, 2015a). In this study, an image steganography technique is being discussed. Here, first, DCT of cover image is obtained along with the secret text message, which has been taken as a binary stream. Then, that secret message is embedded using histogram of DCT matrix by multiple LSB replacement method. This method is different from the existing multiple LSB replacement methods in a way that it is applied on the histogram of DCT of the cover image, whereas in general, multiple LSB replacement is applied in spatial domain of cover image.

Gupta Banik & Bandyopadhyay (2015b) explained that human eyes cannot detect changes up to four least significant bits of an image. By exploring such perceptual inability, in this study, four least significant bits are used to embed the binary stream of message.

2 LITERATURE SURVEY

2.1 *Histogram of image*

Histogram is a graphical representation of numeric data distribution (see NIST). It gives probability distribution of continuous variable. The histogram of a digital image, that is, the distribution of gray level can be denoted by the following discrete function (1):

$$H\left(g_k\right) = n_k; k = 0,1,..,M-1 \qquad (1)$$

where g_k is the kth gray level and n_k is the number of pixels in the image having gray level g_k, and M is the number of gray levels.

Histogram is an important technique to increase the contrast (Wang, Fan & Yu, 2009). Yalman & Erturk (2009) proposed a method using histogram of cover image to embed data. Wang et al. (2013) proposed another method of information hiding using histogram shifting.

2.2 *Discrete Cosine Transformation (DCT)*

Discrete cosine Transformation is a successful signal transformation technique. Different types of DCT are available, among which two-dimensional (2D) DCT is often used to transform an image from spatial domain to frequency domain (Papakostas, Karakasis & Koulouriotis, 2008). The formulation of the 2D DCT for an input image C

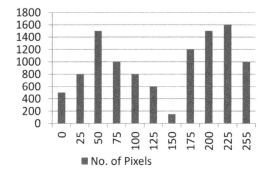

Figure 1. Gray-level representation in histogram of an image.

with i rows and j columns and the output image D is given in equation (2):

$$D_{xy} = a_x a_y \sum_{i=0}^{M-1}\sum_{j=0}^{N-1} C_{ij} \cos\frac{\pi(2m+1)x}{2M}\cos\frac{\pi(2n+1)y}{2N}$$

$$(2)$$

where

$$a_x = \begin{cases} \dfrac{1}{\sqrt{M}}, x=0 \\ \sqrt{\dfrac{2}{M}}, 1\le x \le M-1 \end{cases}$$

$$a_y = \begin{cases} \dfrac{1}{\sqrt{N}}, Y=0 \\ \sqrt{\dfrac{2}{N}}, 1\le y \le N-1 \end{cases}$$

Inverse 2D DCT is also available to reverse the application of 2D DCT on any image. This is defined in equation (3):

$$C_{ij} = \sum_{i=0}^{M-1}\sum_{j=0}^{N-1} a_i a_j D_{xy} \cos\frac{\pi(2m+1)x}{2M}\cos\frac{\pi(2n+1)y}{2N} \quad (3)$$

where $0\le i \le M-1$ and $0\le j \le N-1$

Gupta Banik & Bandyopadhyay (2015c) proposed a method of image steganography using DCT, where at first DCT is performed on cover image followed by secret embedding in the midband frequency coefficients of DCT. Patel & Dave (2012) used one, two, and four LSB replacements to embed secret data in the DCT coefficient values using threshold selected by the authors themselves. Raja et al. (2005) used LSB technique to embed secret followed by DCT to quantize and finally performed run-length encoding to create compressed stego image.

3 PROPOSED METHOD

3.1 Embedding Technique

The proposed technique is based on histogram of DCT of the cover image, which is a unique approach as none of the existing works show any usage of histogram over DCT coefficient values. The binary stream of data has been embedded in the histogram. The image embedded with the embedded text will be treated as stego image.

Here, a 4×4 transformation equation is used, which is given by:

$$t_{ij} = f(x) = \begin{cases} \dfrac{1}{\sqrt{N}} & if\ i=0 \\ \dfrac{\sqrt{2}}{\sqrt{N}}\cos\left[(2j+1)i\pi)/2N\right] & if\ i>0 \end{cases} \quad (4)$$

The various steps involved in the generation of stego image are shown in Figure 2.

From the cover image, each of the pixel values are taken, 128 is subtracted from each of these values, and an image block of 4×4 is taken after equation 4 was applied to obtain DCT of cover image as in equation (2).

Then, histograms of the 1st 16 pixels in eight different gray levels are considered in such a way that the range of histogram should lie between the lowest and highest values of gray level. A matrix of histogram containing eight different levels is shown in Table 1. If the value of histogram is higher than 14, then it should be set to 14 using four-LSB modification of cover image, and secret message can be embedded.

After creating histogram matrix, message stream must be considered. If the message is "ABC......", then the first character should be obtained and converted into binary stream as shown in Table 2.

Now the binary stream of Table 2 should be added to the histogram of Table 1 as shown in Table 3.

Table 3. Addition of binary stream of message upon histogram of DCT of cover image.

Histogram, which has been obtained after embedding binary stream of message, is shown in Table 3.

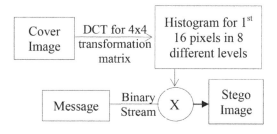

Figure 2. Steps involved in the generation of stego image.

After performing this, pixel values of cover image are considered and transferred into binary stream as shown in Table 5.

After obtaining pixel values of cover image, each of the pixels is transferred into binary stream as shown in Table 6

The binary stream of histogram shown in Table 4 is obtained and the pixel values of cover image shown in Table 6 are added to that to obtain the stego image.

Thus, Table 8 is generated for stego image.

Data of Table 8 are taken and transferred to decimal form to obtain pixel value of stego image as shown in Table 9.

3.2 Extraction Technique

The extraction technique is just the reverse of the embedding technique. The detailed steps of extraction technique are shown in Figure 3.

Here, using equation 5, a 4×4 transformation matrix is created:

Table 1. Histogram values of eight different gray levels.

No of pixel	Gray level 0	Gray level 1	Gray level 2	Gray level 3	Gray level 4	Gray level 5	Gray level 6	Gray level 7
0–15	2	5	1	0	2	4	0	2
16–31	5	0	3	2	0	0	6	0
.	.	.	.	.	.	.	.	.
496–511	3	0	0	8	4	1	0	0

Table 2. Binary stream for character "A".

Character	Binary stream of character							
A	0	1	0	0	0	0	0	1

Table 3. Addition of binary stream of message upon histogram of DCT of cover image.

No of pixel	Gray level 0	Gray level 1	Gray level 2	Gray level 3	Gray level 4	Gray level 5	Gray level 6	Gray level 7
0–15	2	5	1	0	2	4	0	2
.	.	.	.	.	.	.	.	.

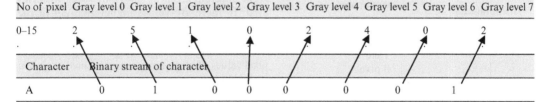

Character	Binary stream of character							
A	0	1	0	0	0	0	0	1

Table 4. Histogram obtained after embedding text.

No of pixel	Gray level 0	Gray level 1	Gray level 2	Gray level 3	Gray level 4	Gray level 5	Gray level 6	Gray level 7
0–15	2	**6**	1	0	2	4	0	**3**
.	.	.	.	.	.	.	.	.

Table 5. Pixel values of cover image.

162	150	160	155	130	145	160	180	.	.
185	100	155	140	130	148	162	155	.	.
138	120	160	168	150	125	132	162	.	.
.	.	.	.	.	.	.	.	.	.
.	.	.	.	.	.	.	.	.	.

Table 6. Pixel values of cover image in binary format.

10100101	10010110	10100000	10100000	10011011	10010001	10100000	10110100	.	.
10111001	01100100	10011011	10001100	10000010	10010100	10100010	10011011	.	.
.	.	.	.	.	.	.	.	.	.
.	.	.	.	.	.	.	.	.	.

Table 7. Binary stream of Histogram.

No of pixel	Gray level 0	Gray level 1	Gray level 2	Gray level 3	Gray level 4	Gray level 5	Gray level 6	Gray level 7
0–15	0010	0110	0001	0000	0010	0100	0000	0011
.	.	.	.	.	.	.	.	.

Table 8. Binary stream obtained after embedding message.

10100010	10010110	10100001	10100000	10010010	10010100	10100000	10110011	.	.
10111001	01100100	1001101	10001100	10001100	10010100	10100010	10011011	.	.
.	.	.	.	.	.	.	.	.	.
.	.	.	.	.	.	.	.	.	.

Table 9. Pixel value of stego image

162	150	161	160	146	148	160	179	.	.
185	100	155	140	130	148	162	155	.	.
.	.	.	.	.	.	.	.	.	.

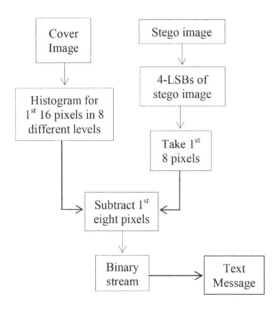

Figure 3. Steps involved in extraction.

$$t_{ij} = f(x) = \begin{cases} \dfrac{1}{\sqrt{2}} & if\ i = 0 \\ \dfrac{2}{N}\cos\left[(2j+1)i\pi\right]/2N] & if\ i > 0 \end{cases} \quad (5)$$

Now considering the cover image, 128 is subtracted from each of the pixel values and then a block of 4×4 from modified pixel value is taken, on which a transformation matrix was applied to obtain DCT of image.

Then, the histogram of the 1st 16 pixels is created in eight different gray levels in such a manner that range of histogram should lie between the lowest and highest values of gray level. Now a matrix of histogram containing eight different levels as shown in Table 10 is created. If there are values higher than 14 in the histogram, then set it to 14 as similar to the process of embedding.

After doing this, stego image is considered and pixel values of stego images are obtained as shown in Table 11.

Afterward, each of the pixels is transferred in binary form as shown in Table 12.

Four least significant bits of each of pixels in Table 12 are taken, as they contain binary stream of message as shown in Table 13.

After obtaining this, it is converted to decimal as shown in Table 14. Then, consider histogram of Table 10 to rewrite that in Table 15, and then subtract both of them to obtain binary message stream.

Binary stream of message, which has been obtained after subtraction, is shown in Table 16.

Now this binary stream is converted into character stream to get the embedded text. This conversion gives result in ASCII values—for letter "A", which was embedded, the above binary stream will generate result as 65, which is the ASCII value for "A".

3.3 Embedding Algorithm

STEP 1: Obtain the binary stream of the given text.

STEP 2: Construct a 4×4 matrix of cover image for DCT

STEP 3: Perform Discrete Cosine Transform of the given cover image.

STEP 4: Construct Histogram of Discrete Cosine Transform matrix of the cover image. Take 16 pixels and form eight different levels—if the value of histogram in a given level is higher than 14, then set it to 14, so that four Least Significant Bits of cover image can be used to embed text.

STEP 5: Take first eight bits of the binary stream of message and add it to the given eight levels of histogram.

STEP 6: Take 1st four Least Significant Bits of cover image and embed histogram containing message there.

STEP 7: Make four Least Significant Bits of eight successive pixels to 0; this will be used as terminating condition.

3.4 Extraction Algorithm

STEP 1: Construct a 4 × 4 matrix of cover image for Discrete Cosine Transform.

STEP 2: Perform Discrete Cosine Transform of the cover image.

STEP 3: Construct Histogram of Discrete Cosine Transform matrix of the cover image. Take 16 pixels and form eight different levels – if the value of histogram in a given level is higher than 14, make it 14, so that four Least Significant Bits of cover image can be used to transfer histogram embedded text.

STEP 4: From the stego image, obtain four LSBs of each pixel and convert it to decimal.

STEP 5: Subtract each of the value from histogram of the cover image and obtain binary stream

STEP 6: Convert binary stream into text stream.

STEP 7: Repeat this process until it generates four 0's in Least Significant Bit of eight consecutive pixels.

The processes of embedding and extraction are shown in Figures 4 and 5, respectively. Here, all of the tests were performed using C language in Windows platform on 512 × 512 gray scale image.

Table 10. Histogram of cover image.

No of pixel	Gray level 0	Gray level 1	Gray level 2	Gray level 3	Gray level 4	Gray level 5	Gray level 6	Gray level 7
0–15	2	5	1	0	2	4	0	2
16–31	5	0	3	2	0	0	6	0
.	.	.	.	.	.	.	.	.
496–511	3	0	0	8	4	1	0	0

Table 11. Pixel values of stego image.

162	150	161	160	146	148	160	179	.	.
185	100	155	140	130	148	162	155	.	.
.	.	.	.	.	.	.	.	.	.

Table 12. Binary stream of pixel values of stego image.

10100010	10010110	10100001	10100000	10010010	10010100	10100000	10110011	.
10111001	01100100	1001101	10001100	10001100	10010100	10100010	10011011	.
.	.	.	.	.	.	.	.	.
.	.	.	.	.	.	.	.	.

Table 13. Four-LSB bits of the stego image containing message.

No of pixel	Gray level 0	Gray level 1	Gray level 2	Gray level 3	Gray level 4	Gray level 5	Gray level 6	Gray level 7
0–15	0010	0110	0001	0000	0010	0100	0000	0011
16–31	.	.	.	.	.	.	.	.
.	.	.	.	.	.	.	.	.

Table 14. Histogram in decimal containing message.

No of pixel	Gray level 0	Gray level 1	Gray level 2	Gray level 3	Gray level 4	Gray level 5	Gray level 6	Gray level 7
0–15	2	6	1	0	2	4	0	3

Table 15. Histogram of Cover image.

No of pixel	Gray level 0	Gray level 1	Gray level 2	Gray level 3	Gray level 4	Gray level 5	Gray level 6	Gray level 7
0–15	2	5	1	0	2	4	0	2

Table 16. Binary stream of obtained message.

Binary stream of character							
0	1	0	0	0	0	0	1

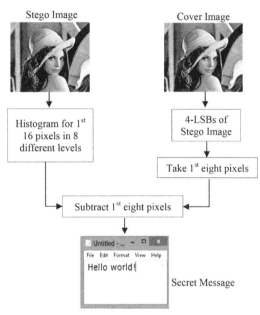

Figure 5. Extraction of secret message from stego image.

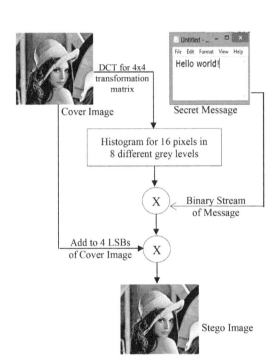

Figure 4. Steps involved in generating stego image.

4 RESULT AND ANALYSIS

PSNR stands for Peak Signal-to-Noise Ratio. It computes the difference between two images in decibels. This ratio is used as the quantity measurement between original and stego image (see NI, 2013). It is given by the below equation:

$$PSNR = 10\log_{10}\left(\frac{255^2}{MSE}\right) \qquad (6)$$

$$MSE = 1/MxN \sum_{i=0}^{M-1}\sum_{j=0}^{N-1}\left|I_{i,j} - k_{i,j}\right| \qquad (7)$$

where $M \times N$ is the size of M rows and N columns; $I_{i,j}$ is the value of (i,j) cell for the cover image; and $K_{i,j}$ is the value of (i,j) cell for the stego image.

MSE stands for Mean Square Error, which is the cumulative square error between the stego and original image.

Table 17. Value of PSNR and BER for secret message.

No. of characters in secret message	PSNR	BER
30	61.288	.000412
40	59.678	.000587
50	58.346	.000700
60	57.342	.008890

Table 18. Table for comparison.

No. of hidden characters in the proposed method	BER of the proposed method	Depth of Hiding of the existing method	BER of the existing method
30	.0004120	1	2.780208
40	.0005870	2	8.330555
50	.0007000	3	8.315001
60	.0088900	4	8.326388
100	0.282738	5	8.347917
500	0.500095	6	8.265972
1500	0.500961	7	8.359446

The Bit Error Rate (BER) is the number of bit error to the total number of transmitted bit (Breed, 2003).

As an example, assume that the transmitted bit sequence is:

0 1 1 0 0 0 1 0 1 1

And the following are the received sequence of bits:

0 0̲ 1 0 0 0 0̲ 0 1 0̲

Here, bit error obtained is 3. Therefore, BER in this case is 3/10, that is, 0.3.

Here, PSNR and BER are calculated to evaluate the quality of the proposed method, the values of which are shown in Table 17.

In this proposed method, data to be hidden are text and cover is gray scale image. Here, BER value of the proposed method has been used for comparison with the steganographic method published by Raja et al. (2005). However, this method is lossy, as here some of secret image data are lost, but the proposed method is lossless as no secret data is lost during embedding or extraction. The comparison results are shown in Table 18.

5 CONCLUSION

In this study, the application of four Least Significant Bits on Histogram of Discrete Cosine Transformation is proposed in an innovative way. The result of quality metrics obtained from the proposed method proves the efficiency of this steganographic technique; furthermore, it shows better result while comparing with the existing method.

REFERENCES

Breed G. (2003) "Bit Error Rate: Fundamental Concepts and Measurement Issues", High Frequency Electronics, Summit Technical Media, LLC. Retrieved from http://www.highfreqelec.summittechmedia.com/Jan03/HFE0103_Tutorial.pdf.

Gupta Banik B and Bandyopadhyay S. K (2015a), "Review on Steganography in Digital Media", International Journal of Science and Research, Volume 4, Issue 2, pp. 265–274. Retrieved from https://www.ijsr.net/archive/v4i2/SUB151127.pdf.

Gupta Banik B. and Bandyopadhyay S. K. (2015b) "AnImage Steganography Method on Edge Detection Using Multiple LSB Modification Technique", Journal of Basic and Applied Research International of International Knowledge Press, Volume 9 Issue 2 pp. 75–80. Retrieved from http://www.ikpress.org/abstract/4140.

Gupta Banik B. and Bandyopadhyay S. K. (2015c), "Implementation of Image Steganography Algorithm using Scrambled Image and Quantization Coefficient Modification in DCT", Proceedings of 2015 IEEE International Conference on Research in Computational Intelligence and Communication Networks pp. 400–405. doi: 10.1109/ICRCICN.2015.7434272.

National Institute of Standard and Technology (NIST), Retrieved from http://www.itl.nist.gov/div898/handbook/eda/section3/histogra.htm.

National Instruments (NI, 2013), "Peak Signal-to-Noise Ratio as an Image Quality Metric" Retrieved from http://www.ni.com/white-paper/13306/en/

Papakostas G.A., Karakasis E. G., Koulouriotis D. E. (2008) "On accelerating the computation of 2-D Discrete Cosine Transform in image processing", Proceedings of International Conference on Signals and Electronic Systems, pp.7–10. doi: 10.1109/ICSES.2008.4673343.

Patel H, Dave P. (2012) "Steganography Technique Based on DCT Coefficients", International Journal of Engineering Research and Applications (IJERA), Vol. 2, Issue 1 pp. 713–717. Retrieved from http://www.ijera.com/papers/Vol2_issue1/DK217137 17.pdf.

Raja K.B., Chowdary C.R., Venugopal K.R., Patnaik L.M. (2005) "A Secure Image Steganography using LSB, DCT and Compression Techniques on Raw Images", Proceedings of 3rd International Conference on Intelligent Sensing and Information Processing, pp.170–176. doi: 10.1109/ICISIP.2005.1619431.

Wang S, Fan Y, Yu P (2009) "A Watermarking Algorithm of Gray Image Based on Histogram", Proceedings of 2009 2nd International Congress on Image and Signal Processing. doi: 10.1109/CISP. 2009.5303871.

Wang W., Zhang Y, Huang C, Wang S. (2013), "Steganography of Data Embedding in Multimedia Images Using Interpolation and Histogram Shifting", Proceedings of Ninth International Conference on Intelligent Information Hiding and Multimedia Signal Processing, pp. 387–390 doi: 10.1109/IIH-MSP.2013.103.

Yalman Y, Erturk I. (2009), "A New Histogram Modification Based Robust Image Data Hiding Technique", Proceedings of 24th International Symposium on Computer and Information Sciences pp.39–43 doi: 10.1109/ISCIS.2009.5291922.

Computer, Communication and Electrical Technology – Guha, Chakraborty & Dutta (Eds)
© *2017 Taylor & Francis Group, ISBN 978-1-138-03157-9*

A parallel cell based architecture for real time morphological edge detection

Pradipta Roy & Binoy Kumar Das
Integrated Test Range, DRDO, Chandipur, India

ABSTRACT: Video Segmentation is indispensable task in any level of video processing. Edge or Gradient image extraction is the starting point in many segmentation algorithms like watershed or Active Contour Model. It is very difficult to perform real time edge detection for high resolution video in general purpose microprocessors. Moreover gradient operator requires square and square root operation which is extremely hardware consuming. To alleviate this problem we have proposed a simple cell network based configurable parallel architecture which performs Morphological gradient detection in few clock cycles. Due to its regular architecture it is also very easy to implement in VLSI. We have implemented the architecture in XILINX VIRTEX II P, FPGA and achieved a 4 clock cycle operations for each morphological operators like dilation and erosion. With a 400 MHz processing clock we have achieved a throughput of 40 MPS (Mega Pixel per Second), which is more than the required real time throughput of PAL standard video. Since most of the core operations in video segmentation can be mapped to morphological operations it can be used as configurable segmentation engine to perform a wide variety of segmentation task.

1 INTRODUCTION

Video segmentation is a critical and essential task for many computer vision applications like tracking, surveillance; object based video coding (MPEG4 and MPEG7), robotics and different medical visual applications. Success of many subsequent processing and decision depends on the efficiency of segmentation step. Video segmentation presents challenges in multi-dimensional aspects. First and foremost criterion is real time implementation obeying inter-frame timing constraint. Though wide varieties of algorithms for video segmentation are evolving over the years most of them are meant for software implantation. As software implementations are essentially sequential processing, it puts a severe restriction for processing area in real time. Mapping well defined and fixed-function algorithms to hardware modules using custom and semi custom VLSI technologies can results in the highest efficiency in cost and performance. An example can be sited in support that a segmentation algorithm involving background subtraction and watershed algorithm to a QCIF format frame takes 19.31 ms in a Field Programmable Gate Array (FPGA) platform compared to 233 ms in software implementation in SGI 02 workstation (Chien, Huang, & Chen 2002). Thus a hardware solution to video segmentation task is preferred in most of the time consuming cases. A reconfigurable attribute in the hardware architecture is also sought to implement a wide variety of segmentation algorithm in a basic core processing engine.

In this paper we propose a cell network based architecture which performs two basic morphological operations, namely Dilation and Erosion in 4 clock cycles. As we can show different segmentation algorithms can be decomposed using basic morphological operations the circuit can be used for the same. In this paper we have taken Gradient image calculation as the implementation task with our system. Gradient image or edge map is starting point for many segmentation algorithms like Watershed segmentation, active contour model etc. But calculation of gradient for complete image is time consuming in software implementation and resource consuming in hardware implementation as it involves square and square root calculation. We have implemented a simple cell network based configurable parallel architecture which performs morphological gradient detection in few clock cycles with simple cell network architecture.

The organization of this paper is as follows. First we will introduce related works and how different core operations can be mapped in morphological operations. The following sections describe the algorithm and our proposed architectural implementation. Finally we conclude with results and future scope of work.

2 MORPHOLOGICAL OPERATIONS AND EXISTING IMPLEMENTATIONS

Mathematical morphology is a useful tool for various image processing tasks (Gonzalez & Woods

2012). The morphological operations can be extended to video processing algorithms quite easily. The basic operations of morphology are dilation and erosion. We here describe these morphological operations with respect to binary frame which can be later extended to gray or color scenes. Lixia et al. have proposed complex morphological edge detection applicable to noisy images (Lixia 2010). Wang put forward improved morphological operators about edge detection, but their results either cannot detect the details of the edge well, or cannot filter out the noise well (Wang, Zhang, & Gao 2013).

Let A and B are two sets in Z^2 space. We define two operations namely reflection and translation of set B and A respectively as

Reflection of set B,

$$\hat{B} = \{d \mid d = -b \; for \; b \in B\} \qquad (1)$$

Translation of set A,

$$(A)_z = \{c \mid c = a + z \; for \; a \in A) \qquad (2)$$

Now let us define two basic operations Dilation and Erosion. Dilation is the operation that thickens or grows specific pixels (pixels labeled by '1' in binary images), controlled by a mask called structuring element. Mathematically it is defined as follows.

Dilation of A with *structuring element* B,

$$A \oplus B = \{z \mid (\hat{B})_z \cap A \neq \Phi\} \qquad (3)$$

where, Φ is empty set.

On the other hand erosion thins or erodes the object in binary scene, controlled by structuring element. Erosion of set A by structuring element B is defined as:

$$A \ominus B = \{z \mid (\hat{B})_z \cap A^c \neq \Phi \qquad (4)$$

where, A^c is the complement of set A

The morphological gradient of a image A is defined as:

$$Grad(A) = (A \oplus B) - (A \ominus B) \qquad (5)$$

Chien et al. had analyzed different video segmentation operations and classified them in five groups (Chien et al. 2002). They are morphological operations, region growing operations, motion detection operation, pixel operations and other operations like relaxation, Canny edge detection etc. They have shown that all the operations can be mapped into basic morphological opera-

tions. A programmable Processing Element (PE) array structure is proposed in their paper, though detailed functional and architectural description is not given. They have proposed a hardware accelerator for video segmentation and total task is divided between hardware and software. Diamantaras et al. proposed a systolic architecture for Morphological operations though their hardware burden is much higher (Diamantaras & Kung 1997). Sheu et al proposed data reuse architecture for morphological operations but their implementation lacks programmability feature (Sheu et al. 1993).

3 PROPOSED ARCHITECTURE

We have designed hardware architecture capable of implementing basic morphological operations namely Dilation and erosion. The core architecture is a programmable cell array implemented using small FSM. Overall system is implemented in an FPGA with PCI interface. The PCI interface is used to load different combination of operation sequence and different structural element from PC to suite user requirement. With different sequence of Morphological operations a variety of video segmentation task can be performed. Here we will describe the Morphological Gradient detection using a sequence of Dilation and Erosion. The overall block diagram of the system is shown in Figure 1.

The system utilizes two Memory blocks (implemented in distributed RAM) to store input image frame and output frame. The input frame buffer can be filled by real time video data coming from ADC data or by intermediate image data from Image Output buffer. The main sequence of operations is controlled by a centralized Control engine which generates different control, address signals for proper operations of all the sub blocks and memory. The control Unit also accepts user data from PCI interface and sends them to core cell

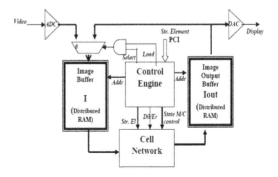

Figure 1. Architectural block diagram of the proposed system.

network with appropriate timing. The loading of input buffer is controlled using the signal load and select from control engine. Select is high when intermediate data has to be loaded into input buffer. The output buffer can be reloaded to input buffer as well as can be sent to DAC for display in monitor.

3.1 *Cell network*

The cell network consists of parallel cell array implemented using FSM. Each cell represent individual pixel. As morphological operations are regular in nature, i.e. same operation is applied to all the cells simultaneously; a global FSM is used instead of local FSMs to save state storage memory. The cells are interconnected to their neighboring cells in eight neighborhoods with a particular fashion. The global connection for loading SE, global control signals and clock is connected separately. An individual cell structure is shown in Figure 2.

Each cell is associated with corresponding input and output frame memory element ($I(j,k)$ and $Iout$ (j,k) for cell (j,k)). The global control signal determines the operation of the cell either in dilation mode or erosion mode. The three control signals namely latch en, transmit and receive sequence the internal operation of each cell. These signals are generated from global state machine. These signals are used by each cell to generate output and intermediate variables, equivalent to local state machines running as shown in Figure 3.

Each state machine has four states, *Wait, Latch En, Transmit* and *Receive*. The machine runs in idle condition in *Wait* state during frame initialization and loading. In *Latch En* state the input image is loaded into cell memory as well as structuring element for operation is also loaded. The connector outputs to neighboring cells are also kept at '0' value. In *Transmit* state the connector values are generated depending on condition of control signal

(dilation/erosion) and latched intensity value. These connection values are latched in respective cell during *Receive* state. Then the machine again goes to *Wait* cell for next operation sequence to appear.

The inter cell connections are shown in Figure 4. The latched values of connections from neighboring cells and intensity value of each cell is used in a combinatorial circuit to generate Dilated/ Eroded output according to the control signal. This circuit is shown in Figure 5.

4 OPERATION SEQUENCE AND TIMING DIAGRAM

To achieve morphological gradient we have to perform dilation followed by erosion. We load

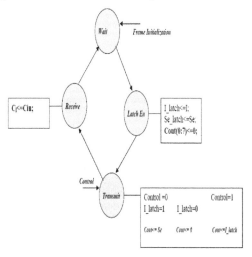

Figure 3. State machine controlling the operation of each cell.

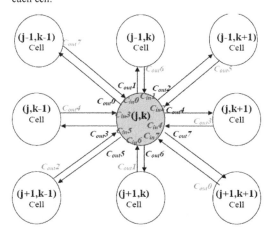

Figure 4. Connection diagram between in and out connectors of a cell and its neighbors.

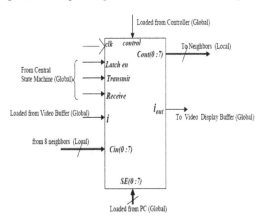

Figure 2. A local processing cell.

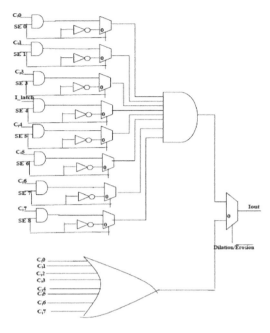

Figure 5. Cell architecture for dilation and erosion.

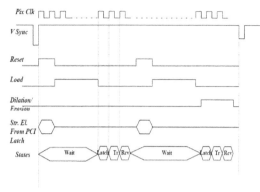

Figure 6. Overall timing diagram.

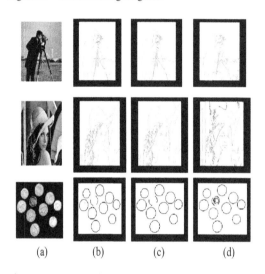

Figure 7. Detected edge of Cameraman, Lena and coins. (a) original image, (b) Sobel output, (c) Prewitt output and (d) Proposed output.

the structural elements from PC through PCI Bus. The intermediate dilated and eroded image of current frame is stored in intermediate buffers implemented in distributed RAM. Though output buffer can be reloaded into input buffer it is not used for gradient detection. Initially the dilated output is stored in *Iout* Buffer and when eroded data is available in each individual cell output the output buffer is updated by subtracting the eroded data from previously stored dilated data. For this operation a special memory interface circuit is designed which is controlled by main control unit. The overall timing diagram is shown in Figure 6.

5 RESULTS AND DISCUSSIONS

We have first simulated our design in MATLAB environment. We have compared the detected edge of our proposed methods with standard edge detector output like Sobel and Prewitt. The result is displayed in Figure 7 for three standard images namely *Cameraman, Lena and Coins*. We can see from the figure that proposed edge detector performs better than Sobel and Prewitt as it detects more edge details from the images. The quantitative evaluation of the method is done using the metric Peak Signal to Noise Ratio (PSNR). The values are summarized in Table 2, which shows an improved behavior of proposed method.

Table 1. Operation sequence and buffer status.

Operation sequence	Input buffer	Output buffer
Load	Input image	–
Dilate	Input image	Dilated image, I_{dil}
Load	Input image	Dilated image, I_{dil}
Erode	Input image	$I_{out} = I_{dil} - I_{out}$

We have implemented the design in VirtexII P FPGA. The logic utilization of combined cell architecture for 768×576 (PAL resolution) array is about 60% of XC2VP30 FF1152 Chip. The maximum frequency achieved for a pipelined design is 533 MHz corresponding to a critical path of 1.87 ns. Though, we have operated the design with 400 MHz clock (a 100 MHz crystal oscillator is frequency multiplied by 4 using Digital Clock

Table 2. PSNR of different edge detectors.

Image	Sobel	Prewitt	Proposed
Cameraman	5.054	5.321	6.276
Lena	9.632	9.468	10.389
Coins	14.054	15.321	15.776

Table 3. Hardware implementation comparison.

	Chien et al. (2002)	Sheu et al. (1997)	Proposed
Hardware Utilization	64764 Gates	1800 Gates	18483 Logic Cells
Image Size	352 × 288	512 × 512	768 × 576
Clock Frequency	40 MHz	33 MHz	400 MHz
Throughput	9.97 MPS	7.864 MPS	40 MPS

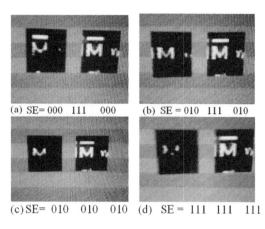

(a) SE= 000 111 000   (b) SE = 010 111 010

(c) SE= 010 010 010   (d) SE = 111 111 111

Figure 9. Eroded images with different structuring elements (right image is original image whereas left image is processed image).

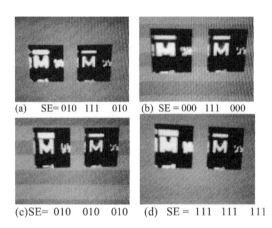

(a) SE= 010 111 010   (b) SE = 000 111 000

(c) SE= 010 010 010   (d) SE = 111 111 111

Figure 8. Dilated images with different structuring elements (right image is original image whereas left image is processed image).

Figure 10. Real time video: edge detection.

Manager available inside FPGA). For edge detection we have two operations erosion and dilation consumes 2 × 4 clock cycles or 8 clock cycles per pixel. Another two clock cycles are required for set difference and registering the output. Hence 10 clock cycles are required to process each pixel. Thus, a throughput of 40 Million Pixels per Second (MPS) is possible with this scheme. This rate is much higher than real time PAL (768 × 576) video rate at 25 frames per second, where the required throughput is 11.05 MPS. Performance comparison with hardware implemented morphological edge detectors available in literature is done in Table 3.

From Table 3 we can see that, though consuming a sizeable hardware resource, our design operates at higher speed and provides almost four times faster output. It also operates on a bigger size of image (PAL resolution).

We have simulated our result in Xilinx ISE Simulator and found the performance satisfactory. The real time morphological operations with different structural element are shown in Figures 8, 9 and 10 which are the captured video monitor output.

6 CONCLUSION

We have implemented a configurable Morphological Engine for performing gradient detection from a video scene. The real time performance is satisfactory as each operation takes 4 clock cycles due to parallel structure. Due to regular architecture VLSI implementation is very attractive for this architecture. Furthermore this architecture due to its configuration capability can be extended to other video processing application. The architecture can also be scaled for gray scale morphological operations.

REFERENCES

Chien Shao-Yi, Huang Yu-Wen, & Chen Liang-Gee (2002). A hardware accelerator for video segmentation using programmable morphology PE array. In *Proc. IEEE International Symposium on Circuits and Systems. IV,* 341–344.

Diamantaras K. I., & Kung S. Y. (1997). A Linear Systolic Array for Real-Time Morphological Image Processing. *Journal of VLSI Signal Processing Systems.* 17(1), 43–55.

Gonzalez R. C., & Woods R. E. (2012), Digital Image Processing, Pearson Education Inc. (3rd edition).

Lixia X., Tao L., Zuo-cheng W. (2010). Adaptive Edge Detection Algorithm Based on Morphology. *Computer Engineering*, 36,. 214–216.

Sheu Ming-Hwa, Wang Jhing-Fa, Chen Jer-Sheng, Suen An-Nan, Jeang Yuan-Long, & Lee Jau-Yien (1992). A Data-Reuse Architecture for Gray-Scale Morphologic Operations. *IEEE Transactions on Circuits and system-II: Analog and digital signal processing*, 39(10).

Xiufang W., Xingyuan Z., Running G. (2013). Research of Image Edge Detection Based on Mathematical Morphology. *International Journal of Signal Processing, Image Processing and Pattern Recognition.* 6. 227–236.

Computer, Communication and Electrical Technology – Guha, Chakraborty & Dutta (Eds)
© 2017 Taylor & Francis Group, ISBN 978-1-138-03157-9

A study on Bangla handwritten compound character recognition using LDP

M. Das
Department of Computer Science and Engineering, Murshidabad College of Engineering and Technology, West Bengal, India

S. Halsana, A. Banerjee & A. Sahoo
Department of Information Technology, Murshidabad College of Engineering and Technology, West Bengal, India

ABSTRACT: In this paper, Bangla handwritten compound character recognition by using simple yet robust Local Directional Pattern (LDP) features is studied and reported. LDP is robust as it is gradient based and resistant to non-monotonic illumination changes. No single algorithm can recognize the alphabets of different script languages. Besides, Bangla script contains a higher order of alphabet set, and recognition of these alphabets with high accuracy is a challenging task. Here, the character images are divided into blocks and sub-blocks based on the Centre of Gravity (COG) and features are extracted from each block. A total of 1176 dimension feature vectors are extracted. The Bangla data set, CMATERdb3, is used here in the experiment by considering 50 randomly selected character classes. In each class, 100 samples are taken and 2390 images are tested. Experimentally, 98.8% average accuracy is achieved by using LDP feature.

1 INTRODUCTION

Bangla ("*Bengali*") is the fourth most widely spoken language in India and fifth in the world. It is also the national language of Bangladesh. Nowadays, it is very important to automate business document processing, bank cheque processing, postal address processing, and many more with portable devices such as mobile phone and notepad. The automation process requires the recognition of characters known as Optical Character Recognition (OCR).

The real challenging issue in Bangla script is that it contains several alphabets with complex shape and structure. It becomes more challenging when handwritten Bangla alphabets are considered with different stylus of representation of individual writer. Bangla script contains 51 basic characters (11 vowels, often called modifier, and 40 consonants), one special character called Diacritic, and about 334 compound characters. The basic character sets in Bangla script are shown in Figures 1 and 2. Table 1 shows the formation of Bangla compound characters from basic characters.

Figure 1. Vowels of Bangla script.

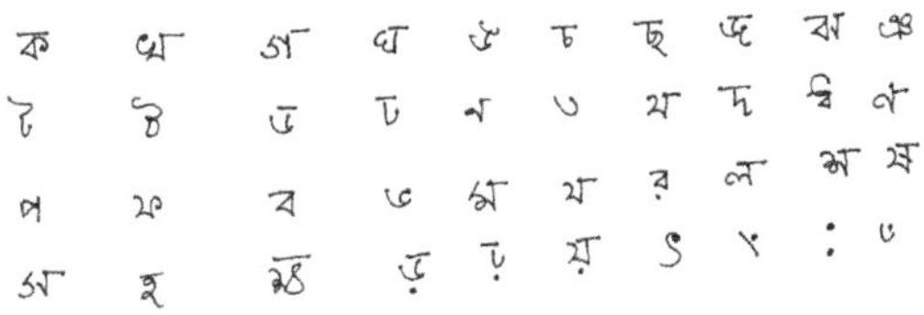

Figure 2. Consonants of Bangla script.

Table 1. Illustration of the formation of Bangla compound characters from basic characters.

Basic characters		Compound character
ক + র	=	ক্র
ঙ + শ	=	ঙ্শ
চ + ছ	=	চ্ছ
ন + ট	=	ন্ট
ক + ম + ম	=	ক্ম্ম
স + ত + র	=	স্ত্র

33

2 BRIEF DESCRIPTION OF PREVIOUS WORK

Several papers concerning handwritten character recognition in English or Roman languages exist in the literature, but only few papers have been found for Bangla handwritten character recognition (Bhowmik et al. 2004, Basu et al. 2005, Roy et al. 2005, Pal et al. 2008, among others) to date. Furthermore, most papers deal with Bangla numerals (Das 2012, Das 2012) or Bangla basic character set (Das 2009, Rahman 2002, Bhattacharya 2006). One of the important contributions considering a full set of handwritten compound characters was made by N. Das et al. (2012). The features are used based on Convex Hull and Longest Run. A two-pass classifier was used with Genetic Algorithm to recognize compound characters. A total of 240 classes, consisting of 33,282 train and 8254 test character images, are considered here. The recognition rates determined by the authors using SVM and MLP are 84.66% and 78.92%, respectively. The lowest recognition rate found in this paper for class ৲খ is 36%. Considering OCR on Bangla characters, two major approaches were found: single stage (Bhowmik 2004, Basu 2005) and multistage (Roy 2005, Das 2012, Rahman 2002, Bhattacharya 2005). Multi-layer perceptions (MLPs) (Bhowmik 2004, Basu 2005, Das 2010, Bhattacharya 2005), Support Vector Machines (SVMs) (Das 2012, Das 2010), and Quadratic Discriminate Function (QDF) (Bhattacharya 2008, Pal 2008) are the different classifiers involved with OCR on handwritten Bangla alphabetic character recognition. Pal et al. (2008) used Modified Quadratic Discriminant Function (MQDF) for recognition procedure based on directional information by arc tangent on gradient. Fivefold cross-validation technique has accuracy of approximately 85.90%. U. Bhattacharya et al. (2008) used MQDF on online handwritten compound character recognition over word level and achieved an accuracy of 82.34% on test set. To the best of the authors' knowledge, very few papers concerned with compound characters, which constitutes 85% of the total character set. The next section describes our approach on Bangla handwritten compound character recognition using LDP features and SVM classifier.

3 PRESENT WORK

3.1 Data set

In this paper, the CMATERdb3 data set consisting of 171 classes of isolated compound character sets is considered. Among these sets, we choose 50 classes randomly, presented in Table 2, and extract features followed by some preprocessing.

Table 2. Recognition accuracy of different class of Bangla character.

Class label	Bangla character class	Count of test samples	Count of positive classification	Percentage of average accuracy
1	পর	52	27	0.982845
2	ওগ	43	38	0.995816
3	কষ	49	40	0.991213
4	তর	53	39	0.985774
5	নদ	51	41	0.991213
6	চছ	49	36	0.993305
7	নত	51	33	0.983682
8	নদর	49	27	0.986611
9	নতু	55	35	0.98159
10	গর	51	33	0.985774
11	সট	61	33	0.980335
12	মব	48	37	0.990377
13	তত	52	34	0.985356
14	কত	51	48	0.995397
15	নট	40	27	0.986611
16	লপ	52	19	0.975314
17	ষট	44	30	0.986192
18	নতর	50	28	0.979916
19	কর	54	44	0.992887
20	নন	51	42	0.990377
21	দধ	51	44	0.996234
22	নধ	49	33	0.98954
23	ওক	41	33	0.994142
24	নড	51	26	0.980335
25	ফর	51	37	0.991213
26	জঞ	41	31	0.992469
27	কট	42	33	0.992469
28	শচ	53	41	0.991632
29	টর	44	30	0.990795
30	তব	55	43	0.991632
31	লল	52	36	0.985774
32	বর	52	47	0.995397
33	ণড	40	33	0.994979
34	শর	50	35	0.981172
35	দর	51	32	0.988285
36	সপ	26	15	0.983264
37	ঞজ	40	28	0.993305
38	নস	50	40	0.985774
39	মভ	50	33	0.982845
40	শব	51	40	0.988285
41	বদ	50	41	0.991632
42	শন	50	33	0.988703
43	পপ	45	30	0.986611
44	পত	51	37	0.990795
45	লট	41	28	0.985774
46	মফ	25	14	0.984100
47	পট	48	25	0.986611
48	মর	50	29	0.980753
49	হম	43	36	0.993724
50	মফ	41	22	0.983682

3.2 Preprocessing

Before going to feature extraction procedure, the character images are processed by the following steps.

3.2.1 Binarization
The character images are binarized by Ostu's method, which is based on local adaptive-based thresholding method.

3.2.2 Region of interest detection
After binarization, the region enclosed by the rectangle to which the compound character belongs is extracted.

3.2.3 Linear size normalization
All character images are normalized to a fixed size of 64×64 by experiment with an aspect ratio preserve linear mapping technique.

3.3 Feature extraction

Local Directional Pattern (LDP) features (Jabid 2010) on Bangla handwritten character recognition were used. LDP computes the edge response of an image in 8-direction using Kirch's mask and generates a coded image I_L. The masks are shown in Figure 3. For each image pixel I(x, y), m_i, i = 0–7 is calculated, which gives the response in 8-direction. LDP chooses K highest response value to determine the presence of corner or edges in a particular direction by setting top K number of values of $|m_i|$ to 1 and others to 0, as shown in Figure 4. If K is set to 3, the encoded image will consist of 56 distinct values. The histogram of LDP image I_L(x, y) can be defined as:

$$H_{LDP} = \sum_{x,y} P(I_L(x,y) = C_i)$$

$$C_i = i - th\ LDP\ code\ (0 \le i \le 56).$$

$$P(A) = \begin{cases} 1 & if\ A\ is\ TRUE \\ 0 & otherwise \end{cases}$$

$$\begin{bmatrix} -3 & -3 & 5 \\ -3 & 0 & 5 \\ -3 & -3 & 5 \end{bmatrix}$$
East m_0

$$\begin{bmatrix} -3 & 5 & 5 \\ -3 & 0 & 5 \\ -3 & -3 & -3 \end{bmatrix}$$
North East m_1

$$\begin{bmatrix} 5 & 5 & 5 \\ -3 & 0 & 5 \\ -3 & -3 & -3 \end{bmatrix}$$
North m_2

$$\begin{bmatrix} 5 & 5 & -3 \\ 5 & 0 & -3 \\ -3 & -3 & -3 \end{bmatrix}$$
NorthWest m_3

$$\begin{bmatrix} 5 & -3 & -3 \\ 5 & 0 & -3 \\ 5 & -3 & -3 \end{bmatrix}$$
West m_4

$$\begin{bmatrix} -3 & -3 & -3 \\ 5 & 0 & -3 \\ 5 & 5 & -3 \end{bmatrix}$$
SouthWest m_5

$$\begin{bmatrix} -3 & -3 & -3 \\ -3 & 0 & -3 \\ 5 & 5 & 5 \end{bmatrix}$$
South m_6

$$\begin{bmatrix} -3 & -3 & -3 \\ -3 & 0 & 5 \\ -3 & 5 & 5 \end{bmatrix}$$
South East m_7

Figure 3. Kirch's mask.

85	32	26
53	50	10
60	38	45

↓

	m_7	m_6	m_5	m_4	m_3	m_2	m_1	m_0
Mask Index	m_7	m_6	m_5	m_4	m_3	m_2	m_1	m_0
Mask value	161	97	161	537	313	97	-503	-393
Rank	6	7	5	1	4	8	2	3
Code Bit	0	0	0	1	0	0	1	1
LDP Code	19							

Figure 4. LDP code generation.

In this paper, the following steps are taken to extract LDP features from the character images

Step 1: Divide the character image based on Center of Gravity (COG) into four blocks, which are further divided into 16 sub-blocks. COG is computed using the following equations:

$$COG_x = \frac{\sum_{x,y} x \bullet I(x,y)}{\sum_{x,y} I(x,y)} \tag{1}$$

$$COG_y = \frac{\sum_{x,y} y \bullet I(x,y)}{\sum_{x,y} I(x,y)} \tag{2}$$

Equations 1 and 2 give COG values of X-coordinate and Y-coordinate, respectively,

$$where\ I(x,y) = \begin{cases} 1 & for\ ON\ pixel \\ 0 & otherwise \end{cases}$$

Step 2: Obtain the LDP features $\{H^i_{LDP}\}^{16}_{i=1}$ from each of these sub-blocks and concatenate the results of the length ($56 \times 16 = 896$) of feature vector. In this paper, we neglect the 1st bin of H^i_{LDP}, as it covers mostly the background or the region in strokes. Thus, we obtain a reduced feature vector of length $55 \times 16 = 880$. LDP features are also extracted from the four main blocks and from the entire image considering it as a single block. After concatenation, we obtain the total feature length of 1155. This process is depicted in Figure 5.

3.4 Classification

Support Vector Machine (SVM) is a widely accepted supervised classifier in pattern recognition. It is now equally accepted in recognizing Bangla handwriting over MLP (Das 2010).

35

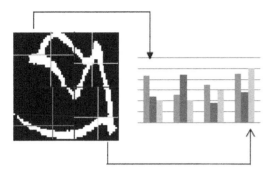

Figure 5.   Feature vector generation from block image.

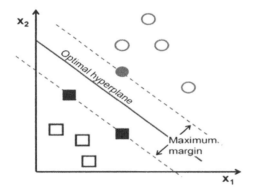

Figure 6.   Optimal hyper-plane formation by SVM.

Basically, SVM is designed to solve two class problems by creating a hyper-plane between positive and negative class data set and maximizing the gap of separation between them as shown in Fig. 6. Labeled data (D_i, L_i), i = 1, ..., m, where $D_i \in R^n$ is the training set and $L_i \in \{1, -1\}$ is the label, are fed into the system and SVM attempts to put the hyper-plane that gives the largest minimum distance to the training examples.

Maximal separable hyper-plane is obtained by solving the optimization equation 3:

$$\min_{w,b,\tau} \frac{1}{2} w^T w + C \sum_{i=1}^{m} \tau_i \qquad (3)$$

Under the condition,

$$L_i \left(w^T \varnothing (D_i) + b \right) \geq 1 - \tau_i$$

$$\tau_i \geq 0$$

where $\varnothing$, is the function, D_i is mapped into higher dimension, and C(>0) is the penalty parameter of the error term. The Kernal function for SVM is given by:

$$K\left(D_i, D_j\right) \equiv \varnothing \left(D_i\right)^T \varnothing \left(D_j\right) \qquad (4)$$

For Radial Basis Function (RBF):

$$K(D_i, D_j) = \exp(-\gamma \| D_i - D_j \|^2), \gamma > 0 \qquad (5)$$

is the Kernal parameter. Here, the RBF kernel is used, as it gives better result in handwritten character recognition applications (Das 2010).

4   EXPERIMENT AND RESULT

We conduct our experiment using SVM classifier in MATLAB on CMATERdb3 data set. For training, we choose 50 classes randomly using 100 images from each class from train set and 2390 images from test set. The number of test samples in each class is varying with respect to the availability. Average recognition accuracy of each class is calculated from the confusion matrix using equation 6:

$$\text{Recognition accuracy} = \frac{TP + TN}{TP + FP + FN + TN} \qquad (6)$$

where TP (True Positive) measures true acceptance in class and FP (False Positive) measures false acceptance in class.

Using these, the recognition accuracies found in test data set are given in Table 2.

A comparison result with Histogram of Oriented Gradient (HOG) features extracted from Bangla data set is shown in Figures 7 and 8 in terms of accuracy and true positive response,

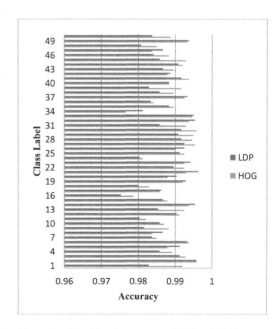

Figure 7.   Classification accuracy LDP vs. HOG.

36

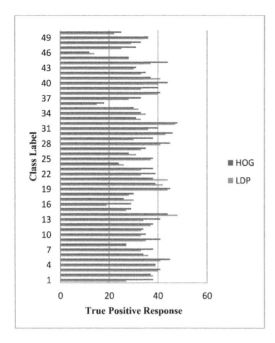

Figure 8. Recognition rate based on true positive response HOG vs. LDP.

Table 3. Some misclassified character images of test samples.

Image sample	True class	Misclassified class
প্র	প্র	গ্র
ক্ষ	ক্ষ	ম্ফ
ত্র	ত্র	ক্র
ন্দ	ন্দ	ব্দ
ন্ত	ন্ত	ন্ড
ন্দ্র	ন্দর	ন্ত্র
ন্ট	ন্ট	ন্ড

respectively. An average accuracy of 98.6% is observed using LDP-based features and 98.7% by using HOG. The best true positive response was observed in class 14 (ড্গ) and lowest in class 16 (ঞ) from LDP feature-based recognition. Some incorrectly classified samples are shown in Table 3.

5 DISCUSSION

The recognition accuracy that we obtained using LDP feature is comparable to HOG feature. The average true positive response is 72% using only LDP feature. Improvement by combining these LDP and HOG together with some other topological features will be studied in future. An average accuracy of 96.6% was found by using PCA on data set and reduced to 272 dimension feature vectors by experiment.

REFERENCES

Basu, S., N. Das, R. Sarkar, M. Kundu, M. Nasipuri & K. Basu (2005), Handwritten Bangla alphabet recognition using an MLP based classifier, in: Proceedings of the 2nd National Conference on Computer Processing of Bangla, Dhaka, Bagladesh, pp. 285–291.

Basu, S., N. Das, R. Sarkar, M. Kundu, M. Nasipuri & D. Basu (2009), A hierarchical approach to recognition of hand written Bangla characters, Pattern Recognition. 42, 1467–1484.

Bag, S., P. Bhowmick & G. Harit (2011), Recognition of Bengali hand written characters using skeletal convexity and dynamic programming, in: Proceeding soft the Second International Conference on Emerging Applications of Information Technology (EAIT), pp. 265–268.

Bag, S., G. Harit & P. Bhowmick (2014), Recognition of Bangla compound characters using structural decomposition, Journal of Pattern Recognition, pp. 1187–1201.

Bhowmik, T., U. Bhattacharya & S. Parui (2004), Recognition of Bangla hand written characters using an MLP classifier based on stroke features, in: Pal. N, N. Kasabov, R. Mudi, S. Pal. & S. Parui (Eds.), Neural Information Processing, Springer, Berlin, Heidelberg, pp. 814–819.

Bhattacharya, U., S. K. Parui, M. Shridhar & F. Kimura (2005), Two-stage recognition of hand written bangle alphanumeric characters using neural classifiers, in: B. Prasad(Ed.) 2nd Indian International Conference on Artificial Intelligence, Pune, India, pp. 1357–1376.

Bhattacharya, U., M. Shridhar & S. Parui (2006), On recognition of hand written Bangla characters, in: Kalra. P & S. Peleg (Eds.), Computer Vision, Graphics and Image Processing, Springer, Berlin, Heidelberg, pp. 817–828.

Bhattacharya, U., A. Nigam, S. Y. Rawat & P. K. Parui (2008) An Analytic Scheme for Online Handwritten Bangla Cursive Word Recognition, Proc. of the 11th ICFHR, pp. 320–325.

Das, N. & Basu. S & Sarkar. R & Kundu. M & Nasipuri. M & Basu. D, (2009), "An Improved Feature Descriptor for Recognition of Handwritten Bangla Alphabet," in International conference on Signal and Image Processing, Mysore, India, pp. 451–454.

Das, N., S. Basu, R. Sarkar, M. Kundu, M. Nasipuri, & D. Basu (2009), "Handwritten Bangla Compound character recognition: Potential challenges and probable solution," in 4th Indian International Conference on Artificial Intelligence, Bangalore, pp. 1901–1913.

Das, N., D. Das, R. Sarkar, S. Basu, M. Kundu & M. Nasipuri (2010), Hand written Bangla basic and compound character recognition using MLP and SVM classifier, J. Comput. 2109–115.

Das, N., R. Sarkar, S. Basu, M. Kundu, M. Nasipuri & D. Basu (2012), "A genetic algorithm based region sampling for selection of local features in handwritten digit recognition application," Applied Soft Computing, vol. 12, pp. 1592–1606.

Das, N., J. Reddy, R. Sarkar, S. Basu, M. Kundu, M. Nasipuri & D. Basu (2012), "A statistical & topological feature combination for recognition of handwritten numerals," Applied Soft Computing, vol. 12, pp. 2486–2495.

Das, N., K. Acharya, R. Sarkar, S. Basu, M. Kundu & M. Nasipuri (2012), "A Novel GA-SVM Based Multistage Approach for Recognition of Handwritten Bangla Compound Characters," Proceedings of the International Conference on Information Systems Design and Intelligent Applications 2012 (INDIA 2012) held in Visakhapatnam, India, January 2012." vol. 132, S. Satapathy, et al., Eds., ed: Springer Berlin / Heidelberg, pp. 145–152.

Das, N., R. Sarkar, S. Basu, P. Saha, M. Kundu & M. Nasipuri (2015), Handwritten Bangla character recognition using a soft computing paradigm embedded in two-pass approach, Journal Pattern Recognition, pp-2054–2071.

Das, N., K. Acharya, R. Sarkar, S. Basu, M. Kundu & M. Nasipuri, "A Benchmark Data Base of Isolated Bangla Handwritten Compound Characters" IJDAR (Revised version communicated).

Jabid, T., MD. Kabir, O. Chae, (2010), Local Directional Pattern (LDP) for Face Recognition, Proceedings of the IEEE International Conference on Consumer Electronics, 329–330.

Pal, U., T. Wakabayashi & F. Kimura (2008), MQDF Based Recognition of Off-line Bangla Handwritten Compound Character, Journal Advances in Engineering Science, pp. 1–8.

Rahman, A. F. R, R. Rahman & M. CFairhurst (2002), Recognition of hand written Bengali characters: a novel multistage approach, Pattern Recognit. 35997–1006.

Roy, K., U. Pal & F. Kimura (2005), Bangla hand written character recognition in: Prasa. B, d (Ed.) 2nd Indian International Conference on Artificial Intelligence, Pune, India, pp. 431–443.

Computer, Communication and Electrical Technology – Guha, Chakraborty & Dutta (Eds)
© 2017 Taylor & Francis Group, ISBN 978-1-138-03157-9

Handwritten signature recognition and verification using artificial neural network

A. Roy, S. Dutta & M. Das
Department of Computer Science and Engineering, Murshidabad College of Engineering and Technology, West Bengal, India

ABSTRACT: Handwritten signature is an important biometric attribute of a human being and the most natural method of authenticating a person's identity. Verification of handwritten signature can be performed either off-line or online based on application. In this paper, offline handwritten signature recognition and verification using neural network is projected, where the signature is written on a paper and is obtained using a scanner and presented in an image format. There are various approaches to signature recognition with a lot of scope of research. The method presented in this paper consists of global feature extraction, neural network training with extracted features and verification. A verification stage includes applying the extracted features of test signature to a trained neural network. Signatures are verified based on parameters extracted from the signature using various global features. The method is implemented using MATLAB.

1 INTRODUCTION

The hand written signature is regarded as the primary means of identifying the signer of a written document based on the implicit assumption that a person's normal signature changes slowly and is very difficult to erase, alter or forge without detection. Signatures are composed of special characters and flourishes and therefore most of the time they can be unreadable. Also intrapersonal variations and interpersonal differences make it necessary to analyse them as complete images and not as letters and words put together, referring for a detailed description to Velez et al. (2003).

Signature recognition is the process of verifying the writer's identity by checking the signature against samples kept in the database.

There are several approaches of verifying the authenticity of a signature. Approaches for signature verification fall into two categories according to the acquisition of data i.e. On-line and Off-line (Faunder-Zanuy 2007, Jain et al. 2002). The difference of on-line and off-line lies in how data are obtained. In on-line, data are obtained using an electronic tablet and other devices. In the off-line, images of the signature written on a paper are obtained using a scanner or a camera (Velez et al. 2003).

Off-line data is a 2-D image of the signature. Processing Off-line signature is complex due to the absence of stable dynamic characteristics.

The non-repetitive nature of variation of the signatures, because of age, illness, geographic location and perhaps to some extent the emotional state of the person, accentuates the problem. All these coupled cause large intrapersonal variation.

The system for signature verification should be neither too sensitive nor too coarse. It should have an acceptable trade-off between a low False Acceptance Rate (FAR) and a low False Rejection Rate (FRR) (Coetzer et al. 2004).

In this paper the various approaches have been taken for offline signature verification using neural network. The main reasons for the widespread usage of Neural Networks (NNs) in pattern recognition are their power and ease of use (Pansare & Bhatia 2012). A simple approach is to firstly extract a feature set representing the signature (details like length, height, duration, etc.), with several samples from different signers. The second step is for the NN to learn the relationship between a signature and its class (either "genuine" or "forgery"). Once this relationship has been learned, the network can be presented with test signatures that can be classified as belonging to a particular signer. NNs therefore are highly suited to modelling global aspects of handwritten signatures (Blumenstein et al. 2006).

We approach the problem in two steps: (i) Training signatures, (ii) recognition or verification of given signature. The block diagram of the system is given in Figure 1.

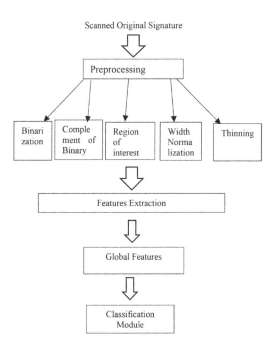

Figure 1. Block diagram of proposed system.

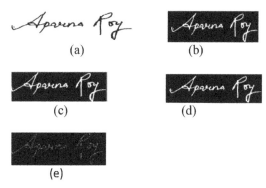

Figure 2. Preprocessing steps: (a) Binarization (b) Complement of binary image (c) Region of Interest Detection (d) Width Normalization (e) Thinning.

2 PROPOSED METHODOLOGY

In this section, block diagram of system is discussed. Fig. 1 gives the block diagram of proposed signature verification system which verifies the authenticity of given signature of a person.

These steps are

2.1 Image pre-processing
2.2 Feature extraction
2.3 Classification of Neural network

2.1 *Pre-processing*

The pre-processing step is applied both in training and testing phases. Signatures are scanned in RGB. The purpose of this phase is to make signatures standard and to improve accuracy for feature extraction. The pre-processing stage involves some of the following steps:

2.1.1 *Binarization*
The signature image in RGB has been converted to binary image to make features extraction simpler. Figure 2(a).

2.1.2 *Complement of binary image*
Complement of binary image is obtained for computational simplification by changing the background into black and foreground of image into white. Figure 2(b).

2.1.3 *Region of interest*
Region of Interest have to be identified from both Sample Signature and corresponding Test Signature. Cropping is done with respect to bounding box of image by calculating minimum and maximum value of X and Y coordinates of image and using Matlab function. Normally all the signatures in the database are made to fit inside a rectangle of same height and width. Figure 2(c).

2.1.4 *Width normalization*
Signature dimension may have intrapersonal and interpersonal differences. So the image width is adjusted to a default value. Figure 2(d).

2.1.5 *Thinning*
The goal of thinning is to eliminate the thickness differences of pen by making the image one pixel thick. In this system Hilditch's Algorithm is used Figure 2(e).

2.2 *Features extraction*

Feature extraction is the second major step in signature recognition and verification. An ideal feature extraction technique uses a minimal feature set that is used to maximize interpersonal distance between signature samples of different individuals while minimizing intrapersonal distances for those belonging to the same individual [4] (Batista et al. 2007). Various features used in the feature set of the proposed method are explained as follows:

2.2.1 *Aspect ratio*
It is the ratio of width of signature image to the height of the image. This is done because width or height of person's signature may vary but its ratio remains approximately equal.

2.2.2 *Area of normalization*

Normalized Area (NA) is the ratio of the area occupied by signature pixels to the total area of the image.

2.2.3 *White pixel density*

White pixel density can be defined total number of white pixel divided by total number of pixels of an image.

2.2.4 *Euler number*

The Euler number is the total number of objects in the image minus the total number of holes in those objects. Objects are connected sets of on pixels, that is, pixels having a value of 1.

2.2.5 *Centroid*

Returns a 1-by-Q vector that specifies the centre of mass of the region. The first element of Centroid is the horizontal coordinate (or x-coordinate) of the centre of mass, and the second element is the vertical coordinate (or y-coordinate).

2.2.6 *Entropy*

Entropy is a statistical measure of randomness that can be used to characterize the texture of the input image.

Entropy is defined as sum (p.*log2(p))
where p contains the histogram counts.

2.2.7 *Maximum horizontal histogram and vertical histogram*

The horizontal histograms are calculated for each row and the row which has the highest value is taken as maximum horizontal histogram. The vertical histograms are calculated for each column and the column which has the highest value is taken as maximum vertical histogram.

2.2.8 *Number of objects*

Number of objects of an image can be calculated by finding the connected components of the image and then we can apply property NumObjects in getfield method.

2.2.9 *Solidity*

Scalar specifying the proportion of the pixels in the convex hull that are also in the region. Computed as Area/Convex Area. This property is supported only for 2-D input label matrices. Also some statistical features like mean, variance, and standard deviation are also used.

2.3 *Classification of neural network*

The simplest definition of a neural network is provided by the inventor of one of the first neuron computer, Dr. Robert Hecht-Nielsen. He defines a neural network as:

"A computing system made up of a number of simple, extremely interrelated processing elements, which practice information by their dynamic state response to peripheral inputs" details at http://www.psych.utoronto.ca

There are several algorithms that can be used to create an artificial neural network, but the Back propagation (Vapnik 1995) was chosen because it is probably the easiest to implement, while preserving efficiency of the network. Backward Propagation Artificial Neural Network (ANN) (Mitchell 1997) use more than one input layers (usually 3). Feed forward back propagation neural network use to classify signature according to feature vector characteristic. Input vectors and the corresponding target vectors are used to train feed forward back propagation neural network.

Neural network train until it can classify the defined pattern. The training algorithms use the gradient of the performance function to determine how to adjust the weights to minimize performance. The gradient is determined using a technique called back propagation, which involves performing computations backwards through the network. The back propagation computation is derived using the chain rule of calculus.

3 SIGNATURE DATABASE

In this paper, for training and testing of signature recognition and verification 126 signature images are taken from 7 individuals.

For training 10 signatures (5 genuine & 5 random forge) are taken from each individual. And for testing 8 signatures (4 genuine & 4 random forge) are taken from each individual.

4 TRAINING AND TESTING

The recognition phase consists of two parts, training and testing respectively which is accomplished by Neural Network.

4.1 *Training phase*

4.1.1 *Neural network training*

In this proposed work, the neural network is trained with 70 signatures from 7 different individuals and some forged signatures are also used for training.

After pre-processing the signature images thinning is applied on the signature images to make it single pixel width to get better result. Then from this thinned images 11 features are extracted, that include energy, entropy, aspect ratio, histogram, solidity etc. this feature set is used to train Neural

Table 1. Comparison of the recognition results.

	True classification ratio	False classification ratio
ANN	89.28%	7.14%

Table 2. Comparison of verification results.

	FAR	FRR
ANN	7.14%	10.71%

Network (NN), the type of NN used is feed forward back propagation neural network. The NN is trained using this feature set on 70 signature images and also on some sample images testing have been performed that does not existed in the database.

4.2 Testing phase and results

Testing Phase include recognition and verification of signature which is done using Neural Network. Recognition is the process of finding the identification of the signature owner. Verification is the decision about whether the signature belongs to the authentic user.

4.2.1 Recognition

The original signatures are used for recognition. Table 1 shows the recognition performance of ANN (Artificial Neural Network).

4.2.2 Verification

In this system for each person 4 original and 4 forgery signature are tested. The possible cases in verification are False Acceptance (FA), False Rejection (FR). (FAR): It is the rate of number of forgeries accepted as genuine to the total number of forgeries submitted. (FRR): It is the ratio of the number of genuine test signatures rejected to the total number of genuine test signature submitted in verification the same 11 features are used.

The verification results of ANN's Back propagation method are given in Table 2.

5 CONCLUSION

The extracted features are used to train a neural network using Feed Forward Neural Network (FFNN) with error back propagation training algorithm. When the network was presented with signature samples for testing, it could recognize 25 signatures out of 28 signatures that were provided. Hence, the correct classification rate of the system is 89.28%.

However, it exhibited poor performance when it was presented with signatures that it was not trained for earlier. We did not consider this a "high risk" case because recognition step is always followed by verification step and these kinds of false positives can be easily caught by the verification system. Generally the failure to recognize/verify a signature was due to poor image quality and high similarity between two signatures. Recognition and verification ability of the system by using local features in combination with global features in the input data set will be studied in future. Moreover SVM (Support Vector Machine) classifier can be used to obtain better results.

REFERENCES

Batista, L., D. Rivard, R. Sabourin, E. Granger & P. Maupin (2007). State of the art in off-line signature verification. In: Verma B., Bluemenstein M. (eds.), Pattern Recognition Technologies and Applications: Recent Advances, (le). IGI Global, Hershey (2007).

Blumenstein, M., S. Armand & V. Muthukkumarasamy (2006). Off-line Signature Verification using the Enhanced Modified Direction Feature and Neural based Classification. International Joint Conference on Neural Networks, 1663–1669.

Coetzer, J., B. Herbst & J. du Preez (2004). Off-line signature verification using discrete random transform and a hidden Markov model. EURASPI Journal on Applied signal Processing, 559–571.

Faunder-Zanuy, M. (2007). Online signature based on vq-dtw. Pattern Recognition 40(3), 981–992.

Jain, A., F. Griess & S. Connell (2002). Online signature verification. Pattern Recognition 35(12), 2963–2972.

Mitchell, T. (1997). Machine Learning, McGraw-Hill, 81–126.

Pansare, A. & S. Bhatia (2012). Handwritten signature verification using neural network, International Journal of Applied Information Systems (IJAIS), 1(2), 2249–0868.

Velez, J.F., A. Sanchez & A.B. Moreno (2003). Robust Off-line Signautre Verification Using Compression Networks And Positional Cuttings, Proc. IEEE Workshop on Neural Networks for Signal processing, 1, 627–636.

Vapnik, V.N. (1995). The Nature of Statistical Learning Theory, Springer, 122–126.

Computer, Communication and Electrical Technology – Guha, Chakraborty & Dutta (Eds)
© *2017 Taylor & Francis Group, ISBN 978-1-138-03157-9*

Breast cancer detection using feature selection and active learning

S. Begum & S.P. Bera
Department of Computer Science and Engineering, Government College of Engineering and Textile Technology Berhampore, Murshidabad, West Bengal, India

D. Chakraborty
Department of ECE, Murshidabad College of Engineering Technology, Berhampore, Murshidabad, West Bengal, India

R. Sarkar
Department of Computer Science and Engineering, Jadavpur University, Kolkata, India

ABSTRACT: In any data classification problem where supervised learning is used, it is presumed that the labelled data can be obtained at ease. In fact, there are several research problems, because it is very expensive and tedious to obtain the labelled data. The situation can be conquered with a new theme called active learning. In this paper, we proposed pool-based active learning method, where user observes the pool of non-labelled instances to access the breast cancer data set. After selecting few samples from the pooled data, the user needs to label them. Here, we suggested three active learning methods with Support Vector Machines (SVMs) as a classifier and three methods, namely Entropy (Entrp), Smallest Margin (SM) and Least Confidence (LC), for choosing uncertain samples from the pooled data. In addition, to avoid redundancy and unwanted samples, we incorporated three feature selection algorithms, *namely* Fuzzy Preference-Based Rough Set (FPRS), Signal-to-Noise Ratio (SNR) and Neighbourhood Rough Set-based Feature Evaluation and Reduction (fs_con_N) to obtain the optimal number of features from the microarray data set.

1 INTRODUCTION

In supervised learning, algorithm providing label for the samples of training set is expensive and time consuming. Hence, active learning is worthwhile so as to minimise the size of the training set. Here, very few samples are assigned with label and merged with the original training set. Hence, in such paradigm, the actual training set is successively enlarged following an interactive procedure, which incorporates an expert (usually a human supervisor) to assign a correct label to any queried sample. In active learning, query function repeatedly asks for the label of the sample, which seems to be the most informative for a fruitful training of the classifier. The supervisor assigns label to the selected sample and again retrains the classifier using the refreshed training set. In this way, unwanted labelling of non-informative samples can be avoided, which is beneficial by reducing the time and cost of training samples. The pool-based active learning (Persello & Bruzzone 2014) method, where we observe pool of unlabeled samples, provides a query for selecting some samples from the pooled data. Then, it asks for providing label for these selected samples. The query for selecting the

most uncertain sample from the job pool plays an important role in active learning method.

In this paper, we applied active learning approach to obtain an efficient classifier for microarray data set. We used three methods, *namely* Entrp, SM and LC, to obtain the most uncertain samples from the pooled data set, as microarray data set is characterised by a large number of features. Hence, we proposed three feature selection algorithms, FPRS (Nanda & Majumder 1992), SNR (Maulik, Mukhopadhyay, & Chakraborty 2007) and fs_con_N (Swiniarski 2001), to obtain optimal number of features from the microarray data set. In addition, we compared its performances with supervised learning algorithm in terms of percentage of accuracy. The details of supervised learning algorithm are presented in Cunningham, Cord & Delany (2007).

2 BACKGROUND

2.1 *Support vector machine*

SVM (Burges 1998) is invented by Vapnik based on the concept of Vapnik–Chervonenkis (VC) theory and Structural Risk Minimization (SRM) principle, which attempt to enhance the margin

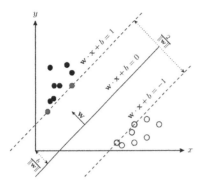

Figure 1. SVM for binary classification in linearly separable cases.

and reduce the training set error to achieve the best performance. Now we elaborate the basic SVM with binary classification problems.

Consider a binary classification problem: $\{x_i, y_i\}$ where $i = \{1, 2, ..., l\}$, $y_i \in \{-1, 1\}$ and $x_i \in R^d$, where x_i are the objects and y_i are the labels. The hyperplane is given by $w^t x + b = 0$, where w is an n-dimensional coefficient vector, which is normal to the hyperplane, and b is the offset with respect to origin. SVM (Karatzoglou, Meyer, & Hornik 2006) produces an optimal separating hyperplane, which maximises the separating margin between the two classes of data. Let us define two hyperplanes $P1 : w^T x + b = 1$, which is closest to one side of samples and $P2 : w^T x + b = -1$, which is closest to another side of data.

Now the maximum margin between the two separating hyperplanes is $d = (x^+ - x^-) \cdot \dfrac{w}{\|w\|} = \dfrac{2}{\|w\|}$. Maximising the distance means

Minimise

$$L(w) = 1/2\|w\|^2 \tag{1}$$

so that

$$y_i(w^T x_i + b) \geq 1, \forall i \tag{2}$$

The above equation can be written in terms of Lagrangian multiplier as:

Minimise

$$Lr_p(w, b, \beta_i) = 1/2\|w\|^2 - \sum_{i=1}^{i=n} \beta_i \left(y_i\left(w^T x_i + b\right) - 1\right) \tag{3}$$

where β_i are Lagrangian multipliers, $\beta_i > 0$. Now begin equation

$$\frac{\partial Lr_p}{\partial b} = 0 \Rightarrow \sum_{i=1}^{i=n} \beta_i y_i = 0 \tag{4}$$

$$\frac{\partial Lr_p}{\partial w} = 0 \Rightarrow w = \sum_{i=1}^{i=n} \beta_i y_i x_i \tag{5}$$

Now we substitute b and w in equation 3, which is transformed into its dual problem
Maximise

$$\sum_{i=1}^{i=n} \beta_i - 1/2 \sum_{i=1}^{i=n} \sum_{j=1}^{j=n} \beta_i \beta_j y_i y_j x_i^T x_j \tag{6}$$

so that

$$\sum_{i=1}^{i=n} \beta_i y_i = 0, \beta_i \geq 0 \tag{7}$$

From the Karush–Kuhn–Tucker (KKT) condition, $\beta_i(y_i(w^T x_i + b) - 1) = 0$. Hence, the support vector (sv) $\beta_i \neq 0$ carries the information regarding classification problem. Hence, the solution is

$$w = \sum_{i=1}^{i=n} \beta_i y_i x_i = \sum_{i \in sv} \beta_i y_i x_i \tag{8}$$

The equation $y_i(w^T x_i + b) - 1 = 0$ gives the value for b, where x_i is sv. Hence, the linear function can be expressed as:

$$L(x) = w^T x + b = \sum_{i \in sv} \beta_i y_i x_i^T x + b \tag{9}$$

The kernel function $k(\cdot, \cdot)$ maps the data from input vectors to a higher dimensional feature space $k(x_i, x_j) = (\phi_i, \phi_j)$. Kernel function (Yekkehkhany, Safary, Homayouni, & Hasanlou 2014) helps to develop different types of SVM. The different kernel functions are as follows:

(i) Linear kernel: $k(a_i, a_j) = a_i^T a_j$
(ii) Polynomial kernel: $k(a_i, a_j) = (1 + a_i^T a_j)^d$
(iii) RBF kernel: $k(a_i, a_j) = exp(-\|a_i - a_j\|^2 / 2\sigma^2)$
(iv) Sigmoid kernel: $k(a_i, a_j) = tanh(\beta a_i \cdot a_j + b)$

SVM constructs an optimal separating hyperplane. In linear kernel, $a_i \cdot a_j$ computes the dot product between two vectors. The degree of the polynomial kernel is measured by d, which is a positive constant. The similarity between two vectors a_i and a_j is measured by sigmoid kernel. Here, we considered RBF kernel, as it is beneficial, when there is no prior knowledge about the data set.

3 METHODOLOGY AND EXPERIMENTS

In this section, we briefly summarise the essential steps of the feature selection method and describe our proposed method.

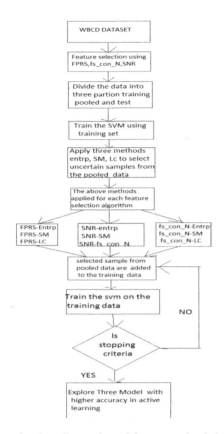

Figure 2.   Overall procedure of the proposed technique.

3.1 *Proposed technique*

We have a set of training samples and a pool of unlabeled data. The purpose is to increase the efficiency of the classifier using active learning method by repeatedly asking for the class label of the unlabeled data. We recommended algorithm executes query based on Entrp, SM and LC methods, respectively. The algorithm can be described as follows:

Step 1:   Preprocess the data in view of shorter training time and increased generalisation by decreasing overfitting. Three feature selection methods, namely Fuzzy Upward Consistency (FUC) model of FPRS, SNR and fs_con_N, are employed to preprocess the data set.

Step 2:   Train the classifier with the labelled data.

Step 3:   Data selection using active learning:

a. Entrp-Based Active Learning: Query the sample x, which the learner is most uncertain about. The entropy (Knuth 2014) of a sample x is denoted by:

$$\psi_{entrp}(x) = -\sum_y P_\theta(y\,|\,x) log_2 P_\theta(y\,|\,x) \qquad (10)$$

$P(y|x)$ is the probability of sample x for class y. The entropy of any sample x can be obtained for each class y_i. The maximum entropy designates the maximum uncertain sample. Then, the entropy of each sample is evaluated. Now five samples are selected from the pooled data with the highest entropy value and then removed from the pooled data (we provide the class label of the unlabeled data from our original data set as if it is given by the human annotator). Then, it is added to the original training data. In the following iteration, next five samples are chosen with the highest entropy value from the pooled data and added to the training data. This is continued for nth iteration.

b. SM (between most likely and second most likely labels)-based active learning: Query the sample x, which the learner is most uncertain about. The smallest margin (Schapire & Bartlett 1997) of a sample x is denoted by:

$$\psi_{SM}(x) = P_\theta(y_1^*|x) - P_\theta(y_2^*|x) \qquad (11)$$

where $P(y_1\,|\,x)$ is the probability of the sample x for class y_1 and $P(y_2\,|\,x)$ is the probability of the sample x for class y_2. The smallest margin value indicates the most uncertain sample. Then, the smallest margin value for each sample is evaluated. Then, five samples from the pooled data with the smallest margin value are selected and removed from the pooled data (we provide the class label of the unlabeled data from our original data set, as if it is given by the human annotator). Then, it is added to the original training data. In the following iteration, next five samples with the smallest margin value are chosen from the pooled data and added to the training data. This is continued for nth iteration.

c. LC-Based active learning: Query the sample x, which the learner is most uncertain about. The LC (Shafer & Vovk 2008) of a sample x is denoted by:

$$\psi_{LC}(x) = 1 - P_\theta(y^*\,|\,x) \qquad (12)$$

where $P_\theta(y^*\,|\,x)$ is the probability of the sample x for all classes of y_i. The LC value indicates the most uncertain sample. Then, the LC value for each sample is evaluated. Then, five samples from the pooled data are selected with LC values and removed from the pooled data set (we provide the class label of the unlabeled data from our original data set, as if it is given by the human annotator). Then, it is added to the original training data. In the following iteration, next five samples with the smallest LC value from the pooled data are selected and added to the training data. This is continued for nth iteration.

Step 4: The above steps continued for nth iteration, where at each iteration, five samples from the pooled data are added to the training data set, again retrain the classifier.

Step 5: The algorithm stops when the users find that the performance of the classifier increased with active learning. The highest classification accuracy from the final iteration is defined as the final accuracy.

The implementation is carried out via LIBSVM software, which is originally designed by Chang and Lin. The evaluation is performed on Intel Quad-Core Xeon 5130 CPU (2.0 GHz) with 4GB of RAM.

3.2 Feature selection method

Suppose that we have a microarray data set D containing m samples from different cancer subtypes. Each sample has n genes as its feature. The presupposition here is that not all genes are essential to diagnose the disease. Some genes are immaterial and some are dispensable. This may be a bottleneck to get a good classification accuracy of some machine learning algorithms. Hence, we used three feature selection (Wang & Zhou 2013) algorithms, namely FPRS, SNR and fs_con_N, in our proposed method.

FPRS: In this paper, we have used FPRS as a feature selection method applied onto the breast cancer data set. The following is the explanation of fuzzy preference relations: A fuzzy product set $u \times u$ can be outlined using a membership function $\mu_r : u \in [0,1]$. A fuzzy preference relation is denoted by a $p \times p$ matrix $(r_{ij})_{p \times p}$, where r_{ij} denotes the preference of r_i against r_j. $r_{ij} = 1/2$ denotes r_i and r_j are the same, $r_{ij} > 1/2$ shows that r_i has higher preference than r_j. $r_{ij} = 1$ shows that r_i has an absolute preference over r_j and $r_{ij} < 1/2$ denotes r_j is preferred over r_i. The preference matrix is $r_{ij} + r_{ji} = 1, \forall i, j \in 1, 2, ..., p$, where the cardinality of u is finite.

u can be considered as a finite number of elements $u = \{o_1, o_2, \cdots, o_p\}$. The feature value of any object o can be measured by $f(o,l)$, where l is the feature of the object. The upward and downward fuzzy preference relations over u are denoted by:

$$r_{ij} >= \frac{1}{1 + e^{-\beta(f(o_i,l)-f(o_j,l))}}$$

and

$$r_{ij} <= \frac{1}{1 + e^{-\beta(f(o_i,l)-f(o_j,l))}}$$

where β is a positive constant. Hence, FPRS is an ammulgumation between fuzzy preference relation

and Rough Set (RS) (Pawlak 2002), (Pawlak 1982) theory, which is shaped to evaluate the fuzzy preference. Let (U, F) denote an information system, where $U = \{o_1, o_2, \cdots, o_n\}$ is a non-empty finite set of objects and $F = \{f_1, f_2, \cdots, f_n\}$ is a finite set of attributes to classify the object. A Decision Table (DT) can be denoted by (U, C, D), where the set of features are put in the group condition C and decision D. Conditions (attributes) are given, now we have to forecast the decision for the objects of U. Let $\{dc_1, \cdots, dc_n\}$ be the N decision class labels, where $dc_1 \leq dc_2 \leq \cdots \leq dc_n$. Lower approximation and upper approximation are two fundamental operations in fuzzy rough set theory. Let $R^>$ and $R^<$ be the fuzzy preference relation designated by $P \subseteq C$. The fuzzy preference approximation qualities of the decision D in terms of P can be defined by:

$$\alpha_p(D) = \frac{\sum_j \left(\sum_{x \in dc_j^\leq} \overline{R^< dc_j^\leq(x)} + \sum_{x \in dc_j^\geq} \overline{R^> dc_j^\geq(x)} \right)}{\sum_j \left(\|dc_j^\geq\| + \|dc_j^\leq\| \right)}$$

Obviously, $0 \leq \alpha_p^>(D^\geq) \leq 1, 0 \leq \alpha_p^<(D^\leq) \leq 1$ and $0 \leq \alpha_p(D) \leq 1$. We say that D is upward consistent if $\alpha_p^>(D^\geq) = 1$. The Fuzzy Upward Consistency (FUC) model of FPRS is explained in Hu, U & Gua (2010).

Signal-to-Noise Ratio (SNR): Signal-to-noise ratio can be explained by:

$$SNR = \frac{\mu_1 - \mu_2}{\sigma_1 + \sigma_2} \tag{13}$$

where μ_i and σ_i, $i \in (1, 2)$, respectively, denote the mean and standard deviation of class i with respect to corresponding features. Higher absolute value for any particular gene shows that the gene expression level is high for that particular class and low in another class. Hence, this biasness is beneficial in differentiating the gene that exploits its versatility in the two classes of samples. After evaluating the SNR value for each gene, the genes are sorted in descending order according to their SNR value.

3.3 Data set description

Breast cancer is an untrammelled extension of breast cells. Tumour can be of two types, *viz.* benign (not harmful to health) and malignant (has the potential to be dangerous). Benign tumours are not harmful: their cells resemble normal cell; they grow at a slow pace and do not disrupt nearby tissue or affects other parts of the body. Malignant tumours are treated as cancerous. Malignant tumour cell spreads throughout the body beyond any control. Usually breast cancer seems to be

Table 1. The details the nine features of breast cancer data.

Label	Attribute	Domain
C1	Clump Thickness	
C2	Uniformity of Cell Size	
C3	Uniformity of Cell Shape	
C4	Marginal Adhesion	
C5	Single Epithelial Cell Size	1–10
C6	Bare Nuclei	
C7	Bland Chromatin	
C8	Normal Nucleoli	
C9	Mitoses	

Table 2. The details partitions of breast cancer data.

Total samples	Training set	Pooled data	Test set
683	60	149	434

malignant tumour, which has emerged from the cells in the breast.

In the proposed method, we conducted our experiment on Wisconsin Breast Cancer Data set (WBCD) (UCI Repository of machine learning databases). The WBCD data set contains 699 instances, among which 16 samples have missing values. Here, we worked on 683 samples, of which 444 belong to benign class and 239 samples belong to malignant samples. Each instance is characterised by nine features. The details of attributes are shown in Table 1. Class 2 represents benign tumour and class 4 represents malignant tumour (http://www.breastcancer.org/symptoms/understand_*bc*/what_*is*_bc.*jsp*).

4 EXPERIMENTAL RESULTS

To evaluate the effectiveness of our proposed technique, we conduct our experiment on WBCD data set. Here, we observed the performance of SVM against Naive Bayes, Multi Layer Perceptron (MLP) and Decision tree classifiers on the test data of size 482, which is 70% of the total data set size. We conducted the experiment by taking all the attributes of the data set into account, with the training set size of 204. The performances of SVM and other classifiers are shown in Table 3. As the performance of SVM is better than that of the other classifiers, we have chosen SVM as a learner in our proposed technique. Significant feature subsets are selected by FUC, fs_con_N and SNR methods. The classification accuracy on the test data following the method of Entrp-based active learning is shown in Table 4. Similarly, the results of SM- and LC-based active learning

Table 3. The Performance of the Classifiers in terms of accuracy.

SVM	Naive Bayes	MLP	Decision tree
97.51	96.05	95.85	92.94

Table 4. Entropy based active learning: Accuracy (%).

Model	Feature selection algorithm		
	FUC	fs_con_N	SNR
Supervised Learning	91.24	89.86	91.24
Active Learning	**94.93**	92.16	91.24

Table 5. Smallest margin based active learning: Accuracy (%).

Model	Feature selection algorithm		
	FUC	fs_con_N	SNR
Supervised Learning	91.24	89.86	91.24
Active Learning	**94.70**	90.78	91.70

Table 6. Least confidence based active learning: Accuracy (%).

Model	Feature selection algorithm		
	FUC	fs_con_N	SNR
Supervised Learning	91.24	89.86	91.24
Active Learning	**94.23**	92.62	91.93

methods are shown in Tables 5 and 6, respectively. It can be observed from the tables that Entrp-based FUC for feature selection active SVM provides 94.93% accuracy, which is the best empirical performance. On the contrary, the active SVM yields 94.70% accuracy using the selected features obtained by the SM-based FUC. As Active learning produces better result than supervised learning, we have given the statistical measurement of the proposed approach for the active SVM with FUC model. The values of recall, precision and F_measure for FUC model with three respective models are presented in Table 7. It is evident from the table that the proposed method exhibits better performance than supervised learning methods. Hence, it is clear that active learning is advantageous when only few labelled samples are available. The experimental result also shows that active SVM with FUC model outperforms the supervised learning methods.

Table 7. Performance measure (%) of FUC model.

Model	FUC		
	Precision	Recall	f_measure
Entrp-based FUC	92.19	100	95.94
SM-based FUC	92.19	99.61	95.76
Lc-based FUC	92.55	98.49	95.42

5 CONCLUSIONS

The present research delegates to explore the small labelled sample size problem in microarray data-based outcome foreshadow for breast cancer detection. This mainly tries to focus on the effectiveness of active learning rather than supervised learning method. As it is very difficult to have labelled sample, active learning has a great impact on the prediction of human cancer. Our result manifests noteworthy prospective of active learning in clinical diagnosis. We believe that the promising results obtained by the proposed method (active SVM with FUC) in classifying the breast cancer can ensure that the physicians can make very accurate diagnosis. In future work, we will explore our method on other data sets such as multispectral and hyperspectral data, gene microarray data, web data and so on, to demonstrate its potentiality (15–17 November 2014).

REFERENCES

Burges, C. J. C. (1998). A tutorial on support vector machine for pattern recognition. *Data Mining and Knowledge Discovery. 2*, 121–127.
Cunningham, P., M. Cord, & S. J. Delany (2007). *Machine Learning Technique for Multimedia.*
Hu, Q., D. U, & M. Gua (2010). Fuzzy preference rough set, information sciences. *Information Sciences 180*, 2003–2022.
Karatzoglou, A., D. Meyer, & K. Hornik (2006). Support vector machine in r. *Journal of Statictical Software. 15.*
Knuth, K. H. (2014). Entropy 2014. *J. Geophys. Res.—Earth Surface doi: 10.3390/e16020726*, 726–728.
Maulik, U., A. Mukhopadhyay, & D. Chakraborty (2007). Gene expression based cancer subtypes prediction through feature selection and transductive svm. *Journal of Latex class Files. 6.*
Nanda, S. & S. Majumder (1992). Fuzzy rough sets, fuzzy sets and systems. *45*, 157–160.
Pawlak, Z. (1982). Rough sets. *International Journal of Computer and Information Sciences 11.*
Pawlak, Z. (2002). Rough set theory and its applications. *Journal of Telecommunication and Information Technology.*
Persello, C. & B. Bruzzone (2014). Active and semisupervised learning for the classification of remote sensing images. *IEEE Transactions on Geoscience and Remote Sensing. 52.*
Schapire, R. E. & W. Bartlett, P. Lee (1997). Boosting the margin: A new explanation for the effectiveness of voting methods. *Machine Learning: Proceedings of the Fourteenth International Conference.*
Shafer, G. & V. Vovk (2008). A tutorial on conformal prediction. *Journal of Machine Learning Research. 9*, 371–421.
Swiniarski, R. W. (2001). Rough set methods in feature reduction and classification. *Int.l J. Appl. Math. Comput. Sci. 11*, 565–582.
Wang, G. & Y. Zhou (2013). A feature subset selection algorithm automatic recommendation method. *Journal of Artificial Intelligence Research. 47*, 1–34.
Yekkehkhany, B., A. Safary, S. Homayouni, & M. Hasanlou (2014). A comparison study of different kernel functions for svm-based classification of multitemporal polarimetry sar data. *The 1st ISPRS International conference on Geospatial Information Research.*

Computer, Communication and Electrical Technology – Guha, Chakraborty & Dutta (Eds)
© 2017 Taylor & Francis Group, ISBN 978-1-138-03157-9

Image contrast enhancement using a crossover-included hybrid artificial bee colony optimization

S.K. Mondal
Department of Instrumentation and Control Engineering, Haldia Institute of Technology, Haldia, India

A. Chatterjee
Department of Printing Engineering, Jadavpur University, Kolkata, India

B. Tudu
Department of Instrumentation and Electronics Engineering, Jadavpur University, Kolkata, India

ABSTRACT: Histogram Equalization (HE) is a popular approach toward improvement of low-contrast images. Many developments have been reported to address several limitations of the conventional HE. However, such algorithms also frequently fail to retain important image characteristics such as brightness, noise characteristics, and textural contents. Optimization of well-formulated objective function may address such problems. In this paper, natural behavior-inspired optimization techniques have been adopted in a hybrid manner, where the search dynamics of Artificial Bee Colony (ABC) is clubbed with the crossover operation of Genetic Algorithm (GA). The hybridization is adopted to avoid the complexity of multi-objective optimization, which is mathematically expensive. The hybrid optimization is employed with the objective function formulated with different image quality metrics conveying different image characteristics. The implementations are made with standard database and found to be potential over conventional techniques in terms of both visual and objective comparisons.

1 INTRODUCTION

Image enhancement is an important process for betterment of image quality in the field of image processing system, such as medical image processing, space image processing, and remote sensing (Zimmerman et al. 1988, Patil & Patil 2015). These operations are applied on digital images to improve the human perception of information. Contrast enhancement is one of the most vital parts of image enhancement systems. A contrast enhancement technique can obtain better-quality image for image-processing application (Gupta 2016).

Classical Histogram Equalization (CHE) is a common way to improve image contrast, which probabilistically remaps the existing image intensity levels to the available intensity levels [Ting et al. 2015]. In addition to conventional HE, many algorithms have been presented in this regard, such as Brightness Preserving Bi-Histogram Equalization (BBHE) and Dual Sub Image Histogram Equalization (DSIHE) (Wang & Ye 2005, Butola et al. 2016). However, many of such techniques suffer from problems like overenhancement, whitening of the image, non-preservation of image brightness, and false contouring.

Optimization-based HE may be a possible alternative to such problems with conventional techniques. Such types of techniques are Genetic Algorithm, Particle Swarm Optimization, and Artificial Bee Colony. These optimization techniques search an optimal solution, which will enhance the image contrast as well as preserve the mean brightness of the tested images to provide better result.

This paper presents a hybrid optimization approach, where the search dynamics of ABC and GA are combined. Artificial Bee Colony algorithm was proposed by Karaboga and Basturk (Karaboga 2005). It is observed that ABC algorithm is superior to other optimization algorithms such as GA and PSO (Draa & Bouaziz 2014). The Artificial Bee Colony (ABC) algorithm is a relatively new swarm intelligent technique, which is designed by the natural behavior of real honey bees in food foraging. In ABC algorithm, there are three types of bees: employed bees for searching food source, onlooker bees that are waiting on the dance area to choose a food source, and scouts that carry out a random search for new food source (Akay 2013). It is observed that ABC algorithm is very much effective for searching new solutions, but not so effective for generation-desired final solution (Zhu 2010).

The term Genetic Algorithm was introduced by John Holland (Maheshwari et al. 2016). This is basically the natural selection process invented by Charles Darwin by taking one input and computing an output from multiple solutions. One of the most potential operations in GA is crossover. It is a genetic parameter that combines two chromosomes (can also be called as Parents) to produce a new chromosome (also called as Children). The output of crossover is expected to provide the new chromosomes with betterment than both of the parents if it takes the best characteristics from each of the parents (Hashemi et al. 2010). The approach in this paper performs the search using ABC algorithm while performing the crossover operations on the solutions to find one solution as optimal.

The rest of the paper is organized as follows: section 2 describes the search dynamics of ABC and crossover of GA as well as the development of hybrid dynamics. The objective function designed for this work is been presented in section 3. The results obtained with various standard test images and related discussions are described in section 4, while the concluding remarks including major findings and future prospects are stated in section 5.

2 HYBRID ABC OPTIMIZATION

2.1 Artificial Bee Colony

In ABC algorithm, each food source assigned by employed bees represents a possible solution for optimization problem and the nectar amount represents the fitness of that solution (Karaboga & Basturk 2007). Let N be the food source number and the position of the ith food source be X_i ($i = 1$, 2, ..., N). The N number of food sources are randomly generated and assigned to N employed bees. These employed bees associated to the ith food source search for new solution by using Eq. (1) (Karaboga & Akay 2009):

$$V_{ij} = X_{ij} + \theta_{ij}(X_{ij} - X_{kj}) \quad (1)$$

where, $j = 1, 2...D$, D is the dimension of the optimized problem, θ_{ij} is a random real number within the range $[-1, 1]$, and k is a randomly selected index number in that colony. Then, each onlooker chooses a food source searched by employed bees based on the probability and produces a new food source. The probability function can be calculated using Eq. (2) (Draa & Bouaziz 2014):

$$p_i = fit_i \bigg/ \sum_{j=1}^{N} fit_j \quad (2)$$

where fit_i is the fitness of the solution X_i, fit_j is the total fitness of N number of food sources. The food source, which is not accepted according to the fitness value, becomes a scout and makes a random search for new solution by Eq. (3) (Draa & Bouaziz 2014):

$$Y_{ij} = X_j^{min} + r(X_j^{max} - X_j^{min}) \quad (3)$$

where r is a randomly generated real number within the range $[0, 1]$ and X_j^{min} and X_j^{max} are the lower and upper limits in the jth dimension of that problem, respectively.

2.2 Crossover of Genetic Algorithm

Genetic Algorithm (GA) is a process used to solve various optimization problems based on a natural selection process that creates biological evolution. One of the major steps in genetic algorithm is crossover. Crossover operation is a genetic operation used to produce new offspring (children) from the existing offspring (parent). Figure 1 shows the crossover process to produce children from parents (Hole et al. 2013).

2.3 Hybrid ABC algorithm

In this method, ABC algorithm is executed first and continues according to the number of iterations. For each iteration process, one best solution is achieved. At the end of iterations, a number of best solutions are generated. These solutions will act as the randomly generated solutions for crossover operation. Then, crossover is executed to find the optimal values, which will provide much better quality of image for all aspects. The pseudo-code of this method is shown below.

1. Take low-contrast image
2. Compute histogram and the number of unique intensity levels.
3. Initialize random solutions for ABC operation.
4. Execute ABC algorithm and compute various solutions through optimizing the objective function (as described in section 3) according to the number of iterations.

Figure 1.   Genetic Algorithm crossover techniques.

5. Take those optimized solutions as the random solutions for crossover operation of GA.
6. Execute GA crossover operation.
7. Generate new solution.
8. Evaluate it against the optimization target.
9. Stop once the set target is reached.
10. Reconstruct the image.

The flowchart of this method is shown in Figure 2.

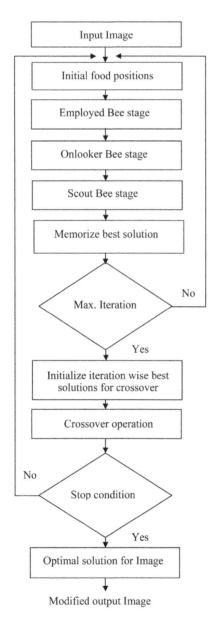

Figure 2. Flowchart of a hybrid crossover-based ABC algorithm.

3 OBJECTIVE FUNCTION FORMULATION

In this paper, the objective function is designed using three important image quality metrics, namely Absolute Mean Brightness Error (AMBE), Peak Signal-to-Noise Ratio (PSNR), and entropy. The brief mathematical descriptions of these are shown below.

AMBE is defined as the absolute difference between the input image mean and the output image mean. It is used to preserve the original brightness of the input image. It is calculated by Eq. (4) (Zadbuke 2012):

$$AMBE = |E(X) - E(Y)| \qquad (4)$$

where X and Y denote the input and output images, respectively, and E (.) denotes the expected value of statistical mean. Lower AMBE indicates better brightness preservation of the image.

Entropy is an important parameter to measure the richness of the details in the output image. The maximum value of entropy indicates high quality of image. Entropy is calculated by Eq. (5) (Wang & Ye 2005):

$$Ent[P] = -\sum_{k=0} P(k) \log_2 P(k) \qquad (5)$$

where P indicates probability, which is the difference between two adjacent pixels.

Peak Signal-to-Noise Ratio, often abbreviated as PSNR, is defined as the ratio of the maximum possible power of a signal to the power of noise that affects the efficiency of its final output. PSNR is commonly used to measure the quality of reconstruction for image compression purpose. The signal in this case is the original data, and the noise is the error introduced by that compression. To calculate PSNR, Mean Square Error, often abbreviated as MSE, should be calculated first.

Assuming that N is the total number of pixels for input image (X) and output image (Y). Then, MSE (Mean Squared Error) is calculated by Eq. (6) (Kaur & Chand 2012):

$$MSE = \sum_i \sum_j |X(i,j) - Y(i,j)|^2 / N \qquad (6)$$

where i and j denote pixel positions. Low value of MSE indicates betterment of image quality.

Depending on MSE (Mean Square Error), PSNR is then calculated by Eq. (7) (Kaur & Chand 2012):

$$PSNR = 10 \log_{10} (L-1)^2 / MSE \qquad (7)$$

where L denotes the total intensity levels. The higher value of PSNR indicates better output image quality.

The objective function in this paper is designed in Eq. (8), where PSNR and entropy are used to divide the AMBE, as both of them indicate betterment with higher values while AMBE shows betterment with lower values. It also enables the optimization toward minimization of the objective function:

$$\frac{AMBE(X,Y)}{PSNR(X,Y)+Entropy(Y)} \tag{8}$$

4 RESULTS AND DISCUSSIONS

The proposed method is tested with a number of standard test images. Three of the results are presented here for visual comparison with the results of conventional techniques. Figures 3, 4, and 5 represent the enhance results of various methods including the presented method for "Nut", "Solar System", and "Man" image, respectively. All the images are shown in gray scale format. The results show that the presented method shows significant visual improvement over the other techniques. It is worth noting that the presented method not only improves contrast, but also tries to preserve the visual information of the image and thus the gray tones are better retained while the other techniques tend toward more black and white images with lesser gray tonal effects. Furthermore, the presented technique overcomes

    (a)   (b)   (c)   (d)   (e)

Figure 3. Nut image: (a) original image, (b) CHE-based enhancement, (c) BBHE-based enhancement, (d) DSIHE-based enhancement, (e) crossover-included hybrid ABC-based enhancement.

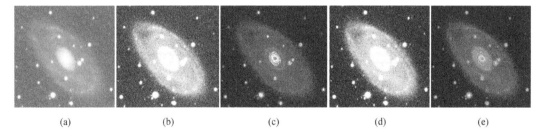

    (a)   (b)   (c)   (d)   (e)

Figure 4. Solar system image: (a) original image, (b) CHE-based enhancement, (c) BBHE-based enhancement, (d) DSIHE-based enhancement, (e) crossover-included hybrid ABC-based enhancement.

    (a)   (b)   (c)   (d)   (e)

Figure 5. Man image: (a) original image, (b) CHE-based enhancement, (c) BBHE-based enhancement, (d) DSIHE-based enhancement, (e) crossover-included hybrid ABC-based enhancement.

Table 1. Average mean values of PSNR, AMBE, and entropy for different methodologies.

Techniques metrics	CHE	BBHE	DSIHE	Hybrid ABC
PSNR	11.12 dB	14.1 dB	10.21 dB	16.55 dB
AMBE	42.96	24.01	32.73	8.41
Entropy	5.44	5.36	5.38	5.46

the false contouring and patches effects, which are present in conventional techniques and give sort of synthesized image appearance. The objective comparison using different image quality metrics using the standard database (Sipi Image Database) is shown in Table 1.

In Table 1, average mean values of PSNR, AMBE, and Entropy are indicated for different methodologies by considering different types of gray-scale images. All the tested images are considered using the standard database [Sipi Image Database]. Here the first column indicates metric parameters for CHE method and the last column shows the presented hybrid ABC method.

To obtain high-quality image for visual purpose as well as informational purpose, the modified output images from different methodologies should have low AMBE, maximum value of Entropy, and high value of PSNR. It is observed that for the case of presented hybrid ABC method, the desired values are comparatively much better with respect to other methodologies indicated here.

5 CONCLUSIONS

In this paper, we presented a crossover-included hybrid ABC methodology that integrates linear crossover operation from GA with search dynamics of ABC. The hybrid technique is implemented to minimize the objective function designed with important image characteristics. The method is tested with different images and found to be competitive in comparison with CHE, BBHE, and DSIHE. The work can be further extended to the implementation of color images, a more robust optimization function, implementation with multi-objective optimization algorithms, etc. The major limitation of the presented method is the processing time, which is higher than the conventional techniques, because of the iterative nature of the presented technique. The visual as well as objective comparison shows that the presented method can be a potential alternative for image contrast enhancement.

REFERENCES

Akay, B. (2013). A study on particle swarm optimization and artificial bee colony algorithms for multi level thresholding. *Appl. Soft Computing. 13*, 3066–3091.

Butola, R., S. Pratik, & U. Kumar (2015). A Comparison of Thresholding Based Image Enhancement Techniques. *International Journal of Computer Science and Mobile Computing. 4(1)*, 314–319.

Draa, A. & A. Bouaziz (2014). An artificial bee colony algorithm for image contrast enhancement. *Swarm and Evolutionary Computation. 16*, 69–84.

Gupta, P. (2016). Contrast Enhancement for Retinal Images using Multi-Objective Genetic Algorithm. *International Journal of Emerging Trends in Engineering and Development. 1(6)*, 8–10.

Hashemi, S., S. Kiyani, N. Noroozi, & M. E. Moghaddam (2010). An image contrast enhancement method based on genetic algorithm. *Pattern Reorganization Literature. 31*, 1816–1824.

Hole, K. R., V. S. Gulhane, & N. D. Shellokar (2013). Application of Genetic Algorithm for Image Enhancement and Segmentation. *International Journal of Advanced Research in Computer Engineering & Technology (IJARCET). 2(4)*, 1342–1346.

Karaboga, D. (2005). An Idea Based on Bee Swarm for Numerical Optimization. *Tech. Rep.-TR06.*

Karaboga, D. & B. Akay (2009). A comparative study of Artificial Bee Colony algorithm. *Applied Mathematics and Computation. 214*, 108–132.

Karaboga, D. & B. Basturk (2007). Artificial Bee Colony Optimization Algorithm for Solving Constrained Optimization Problems. *IFSA. 4529*, 789–798.

Kaur, J. & O. Chand (2012). Comparative analysis for contrast enhancement using histogram equalization techniques. *JBRCS. 3(5).*

Maheshwari, A., R. Garg, & N. Sharma (2016). A Review Paper on Brief Introduction of Genetic Algorithm. *International Journal of Emerging Research in Management & Technology. 5(2)*, 87–89.

Patil, P. & A. M. Patil (2015). Contrast Enhancement Technique for Remote Sensing Images. *International Journal of Emerging Trends & Technology in Computer Science. 4(4)*, 57–61.

Sipi Image Database. *http://www.sipi.usc.edu/database.*

Ting, C. C., B. F. Wu, M. L. Chung, C. C. Chiu, & Y. C. Wu (2015). Visual Contrast Enhancement Algorithm Based on Histogram Equalization. *Sensors. 15*, 16981–16999.

Wang, C. & Z. Ye (2005). Brightness Preserving Histogram Equalization with Maximum Entropy: A Variational Perspective. *IEEE Transactions on Consumer Electronics. 51(4).*

Zadbuke, A. S. (2012). Brightness Preserving Image Enhancement Using Modified Dualistic Sub Image Histogram Equalization. *International Journal of Scientific & Engineering Research (IJSER). 3(2)*, 1–6.

Zhu, G. (2010). Gbest-guided artificial bee colony algorithm for numerical function optimization. *Applied Mathematics and Computation. 217(7)*, 3166–3173.

Zimmerman J. B., S. M. Pizer, E. V. Staab, J. R. Perry, W. Mccartney, & B. C. Brenton (1988). An evaluation of the effectiveness of adaptive histogram equalization for contrast enhancement. *IEEE Transactions on Medical Imaging. 7(4)*, 304–312.

Computer, Communication and Electrical Technology – Guha, Chakraborty & Dutta (Eds)
© 2017 Taylor & Francis Group, ISBN 978-1-138-03157-9

Supervised machine learning vs. Lexicon-based text classification for sentiment analysis: A comparative study

Sumit Gupta & Santanu Mandal

Department of Computer Science and Engineering, University Institute of Technology,
The University of Burdwan, West Bengal, India

ABSTRACT: Textual data available online in the form of tweets, posts, messages, blogs, reviews, comments etc. have proven to be instrumental in comprehending the sentimental needs and necessities of an individual. With the increase in the number of users joining the social communities, there has been a whopping demand among researchers to analyze the online content by classifying text to predict sentiments and emotional traits as much consummately as possible. In this paper, we have proposed our text classification algorithms based on the Machine Learning approaches and the other on Lexicon-Based approach to observe how they work. Further, we have specified the performance-wise differences between the three classifiers viz. SVM classifier, Naïve Bayes Classifier and Dictionary-Based classifier to prove the supremacy of one over the other.

Keywords: sentiment analysis, SVM classifier, Naïve Bayes classifier, Dictionary-Based approach

1 INTRODUCTION

Opinion Mining or Sentiment Analysis (SA) has been implemented in many research areas like mood detection, speech recognitions, image classifications, linguistics, psychology, textual data analysis and so on. Virtual communities have proven to be an ideal resource of textual data that can be taken as input and fed to SA systems to predict the correct sentiments and emotional polarity levels of a user.

An opinion is basically a text, a phrase or an expression that consists of two fundamental components-target (or topic) and sentiment. An entity is considered as the target and the judgment made by the target is considered as the sentiment. For example, in the text- "The boys were having fun at the park," the noun "boys" is the target and the sentiment (as conveyed in the text by the verb "fun") is positive.

The basic working behind SA systems is to classify the polarity of a given text at document level and sentence level and check whether the expressed opinion in a document, a sentence or an entity feature is positive, negative, or neutral.

This paper mainly deals with the text classification of textual data based on SVM classifier & Naïve Bayes Classifier (Supervised Machine Learning Approaches) and Dictionary-Based classifier (Lexicon-Based Approach) to observe which

one of these performs better and can be used for sentiment prediction.

The rest of the paper is organized as follows: Section II discusses about the related works, the proposed architecture and algorithm of classifiers following the Supervised Machine Learning Approach. Section III elaborates the related works, the proposed architecture and algorithm of the Dictionary-Based classifier. Section IV shows the results highlighting the performance level comparison between the two classification approaches. Finally, we have concluded this paper by suggesting some tasks that can be taken up in future.

2 SUPERVISED MACHINE LEARNING APPROACHES

Supervised algorithms generally work based on the training data and what has been learned in the past so that new data can be classified accordingly. Support Vector Machine (SVM) classifier is a linear classifier which makes the classification decision based on the value of a linear combination of the characteristics. It is easy to calculate and very simple in nature. The Naïve Bayes classifiers are a family of simple probabilistic classifiers based on Bayes theorem of probability with strong (naive) independence assumptions between the features to predict the class of unknown data set. They can be extremely fast relative to other classification algorithms.

There has been a lot of research work done previously on SA using SVM & Naïve Bayes classifiers. Gautam and Yadav (2014) have measured the effect of varying the training set size on the classification Accuracy and F-score. The (Accuracy, F-score) pair values for the complete training dataset were calculated as- SVM (76.47%, 0.7370), Naïve Bayes (74.68%, 0.6892), Ensemble 1 using 'AND-type' fusion (76.21%, 0.7365) and Ensemble 2 using 'OR-type' fusion (74.94%, 0.6882). Abdelwahab et al. (2015) have used tweets for performing Sentiment Analysis. Their approach dealt with pre-processing of data, extraction of meaningful adjective (feature vector) and selection of feature vector list. The accuracy calculation of various classifiers used were- Naïve Bayes (88.2%), Maximum entropy (83.8%), SVM (85.5%) and WordNet based Semantic Analysis (89.9%).

Let us now understand the working of the Supervised Machine Learning classification approach through a proposed architecture (Figure 1) and an algorithm.

The SA system consists of extracting the raw data from various texts available on social media and arranging them as per polarity traits. The oriented data is then fed to the database processing section where the words are matched with the content of Database. The Database consists of two tables (positive table and negative table) containing emotional words (based on emotional dictionary). We have initialized the value of emotional traits as +1 for positive and −1 for negative. If the sentence contains any emotion word then the extraction of emotion features in the sentence is done by analysing the sentence structure. Then text classification is done based on the emotion features. The last step is the polarity classification to generate the type of emotional polarity-positive, negative or neutral.

The proposed algorithm is as follows:

Input: Unstructured text $T := (m_1, m_2 ..., m_n)$ [where m_i represents the i-th message]
Database: Training Dataset
Positive Dataset: $pword := (p_1, p_2, ..., p_n)$
[where p_j represents the j-th positive emotional word]
Negative Dataset: $nword := (n_1, n_2, ..., n_n)$
[where n_j represents the j-th negative emotional word]
Output: Polarity Levels: $posp, negp, neup$
[where $posp, negp$ and $neup$ represents the positive polarity level, negative polarity level and neutral polarity level respectively]

Step 1: [Apply text mining process to T to produce structured words pertaining to each i-th message and set the variable count to the message count that counts the number of messages in T.]
$count := msg\_count(T); m_i := (w_{i1}, w_{i2}, ..., w_{in})$
[where w_{ik} is the k-th word of the i-th message]

Step 2: [Initialize all emotional counter variables to 0.] $pos := 0, neg := 0, total1 := 0; total2 := 0;$

Step 3: [Take each i-th message m_i for i: = 1 to count and set variable $total1$ to the word count of m_i] $total1 := msg\_word\_count(m_i);$

Step 4: Load into the application all the positive emotional words from the Positive dataset $pword$.

Step 5: Fetch each word w_{ik} of the message for k: = 1 to $total1$ from the application and proceed for matching with the already loaded dataset $pword$.

Step 6: If there is any positive word matching, increment the pos counter value by 1. $pos := pos + 1;$

Step 7: Increment the value of k and repeat steps 3 to 5 until all the words of the i-th message are checked for positive word matching. The iteration stops when k becomes greater than $total1$.

Step 8: Similarly, load into the application all the negative emotional words from the Negative dataset $nword$.

Step 9: Fetch each word w_{ik} of the message for k: = 1 to $total1$ from the application and proceed for matching with the already loaded dataset $nword$.

Step 10: If there is any negative word matching, increment the neg counter by 1. $neg := neg + 1;$

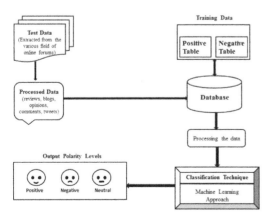

Figure 1. Our proposed architecture based on the Supervised Machine Learning classification approach.

Step 11: Increment the value of k and repeat steps 8 to 10 until all the words of the i-th message are checked for negative word matching. The iteration stops when k becomes greater than *total1*.

Step 12: Calculate the total number of words that are matched. *total2: = total2 + total1;*

Step 13: Increment the value of i and go to step 3 to continue the iteration process for the next i-th message until all the messages are taken as input. The iteration stops when i becomes greater than count.

Step 14: [SVM]: [Compute the polarity levels for each type of polarities viz. positive, negative and neutral.] *posp: = pos/total2; negp: = neg/total2; neup: = posp – negp;*

Step 15: (a) If *posp* and *negp* result in zero, then return a prompt message to the user that the message has no emotion word as its content and exit.
(b) If *neup* results in zero, then return a prompt message to the user that the user has mixed emotional state and is an ambivalent person and exit.

Step 16: Predict the sentiment as "Positive" if *posp* is greater than *negp* and print that the person is a optimist, or else predict the sentiment as "Negative" and print that the person is a pessimist.

Moreover, we have modified the same algorithm to perform the simplified version of Naïve Bayes Classifier by altering the Step 14 of the algorithm as:

Step 14 [Naïve Bayes]: [Compute the polarity levels for each type of polarities viz. positive, negative and neutral.] *posp: = log(pos)/total2; negp: = log(neg)/total2; neup: = posp – negp;*

3 DICTIONARY BASED APPROACH

In Lexicon-Based approach, human beings play a crucial role in making text classification. It is a method that focuses on fixed expressions that occur frequently in speeches, lectures or dialogues, which make up a larger part of discourse than unique phrases and sentences. Although it yields more precise results compared to its Machine learning counterpart, yet it suffers from time overhead because it requires manual intervention. The Dictionary-Based classifier is based on dictionary entries, which means that the words will be translated just as it is done by a dictionary—word by word, usually without much correlation of meaning between them.

Let us understand about this classifier from previous related works. Palanisamy et al. (2013) have proposed a hybrid Machine learning &

lexicon-based classification approach that works by replacing some words with its synonyms using the domain dictionary. The classification task when performed using the hybrid approach yielded better accuracy—Naïve Bayes (91%), SVM (87%) and K-Nearest Neighbour (85%). Zamin et al. (2013) have designed a simple lexicon based system for the SemEval-2013 Task 2 for doing Sentiment Analysis on Twitter data using the Serendio's Sentiment engine. The Serendio's approach consists of positive, negative, negation, stop words and phrases. The system classifies the tweets as positive or negative by identifying and extracting sentiments from emoticons and hash tags yielding an F-score of 0.8004. Khalif and Omar (2014) have proposed an unsupervised approach to apply Part-of-Speech (POS) tags to Malay and English words. They have used the Statistical Dictionary-based Word Alignment Algorithm that acts as an efficient tool for extracting information from a resource-poor language.

Let us now understand the working of the Dictionary-Based classifier through a proposed architecture and an algorithm.

The Dictionary-based SA consists of extracting the raw data from various texts available from online communities and arranging them as per polarity traits. The oriented data is then passed to the database processing section where the words are compared with the content of Emotional Database. The Emotional Database consists of six different tables. The tables are initially classified into Positive and Negative tables based on the polarity of the words. Each of these tables has been further classified into three sub-tables, each of which comprises words based upon their degree of

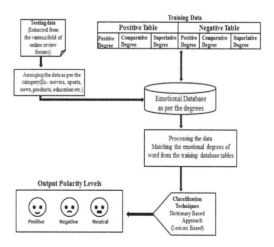

Figure 2. Our proposed architecture based on the Dictionary-Based classification approach.

comparison viz. positive, comparative and superlative. We have used our proposed dictionary-based text classification algorithm to calculate the sentiment value and thus predict the polarity levels of a user.

The proposed Dictionary-based classification algorithm is as follows:

Input: Unstructured text T: $= (m_1, m_2, ..., m_n)$
[where m_i represents the i-th message]
Database: Training Dataset
For Positive Dataset:
positive degree, comparative degree, superlative degree
For Negative Dataset:
positive degree, comparative degree, superlative degree
Output: *sentival*
[which is the value of the sentiment] &
Polarity Levels: *positive*, *negative* and *neutral*

Step 1: [Apply text mining process to T to produce structured words pertaining to each i-th message and set the variable count to the message count that counts the number of messages in T.]
count: $= msg\_count(T); m_i := (w_{i1}, w_{i2}, ..., w_{in})$ [where w_{ik} is the k-th word of the i-th message]

Step 2: [Initialize all sentiment counter variables to 0.] *degree:* $= 0$, *scale:* $= 0$, *sentival:* $= 0$;

Step 3: [Take each i-th message m_i for i = 1 to count and set variable *total1* to the word count of m_i]
total1: $=$ msg_word_*count(m_i)*;

Step 4: Load into the application all the emotional database tables i.e., positive and negative.

Step 5: Fetch each word w_{ik} of the message for k: $= 1$ to *total1* from the application and proceed for matching with the emotional database words already loaded into the dataset.

Step 6: Compare the polarity and the degree of each word as follows:
a) If polarity of w_{ik} is positive and the degree of comparison is positive then set degree: $= +0.22$
b) If polarity of w_{ik} is positive and the degree of comparison is comparative then set degree: $= +0.335$
c) If polarity of w_{ik} is positive and the degree of comparison is superlative then set degree: $= +0.445$
d) If polarity of w_{ik} is negative and the degree of comparison is positive then set degree: $= -0.22$
e) If polarity of w_{ik} is negative and the degree of comparison is comparative then set degree: $= -0.335$

f) If polarity of w_{ik} is negative and the degree of comparison is superlative then set degree: $= -0.445$
g) Otherwise set degree: $= 0$

Step 7: Check for the existence of scale value that enhances the degree of comparison as follows:
a) If w_{ik} is equal to "more" or "very", then set scale: $= +0.115$ when $w_{ik}+_1$ is a word that belongs to Positive sentiment table or else set scale: $= -0.115$ when $w_{ik}+_1$ is a word that belongs to Negative sentiment table.
b) If w_{ik} is equal to "most" or "much", then set scale: $= +0.225$ when $w_{ik}+_1$ is a word that belongs to Positive sentiment table or else set scale: $= -0.225$ when $w_{ik}+_1$ is a word that belongs to Negative sentiment table.

Step 8: Calculate *sentival* using the following formula: *sentival:* $=$ *sentival* $+$ *degree* $+$ *scale*;

Step 9: Increment the value of k and repeat steps 5 to 8 until all the words of the i-th message are checked.

Step 10: Increment the value of i and go to step 3 to continue the iteration process for the next i-th message until all the messages are taken as input. The iteration stops when value of i becomes greater than count.

Step 11: The sentiment polarity level of a user is predicted based on the following conditions:
a) The sentiment is "Positive" if sentival is greater than 0. Also prompt a message that the person is an optimist.
b) The sentiment is "Negative" if sentival is less than 0. Also prompt a message that the person is a pessimist.
c) Otherwise predict the sentiment as "Neutral" and print that the person is an ambivalent and has mixed emotions.

4 IMPLEMENTATION AND RESULT

Datasets: The training datasets applied in Sentiment Analysis are a relevant item in this field. Some of the well-known text-based datasets available online are the SemEval-Task2 development-test set (consisting of Tweets), the International Survey of Emotion Antecedents and Reactions (ISEAR) dataset which is a source of emotion-related words, the Digg dataset from Cyber emotion that contains data about stories promoted to Digg's front page, the SemEval Affective Text-2007 that consists of news headlines drawn from major newspapers such as New York Times, CNN, and BBC News,

58

Table 1. Performance comparison of supervised machine learning and dictionary based classification approaches.

| Measure | Formula | Machine learning approach | | Dictionary based approach |
		SVM	Naive bayes	
Accuracy	(TP + Tn)/(P + N)	0.89	0.98	0.81
Error rate	(Fp + Fn)/(P + N)	0.11	0.02	0.19
Recall	Tp/P	0.84	0.96	0.92
Precision	Tp/(Tp + Fp)	0.92	0.98	0.84
F-score	(2 * Precision * Recall)/(Precision + Recall)	0.88	0.98	0.87

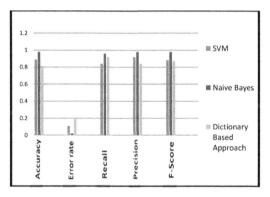

Figure 3. Comparison result of classifiers using column-chart.

as well as from the Google News search engine and the standard Sentiment 140 dataset to name a few. Moreover we have manually collected data from several online communities and forums to build our database with approximately 7000 positive and negative emotional words. Then we have run 200 sentences (containing both positive as well as negative words) on the system as test cases to calculate the performance and accuracy of our proposed SA systems. The datasets are made available for non-commercial and research purposes only.

Result & Analysis: We have implemented our algorithm and fed different test cases to analyze the difference in performance characteristics between the linear, probabilistic and lexicon-based classification approaches. Metrics such as Accuracy, Error rate, Recall, Precision and F-score (or harmonic mean of precision and recall) are calculated and listed in Table 1 and represented using a column-chart in Figure 3.

The accuracy value of 98% easily shows that text classification using probability-based Naïve Bayes classifier is far better than the other two classifiers that we have used while building our SA system.

5 FUTURE SCOPE AND CONCLUSIONS

In this paper we have shown a comparative study of only three classifiers- two Supervised Machine Learning-Based (SVM & Naïve Bayes) and the other Dictionary-Based. In future we tend to study and implement other well-known classifiers so that we can present a complete performance-wise comparison statistics and identify the best classifier for text classification in a SA system.

REFERENCES

Abdelwahab, O., M. Bahgat, C. J. Lowrance, & A. Elmaghraby (2015). Effect of Training Set Size on SVM and Naïve Bayes for Twitter Sentiment Analysis. *Proc. of the IEEE Intl. Sympo. on Sig. Process. and Info. Tech. (ISSPIT), 15,* 46–51. doi: 10.1109/ISSPIT.2015.7394379.

Gautam, G., & D. Yadav (2014). Sentiment Analysis of Twitter Data Using Machine Learning Approaches and Semantic Analysis. *Proc. of the IEEE Intl. Conf. on Cont. Computing (IC3), 7,* 437–442. doi: 10.1109/IC3.2014. 6897213.

Khalif, K., & N. Omar (2014). A Hybrid Method Using Lexicon-Based Approach and Naive Bayes Classifier for Arabic Opinion Question Answering. *J. Comput. Sci., 10,* 1961–1968. doi: 10.3844/jcssp.2014.1961.1968.

Palanisamy, P., V. Yadav, & H. Elchur (2013). Serendio: Simple and Practical Lexicon Based Approach to Sentiment Analysis. *Proc. of the Joint Conf. on Lexical and Computational Semantics, 2,* 543–548. Retrieved from http://www.aclweb.org/anthology/S13-2091.pdf.

Zamin, N., A. Oxley, Z. A. Bakar, & S. A. Farhan (2013). A Statistical Dictionary-based Word Alignment Algorithm: An Unsupervised Approach. *Proc. of the IEEE Intl. Conf. on Comput. & Info. Sci. (ICCIS), 1,* 396–402. doi: 10.1109/ICCISci.2012.6297278.

Computer, Communication and Electrical Technology – Guha, Chakraborty & Dutta (Eds)
© 2017 Taylor & Francis Group, ISBN 978-1-138-03157-9

Network anomaly detection using a fuzzy rule-based classifier

Soumadip Ghosh & Arindrajit Pal
Department of Computer Science and Engineering, Academy of Technology, Hooghly, West Bengal, India

Amitava Nag
Department of Information Technology, Academy of Technology, Hooghly, West Bengal, India

Shayak Sadhu & Ramsekher Pati
Department of Computer Science and Engineering, Academy of Technology, Hooghly, West Bengal, India

ABSTRACT: Network anomaly detection is one of the important approaches in the field of Internet security to detect threat to network resources. In this work, we address the problems related to network anomaly detection. We use an adaptive method to construct a fuzzy rule-based classification system for classifying these types of problems. The classification of network anomaly is an effervescent research area. The proposed method consists of an error correction-based learning procedure. The error correction-based learning procedure regulates the rank of confidence of each fuzzy rule by its classification performance. In this work, we also compare the performance of the fuzzy rule-based classifier with Support Vector Machine (SVM) and K-Nearest Neighbor (KNN) classifiers. This performance evaluation is done to prove the superiority of the proposed classifier.

1 INTRODUCTION

The computer is used to store vital data, manipulate that data to turn it into knowledge, and perform many other basic and complex mathematical operations. Intrusion Detection Systems (IDSs) and firewalls have been extensively used to make our computers safe from intrusion. The intrusion detection systems (Fayyad 1996) aim to detect network anomalies, but firewalls take action according to a predefined set of rules. However, distinguishing a specific anomaly provides us with valuable information about the attacker that may be used to further protect the system, or to respond accordingly. In this way, detecting network intrusion is a current challenge due to the rapid development of the Internet and the number of potential intruders.

The perfect detection and classification of network anomalies based on traffic attribute distributions is still a major challenge (Lunt 1993). Together with volume metrics, traffic feature distributions are the main source of information of approaches scalable for large-scale networks. The degree of intelligence of firewalls and other simple boundary devices is not sufficient to observe, recognize, and identify the attack signatures that may be present in the traffic monitored by them and the log files collected by them. Naturally, Intrusion Detection Systems (IDSs) are gaining more importance to maintain the network security due to this inefficiency of firewalls and other simple boundary devices.

The IDS can be simply described as a specialized tool that understands how to read and interpret the log file contents from servers, routers, and other network devices. Moreover, the IDS stores a knowledgebase containing information about the known attack signatures. When it recognizes any signature that has a close relation to the attack signatures kept in its knowledgebase, it generates alarm or alert and also performs various kinds of automatic actions like disconnecting Internet links or shutting down specific servers, as well as tries to actively identify the attacker. In brief, the IDS performs the function of an antivirus for a network.

Intrusion detection systems are generally classified in two ways (Wespi 2002, Hoglund 2000). First, we can classify IDSs according to the devices or system they monitor. In this case, we can categorize IDSs into three types namely:

- Network IDS: network IDS monitors the entire network traffic without adding any significant overhead to the network.
- Host IDS: host IDS is host specific and monitors whether that specific host is attacked by any malicious signature.
- Application IDS: application IDS is application specific and monitors the events occurring in some specific applications.

Second, intrusion detection systems can also be categorized according to their differing approaches to event analysis. Some IDSs use a signature detection technique to search and match patterns to detect malicious signatures in the network. These IDSs are known as signature-based IDSs. There are IDSs that check the traffic, transmissions, or behaviors for anomalies in the network that may indicate the attack. These IDSs are called anomaly-based IDSs. The basic principle used in anomaly-based IDSs is that the "intruder behavior" is significantly different from the "normal user behavior" that can easily be recognized by identifying the differences involved (Durst 1999). One of the main problems of anomaly-based IDSs is that the normal user behavior is not static and varies over time, which makes the anomaly-based IDSs prone to a high false positive rate, i.e., they will detect a normal behavior to be an intruder behavior. As a result, there is a necessity of using several classification techniques such as KNN and SVM for a better training of the anomaly-based IDSs (Mukkamala 2002, Northcutt 2002).

This paper is organized as follows: Section 2 presents the related works done in this field; Section 3 provides a brief introduction to the classification procedure; Section 4 describes the proposed fuzzy rule-based classification method; Section 5 explains the methodology in terms of the proposed fuzzy rule-based approach, KNN, and SVM; Section 6 provides the results and discussion related to the performance analysis of the classifiers; finally, Section 7 concludes this research study.

2 REVIEW WORK

Classification is an important data mining technique that has been applied to anomaly detection to successfully identify the network anomaly pattern. Lee and Stolfo (Chi 1995) proposed a systematic data mining framework for intrusion detection. This framework consists of classification, association rules that can be used to construct detection models. Lee (1998) presented PNRule for a multi-class classification problem. The modeling of network behavior was proposed by Agarwal (2000). Traffic models such as the self-similar models introduced by Norros (Ye 2001) and the cascade models originally developed by Crousse et al. (Norros 1994) accurately capture the fractal scaling properties. Soft computing paradigms, like fuzzy rule-based classifiers to model efficient systems, were investigated by Northcutt (2002). Ishibuchi et al. (Reidi 1999) presented a fuzzy rule-based classification system. Poojitha et al. also illustrated the use of classification in the intrusion detection system. They have used the artificial neural network as the classification algorithm to detect anomaly in the network.

Manikopoulos (Ishibuchi 2001), Yeung and Chow (Manikopoulos 2002) developed different techniques for network intrusion and fault detection. There have been many earlier studies of network fault recognition methods (Yeung 2002). Feather et al. (Ward 1998) used statistical deviations from network traffic behavior to classify faults. Another paper (Feather 2002) used a fault detection technique in an Ethernet network using anomaly signature matching. These research works (Poojitha 2010, Kaushik 2011, Revathi 2013, Kumar 2013) also contributed immensely to network anomaly detection.

3 CLASSIFICATION

The proposed method is based on the concept of the classification technique. Classification is the process of using a mathematical model or a classifier that is able to describe and differentiate among various data classes so that the model can successfully predict the class of entities, objects, or tuples to which the target class belongs. Data classification is a two-step procedure (Han 2000). In the first step, a classifier is constructed based on a set of predefined mathematical concepts and data structures. Next, this model needs to be trained so that it can adapt to the current problem and successfully classify tuples with an unknown class label. This phase is called the training phase. In this phase, the classifier learns from a training dataset and its associated class attributes. In the next phase, testing is performed. In this stage, the model is used to predict tuples with an unknown class label. This output class label is the predicted output of the classifier that will later be used for performance evaluation. The NSL-KDD dataset (NSL KDD) of UCI, which is a refined version of the KDD cup99 dataset, is used in our work.

4 FUZZY RULE-BASED CLASSIFIER

Fuzzy rule-based classification system demonstrates good generalization ability in a high-dimensional aspect and has been a dynamic research topic for a long time (Northcutt 2002). Classification problems involve transferring a class C_j from a predefined class set $C = \{C_1, \ldots, C_M\}$ to an object that is denoted as an element in a feature space $x \in S^N$. We need to find a mapping criterion for designing a classifier like D: $S^N \rightarrow C$.

This is optimal for a certain criterion (D) that decides the classifier performance. The final goal is to develop a classifier that allocates class labels

Figure 1. Fuzzy rule-based classification system.

with the smallest possible error across the total feature space. The classifier possibly has a set of fuzzy rules, a neural network, a decision tree, etc. If a classifier is based on a set of fuzzy rules, it is called the Fuzzy Rule-Based Classification System (FRBCS). The FRBCS consists of a Knowledge-Base (KB) and a Fuzzy Reasoning-based Method (FRM). The KB consists of two parts, namely the Rule Base (RB) and the Data Base (DB). They describe the semantic of the fuzzy subsets related to the class labels that form the condition part in *if...then...else* rules of the FRM. The FRM uses the information from the KB to establish a label class for all admissible models. This structure is shown in Fig. 1.

To build an FRBCS, we first use a set of pre-classified examples, from which we must determine:

- the method which constructs a set of fuzzy rules for the considered classification problem;
- the fuzzy reasoning method which classifies an unknown pattern that is taken as the input.

The well-known fuzzy rule-based classification algorithm, namely RIPPER (Cohen 1995), is an inductive rule learner. In particular, it can develop an efficient FRBCS-based system. Thus, we use the RIPPER algorithm in this work. This algorithm generates the classification model based on a set of fuzzy rules that can successfully detect malicious executables. This algorithm uses libBFD information as features.

5 METHODOLOGY

In this work, we use a fuzzy rule-based classifier (RIPPER) as the primary classifier and compare its performance with two more classifiers, namely K-Nearest Neighbor (KNN) (Altman 1992) and Support Vector Machine (SVM) (Cortes 1995), which have been chosen as secondary classifiers. KNN uses Euclidean distance to calculate the distance among the neighbors, and uses the nearest distance to classify data. SVM uses dot product as the kernel function, and uses sequential minimal optimization to separate the hyperplane. These two classifiers are chosen to compare its performance with the fuzzy rule-based classifier. KNN is

Table 1. Some basic features of the dataset.

Feature name	Description	Type
Duration	Length (number of seconds) of the connection	Continuous
protocol_type	Type of the protocol, e.g., tcp, udp, etc.	Discrete
Service	Network service on the destination, e.g., http, telnet, etc.	Discrete
src_bytes	Number of data bytes from source to destination	Continuous
dst_bytes	Number of data bytes from destination to source	Continuous
Flag	Normal or error status of the connection	Discrete
land	1 if connection is from/to the same host/port; 0 otherwise	Discrete
wrong_fragment	Number of "wrong" 'fragments	Continuous
urgent	Number of urgent packets	Continuous

a lazy classifier that produces good result when the available dataset is large. The dataset that we have used is fairly large (containing more than 25,000 tuples), and thus KNN is selected. SVM is a supervised classification model that works well with large datasets. SVM is built based on the theory of hyperplane, in which it creates an n-dimensional space. In this case, n is the number of attributes in the dataset. As we have a small number of attributes, the use of this classifier would produce good result.

In this work, we use 10% of the given NSL-KDD dataset, to which we apply the classification techniques of data mining. In this dataset, the classifiers classify the dataset into two classes, namely 'normal' and 'anomaly'. We have chosen this dataset because it is free from any redundant record and exhaustive data cleaning will not be required. Moreover, this dataset does not contain any missing value of any attribute. The meanings of the basic attributes that we have used from this dataset are described in Table 1.

6 RESULTS AND DISCUSSION

We applied a 10-fold cross-validation method for data distribution with respect to training and testing. The simulation was performed in a MATLAB environment (version R2013a) installed on a PC with AMD FX-4300 processor having a clock speed of 3.80 GHz and 8 GB RAM. Several standard statistical measures were used to measure the performance of each of the three classifiers and compared among them. These measures (Fawcett 2006, David 2011) were also used to evaluate the performance of each classifier. The statistical measures are as follows:

- Accuracy: It is the percentage of correctly classified data. It is represented by the following equation:

$$Accuracy = \frac{TP + TN}{TP + TN + FP + FN} \qquad (1)$$

- Precession: It is the proportion of the predicted positive cases that are correct, which can be calculated using the equation:

$$Precision = \frac{TP}{TP + FP} \qquad (2)$$

- FP-Rate: It is the proportion of negative cases that are incorrectly classified as positive, which can be calculated using the equation:

$$FP - Rate = \frac{FP}{FP + TN} \qquad (3)$$

- TP-Rate or Recall: It is the percentage of positive cases that are correctly identified, which can be calculated using the equation:

$$Recall = TP - Rate = \frac{TP}{TP + FN} \qquad (4)$$

- F-Measure: It is the harmonic mean of precession and recall, which is given by the following equation:

$$F - Measure = \frac{2 * Precision * Recall}{Precision + Recall} \qquad (5)$$

After simulation of the experiment, it was observed that the fuzzy-rule-based classifier produced a far superior result compared with the other two classifiers, as outlined in Table 2. Simulation of the fuzzy rule-based classifier and KNN was moderately faster, whereas SVM took longer time to classify. The confusion matrix for each of the classifiers was also generated. This matrix was used to calculate the True Positive Rate (TP) or Recall, False Positive (FP) Rate, Precision, and

Table 2. Classification performance.

Classifier	Accuracy	TP Rate	FP Rate	Precession	F-Measure
RIPPER	99.4%	99.4%	3.0%	99.4%	99.4%
SVM	93.6%	93.7%	6.3%	93.6%	93.6%
KNN	95.8%	95.8%	4.2%	95.9%	95.8%

F-Measure values. Table 2 presents the classification performance of these classifiers.

Among these classifiers, RIPPER has the highest F-Measure and accuracy. This proves that RIPPER works significantly better than SVM and KNN for the given dataset.

7 CONCLUSION

This study proposed a fuzzy rule-based classification method for anomaly detection and proved its efficiency successfully using a given NSL-KDD dataset. Its aim was to analyze and investigate the operational aspects of the proposed technique in comparison with KNN and SVM classifiers. The proposed classification method had an accuracy of 99.4% for the given dataset. These results were better than those of MLP and SVM. The proposed classifier also had the lowest FP Rate value and the highest F-Measure value compared with its other counterparts. Thus, it can be concluded that the proposed technique has a great potential in detecting network anomaly.

REFERENCES

Agarwal, R. & M. V. Joshi (2005), PNRule: A New Framework for Learning Classifier Models in Data Mining (A Case-Study in Network Intrusion Detection), *Technical Report. No. RC 21719*, IBM Research Division.

Altman, N. S. (1992), An introduction to kernel and nearest-neighbor nonparametric regression, *J. The American Statistician*, 46 (3), 175–185.

Chi, Z., J. Wuand H. Yan (1995), Handwritten numeral recognition using self-organizing maps and fuzzy rules, *Pattern Recognition 28 (1)*, 59–66.

Cohen, W. W. (1995), Fast Effective Rule Induction, *Proc. 12th International Conference on Machine Learning (ICML)*, 115–123.

Cortes C. & V. Vapnik (1995), Support-vector networks, *J. Machine Learning 20 (3)*, 273–297, doi:10.1007/BF00994018.

David, M. W. (2011), Evaluation: From Precision, Recall and F-Measure to ROC, Informedness, Markedness & Correlation, *J. Machine Learning Technologies*, 2 (1), 37–63.

Durst, R., T. Champion, B. Witten, E. Miller & L. Spagnuolo (1999), Testing and evaluating computer intrusion detection systems. *Comm. of the ACM 42*, 53–61.

Fawcett, T. (2006), An Introduction to ROC Analysis. *Pattern Recognition Letters,* 27 (8), 861–874.

Fayyad, U., G. Piattetsky-Shapiro & P. Smyth (1996), Fromdata mining to knowledge discovery: an overview, *Advances in knowledge discovery and data mining,* American Association for Artificial Intelligence, USA.

Feather, F., D. Siewiorek & R. Maxion (2000), Fault detection in an Ethernet network using anomaly signature matching, *In Proceedings of ACM SIGCOMM '93,* San Francisco, CA.

Han, J. & M. Kamber (2000), Data Mining: Concepts & Techniques, *Morgan & Kaufmann.*

Hoglund, A., K. Hatonen & A. Sorvari (2000), A computer host-based user anomaly detection system using the self-organizing map. *Proc. IJCNN 2000,* 5, 411–416.

Ishibuchi H. & T. Nakashima (2001), Effect of rule weights in fuzzy rule-based classification systems, *IEEE Trans. Fuzzy System* 9, 506–515.

Kaushik, S. S. & P. R. Deshmukh (2011), Detection of Attacks in an Intrusion Detection System, *International J. Computer Science and Information Technologies,* 2 (3), 982–986.

Kumar, V., H. Chauhan & D. Panwar (2013), K-Means Clustering Approach to Analyze NSL-KDD Intrusion Detection Dataset, *International Journal of Soft Computing & Engineering,* 3 (4).

Lee, W. & S. J. Stolfo (1998), Data mining approaches for intrusion detection, *Proc. of the 7th conf. on USENIX Security Symposium* 7, 1–16.

Lunt, T. F. (1993), A survey of intrusion detection techniques. *J. Computers and Security 12,* 405–418,.

Manikopoulos, C. & S. Papavassiliou (2002), Network intrusion and fault detection: a statistical anomaly approach. *IEEE Communications Magazine,* 40, 76–82.

Mukkamala, S., G. Janoski & A. Sung (2002), Intrusion detection using neural networks and support vector machines. *Proc. IJCNN 2002,* Honolulu, 2, 1702–1707.

Norros I, A storage model with self-similar input (1994), *J. Queueing Syst.* 16, 387–396.

Northcutt, S. & J. Novak (2002), Network Intrusion Detection. 3rd edn. New Riders, Indianapolis.

NSL KDD dataset, *UCI machine language repository,* http://nsl.cs.unb.ca/NSL-KDD.

Poojitha, G., K. K. Naveen & R. P. Jayarami (2010), Intrusion Detection using Artificial Neural Network, *Proc. Computing Communication and Networking Technologies (ICCCNT),* 1–7.

Reidi, R. H., M. S. Crouse, V. J. Ribeiro & R. G. Baraniuk (1999), A multifractal wavelet model with application to network traffic, *IEEE Trans Information Theory,* 45 (3), 992–1018.

Revathi S. & Malathi A. (2013), A Detailed Analysis on NSL-KDD Dataset Using Various Machine Learning Techniques for Intrusion Detection, *International Journal of Engineering Research and Technology (IJERT),* 2 (12).

Ward, A., P. Glynn & K. Richardson (1998), Internet service performance failure detection, Performance Evaluation Review.

Wespi, A., G. Vigna & L. Deri (2002), Recent Advances in Intrusion Detection. *Proc. LNCS 2516.* Springer Verlag.

Ye, T., S. Kalyanaraman, D. Harrison, B. Sikdar, B. Mo, H. T. Kaur, K.Vastola & B. Szymanski (2001), Network management and control using collaborative on-line simulation, *Proc. CNDSMS,* DOI: 10.1109/ICC.2001. 936304.

Yeung, D, Y. & C. Chow (2002). Parzen-window network intrusion detectors. *Proc. International Conference on Pattern Recognition (ICPR'02)* IEEE Computer Society, 4, 385–388.

Computer, Communication and Electrical Technology – Guha, Chakraborty & Dutta (Eds)
© 2017 Taylor & Francis Group, ISBN 978-1-138-03157-9

Hybrid VVRKFA-SLFN classifier for improved pattern classification

S. Ghorai
Department of AEIE, Heritage Institute of Technology, Kolkata, India

ABSTRACT: In this work, recently proposed learning algorithm Vector-Valued Regularized Kernel Function Approximation (VVRKFA) and Single-hidden Layer Feedforward Neural Network (SLFN) are combined together to form a hybrid classifier called VVRKFA-SLFN. In VVRKFA method, multiclass data are mapped to the low dimensional label space through regression technique and classification is carried out in this space by measuring the Mahalanobis distances. On the other hand, SLFN has exceptionally high-speed and can achieve high accuracy on unknown samples. In the proposed hybrid learning system, VVRKFA is used to map the patterns from high dimensional feature space to a subspace and SLFN is trained on these low dimensional vectors to determine the class labels. The presented technique is verified experimentally both on benchmark and gene-microarray data sets to show the effectiveness of VVRKFA-SLFN classifier with enhanced correctness and testing time.

1 INTRODUCTION

Most of the practical classification problems deal with multi-category samples and are tricky compared to binary data classification. In our earlier research work, we have presented a new technique of fast multiclass data classification method called Vector-Valued Regularized Kernel Function Approximation (VVRKFA) (Ghorai, Mukherjee, & Dutta 2010). VVRKFA eliminates the drawbacks of multiclass data classification algorithms that use to decompose a multiclass problem into a number of binary classification problems (Allwein et al. 2000, Dietterich and Bakiri 1995, Hsu and Lin 2002, Kreßel 1999). VVRKFA is a unusual type classifier that uses regression technique to classify multiclass data by mapping the feature vectors into a low-dimensional subspace. The dimension of this subspace is equivalent to that of the category of data in a problem. The label of a test sample is determined in the low-dimensional subspace by the minimum Mahalanobis distance measured from the centroids of different classes. VVRKFA method of classification has an improved training and testing time complexity for large data sets in comparison to multicategory SVM classifier (Ghorai et al. 2010). This research work is an effort to improve the classification accuracy of VVRKFA.

Classification by a Single-hidden Layer Feedforward Neural Network (SLFN) was introduced long ago by Broomhead and Lowe (Broomhead & Lowe 1988) and subsequently elaborated by Lowe (Lowe 1989). Broomhead and Lowe established that computationally expensive gradient based learning techniques of neural networks can be avoided if one merely considers the hidden Radial Basis Function (RBF) neurons uniformly over a grid and train only the output weights. In such cases, the network training reduces to a simple least square minimization problem and its output weights are simply obtained by computing the inverse of hidden layer output matrices. Broomhead and Lowe also introduced the idea of randomizing the RBF center selection to choose very fewer centers than data points. Several other authors later developed different networks with randomly selected hidden neurons (Igelnik and Pao 1995, Kaminski and Strumillo 1997, Pao et al. 1994, Wettschereck and Dietterich 1992). In a slight different way, Huang et al. (Huang, Zhu, & Siew 2004, Huang & Siew 2004, Huang & Siew 2005), have presented Extreme Learning Machine (ELM) for training of SLFNs. According to this method, both input weights and hidden layer biases are initialized at random using any nonlinear activation function. In their recent works, Huang and others (Huang, Chen, & Siew 2006, Huang, Zhu, & Siew 2006, Huang & Chen 2007, Huang & Chen 2008) demonstrated that ELM can be used as universal approximators to learn Incremental Extreme Learning Machine (I-ELM) that outperforms many algorithms. Further, ELM can be employed in regression as well as in multiclass data classification applications (Huang, Zhou, Ding, & Zhang 2012). It has been observed from the research work of Huang et al. (Huang, Zhu, & Siew 2004, Huang & Siew 2004, Huang & Siew 2005) that the learning time complexity of ELM is absolutely small compared to the conventional feedforward neural network learning algorithms with enhanced classification accuracy. This property of ELM has motivated us to employ similar form of SLFN in the VVRKFA framework.

In this work, a hybrid pattern classifier is proposed for improved pattern classification in low dimensional space by combining VVRKFA and SLFN classifier similar to ELM. The VVRKFA classifier is used to map the feature vectors in the low-dimensional subspace. The task of assigning a label to each test pattern is performed in this low-dimensional subspace by employing a SLFN classifier instead of the use of distance measure. Originally, in VVRKFA the class label of a test pattern is determined by computing the Mahalanobis distances of it from the class centroids in the low-dimensional subspace. The use of SLFN for this purpose not only speed up the testing time of VVRKFA classifier but also improves its performance. Further, a small number of hidden neurons in VVRKFA-SLFN classifier can produce better classification accuracies in comparison to SLFN constructed with a lot of hidden neurons which may cause over training. This fact is verified experimentally on several standard data sets.

2 VVRKFA METHOD OF CLASSIFICATION

The concept of VVRKFA method of classification is graphically depicted in Fig. 1. There are three steps of classification in VVRKFA method. These are encoding of class labels by suitable scoring technique, regression of the class labels on the input attributes to extract low-dimensional subspace and decoding of the class labels in that space by training a classifier. Instead of direct regression on input

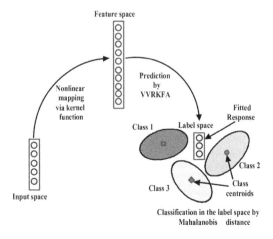

Figure 1. VVRKFA method of multiclass data classification.

space VVRKFA employed kernel trick to map a training pattern into a high-dimensional space. A nonlinear regularized kernel function is then fitted to map the patterns from this high-dimensional space to the low-dimensional subspace. The subspace dimension depends on category of the samples in the problem. The classification is carried out in the subspace with the help of mapped low-dimensional patterns obtained through the vector-valued regression. The testing of a new pattern is also carried out in this label space by mapping it from input space to feature space and then finding its fitted response in the label space.

Let us consider a multiclass problem of N different categories of samples with m features and training data $X\{(x_i, y_i): x_i \in \Re^n, y_i \in \Re^N, i = 1, 2, ..., m\}$. VVRKFA solves the following optimization problem to obtain the mapping function:

$$\text{Min } J(\Theta, b, \xi) = \frac{C}{2} \text{tr}([\Theta \ b]^T [\Theta \ b])$$
$$+ \frac{1}{2} \sum_{i=1}^{m} \| \xi_i \|^2 \quad (1)$$
$$\text{s.t. } \Theta k(x_i^T, B^T)^T + b + \xi_i = y_i, i = 1, 2, \cdots, m.$$

The class label y_i of a sample x_i of jth class is constructed as follows:

$$y_i = [y_{i1}, y_{i2}, ..., y_{iN}]^T \text{ with } y_{ij} = 1 \text{ and } y_{ik} = 0, \atop \text{for } k \neq j \in N. \quad (2)$$

In (1), the matrix $\Theta \in \Re^{N \times \bar{m}}$ contains the regression coefficients for mapping a feature vector $\phi(x_i) = k(x_i^T, B^T)^T$ from the high-dimensional space $\in \Re^{\bar{m}}$ to the low-dimensional subspace $\in \Re^N$, matrix $B \in \Re^{\bar{m} \times n}$ contains $\bar{m}$ random samples from training set $A \in \Re^{m \times n}$ to form the reduced kernel matrix, $k(.,.)$ is nonlinear kernel function, $b \in \Re^N$ is the bias vector, $\xi_i \in \Re^N$ is the slack variable. Equation (2) can be reduced to a unconstrained optimization problem by replacing the value of ξ from the constraint to the objective function. This quadratic unconstrained minimization problem is solved by equating its first derivative to zero (Ghorai, Mukherjee, & Dutta 2010, Ghorai, Mukherjee, & Dutta 2012). Once the solution Θ and b are obtained the vector-valued mapping function can be obtained as:

$$\hat{\rho}(x_i) = \Theta k(x_i^T, B^T)^T + b, \quad (3)$$

where $\hat{\rho}$ is the fitted vector-valued response of input patterns $x \in \Re^n$ in the label space having dimension N. The second stage of VVRKFA classification approach is the decoding of these label vectors. The vector valued responses $\{\hat{\rho}(x_i) \in \Re^N\}_{i=1}^m$ of training patterns $\{x_i\}_{i=1}^m$ are obtained by (3) to obtain the centroid of the

N different classes denoted by $\{\overline{\rho}^{(1)}, \overline{\rho}^{(2)}, ..., \overline{\rho}^{(N)}\}$. A centroid $\overline{\rho}^{(j)}$ of jth class is computed by

$$\overline{\rho}^{(j)} = \frac{1}{m_j} \sum_{i=1}^{m_j} \hat{\rho}(x_i), \tag{4}$$

where m_j represents the number of samples in class j. The decision about the category of a sample $Class\,(x_t)$ is determined by

$$Class(x_t) = \arg \min_{1 \le j \le N} d_M\left(\hat{\rho}(x_t), \overline{\rho}^{(x_j)} \mid \hat{\Sigma}\right) \tag{5}$$

where $d_M(.,.)$ indicates the Mahalanobis distance given by

$$d_M\left(\hat{\rho}(x_t), \overline{\rho}^{(x_j)} \mid \hat{\Sigma}\right) \\ = \sqrt{[\hat{\rho}(x_t) - \overline{\rho}^{(x_j)}]^T \hat{\Sigma}^{-1}[\hat{\rho}(x_t) - \overline{\rho}^{(x_j)}]}, \tag{6}$$

$$\hat{\Sigma} = \sum_{j=1}^{N} (m_j - 1)\hat{\Sigma}^{(j)} / (m - N) \tag{7}$$

is the pooled within class sample co-variance matrix, and

$$\hat{\Sigma}^{(j)} = \frac{1}{m_j - 1}$$

$$\times \sum_{i=1 \| x_i \in Class(j)}^{m_j} \left(\hat{\rho}(x_i) - \overline{\rho}^{(j)}\right)\left(\hat{\rho}(x_i) - \overline{\rho}^{(j)}\right)^T \tag{8}$$

is the co-variance matrix of jth class. In the present work this classification part of the VVRKFA is performed by employing SLFN as explained below.

3 HYBRID VVRKFA-SLFN CLASSIFIER

As discussed in section 2, VVRKFA classifies a test pattern in low-dimensional subspace based on Mahalanobis distance measurement from the class centroids to it. The vector-valued responses $\{\hat{\rho}(x_i) \in \Re^N\}_{i=1}^{m}$ of the training patterns $\{x_i\}_{i=1}^{m}$ are obtained by the approximated function (3) using a nonlinear kernel function k. These vector-valued responses and the target vectors together $(\hat{\rho}_i, y_i) \in \Re^N \times \Re^N$ are used as the training data of the SLFN classifier. According to the theory of SLFN classifier like ELM (Huang et al. 2004, Huang and Siew 2004), a SLFN, having q number of hidden nodes, can estimate m training samples with absolutely no error. This points toward the existence of β_i, a_i and d_i such that

$$f_q(\hat{\rho}_j) = \sum_{i=1}^{q} \beta_i G(a_i, d_i, \hat{\rho}_j) = y_j, j = 1, \dots m. \tag{9}$$

Here, $a_i = [a_{i1}, a_{i2}, ..., a_{iN}]^T$ is the weight vector that connects the ith hidden neuron and the input neurons, $\beta_i = [\beta_{i1}, \beta_{i2}, ..., \beta_{iq}]^T$ is the weight vector that connects the ith hidden neuron and the output neurons, and d_i is the threshold of ith hidden neuron. $G(a_i, d_i, \hat{\rho}_j)$ indicates the ith hidden node's response with respect to the input $\hat{\rho}_j$ and it can be expressed as

$$G(a_i, d_i, \hat{\rho}_j) = g(a_i \cdot \hat{\rho}_j + d_i), \tag{10}$$

where, g is the activation function. The equation (9) can be expressed in a matrix-vector form as

$$W\beta = Y, \tag{11}$$

where,

$$W(a_1, ..., a_q, d_1, ..., d_q, \hat{\rho}_1, ..., \hat{\rho}_m)$$

$$= \begin{bmatrix} G(a_1, d_1, \hat{\rho}_1) & \dots & G(a_q, d_q, \hat{\rho}_1) \\ \vdots & \dots & \vdots \\ G(a_1, d_1, \hat{\rho}_m) & \dots & G(a_q, d_q, \hat{\rho}_m) \end{bmatrix}_{m \times q},$$

$$\beta = \begin{bmatrix} \beta_1^T \\ \vdots \\ \beta_q^T \end{bmatrix}_{q \times N} \text{ and } Y = \begin{bmatrix} y_1^T \\ \vdots \\ y_m^T \end{bmatrix}_{m \times N}. \tag{12}$$

The matrix W contains weights of hidden layer output of the network; the ith column of W is the ith hidden node's output vector with respect to inputs $\hat{\rho}_1; \hat{\rho}_2; ...; \hat{\rho}_m$ and the jth row of W is the output vector of the hidden layer with respect to input $\hat{\rho}_j$. Equation (11) then reduces to a linear system and the output weights β are determined by

$$\hat{\beta} = W^\dagger Y \tag{13}$$

where $W^\dagger$ is the Moore-Penrose generalized inverse (Serre 2002) of the output matrix W of hidden layer. If W and β are known, the class labels of a new test patterns $x_t \in \Re^n$ can be determined from its mapped label vectors $\hat{\rho}(x_t)$.

4 EXPERIMENTAL RESULTS

4.1 Benchmark data set

The efficacy of this present method is evaluated on a varieties of data sets which include eight benchmark data sets from UCI repository (Blake & Merz 1998) and four microarray expression data sets, namely SRBCT (Khan, Wei, & Ringner 2001), Lung cancer (Bhattacharjee & et al. 2001), Brain tumor2 (Nutt & et al. 2003) and Leukemia2 (Armstrong & et al. 2002), for multicategory cancer classification.

The specification of these data sets such as number of classes, number of input features and number of samples are shown in Table 1 and Table 2 separately for the two types of data sets.

4.2 Experimental setup

The performance of VVRKFA-SLFN classifier is compared with VVRKFA and SLFN in the form of ELM. All the methods are implemented in MATLAB. 10-fold cross validation testing method is used to compare the performances of all the three methods. We employed a grid search technique to identify the optimal parameter set for a classifier. A Gaussian kernel $k(x_i, x_j) = \exp(-\mu) \| x_i - x_j \|^2$

Table 1. Specification of benchmark multiclass data sets from UCI repository.

Data set	No. of classes	No. of features	No. of samples
Iris	3	4	150
Wine	3	13	178
Glass	6	9	214
Vehicle	4	18	846
Segment	7	19	2310
Waveform	3	21	5000
Bupa	2	6	345
WPBC	2	32	110

Table 2. Specification of benchmark multiclass micro-array data sets.

Data set	No. of classes	No. of genes	No. of genes selected	No. of samples
selected	4	2308	10	83
SRBCT	4	12625	10	50
Brain-Tumor2				
Lung	5	12600	10	203
Leukemia	3	12582	10	72

where μ is the Gaussian kernel parameter, is selected as kernel. The regularization parameter C of VVRKFA and VVRKFA-SLFN are selected by tuning from the set $\{C = 10^i \mid i = -7, -6, ..., -1\}$. The kernel parameter μ for above two methods is selected from the set $\{\mu = 2^i \mid i = -8, -7, ..., 8\}$. The number of hidden neurons of VVRKFA-SLFN is kept fixed at 10 for all the data sets while it's value is tuned for ELM classifier for best results. The optimal parameter set for both the classifiers are chosen based on the performance of a parameter set on a tuning data set consisting of 30% of total samples. Once the optimal parameter set is chosen we carried out 10-fold Cross Validation (CV) (Mitchell 1997) testing presenting same fold of data to all the classifiers for computing average testing accuracy. Average testing accuracy and standard deviation are reported by performing 10-fold CV 100 times with random permutation of the training data.

4.3 Results on benchmark data sets

Table 3 shows the performance of SLFN, VVRKFA and VVRKFA-SLFN on 8 benchmark data sets. The results represent average percentage 10-fold CV testing accuracy and its standard deviations for 100 replicate runs with random permutation of data. Table 3 also enlists the selected optimal parameters like number of hidden neurons (q), regularization (C) and kernel (μ) parameters for different methods and the number of classes in a data set. The best accuracy obtained by a classifieris in bold face for each data set in this table.

It is observed from the results of Table 3 that the hybrid VVRKFA-SLFN classifier performs better than VVRKFA on six data sets out of eight. Although, VVRKFA-SLFN performs comperbly with VVRKFA for Iris and Wine data sets. On the other hand, VVRKFA-SLFN performs better than both VVRKFA and SLFN classifiers for all the other six data sets compared in Table 3. The vital feature of the VVRKFA-SLFN classifier is

Table 3. 10-fold testing performance of SLFN, VVRKFA and VVRKFA-SLFN classifiers on benchmark data sets.

Data set (No. of class)	SLFN		VVRKFA			VVRKFA-SLFN			
	q	Tst. acc. (std.)	C	μ	Tst. acc. (std.)	C	μ	q	Tst. acc. (std.)
Iris (3)	30	96.04 (1.13)	10^{-5}	2^{-2}	**98.31** (0.53)	10^{-5}	2^{-2}	10	97.04 (0.75)
Wine (3)	50	97.06 (1.14)	10^{-4}	2^{-2}	**98.60** (0.61)	10^{-4}	2^{-2}	10	98.54 (0.93)
Glass (6)	80	63.76 (2.08)	10^{-5}	2^{-3}	65.25 (1.54)	10^{-5}	2^{-3}	10	**67.52** (1.49)
Vehicle (4)	100	82.31 (0.76)	10^{-5}	2^{-1}	84.47 (0.30)	10^{-5}	2^{-1}	10	**84.68** (0.42)
Segment (7)	100	93.66 (0.26)	10^{-4}	2^{-2}	95.32 (0.08)	10^{-4}	2^{-2}	10	**95.89** (0.17)
Waveform (3)	50	85.14 (0.15)	10^{-3}	2^{-8}	85.84 (0.07)	10^{-3}	2^{-8}	10	**86.38** (0.16)
Bupa (2)	50	69.31 (1.47)	10^{-4}	2^{-2}	71.79 (0.72)	10^{-4}	2^{-2}	10	**72.36** (1.05)
WPBC (2)	50	76.90 (1.40)	10^{-5}	2^{-7}	71.74 (1.60)	10^{-5}	2^{-3}	10	**78.60** (1.23)

that a fixed number of hidden neurons (only 10) for SLFN is sufficient for well discrimination of the patterns in the label space. Whereas the optimal number of hidden neurons for SLFN classifier is required to tune for different data sets.

The variation of testing time with the size of the data set for VVRKFA-SLFN and VVRKFA classifiers is shown in Fig. 2. This testing is performed on waveform data set by considering 50 hidden neurons for SLFN and 10 hidden neurons for VVRKFA-SLFN classifiers. The Fig. 2 shows that with the increase of the number of samples, the average 10-fold testing time increases for both the methods but in all the cases the VVRKFA-SLFN takes less time than that of VVRKFA classifiers. This indicates the speeding up of testing time of VVRKFA-SLFN by the proposed hybrid classifier.

4.4 Results on microarray data sets

The VVRKFA-SLFN method is also applied on a practical problem of cancer detection using gene microarray data. The challenge in microarray data classification is that the number of samples is very less than the number of features (genes), i.e., the data sets suffer from curse of dimensionality.

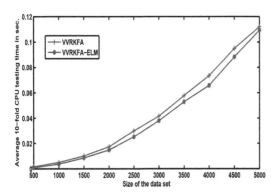

Figure 2. Variation of testing time of VVRKFA and VVRKFA-SLFN classifiers with the increase of number samples.

To facilitate better classification results, this drawback of microarray data is removed first by identifying a small gene subset which causes the disease (Li, Weinberg, Darden, & Pedersen 2001, Guyon, Weston, Barnhill, & Vapnik 2002). We have selected the most relevant 10 genes by MRMR method (Peng, Long, & Ding 2005) as performed in our previous research work (Ghorai, Mukherjee, Sengupta, & Dutta 2011). Incase of microarray data sets, parameter selection is done in the same way as performed for the benchmark data sets but we have evaluated Leave-One-Out Cross Validation (LOO-CV) performance (Mitchell 1997) as there are very less number of samples in each class. In this experiment, the number of hidden neurons of VVRKFA-SLFN is kept fixed at 5 for all the four data sets while it's value is tuned for SLFN classifier for the best results.

Table 4 shows the results obtained on the microarray data sets. From the table it is observed that the performance of VVRKFA-SLFN classifier is better than both VVRKFA and SLFN classifiers on all the four data sets. It also shows that VVRKFA-SLFN classifier can perform better with a fixed and very small number of neurons compared to SLFN classifiers. Thus, VVRKFA-SLFN classifier eliminates the need of tuning of the number of neurons of SLFN classifiers.

4.5 Analysis of results

The above experimental results may be analyzed as follows. In their recent works, Huang et al. (Huang, Zhou, Ding, & Zhang 2012) showed that SLFN (like ELM) has comparable or much improved generalization performance for multiclass data classification with very high training speed (up to thousands times) than conventional SVM. Thus, it is expected that SLFN like ELM classifier will perform better than a simple distance classifier using Mahalanobis distance. This means the SLFN classifier is better able to decode the class labels in the label space than the decoding using Mahalanobis distance. Due to this fact the classification performance of the VVRKFA-SLFN classifier increases than that of the VVRKFA classifier in most of the cases as observed from both the Table 3 and 4.

Table 4. LOO-CV testing performance of SLFN, VVRKFA and VVRKFA-SLFN classifiers on microarray data sets.

Data Set (No. of class)	SLFN		VVRKFA			VVRKFA-SLFN			
	q	Tst. acc. (std.)	C	μ	Tst. acc. (std.)	C	μ	q	Tst. acc. (std.)
SRBCT (4)	10	92.77 (25.90)	10^{-4}	2^{-6}	93.98 (23.79)	10^{-4}	2^{-6}	5	**96.39** (18.66)
Brain tumor2 (4)	10	84.00 (36.66)	10^{-4}	2^{-8}	86.00 (34.7)	10^{-4}	2^{-8}	5	**88.00** (32.50)
Lung (5)	30	92.61 (26.16)	10^{-3}	2^{-2}	94.09 (23.58)	10^{-3}	2^{-2}	5	**94.09** (23.58)
Leukemia (3)	20	94.44 (22.91)	10^{-5}	2^{-7}	94.44 (22.91)	10^{-5}	2^{-7}	5	**95.83** (19.98)

The speeding up of testing time of VVRKFA-SLFN classifier over VVRKFA can be explained with the help of the computational complexity of testing of the two classifiers. In order to test a sample, both VVRKFA and VVRKFA-SLFN use the mapped low-dimensional label space vectors by vector-valued regression technique. The difference in both the classifier is the way of determining the class labels of these mapped patterns. The low-dimensional subspace has a dimension of N if it is a classification problem with N different classes. To test a mapped pattern in N-dimension, SLFN requires to evaluate expressions (10) and (9). Expression (10) calculates the inner product between ith hidden neuron weight $a_i \in \Re^N$ and the input pattern $\hat{\rho}_j \in \Re^N$. This requires N number of multiplications. For q such hidden neurons the total number of multiplication required is qN. Expression (9) calculates the inner product between the weight vector $\beta_i \in \Re^q$ connecting q hidden neurons to an output neuron and output of the ith hidden neuron G_i. For N output neurons, this operation requires another qN number of multiplications. Thus, total complexity of testing by SLFN becomes $O(2qN)$, which is linear to both number of hidden neurons q and number of classes N. On the other hand, testing a mapped sample by VVRKFA method requires Mahalanobis distance measure between the mapped sample $\hat{\rho}(x_t) \in \Re^N$ and the class centroid $\bar{\rho}(x_j) \in \Re^N$ of jth class by the expression (6). This requires $N^2 + N$ multiplications. The total complexity of evaluation of N such Mahalanobis distances becomes $O(N^3 + N^2)$. Since q is the number of hidden neurons in SLFN and itis very small number (~ 5 to 10) in VVRKFA-SLFN classifier the testing complexity of it will be less than the VVRKFA classifier as N increases. This shows the improved testing time complexity of VVRKFA-SLFN classifier over the VVRKFA of classification and it is verified from the Figure 2.

5 CONCLUSION

In this work a recently developed multiclass kernel classifier VVRKFA is combined with a very fast single hidden layer neural network classifier like ELM. The combined hybrid VVRKFA-SLFN classifiers performs better than that of both VVRKFA and SLFN classifiers. In addition, the testing time of the combined classifier is less than the VVRKFA classifier. Further, it is observed that the proposed classifier eliminates the need of tuning of the number of neurons for constructing SLFN classifier. These observations indicate the improvements of classification performance of VVRKFA classifier by the proposed hybrid

VVRKFA-SLFN classifier. This technique can be employed to various real world data classification problems.

REFERENCES

Allwein, E., R. Schapire, & Y. Singer (2000). Reducing multiclass to binary: a unifying approach for margin classifiers. *Journal of Machine Learning Research 1*, 113–141.

Armstrong, S. A. & et al. (2002). Mll translocations specify a distinct gene expression profile that distinguishes a unique leukemia. *Nature Genetics 30*(1), 41–47.

Bhattacharjee, A. & et al. (2001). Classification of human lung carcinomas by mrna expression profiling reveals distinct adenocarcinoma subclasses. *Proc. Nat'l Academy of Sciences, USA, 98*(24), 13790–13795.

Blake, C. & C. Merz (1998). Uci repository for machine learning databases. Technical report, Department of Information and Computer Sciences, University of California, Irvine. [Online]. Available: http://www.ics.uci.edu/mlearn/MLRepository.htmlS.

Broomhead, D. S. & D. Lowe (1988). Multivariable functional interpolation and adaptive networks. *Complex Syst.* (2), 321–355.

Dietterich, T. & G. Bakiri (1995). Solving multiclass learning problems via error-correcting output codes. *Journal of Artificial Intelligence Research 2*, 263–286.

Ghorai, S., A. Mukherjee, & P. K. Dutta (2010). Discriminant analysis for fast multiclass data classification through regularized kernel function approximation. *IEEE Trans. on Neural Networks 101*(6), 1020–1029.

Ghorai, S., A. Mukherjee, & P. K. Dutta (2012). *Advances in Proximal Kernel Classifiers*. Germany: LAP LAMBERT Academic Publishing.

Ghorai, S., A. Mukherjee, S. Sengupta, & P. K. Dutta (2011). Cancer classification from gene expression data by nppc ensemble. *IEEE/ACM Trans. on Computational Biology and Bioinformatics 8*(3), 659–671.

Guyon, I., J. Weston, S. Barnhill, & V. Vapnik (2002). Gene selection for cancer classification using support vector machines. *Machine Learning 46*, 389–422.

Hsu, C.-W. & C.-J. Lin (2002). A comparison of methods for multi-class support vector machines. *IEEE Trans. on Neural Networks 13*(2), 415–425.

Huang, G.-B. & L. Chen (2007). Convex incremental extreme learning machine. *Neurocomputing 70*, 3056–3062.

Huang, G.-B. & L. Chen (2008). Enhanced random search based incremental extreme learning machine. *IEEE Trans. on Neural Networks 71*, 3460–3468.

Huang, G.-B., L. Chen, & C.-K. Siew (2006). Universal approximation using incremental constructive feedforward networks with random hidden nodes. *IEEE Transactions on Neural Networks 17*, 879–892.

Huang, G.-B. & C.-K. Siew (2004). Extreme learning machine: Rbf network case. In *Eighth Int'l Conf. Control, Automation, Robotics, and Vision (ICARCV '04)*.

Huang, G.-B. & C.-K. Siew (2005). Extreme learning machine with randomly assigned rbf kernels. *Int'l J. Information Technology 11*(1).

Huang, G.-B., H. Zhou, X. Ding, & R. Zhang (2012). Extreme learning machine for regression and multiclass classification. *IEEE Transactions on Systems, Man, and Cybernetics – Part B: Cybernetics 42*, 513–529.

Huang, G.-B., Q.-Y. Zhu, & C.-K. Siew (2004). Extreme learning machine: A new learning scheme of feedforward neural networks. In *Int'l Joint Conf. Neural Networks (IJCNN '04)*.

Huang, G.-B., Q.-Y. Zhu, & C.-K. Siew (2006). Extreme learning machine: Theory and applications. *Neurocomputing 70*, 489–501.

Igelnik, B. & Y. H. Pao (1995). Stochastic choice of basis functions in adaptive function approximation and the functionallink net. *IEEE Trans. Neural Networks 6*(6), 1320–1329.

Kaminski, W. & P. Strumillo (1997). Kernel orthonormalization in radial basis function neural networks. *IEEE Trans. Neural Networks 8*(5), 1177–1183.

Khan, J., J. S. Wei, & M. Ringner (2001). Classification and diagnostic prediction of cancers using gene expression profiling and artificial neural networks. *Nature Medicine 7*(6), 673–679.

Kreßel, U. (1999). *Advances in kernel methods-support vector learning*, Chapter Pairwise classification and support vector machines, pp. 255–268. Cambridge: MIT Press.

Li, L., C. R. Weinberg, T. A. Darden, & L. G. Pedersen (2001). Gene selection for sample classification based on gene expression data: Study of sensitivity to choice of parameters of the ga/knn method. *Bioinformatics 17*, 1131–1142.

Lowe, D. (1989). Adaptive radial basis function nonlinearities, and the problem of generalisation. In *1st Inst. Electr. Eng. Int. Conf. Artif. Neural Netw.*, pp. 171–175.

Mitchell, T. M. (1997). *Machine Learning* (1st. ed.). Singapore: The McGRaw Hill Companies, Inc.

Nutt, C. L. & et al. (2003). Gene expression-based classification of malignant gliomas correlates better with survival than histological classification. *Cancer Research 63*(7), 1602–1607.

Pao, Y. H., G. H. Park, & D. J. Sobajic (1994). Learning and generalization characteristics of random vector functional-link net. *Neurocomputing 6*, 163–180.

Peng, H., F. Long, & C. Ding (2005). Feature selection on mutual information: Criteria of max-dependency, max-relevance, and min-redundancy. *IEEE Trans. Pattern Analysis and Machine Intelligence 27*(8), 1226–1238.

Serre, D. (2002). *Matrices: Theory and Applications*. New York: Springer–Verlag.

Wettschereck, D. & T. Dietterich (1992). Improving the performance of radial basis function networks by learning center locations. In *Advances in Neural Information Processing Systems (NIPS)*, Volume 4, San Mateo, CA: Morgan Kaufmann, pp. 1133–1140.

Computer, Communication and Electrical Technology – Guha, Chakraborty & Dutta (Eds)
© 2017 Taylor & Francis Group, ISBN 978-1-138-03157-9

Investigations on user perspective evaluation of some reliability aspects of web services

Subhash Medhi, Abhijit Bora & Tulshi Bezboruah
Department of Electronics and Communication Technology, Gauhati University, Assam, India

ABSTRACT: Web services have emerged as a web based technology for accessing information over the internet using platform neutral web standards and protocols. In this paper we propose to develop and implement a service oriented prototype research service application using. Net technology to study and predict reliability aspects of web services. The uniqueness of our proposed system is the hierarchically designed parent service that authenticates a particular user to invoke services, acting as a service broker and redirects the query to execute children services, acting as a service provider. An automated software testing tool, Mercury LoadRunner is deployed to test and record the specific attributes of the web services and analyse the reliability aspects of the services. The outcome of the experiment will help in adoption and usage of the web services in business applications and integrations. We present here the architecture, procedure of testing, transaction status and reliability estimation of the system under gradual stress of end users.

1 INTRODUCTION

Web Service (WS) technology follows Service Oriented Architecture (SOA) that provides remote procedure call mechanism in distributed environment over the web. Once WS is deployed, various heterogeneous organizations that are running on diverse networks can be interconnected for business processes. The coordination leads to composite services. It is an important issue to assess the trustworthiness of composite WS built on different networks and languages (Duhang et al., 2006). The ever growing reliance on the data and services provided by various WS vendors endorsed these services to be of superior performance and reliable as these services offered over the internet have rapidly permeates our lives due to the convenience and low cost factors (Gokhale Swpana S, Paul J. Vandal and Jijun Lu, 2006). In widespread application of internet technology, WS applications such as online shopping, banking, travel booking, stock trading have been adopted and developed tremendously (Suichang Wang, Fei Ding, 2013). WS is becoming more familiar to the people due to the recent development and applications over the internet as many top companies such as Microsoft, IBM, Oracle and SUN have launched supports for technologies related to WS (Hou et al., 2010). An important quality aspect of software is the capability to which it can perform to its intended operation. The reliability attribute is one of the important qualities of services that affects in overall performance of a WS system. Thus the reli-

ability of software becomes a major concern of the companies as unreliable software may lead to huge economic loss and also may degrade the reputation of an organization. The software failure depends on the software and hardware. Hence, it is essential to assess the reliability of software in parallel to hardware reliability. The software reliability is a probabilistic measurement that can be defined as the probability that the software runs without failure during a specified period of time for a specified environment (Goel, 1985). Many analytical models have been developed to estimate the reliability of software that helps to develop high quality software. In this paper we emphasize on recording the quality attributes of the WS using mercury LoadRunner automated testing tool and used statistical method for data analysis to predict the reliability aspects of the service.

1.1 *Related works*

Abdelkarim et al. (2006) had elaborated some recovery policies to handle and recover from service faults during WS composition. Duhang et al. (2006) had introduced an approach to predict the reliability of WS composition by transformation of Business Process Execution Language (BPEL) specifications into Stochastic Petri Nets (SPN) model. Pat et al. (2007) had identified some parameters that can impact the WS dependability. They had elaborated the methods of dependability enhancement by redundancy in space and time. Gokhale et al. (2006) had proposed Stochastic Reward Net (SRN) based analysis methodology to quantify the performance and reliability tradeoffs. Lizun et al. (2009) had proposed a reliability modeling framework for system reliability and suggested that dynamic modeling and decomposed models reflect the characters of SOA systems and supportive to reliability analysis. Hou et al. (2010) had suggested that WS performance can be improved by optimizing SOAP messages and adopting suitable methods. Bora et al. (2014) presented an empirical study on hierarchical SOAP based WS by implementing WS Security policy. They had elaborated that the response time with security, encryption and signature is more than without security. Medhi et al. (2014) had performed empirical and statistical analysis of hierarchical WS performance by implementing a financial model. Bezboruah et al. (2015) had performed an evaluation of performance of hierarchical WSs using a cluster and non cluster web server. Bora et al. (2015) had proposed a quality evaluation framework for multi service SOAP based WS. The uniqueness of our proposed work from the previous works is that we have performed investigations on user perspective evaluation of some reliability aspects of WS by monitoring HTTP transactions using Microsoft's service oriented application building framework Windows Communication Foundation (WCF). The novelty of this work is that we have monitored the transactional status, reliability analysis and prediction of the system under massive stress of concurrent end user request.

2 PROPOSED WORK AND METHODOLOGY

The objective of the proposed work is to implement a hierarchical electronic Automated Teller Machine (e-ATM) service using WCF technology to study the reliability aspects of the service. The proposed financial prototype service has three tiers compositions as shown in Figure 1, namely: (i) a cli-

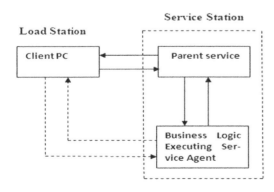

Figure 1. Proposed service architecture.

ent application, (ii) one parent service where users are authenticated and, (iii) a Computing Agents (CA) to execute Business Logic (BL) components. The authenticated user query parameters are redirected to the service agents for executing BL solutions. The Mercury Load Runner is deployed to perform the load test on the proposed model.

The prototype WCF service has been developed to cover all facilities that a bank ATM facilitates to a customer. We implement the service using C# language, Microsoft Internet Information Server (IIS) as Web Server and database with Microsoft Structured Query Language (SQL) 2005, and MS Visual Studio 2012 as Integrated Development Environment (IDE). A client application has also been developed to invoke the service in the same environment. The database size for testing the proposed service is 15,000. The service application has been tested against 200, 400, 600, 900, 1100 and 2000 Virtual User (VU) by deploying it on Mercury LoadRunner to evaluate the reliability attributes of the WS. The statistical analysis is done over the recorded attributes to estimate the reliability of the WS.

3 SOFTWARE AND HARDWARE SPECIFICATION

The software specifications at server side are: (a) IIS 7.5 as Web Server, (b) MS SQL version 2005 as database server, (c) Microsoft Visual Studio version 2012 as Integrated Development Environment (IDE), (d) Internet Explorer as web browser and (e) Windows Server 2008 as Operating System (OS). The software tools such as Microsoft SDK version 7.1 and EasyFit, version 6.5 are used. The hardware configuration includes Intel(R) Xenon(R) CPU E5620 processor with 2.4 GHz speed, 8 GB RAM and 600 GB hard drive. The load generator machine contains the software testing tool such as Mercury LoadRunner. We have created the service script by using the testing tool. The load was given on the WS from a remote desktop PC whose OS is windows XP. The hardware configurations for the remote desktop PC are: (i) Intel(R) Pentium (R) Dual CPUE2200, (ii) Processor speed: 2.2 GHz, (iii) RAM: 1GB and (iv) Hard drive: 150 GB.

4 AUTOMATED TESTING TOOL

The Mercury LoadRunner was deployed to test the services. The Mercury LoadRunner can produce real sense of Virtual User (VU) to stress the system and retrieve information from database servers through web servers (Load Runner, 2016). We have monitored some parameters as displayed in the Load Runner controller. We have recorded the results for statistical analysis. During the testing, a user was set for a think time of 30 s in one transaction and the average steady state period was set for 300 s for each test.

5 TESTING PARAMETERS

The test settings are: (i) think time that a user delayed time for performing successive operations, (ii) work load intensity measured in terms of gradually increasing VU and, (iii) the network speed, which specifies the Bandwidth (BW) of a network. By generating load and gradually increasing stress over the service, we have monitored the HTTP transactions to see how the service performs in real time and its potential bottlenecks. During the testing schedule, we have monitored the number of successful and failure transactions of the service.

6 RESULTS AND ANALYSIS

We have conducted the empirical test at different stress level VU such as: 200, 400, 600, 900, 1100, 1400 and 2000 with maximum BW of 1GBPS. A test case for select operation has been prepared for accessing the ATM service. A data sample against each test is recorded for service metrics. The statistical analysis is carried out over that. The response for 5 minutes duration was considered. The entire load tests were conducted with a ramp up schedule with 1 VU operating for every 15 s. After completion of steady state period of 300 s, all the VU are phased out simultaneously. The different transaction status against massive stress level is given in Table 1. For statistical analysis we have recorded the test for 30 times at 900 VU and the system shows an average connection refusal of 28%. The frequency table and the histogram are shown in Table 2 and Figure 2 respectively. It is observed that the highest density of failure is occurred in the range from 177420.2 to 183430.6. The histogram is skewed left. From these outcomes, we can infer that our Failure Count (FC) distribution is a weibull distribution.

Table 1. HTTP transactions at different VU.

Scenario	VU	Pass	Fail	Total
	200	20489	0	20489
	400	28699	0	28699
SQL	600	34568	0	34568
select	900	486031	175319	661350
operation	1100	443837	195749	639586
	1400	482628	271696	754324
	2000	529208	339527	868735

To further ascertain the distribution, we use EasyFit (EasyFit distribution tool, 2016) parameter estimating statistical software, version 5.6 to estimate the value of shape (α) and scale parameter (β) which is found as $\alpha = 26.11$ and $\beta = 178300$ respectively. The weibull Cumulative Distribution Function (CDF) is calculated using the following formula (1) (Chandra et al, 1981) for each FC.

$$CDF = 1 - exp\left\{(-FC / \beta)^{\alpha}\right\} \qquad (1)$$

6.1 Goodness of Fit (GoF) evaluation using Kolmogornov-Smirnov test

The Kolmogorov Smirnov (KS) GoF test is used to test whether a sample comes from a population with a specific distribution. In the test, the assumed distribution is correct, if the maximum departure between the assumed CDF (F_o) and the empirical CDF (F_n) distributions is small. We have calculated the intermediate values for KS GoF test for weibull distribution. It is seen that the KS GoF test statistic value D_{max} (0.14376) is smaller than the KS table critical value (0.24) for $\alpha = 0.05$ and a sample of size n = 30. Based on these results, we do not reject the hypothesis that the obtained CDF of the FC is distributed weibull ($\alpha = 26.11$, $\beta = 178300$). Hence, the population from where these data were obtained is distributed weibully.

Table 2. Bin and frequency of failure count at 900 VU.

Failure ranges	Frequency
0–159389	1
159389–165399.4	2
165399.4–171409.8	7
171409.8–177420.2	8
177420.2–183430.6	9
>183430.6	3

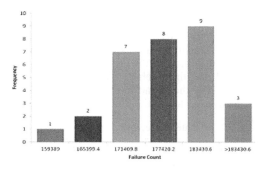

Figure 2. Histogram of failure count.

6.2 Confidence interval of CDF

We estimate the mean value for CDF at 95% confidence interval for 900 VU. The population mean μ can be computed using equation 2 (Spiegel, M. R, 2000).

$$\mu = \bar{x} \pm t_c SD / \sqrt{N} \qquad (2)$$

We consider the mean value of CDF as $\bar{x}$, the critical value from t_c (0.05,29), the standard deviation as SD, sample size as N and the margin of error as $t_c SD/\sqrt{N}$. The estimated population mean μ is calculated and given in Table 3. At 95% confidence interval, the mean value of CDF lies between 0.51 ± 0.12, i.e. 0.63 and 0.39.

6.3 Reliability estimation

Reliability of a system is estimated over phases of time that the system executes without failure in specified period. We have calculated the reliability of our proposed system using average FC and HTTP transactions from 30 sample data i.e FC = 175319, HTTP transaction = 661350. The probability of failure (P_f) is calculated as FC/Total HTTP i.e 175319/661350, P_f = 0.27. The probability of success is the reliability and can be calculated as reliability (R) = 1–P_f = 0.73 (Martin L, Shooman, 2002). Thus the reliability is the probability of no failure within a specified period in a specified environment. Reliability can also be estimated using equation (3) as in (Lizun et al., 2009).

$$R = e^{-\lambda t} \qquad (3)$$

where, λ is failure rate. It is a probability density function for operational time, λ (t). The execution time unit is considered as one day, so we set t = 1, the reliability becomes R = e$^{-\lambda}$. In practice reliability can be measured approximately using equation (4).

$$R = 1 - \lambda \qquad (4)$$

Thus we estimate the following reliabilities at incremented VU level: R(9000) = 0.73, R(1100) = 0.69, R(1400) = 0.64, R(2000) = 0.61. It is observed that the reliability is falling with the rising number of VU. We performed the test at different VU and observed that the reliability at 200, 400 and 600 VU is 1. For these load level, the system will be served as expected to the user's request. However,

Table 3. Estimated values for μ.

N	t_c(0.05,29)	Parameter	$\bar{x}$	SD	$t_c SD/\sqrt{N}$
30	2.045	CDF	0.51	0.31	0.12

in higher load level such as 900, 1100, 1400 and 2000; the reliabilities estimated are 0.73, 0.69, 0.64, and 0.61 respectively. Thus, we can conclude that the reliability falls with the rising number of service users, i.e. the user will get less response from the WCF based service as the web server refuses at massive concurrent request.

7 OVERALL RELIABILITY ASSESSMENT

The overall reliability evaluation for the WCF based service is given in Table 4. In the FC histogram, it is observed that the highest density of FC value is 9 and is occurred in the range from 177420 to 183431 transactions. The FC follows left skewed. Thus we can assume that our FC distribution is a weibull distribution. Using CDF values we have computed the KS statistic using Easy Fit version 5.6. It is a tool for statistical data analysis and simulation that allows fitting probability distributions to make better decisions. In the analysis, it is revealed that the statistics value obtained (0.14386) is smaller than the critical value (0.2417) at α = 0.05 which indicates that the assumed samples were from a population with a specific distribution is weibull. Based on the software analysis results we can conclude that the observed data adequately fits in the weibul distribution. We have evaluated the reliability of our proposed system by increasing VU. For lower

Table 4. Overall reliability evaluation of the service.

Experimental observation	Results
HTTP transaction failure against 200,400,600	Not observed
HTTP transaction failure against 900, 1100,1400,2000	Observed and increases gradually
FC Histogram against 900 VU	Left skewed with highest failure density 9 in the range from 177420 to 183431
Nature of failure distribution K-S test of GoF at 95% confidence level	Weibull Failure distribution fits Weibull
CDF	Mean CDF is 0.51 and lies in between 0.63 to 0.39
Reliability up to 600 VU	R = 1. Strong reliability. Consumer will get expected response
Reliability for 900 VU	R = 0.73, Moderate reliability with the probability of failure occurrences
Reliability more than 900 VU	Service response degrades gradually.

number of VU up to 600, the reliability estimated to be 100% and the reliability decreases at higher number of VU, that is 73%, 69%, 64%, 61% against 900, 1100, 1400 and 2000 VU respectively.

8 CONCLUSION AND FUTURE WORK

In this paper we have conducted user perspective evaluation of some reliability aspects of WS implementing a financial e-ATM model. The proposed model was tested in an automated testing tool Mercury LoadRunner by randomly varying VU. We have observed that the reliability of the system is strong up to 600 VU. Then it decreases gradually with higher number of VU. This occurrence of service failures may be attributed to database server or system resources. The experimental study revealed the strong evidence of availability, scalability and reliability of the proposed service to communicate with massive number of VU. As part of the future work we propose to investigate more reliability aspects of WS which can highlight the reliability of the system.

ACKNOWLEDGEMENT

The acknowledgement goes to the All India Council of Technical Education (AICTE), Govt. of India for the financial support towards the work (F.No. 8023/BOR/RID/RPS (NER)-84/2010–2011 31st March 2011).

REFERENCES

Abdelkarim, E, Maheswary, P., Vladimir, T. (2006). "Recovery Policies for Enhancing Web service Reliability", *IEE International Conference on Web Services* (ICWS'06).

Bezboruah, T and Bora, A. (2015). "Performance Evaluation of Hierachical Soap Based Web Service in Load Balancing Cluster-Based and Non-Cluster Based Web Server", *International Journal of Information Retrieval Research*, Vol. 5, Issue 4, pp. 20–31.

Bora, A and Bezboruah, T. (2015), "Some aspects of QoS for Interoperability of Multi Service Multifunctional Service Oriented Computing", *IEEE International Conference on Research in Computational Intelligence and Communication Network*, pp. 363–368, India.

Bora, A., Bezboruah, T. (2013). "Investigations on Hierachical Web service based on Java Technique",

Proceedings of the World Congress on Engineering, 2013, Vol II, pp. 891–896, WCE 2013, July 3–5, 2013, London, U.K.

Bora, A., Bezboruah, T. (2014). "Investigation on Security Implementation and performance Aspects of MedWS: a Hierarchical SOAP based Web Service", *International Journal of Database Theory and Application*, Vol. 7, No. 4, pp. 169–188, DOI: http:/dx.doi.org/10.14257/ijdta.2014.7.4.13, 2014.

Chandra, M, Singpurwala, N.D and Stephen, M.A. Kolmo-gorov (1981). "Statistics for Test of Fit for the Extreme Value and Weibull distribution", *Journal of the American Statistical Association*, Taylor & Francis, Vol. 76, No. 375, pp. 729–731.

Duhang Zhong and Zhichang Qi. (2006). "A Petri Net Based Approach for Reliability Prediction of Web services", School of Computer Science, National University of Defence Technology, Changsha, China.

EasyFit Distribution tools available at http://www.mathwave.com/easyfit-distribution-fitting.html, Retrieve on 2015.

Goel A. L. (1985). "Software reliability Models: Assumptions, Limitations, and Applicability", *IEEE Transactions on Software Engineering*, Vol SE 11, No. 12, pp. 1411–1423.

Gokhale Swapna S, paul J. Vndal and Jijun Lu: "Performance and Reliability analysis of web server Software Architectures", 12th Pacific Rim International Symposium on Dependable Computing (PRDC, 06).

Hou Zhai-wei, Zhai Hai-xia, Gao Guo-hong (2010). "A Study on Web Services Performance Optimization", *Proceedings of the Third International Symposium on Electronic Commerce and Security Workshops*, pp. 184–188.

Lizun Wang, Xiaoying Bai, Lizhu Zhou and Yinong Chen (2009). "A Hierarchical Reliability Model of Services Based Software System", *33rd Annual IEE International Computer Software and Applications Conference*.

Loadrunner (2015). Available at http://www.en.wikipedia.org//wiki/HP_LoadRunner, Retrieve on 2015.

Martin L. Shooman (2002). "Reliability of computer systems and networks: Fault Tolerance, Analysis, and Design", Polytechnic University and Martin L. Shooman & Associates.

Medhi, S., Bezboruah, T. (2014). "Investigations on implementation of e-ATM Web services based on .NET technique", *International Journal of Information Retrieval Research*, Vol. 4 Issue. 2, pp. 41–56.

Pat. P. W. Chan, Michael R. Lyu and Miroslaw Malek (2007). "Reliable Web Services: Methodology, Experiment and Modeling", IEEE International Conference on Web Services, USA.

Speigel, M. R. (2000). "Theory and problems of probability and Statistics", *Schaum's Outline Series*, MaGraw-Hill Book Company, SI edn, 2000.

Load balancing of the unbalanced cost matrix in a cloud computing network

R.K. Mondal, P. Ray, E. Nandi, P. Sen & D. Sarddar
Department of Computer Science and Engineering, University of Kalyani, Kalyani, India

ABSTRACT: Cloud computing is an incipient Internet accommodation concept that has propagated to provide various accommodations to its users. Cloud computing provides a variety of computing resources to facilitate the execution of sizably voluminous-scale tasks. It is web-predicated distributed computing. Sometimes, many tasks hamper the performance of the network, and the available numbers of servers are not insufficient to execute more than thousand tasks at a time. So, all tasks cannot be executed at the same time. Some nodes are needed to execute and balance the overall load of the network. Load balancing minimizes the completion time and also executes all the tasks in parallel.

It is inconvenient to have equal number servers to execute all tasks. The tasks to be executed in cloud computing would be more than that of the connected servers. Inhibited servers have to execute millions of tasks. Thus, to balance the load and increase the overall computation speedup of the cloud network, we have proposed load balancing with the Hungarian Method to achieve an optimal solution of load balancing.

1 INTRODUCTION

Cloud computing is a web-based service. It has shifted the computing and data away from desktop and portable PCs into large data centers. Cloud computing has also altered the way IT companies buy and design software. As the service is still in its evolving stage, there are many difficulties in cloud computing (Sinha 2016). Some of them are as follows:

a. Ensuring resourceful access control (authentication, authorization, and auditing)
b. Network-level relocation for the requirement of less cost and time to move a job
c. Providing proper data security for the data alteration stage
d. Data availability issues in cloud
e. Data marsh and transitive trust cases
f. The possibility of data race, data security, and inadvertent disclosure of sensitive information and the managing load balancing

In the case of load balancing, various types of information such as the number of jobs waiting in queue, job arrival rate, CPU processing time, etc., at each processor and at neighboring processors as well may be altered for obtaining an efficient and enhanced overall performance. So, many algorithms have been proposed for load balancing. In this paper, we have proposed a new algorithm for proper load balancing to get an improved CPU processing time for issues of better-quality performance.

A. Basic Idea of Load Balancing Algorithms

Load balancing (Mondal et al. 2016) generally chooses the task of how to select the next node and transfers a new request to transfer the load from the overloaded process to the under-loaded process. It enhances the incoming requesting load among the available nodes to improve the performance drastically by the cloud manager. Mainly, there are two types of algorithms based on their implementation method; they are static and dynamic algorithms.

A. *Static Algorithm:* It does not depend on the present state. It decides at the host that the request will be executed before setting up the request.
B. *Dynamic Algorithm:* The load balancer analyses the present state of the load strategies at each available host and executes the request at the suitable host.

Round robin, weighted round robin, least connection scheduling algorithm, etc., are some examples of the existing static algorithms; however, among them, the round robin algorithm is the simplest algorithm. It is used when all the nodes in a cluster have the same processing ability. First-Come-First-Served, Throttled, Honey Foraging Algorithm, etc., are some examples of dynamic scheduling algorithms. These are better than static algorithms and suitable for a large number of requests, which can carry various workload, which would be not able to predict.

2 PERFORMANCE CRITERIA

The proposed algorithm is designed to gather all the scheduling criteria such as maximum CPU utilization, maximum throughput, minimum turnaround time, minimum waiting time, and context switches. We discuss the arrival time, burst time, waiting time, turnaround time, response time, and throughput.

Arrival Time: Arrival time is the time of arrival of a process at the main memory.

Burst Time: The burst time for a process is how long the process holds the CPU.

Waiting Time: The waiting time is how long a process waits in the ready queue.

Turnaround Time: Turnaround time is the time of arrival minus the time of completion of the task.

Response Time: Response time is the time of arrival minus the first response time by the CPU.

Throughput: Throughput is the number of processes completed per unit time period.

3 PROBLEM DEFINITION

In this field, the proposed method has been formulated under a few assumptions:

a. The rules to be implemented are composed of indivisible tasks with no dependency among each other.
b. Each task has no deadline or priority.
c. To estimate the task completion time on each system, a task is submitted for execution by the nodes.
d. The mapping is to be performed in a batch mode.
e. The mapper runs on a disjoined machine and controls the execution of all tasks.
f. Each machine executes a particular task at the same time in the First-Come-First-Served order.
g. The size of the meta-tasks and the number of nodes are known.

4 PROPOSED WORK

Whenever the cost matrix of the problem is not a square matrix, that is, whenever the number of sources is not equivalent to the number of destinations, the matrix is called an unbalanced matrix. In such problems, dummy rows (or columns) are added in the matrix so as to complete it to form a square matrix. The dummy rows or columns will have all costs elements as zeros. The Hungarian Method (Kuhn, 1955) may be used to solve the problem. In our proposed algorithm, we are using the well-known Hungarian method in step 1 to step 5 and step 6 we are using our own step to get optimal solution for the proposed load balancing.

Step 1: *The input of this algorithm is an n*n square.*

Step 2: *Find the smallest element from each row and subtract it from each element in the corresponding row.*

Step 3: *Similarly, find the smallest element and subtract it from each element in the corresponding column.*

Step 4: *Cover all the zeros in the subtracted matrix with a minimum number of horizontal and vertical lines. If the line numbers are n, then an optimal assignment exists. The algorithm stops. Otherwise, if the lines number are less than n then, go to next step.*

Step 5: *Find the smallest element that is not covered by a line in Step 4. Subtract with the smallest uncovered element from all uncovered elements, and add the smallest uncovered element to all elements that are covered twice.*

Step 6: *Now, we have to find unassigned subtasks of that task in this way so that the total task of the unassigned node and other assigned nodes among all nodes would be the smallest.*

Step 7: *Stop.*

5 EXAMPLE

A company has five machines that are used for four jobs. Each job can be assigned to one and only one machine. The completion time (in seconds) of each job on each machine is given in the following table.

Assignment Problem

This is the original cost matrix:

$$
\begin{matrix}
10 & 19 & 8 & 15 \\
10 & 18 & 7 & 17 \\
13 & 16 & 9 & 14 \\
12 & 19 & 8 & 18 \\
14 & 17 & 10 & 19
\end{matrix}
$$

This corresponds to the following optimal assignment in the original cost matrix:

Table 1. Completion time (in seconds) of each job at different machines.

Jobs/Machines	M_{11}	M_{12}	M_{13}	M_{14}
J_{11}	10	19	8	15
J_{12}	10	18	7	17
J_{13}	13	16	9	14
J_{14}	12	19	8	18
J_{15}	14	17	10	19

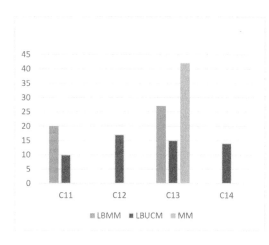

Figure 1. The comparison of the completion time (in seconds) of each task at different nodes.

10	19	8	15
10	18	**7**	17
13	16	9	**14**
12	19	**8**	18
14	**17**	10	19

Now, we have to find unassigned subtasks of a task in this way so that the total task of the unassigned node and other assigned nodes among all nodes would be minimum. Here, the total task of M_3 is minimum when we add 8 to 7, that is, 15 (other values are $M_1 = 22$, $M_3 = 36$, and $M_4 = 32$).

10	19	8	15
10	18	*7*	17
13	16	9	*14*
12	19	*8*	18
14	*17*	10	19

6 COMPARISON

Our proposed method of Load Balancing of an Unbalanced Cost Matrix (LBUCM) can get better load balancing and performance than other algorithms, such as LBMM (Min-You et al. 2000) and MM (Che-Lun et al. 2012), according to the following figure.

7 CONCLUSION

In this paper, we have proposed an efficient algorithm LBUCM for a cloud computing network to allocate tasks to computing nodes according to their resource capability. This proposed method is balancing the working load as well as minimizing the completion time in the cloud computing system. Similarly, LBUCM can get better load balancing and performance than other algorithms, such as LBMM and MM, as shown in the case study.

REFERENCES

Che-Lun H., Hu Y. and Wang H. (2012). "Efficient Load Balancing Algorithm for Cloud Computing Network." *In International Conference on Information Science and Technology*, pp. 28–30.

Kuhn H. W. (1955). The Hungarian method for the assignment problem. *Naval research logistics quarterly*, 83–97.

Min-You W., Shu W. and Zhang H. (2000). "Segmented min-min: A static mapping algorithm for meta-tasks on heterogeneous computing systems." In *hcw, IEEE*, p. 375.

Mondal R. K, Ray P, Sarddar D. (2016). "Load Balancing", *IJ of Research in Computer Applications & Information Technology*, pp. 01–21.

Sinha A. (2016) Cloud Computing in Libraries: Opportunities and Challenges. *Pearl: A Journal of Library and Information Science*, 113–118.

Computer, Communication and Electrical Technology – Guha, Chakraborty & Dutta (Eds)
© 2017 Taylor & Francis Group, ISBN 978-1-138-03157-9

Deadline-driven task scheduling in clustered heterogeneous distributed systems

L. Datta
CEMK, West Bengal, India

ABSTRACT: A distributed system can be viewed as a collection of autonomous computing nodes and communication resources, shared by active users. Because of a large number of job arrivals in some nodes, they may be overloaded while other nodes may remain idle as the response time, storage space, and processing speed differ. Real-time scheduler must consider deadlines of tasks while scheduling. Meeting the deadlines of tasks in a distributed system depends crucially on allocating tasks to idle or lightly loaded and high-speed processors. In this paper, a two-level dynamic scheduling model is proposed by considering deadline of each task. The proposed model is semi-distributed as each cluster is represented by a cluster master, which controls scheduling decisions of worker nodes.

1 INTRODUCTION

A good scheduling protocol allocates tasks among the computing nodes, so that the processing element becomes neither overloaded nor idle. Thus, it balances the load of the nodes and increases the overall system performance. In a clustered distributed computing system, task scheduling can be of two types: local scheduling and global scheduling. In local scheduling, the task is scheduled in a processing node in home cluster while in global scheduling it is allocated to a node outside of home cluster. A task is represented by its arrival time, size, approximate execution time, and resources required. A real-time task also specifies a deadline by which it must complete its execution

The accuracy of a real-time system depends on the outputs of the system as well as the time at which these outputs are produced. Real-time systems, as well as their deadlines, are classified by the consequence of missing a deadline:

Hard—missing a deadline is a total system failure.

Firm—infrequent deadline misses are tolerable, but they may degrade the system's quality of service. The usefulness of a result is zero after its deadline.

Soft—the usefulness of a result degrades after its deadline, thereby degrading the system's quality of service.

Therefore, the real-time scheduler for distributed systems must consider the deadline of each task together with the load-balancing issue while allocating tasks to processor. Real-time tasks may be broadly classified into two categories: periodic processes, which execute on a regular basis, and

aperiodic processes, which are viewed as being activated randomly, for example, a Poisson distribution. Real-time schedulers can be categorized into static and dynamic. Feasibility of application is determined based on current allocation information by dynamic scheduler. A scheduler is static and offline if all scheduling decisions are made before the running of the system. Hence, this scheme is workable only if all the processes are effectively periodic.

A fully distributed system incorporates huge traffic. In this paper, a two-level strategy for aperiodic task scheduling is proposed to meet the deadline constraint and balance workload among the nodes. It reduces the traffic in comparison to fully distributed system. The total system is grouped into clusters consisting of a subset of nodes in the system.

This paper is organized as follows. The status of the considered domain is presented in section 2. Section 3 describes the system model and the responsibilities of each type of node. The 2-Level Task Scheduling (2LTS) policy is discussed in section 4. Section 5 analyzes the communication cost. Simulation results of the experiments are presented in section 6. Section 7 presents the conclusion of this study.

2 RELATED WORK

The theoretical foundation to all modern scheduling algorithms for real-time systems was provided (Liu et al. 1973) for hard real-time tasks executing on a single processor. According to the authors, the upper bound of processor utilization quickly drops to approximately 70% as the number of tasks increases.

Then, Liu and Layland suggested a new, deadline-driven scheduling algorithm, which assigns dynamic priorities to tasks according to their deadlines. Earliest Deadline First is too complex to be implemented in real-time operating system (Li et al. 2009). These algorithms were developed for uniprocessor system. They can be extended for centralized controlled distributed system, which would have several difficulties. The RT-SADS (Atif et al. 1998) algorithm is designed for scheduling aperiodic, non-preemptable, independent, soft real-time tasks with deadlines on a set of identical processors with distributed memory architecture. RT-SADS self-adjusts the scheduling stage duration depending on processor load, task arrival rate, and slack. Epoch scheduling (Karatza et al. 2001) is a special type of scheduling for distributed processor. In this scheme, at the end of an epoch, the scheduler recalculates the priority of each task in the queue using Shortest Task First (STF) criteria. LLF (Mok et al. 1978) assigns priorities depending on laxity, which is the time interval between a task's completion deadline and its remaining processing time requirement. Higher priority was assigned to task with lower laxity. Arabnejad et al. (2014) presented a novel list-based scheduling algorithm called Predict Earliest Finish Time (PEFT) for heterogeneous computing systems. Zeng et al. (2008) presented a modified dynamic critical path algorithm (CBL) to find the earliest possible start time and the latest possible finish time of a task using the distributed nodes network structure. Tasks are sorted in the ascending order of their loads and the processors are sorted in the descending order of their current loads in Zhang et al. (2012). Tasks are assigned to the processors to ensure that the loads assigned to each processor are balanced as much as possible. Ant colony-based task scheduling for real-time operating systems is proposed in Shah et al. (2010). It requires much computation for assigning each task, which itself is time consuming.

3 SYSTEM MODEL

In this paper, a network is considered consisting of N heterogeneous nodes P_1, P_2 ..., P_N connected

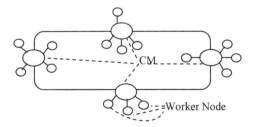

Figure 1. Clustered representation of a distributed network.

by a communication network. Each node has different computational facilities and local memories. The network is organized into L clusters. Each cluster has a specific node designated as the Cluster Master (CM). The distributed system jobs are assumed to arrive at each CM according to Poisson rate λ. The service time of the jobs is exponentially distributed with the mean of $1/\mu$. The jobs are assumed independent and can be executed at any node. Each node maintains a ready queue of jobs to be executed in FCFS manner. The jobs are assumed non-preemptive and aperiodic.

This proposed model is semi-distributed and decentralizes the load-balancing process. It is scalable as it minimizes communication overhead.

4 2-LEVEL TASK SCHEDULING

A set of independent tasks is submitted to the distributed system. The proposed algorithm attempts to assign the task to a processing node such that the deadline of the task can be met. First, CM attempts intra-cluster task scheduling. CM searches for an idle node in home cluster. If no idle node is found, CM searches for a node there whose remaining workload is less than the slack time of a new task. The slack time of the task depends on the processing power of a particular node. If such a node is found, the task is assigned to that node. Otherwise, inter-cluster task scheduling is required. The CM broadcasts a request message to all other CMs. If the receiver CM is able to find a suitable node, it sends response to the initiating CM with its minimum load information. The initiating CM selects

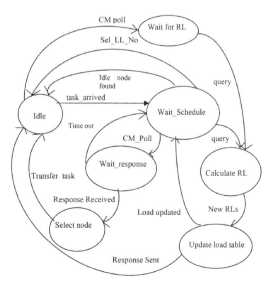

Figure 2. Cluster master state diagram.

the cluster with least minimum load. The new task is transferred to the selected CM. If no response is received, the task is assigned to the least loaded node in home cluster and it misses deadline.

4.1 *Cluster master algorithm*

CM's initial state is idle. Upon new task arrival, the state changes to Wait Schedule. From Idle upon receiving CM poll and from Wait Schedule state, query is sent to the worker nodes of the home cluster and the new state is Calculate RL. New RLs are calculated and accordingly the load table is updated or response is sent to initiating CM. After updating load table, if no suitable node is found, CM poll is sent to all other CMs and the state changes to Wait_response. After receiving responses, the state changes to select node, the most suitable node is chosen, and the task is transferred to the node. The CM returns to Idle state.

Process CM

1. Begin
2. While TRUE
3. Wait for new task arrival or request from other CM
4. Send query to all worker nodes in home cluster
5. Receive remaining load and processing power from each worker node
6. Update new workload in load table
7. If request from other CM is pending
8. If suitable node is found, send response
9. Goto step 3.
10. If new task in home cluster to be scheduled, search for idle node in home cluster
11. If more than one idle node is found, assign the new task to the idle node with the highest processing power
12. If no idle node is found
13. Search for a node whose load is less than slack time of the task considering processing power of each node.
14. If such a node is found
15. Assign the new task to selected node.
16. Else

17. Broadcast CM_Poll to all other CMs.
18. If response is received before time out
19. Select the response with least remaining load.
20. Transfer new task to selected CM.
21. Else
22. Assign new task to the least-loaded node in home cluster.
23. Increment number of missed deadline
24. End

4.2 *Worker node algorithm*

Initial state of worker node is idle. If new tasks are received from CM, it schedules the tasks in its ready queue according to FCFS algorithm. If ready queue becomes empty, it goes back to Idle state. When it receives load query from CM, its state changes to Calculate RL. Then, it sends the remaining load and its power to CM and goes back to the previous state.

Process Worker node

1. Begin
2. While TRUE
3. If a new task is received, add it in ready queue.
4. If ready queue is not empty, schedule tasks in FCFS manner.
5. If load query from CM is received
6. Calculate remaining load
7. Send remaining load and power to CM
8. Calculate each task's turnaround time.
9. End

5 ANALYSIS

Let k, m, and d be the upper bounds on the number of clusters, number of nodes in a cluster, and the diameter of a cluster, respectively.

Theorem 1. The lower bound on total time to assign a task is 2dT + L and the upper bound is (4d + k)T + L, where T is the average message transfer time between adjacent nodes and L is the actual average load transfer time.

Proof: Sending query and receiving load from each node in home cluster requires 2d steps. Therefore, total time to message transfer is 2dT, and load transfer is 2dT+L. If intra-cluster task scheduling is not possible, CM poll to k−1 CMs and their response require maximum k hops. Maximum number of hops to load transfer is (2d + k + 2d) resulting in (4d + k) T+ L time. If the diameter of cluster decreases, this approach produces better result than (Erciyeset al. 2005) inter-cluster task scheduling.

Theorem 2. The total number of messages to assign a task for load balancing is between (2 m) and (2 km+2(k−1))

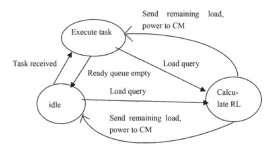

Figure 3. Worker node state machine.

Proof: The total number of messages in intra-cluster scheduling is 2 m. In inter-cluster load balancing, k−1 request messages are sent and most k−1 replies can be received. Each CM sends and receives replies from each worker node resulting (k−1)*2 m messages. Therefore, the maximum number of messages is (2 m + 2(k−1) + (k−1)*2 m). This is much less than that required in Chatterjee et al. (2015).

6 SIMULATION AND RESULT

Simulation experiments evaluated the performance of the proposed 2 LTS policy. The experiments were conducted by varying several performance parameters in the system, namely the number of worker nodes, the number of jobs, and processing power of nodes. It is assumed that on base processing node, a task of size 50 units takes 1 unit of time for execution.

For 1000 tasks in varying number of computing nodes, the number of tasks missed deadline is represented in Figure 5 for three different algorithms. It is found that 2 LTS performs better than the existing algorithms. It is evident from Figure 5 that the proposed algorithm reduces percentage of deadline missed for varying average number of tasks per node compared with unbalanced system and the existing algorithms. Figure 6 shows the Average Turnaround Time (ATAT) for varying number of tasks per node for three different algo-

Table 1. Parameter values.

Parameters	Values
Number of processors	10–50
Number of tasks	100–1000
Task size	1000–2500
Processing power	1–4
Job inter-arrival time	Exponentially distributed with mean 2

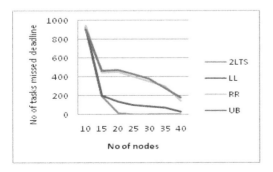

Figure 4. No of tasks missed deadline for 1000 tasks.

rithms and unbalanced system. In LL algorithm for assigning each new task, a large number of message passing are needed, which is time consuming and expensive. This time is not considered while calculating the TAT of each job here. However, it is observed that the performance of 2 LTS is the same as LL.

Figure 7 shows the load on each node of the system for varying average workload. It is observed that, while tasks are assigned using 2 LTS algorithm, load on each node varies slightly. Therefore,

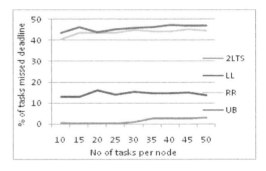

Figure 5. % of deadline miss for varying avg. no. of tasks per node.

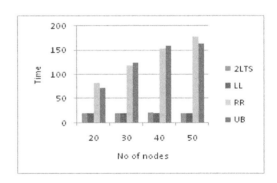

Figure 6. Average turnaround time for varying no. of tasks per node.

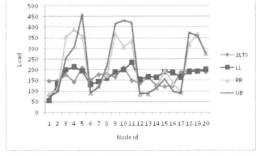

Figure 7. Average load on nodes.

the aim of balancing load for proper resource utilization is assured.

From the graphs, it can be inferred that performance of the system does not vary much with overall system load from average turnaround and load balancing viewpoints. The performance of the system is much better than the existing algorithms in terms of missing deadline. Therefore, considering all variations, it is seen that 2 LTS improves the overall system performance.

7 CONCLUSION

In this paper, a semi-distributed task scheduling method for real-time tasks is proposed for heterogeneous distributed system. A new task is assigned to a processing node, considering the present load status of each node, processing power of the node, and the slack time of the new task. The task assignment method is scalable and has low message and time complexities. The method of partitioning the system into clusters and the method of load transfer are not addressed. It is assumed that the tasks have no precedence. Cluster Master may fail and due to their important functionality in the proposed model, new masters should be elected. Furthermore, a mechanism to exclude faulty nodes from a cluster and add a recovering or a new node to a cluster is needed. These procedures can be implemented using algorithms as in Tulani et al. (2000). Future works should focus on tasks with precedence relations.

REFERENCES

Arabnejad, H. & Barbosa J.G. (2014). List scheduling algorithm for heterogeneous systems by an optimistic cost table, *IEEE Transactions on Parallel and Distributed Systems*, Vol 25, Issue 3.

Atif, Y & Hamidzadeh B. (1998). A Scalable Scheduling Algorithm for Real-Time Distributed Systems, *Proceedings of the 18th International Conference on Distributed Computing Systems*, pp. 352–359.

Chatterjee, M. & Setua, S.K. (2015). A new clustered load balancing approach for distributed systems, *IEEE Conference on Computer, Communication, Control and Information Technology*.

Erciyes, K. & Payli R. U. (2005) A Cluster-Based Dynamic Load Balancing Middleware Protocol for Grids, *Advances in Grid Computing - EGC*, LNCS 3470, 805–812, Springer-Verlag, Berlin.

Karatza H.D. & Hilzer R.C. (2001). Epoch Load Sharing in a Network of Workstations, *IEEE Simulation Symposium*, Proceedings. 34th Annual, 36–42.

Li, Xin, Z. Jia., Li Ma, R. Zhang. & H. Wang. (2009). Earliest deadline scheduling for continuous queries over data streams, *International Conferences on Embedded Software and Systems*.

Liu C.L. & Layland J.W. (1973). Scheduling Algorithms for Multiprogramming in a Hard Real-Time Environment, *Journal of the ACM*, vol. 20, no. 1, pp. 46–61.

Mok, A. & Dertouzos M. (1978). Multiprocessor scheduling in a hard real-time environment, *7th Texas Conference on Computing Systems*.

Shah, A. & Kotecha. K. (2010). Scheduling Algorithm for Real-Time Operating Systems using ACO, *IEEE International Conference on Computational Intelligence and Communication Networks*.

Tunali, T., Erciyes, K., & Soysert, Z. (2000). A Hierarchical Fault-Tolerant Ring Protocol For A Distributed Real-Time System. *Special issue of Parallel and Distributed Computing Practices on Parallel and Distributed Real-Time Systems*, 2(1), 33–44.

Zeng, B., Wei. J. & Liu. H. (2008). Research of Optimal Task Scheduling for Distributed Real-time Embedded Systems, *IEEE-ICESS*.

Zhang, K., Qi. B., Jiang. Q. & Tang. L. (2012). Real-time periodic task scheduling considering load-balance in multiprocessor environment, *3rd IEEE International Conference on Network Infrastructure and Digital Content*.

Computer, Communication and Electrical Technology – Guha, Chakraborty & Dutta (Eds)
© 2017 Taylor & Francis Group, ISBN 978-1-138-03157-9

Survey and analysis of mining requirements in spatio-temporal database

D. Dasgupta & S. Roy
MCA, Asansol Engineering College, Asansol, West Bengal, India

S. Singha Roy
Department of Computer Science, NITTTR, Kolkata, West Bengal, India

A. Chakraborty
Department of Computer Application, Asansol Engineering College, West Bengal, India

ABSTRACT: Data mining is a process of identifying suitable, useful, and reasonable patterns in data. It deals with the analysis of large volume of structured and unstructured data for discovering regularities or relationships between them. Temporal data mining concerns with the investigation of events controlled by one or more dimensions of time, whereas spatial data mining considers the alternative path of embedded and entirely spatial constructs based on several static techniques. Spatio-temporal data mining corresponds to the confluence of a number of fields such as database, statistics, and geographic visualization. It is a user-centric and interactive process, where experts work jointly to achieve solutions for a given problem. Spatio-temporal data mining considers multiple states of the spatio-temporal data to find significant spatio-temporal patterns, which is expensive. This paper covers various data mining areas and several algorithms as well as their application in research.

1 INTRODUCTION

Data mining is a non-trivial process of identifying valid, useful, and understandable patterns in data. It deals with the analysis of large volumes of data, generally unstructured in nature, to discover regularities or relationships between the data. For example, large databases of loan applications can be mined for typical patterns leading to defaults, which can facilitate determining whether a future loan application must be accepted or rejected. This technique is applied in a wide range of business, scientific, and engineering scenarios (Laxman et al. 2006).

Temporal data mining deals with the analysis of events ordered by one or more dimensions of time. If a system contains multiple time lines such as valid time, transaction time, or decision time, then it will be considered as multiple dimensions of time. The spatial data mining considers the alternative path of embedded and exclusively spatial constructs using association rules, clustering, etc.

Spatio-temporal data mining represents the confluence of several fields including spatio-temporal databases, machine learning, statistics, geographic visualization, and information theory. It is a user–centric, interactive process. Mining of spatio-temporal frequent patterns in spatio-temporal data is expensive, because of the complexity in the structure of spatio-temporal data objects and relationships between them (Han et al. 2007).

2 DATA MINING FOUNDATIONS

The last two decades have seen a remarkable increase in the amount of data or information being stored in electronic format. It has been noted that the amount of information in the world doubles every 20 odd months and the size and number of databases are increasing more rapidly. Data mining deals with several technical approaches such as clustering, summarizing data, analyzing changes, learning classification rules, finding dependency networks, and detecting anomalies.

Data mining refers to using a variety of techniques to identify decision-making knowledge. It is used in areas such as decision support, prediction, forecasting, and estimation. A popular data mining toolkit Clementine User Guide is frequently used for data estimation. Basically, data mining is concerned with the analysis of data and finding the patterns and consistencies in any set of data (Han et al. 2011, Pujari 2001).

2.1 Data mining stages

According to Fayyad & Simoudis, the phases of data mining start with raw data and end with the extracted knowledge. Data mining passes through several stages, namely Selection, Preprocessing, Mining, Transformation, Interpretation, Evaluation, Decisions, and use of Discovered Knowledge (Pujari 2001).

2.2 Data mining models

The two types of data mining model or modes of operation used to find the information are as follows:

2.2.1 Verification model

This model acquires some hypothesis from the user and tests its validity against the data, for example, in a marketing division, to launch a new product, it is important to identify the section of the population most likely to buy the new product by formulating a hypothesis to identify potential customers and their purchase information.

2.2.2 Discovery model

This model is based on the system that automatically discovers important information hidden in the data. An example of such a model is a bank database, which is mined to discover the groups of customers to target for a mailing campaign. The data are searched only according to the common characteristics found between the groups of the customers and not based on the hypothesis (Han et al. 2011).

2.3 Data warehousing

A data warehouse is a Relational Database Management System (RDBMS) that is designed particularly to meet the needs of transaction processing systems. It is a centralized data storage area, which can be queried for business benefit. It is a powerful technique that makes the extraction of operational data possible and overcomes inconsistencies between them, so that it suits the high-efficient Decision Support System (Marakas 2003).

2.4 Data mining functions

Data mining methods or functions may be classified by the function they perform. The important functions used in data mining are as follows:

2.4.1 Classification

It is requires the user to define one or more classes. The database contains one or more attributes that denote the class of a tuple, which are known as predicted attributes, whereas the remaining attributes are called predicting attributes. Once the classes are defined, the system infers the rules that govern the classification. A rule is said to be correct if its description covers only all the positive examples of a class.

2.4.2 Association

For a collection of items and a set of records, each of which containing some number of items from a given collection, an association function can be operated. Which actually returns a pattern that

exists among the collection of items? One association can involve any number of items on either side of the rule.

2.4.3 Sequential/temporal pattern

Sequential or temporal pattern functions analyze a collection of records over a period of time to identify the trends. Again, these pattern functions analyze the collections of related records and detect the patterns of products bought over time.

2.5 Data mining techniques

Several data mining techniques exist, which are described as follows:

2.5.1 Cluster analysis

Clustering and segmentation are the procedures of creating a partition so that all the members of each set of the partition are similar according to some metric. A cluster is a set of objects grouped together because of their similarity or closeness. In an unsupervised learning environment, the system has to discover its own classes and the way in which it clusters the data in the database. Many data mining applications make use of clustering according to the similarity.

2.5.2 Deduction and induction

The information inferred from database is more useful. There are two inference techniques: deduction and induction. Deduction technique infers information that has logical consequence of the information in the database, for example, the joint operator applied to two relational tables, where the first is related to employees and departments and the second to departments and managers, deduces a relationship between employees and managers. In general, it follows two techniques: Decision Tree and Rule Induction Method (Mennis et al. 2005).

3 DATA MINING ALGORITHMS

Several algorithms to be used in Data Mining exist, which are described as follows:

3.1 C4.5 algorithm

The C4.5 algorithm generates a classification decision tree for the given data set by recursive partitioning of data. The decision is grown using depth first strategy (Quinlan 2014).

3.2 k-means algorithm

K-means algorithm intends to partition n objects into k clusters, in which each object belongs to the cluster with the nearest mean. This method produces exactly k different clusters of the highest possible distinction. The objective of k-means clustering is to minimize the total intra-cluster variance or the squared error function (Jin et al. 2006).

3.3 *Apriori algorithm*

This is a popular influential algorithm for mining frequent item sets for Boolean association rules. It proceeds by identifying the frequent individual items in the database and extending them to larger item sets as long as those item sets appear sufficiently often in the database (Agrawal et al. 1994).

3.4 *Bayesian algorithm*

This is a simple probabilistic classifier based on Bayes theorem. It is suitable when the dimensions of the inputs are high. Its computational efficiency and classification rate are very high, but they do not work well with small databases (Nikam 2015).

4 SPATIO-TEMPORAL DATABASE AND DATA MINING

An individual is a portion of the space time range that may be extended in both space and time. A physical object may have temporal parts as well, which are generally known as states. The temporal boundaries of states are known as events (West 2002, Andrienko et al. 2006).

4.1 *Spatio-temporal database concepts*

The features involved in managing spatio-temporal data are: (a) elementary features, (b) transactions and multi-user access, (c) programming in database, (d) elements of database administration, (e) scalability [24], (f) performance and VLDB (Very Large Database) issues, (g) distributed databases, (h) special data types, (i) application development and interfaces, (j) reliability, (k) commercial issues, and (l) spatial visualization (Dodge et al. 2005).

4.2 *Spatio-temporal semantics*

The spatio-temporal semantics includes: (a) data types; (b) primitive notions; (c) type of change; (d) the factor evolution in time and space; (e) dimensionality (Pelekis et al. 2004); and (f) space time topology.

4.3 *Spatio-temporal data mining process*

The data mining process usually consists of three phases: (a) pre-processing or data preparation; (b) modeling and validation; and (c) post processing or deployment (Pelekis et al. 2004).

5 CURRENT TRENDS

Consistency of style is very important. It is important to note the spacing, punctuation, and caps in all the examples below.

The immense explosion in developments in Information Technology, Digital Mapping, Remote Sensing, and Geographic Information System demands the development of data-driven inductive approaches to spatial, temporal, and spatio-temporal analysis and modeling. The actual data mining task is the automatic or semi-automatic analysis of large quantities of data in order to extract previously unknown patterns such as groups of data records, unusual records, and dependencies between the records (Yao 2003).

Spatial data mining refers to (a) extraction of knowledge, (b) spatial relationships, and (c) other patterns from spatial data (Han et al. 2007). The domain of temporal mining focuses on the discovery of causal relationships among events that may be ordered in time (Roddick et al. 1999).

Temporal data mining tasks involve: (a) prediction, (b) classification, (c) clustering, and (d) search and retrieval (Laxman et al. 2006). The spatial data mining can be apparently considered as the multi-dimensional equivalent of temporal data mining (Roddick et al. 1999).

Spatio-temporal data mining is carried out to obtain the spatio-temporal knowledge and patterns (Han et al. 2007). It can be characterized in two directions: (a) the embedding of a temporal awareness in spatial systems and (b) the accommodation of space into temporal data mining systems (Roddick et al. 1999, Roddick et al. 2001).

6 RECENT APPLICATIONS

At present, spatio-temporal data mining plays an important role in various fields, such as:

6.1 *Financial data analysis*

In the banking industry, data mining is used heavily in the areas of modeling and predicting credit fraud, evaluating risk, performing trend analyses, analyzing profitability, and so on. In financial market, Bayes classification and decision trees (Sawant et al. 2013) are used to forecast stock prices, options trading, rating bonds, portfolio management, and so on.

6.2 Health care and biomedical research

The past decade has seen an explosive growth in biomedical research, ranging from the development of new pharmaceuticals and advances in cancer therapies to the identification and study of the human genome. Therefore, systems capable of performing temporal abstraction and reasoning become crucial in this context. Data mining has been used in many successful medical applications, including data validation in intensive care, monitoring children's growth, and analyzing the data of a patient with diabetes. In this field, classification and clustering are also used (Jin et al. 2006).

6.3 Crime analysis

Spatio-temporal pattern analysis is a key task in crime analysis. Spatio-temporal Pattern analysis is a process that extracts information and knowledge from geo- and time-referenced data and offers data to crime analysts. The basic objective of spatial crime pattern analysis is to find crime patterns. Use of these patterns makes it possible to identify the root of the crimes (Lee et al. 2006).

Temporal pattern is often considered together with spatial pattern in crime analysis. The key theories behind spatio-temporal crime analysis include routine activity theory and rational choice theory. Basically, crime distribution is determined by the intersection in time and space of suitable targets and motivated offenders (Leong et al. 2015). Clustering, association, and k-means algorithms are used in this field (Zubi et al. 2010).

6.4 Mobility applications

Mobility services are the set of ideas and technologies that transform lives by understanding the geophysical world navigating through the places. Spatio-temporal data mining can also be applied in the fields of Military platforms, Biomass monitoring, Climate change studies, and so on (Norris et al. 2002).

7 RESEARCH DIRECTIONS

Several open issues have been identified dealing with spatio-temporal information to the development of effective methods for interpreting and presenting the results. The mining process for spatio-temporal data is complex in terms of both the mining efficiency and the complexity of patterns that can be extracted from spatio-temporal data sets (Camossi et al. 2008).

The mining needs to consider the following factors: (a) efficiency and scalability, (b) fast response time, (c) enforce the spatial and time constraints, and (d) geographic and time proximity (Han 2003).

The scope of the work on spatio-temporal data mining includes efficient multi-dimensional data model for spatio-temporal database, techniques for spatio-temporal pattern analysis over such model, and realization of spatio-temporal data mining techniques such as spatio-temporal data classification, spatio-temporal data clustering, and spatio-temporal rule mining (Tsoukatos et al. 2001).

In addition to the big data, challenges associated with the propagation of traditional geospatial data sets such as remote sensing imagery are a new challenge. For example, the massive adoption of large-scale video surveillance is imposing some novel challenges due to the exponential growth of the resulting data sets (Norris et al. 2002). Recording of electronic communication such as e-mail logs and web logs has captured human process with respect to security and privacy.

8 CONCLUSION

Data mining in spatio-temporal databases must consider the multiple stages of the spatio-temporal data. It must integrate spatial and temporal information to identify meaningful spatio-temporal patterns. The knowledge of these spatio-temporal patterns allows us to develop more localized or customized business analysis and strategies. It has also become a challenge to identify the temporal and spatial rules taking into account the characteristics of the temporal and spatial dimensions of events on a time and space axis. Further research is also required in the area of compression and sampling. Especially, there is a great need for integrating spatial data mining workflows with modern computing infrastructures such as cloud computing (Klasky et al. 2011) and data spaces (Docan et al. 2012).

REFERENCES

Agrawal, R., & Srikant, R. (1994). Fast algorithms for mining association rules. In Proc. 20th int. conf. very large databases, 1215, 487–499.

Andrienko, G., Parasha, D., May, M., & Teisseire, M. (2006). Mining spatio-temporal data. Journal of Intelligent Information Systems, 27(3), 187–190.

Camossi, E., Bertolotto, M., & Kechadi, T. (2008). Mining spatio-temporal data at different levels of detail. In The European Information Society, Springer Berlin Heidelberg, 225–240.

Docan, C., Parashar, M., & Klasky, S. (2012). DataSpaces: an interaction and coordination framework for coupled simulation workflows. Cluster Computing, 15(2), 163–181.

Dodge, S. & Alesheikh, A. (2005). Evaluating different approaches of spatial database management for moving objects. Proceedings of Map Middle East.

Han, J., Cheng, H., Xin, D., & Yan, X. (2007). Frequent pattern mining: current status and future directions. Data Mining and Knowledge Discovery, 15(1), 55–86.

Han, J., Pei, J., & Kamber, M. (2011). Data mining: concepts and techniques. Elsevier.

Han, J. (2003). Mining Spatiotemporal Knowledge: Methodologies and Research Issues. Urbana, 51, 61801.

Jin, R., Goswami, A., & Agrawal, G. (2006). Fast and exact out-of-core and distributed k-means clustering. Knowledge and Information Systems, 10(1), 17–40.

Klasky, S., Abbasi, H., Logan, J., Parashar, M., Schwan, K., Shoshani, A., & Chacon, L. (2011). In situ data processing for extreme-scale computing. Scientific Discovery through Advanced Computing Program.

Laxman, S., & Sastry, P. S. (2006). A survey of temporal data mining. Sadhana, 31(2), 173–198.

Lee, J. W., & Lee, Y. J. (2006). A knowledge discovery framework for spatiotemporal data mining. Journal of Information Processing Systems, 2(2), 124–129.

Leong, K., & Sung, A. (2015). A review of spatio-temporal pattern analysis approaches on crime analysis. International E-journal of Criminal Sciences, (9), 1–33.

Marakas, G. M. (2003). Modern data warehousing, mining, and visualization: core concepts, Prentice Hall, 100–101.

Mennis, J., & Liu, J. W. (2005). Mining Association Rules in Spatio-Temporal Data: An Analysis of Urban Socioeconomic and Land Cover Change. Transactions in GIS, 9(1), 5–17.

Nikam, S. S. (2015). A Comparative Study of Classification Techniques in Data Mining Algorithms. Oriental Journal of Computer Science & Technology, 8(1), 13–19.

Norris, C., McCahill, M., & Wood, D. (2002). The growth of CCTV: a global perspective on the international diffusion of video surveillance in publicly accessible space. Surveillance & Society, 2(2/3), 110–135.

Pelekis, N., Theodoulidis, B., Kopanakis, I., & Theodoridis, Y. (2004). Literature review of spatio-temporal database models. The Knowledge Engineering Review, 19(3), 235–274.

Pujari, A. K. (2001). Data mining techniques. Universities press.

Quinlan, J. R. (2014). C4.5: programs for machine learning. Elsevier.

Roddick, J. F., & Lees, B. G. (2001). Paradigms for spatial and spatio-temporal data mining. Geographic data mining and knowledge discovery, 33–50.

Roddick, J. F., & Spiliopoulou, M. (1999). A bibliography of temporal, spatial and spatio-temporal data mining research. ACM SIGKDD Explorations Newsletter, 1(1), 34–38.

Sawant, A. A., & Chawan, P. M. (2013). Study of Data Mining Techniques used for Financial Data Analysis. International Journal of Engineering Science and Innovative Technology, 2(3), 503–509.

Sorokine, A. (2003). Multi-scale spatial data models for decision making and environmental modeling. Student Paper Sessions, UCGIS Summer Assembly, 1–9.

Tsoukatos, I., & Gunopulos, D. (2001). Efficient mining of spatiotemporal patterns. In International Symposium on Spatial and Temporal Databases. Springer Berlin Heidelberg, 425–442.

Verhein, F., & Chawla, S. (2008). Mining spatio-temporal patterns in object mobility databases. Data mining and knowledge discovery, 16(1), 5–38.

West M., (2002). A Spatio-temporal model of activity and state, National Science Foundation, ACTOR.

Yao, X. (2003). Research issues in spatio-temporal data mining. In Workshop on Geospatial Visualization and Knowledge Discovery, University Consortium for Geographic Information Science, Virginia, 1–6.

Zubi, Z. S., & Mahmmud, A. A. (2010). Using Data Mining Techniques to Analyze Crime patterns in the Libyan National Crime Data. Recent Advances in Image, Audio and Signal Processing, 79–85.

A verifiable and cheating-resistant secret sharing scheme

Arup Kumar Chattopadhyay
Department of Computer Science and Engineering, Academy of Technology, Hooghly, India

Amitava Nag
Department of Information Technology, Academy of Technology, Hooghly, India

Koushik Majumder
Department of Computer Science and Engineering, MAKAUT, Kolkata, India

ABSTRACT: In Shamir's (k, n)-threshold secret sharing scheme, n shares are distributed among the participants. Each share incorporates an x-value uniquely designated for each participant. If the distribution takes place on insecure channels, malicious users may acquire few shares. If the number of shares obtained by the malicious users is k or more ($\leq n$), the malicious user can reconstruct the secret. We proposed a method for distribution of shares with RSA scheme to encrypt the x-values at insecure network along with verification, such that both the secrecy of shares and secret are maintained.

1 INTRODUCTION

Secret Sharing Schemes (SSS) are defined to share a secret among a group of participants and protect the secret from disloyal access by unauthorized groups. In a (k, n)-threshold secret sharing scheme, a secret S is encoded into n parts called shadows or shares and distributed (by Distributor) among n participants or players. If any k or more ($k \leq n$) shares are obtained, then the full secret S can be reconstructed (reconstruction can be also done by a special entity called Combiner) and no less than k shares can reconstruct the secret or expose any information about the secret. Therefore, any group with $\leq k$ members constitute a disloyal group. Secret sharing was introduced in 1979 by two authors independently – Shamir and Blaklay. Shamir's scheme (Shamir 1979) was based on Lagrange's interpolation and Blacklay's scheme (Blacklay 1979) was based on hyperplane geometry. Mignotte (1982) proposed secret sharing scheme based on Chinese Reminder Theorem (CRT), which was improved by Asmuth–Blooms (Asmuth & Blooms 1983).

The scheme proposed by Shamir is widely used by researchers in different scenarios. Thien and Lin (2002) proposed secret image sharing scheme based on Shamir's scheme. The shares generated by Shamir's scheme are distributed on insecure channels, so the confidentiality of shares becomes volatile and the shares can be misused by malicious users. Furthermore, Zhao et al. (2009) proposed a method to ensure confidentiality of the shares

on insecure channels. The secure key distribution method used by Zhao et al. (2009) implemented secure distribution of shares for medical images proposed by Ulutas et al. (2011), and the authors used Shamir's framework with better authenticity and confidentiality properties.

In (k, n)-threshold secret sharing, when shareholders present their shares in the secret reconstruction phase, dishonest shareholder(s) called cheater(s) can present faked share(s) and thus deceive the other honest shareholders, as they obtain a faked secret as result. Cheater detection and identification are two essential properties for a secret sharing scheme. Harn and Lin (2009) defined a method for the detection and identification of cheaters for Shamir's secret sharing scheme. The method used by Harn and Lin assumed a situation where more than k shares are presented for reconstruction of secret and the redundant shares were used to identify the cheaters.

A number of secret sharing schemes were proposed, which can verify whether the shares received by shareholders are consistent under the condition that both the secrecy of shares and secret are maintained. Harn et al. (2013) proposed a verifiable secret sharing scheme based on CRT and extension of Asmuth–Blooms' secret sharing scheme. Another scheme based on Asmuth–Blooms' scheme was proposed by Liu et al. (2015), and the authors claimed that the proposed scheme is more efficient than the scheme proposed by Harn et al. Liu & Chang (2016) proposed an integrable mechanism for verification with generalized Chinese

Reminder Theorem, Shamir's Secret sharing, and Asmuth–Blooms' secret sharing, and improvised the verification method proposed by Harn et al. (2013) by using one-way hash function.

In our proposed scheme, we ensure that it holds cheating prevention—shares are secure even distributed on public channel and both the shares and secret are verifiable. The remainder of the portions are organized as follows. In section 2, the important entities of threshold secret sharing are briefed. The related schemes and algorithms are discussed in section 3. Our proposed scheme is presented in section 4. In section 5, we discuss the performance of our scheme and the paper concludes in section 6.

2 PRELIMINARIES

The important entities used in a secret sharing scheme are briefed as follows:

Secret: A secret S is the data (can be a text, image, or audio) need to be secured from unauthorized users or unauthorized groups.

Shares: The secret S has to be encoded into n shares or shadows, say $S_1, S_2, ..., S_n$, such that none of them individually reveals any information about the secret.

Dealer: Dealer or distributor D mainly responsible to encode the secret into n shares and distribute them to the participants. Deal mostly the legal owner of the secret or a trusted dealer.

Participants: Participant or players are represented as $P_1, P_2, ..., P_n$ and they are the users seeking for the secret.

Combiner: A combiner C mainly responsible to decode the secret if threshold numbers of shares are obtained from the participants.

In a (k, n)-threshold secret sharing scheme, the *dealer* generates n *shares* and distributes it among n *participants*. If k or more *participants* submit their *shares* to the combiner, then the *secret* can be retrieved in full. If any less than *k shares* are submitted, then no part of the *secret* will be revealed.

3 RELATED STUDY

3.1 *Review Shamir's Secret Sharing Scheme (1979)*

Shamir proposed (k, n)-threshold secret sharing scheme based on Lagrange's interpolation method. For given k points (x_i, y_i), where $i = 1, 2, ..., k$, in the *2D* plane, the Lagrange interpolation polynomial $f^k(x)$ can be constructed by:

$$f^k(x) = \sum_{i=0}^{k} y_i \prod_{j=0, j \neq i}^{k} \frac{x - x_i}{x_j - x_i}.$$

Let the secret is S and the dealer has to generate n shares as $S_1, S_2, ..., S_n$. It allows k or more $(\leq n)$ shares to reconstruct the secret. The solution requires a random $(k-1)$ degree polynomial:

$$F(x) = (S + a_1 x + a_2 x + ... + a_{k-1} x^{k-1}) \, mod \, p$$

where p is a large prime number and the coefficients of polynomial $a_1, a_2, ..., a_{k-1}$ are randomly selected within the range $(1, p-1)$. Dealer computes the shares as follows:

$$S_1 = (1, F(1)), S_2 = (2, F(2)), ..., S_n = (n, F(n)).$$

If k or more $(\leq n)$ shares are obtained, then the polynomial $F(x)$ can be generated by Lagrange interpolation as:

$$F(x) = \sum_{i=0}^{k} S_i \prod_{j=0, j \neq i}^{k} \frac{x - x_i}{x_j - x_i},$$

The secret will be retrieved as $S = F(0)$.

3.2 *Threats in distribution of shares in Shamir's secret sharing*

In Shamir's scheme, each share is distributed as a pair of two integers $(x_i, F(x_i))$, where $x_i \neq 0$. However, in this scheme, the x-values can be easily predicted. Therefore, x-values can be considered as random values within the range $(1, p)$. However still, if the shares are distributed on insecure channels, any knowledge of k or more $\leq n$ shares obtained by malicious users are enough to reconstruct the polynomial $F(x)$. The value of the polynomial at position $x_i = 0$ is the secret $S = F(x_i = 0)$.

For example, in a case where $k = 3$, the polynomial must be a second-order one. Let the polynomial be:

$$F(x) = (9 + 13x + 5x^2) \, mod \, 37$$

All the coefficients are assumed within 0 to $p = 37$, secret $S = F(0)$ and $p = 37$.

The polynomial can be reconstructed (as shown in Figure 1) if 3 or more shares are known. The secret will be recovered where the polynomial

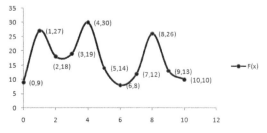

Figure 1. Example of a (3,10)-threshold secret sharing scheme.

98

intersects *Y-axis*, that is, at point $(0, F(0))$. In our example, secret is $S = F(0) = 9$.

3.3 Scheme proposed by Zhao et al. (2009)

The intercommunication between dealer and participants to distribute the shares on an insecure channel was made secure (confidentiality of the shares is not lost) by the cheating proof secret sharing scheme proposed by Zhao et al. (2009). It ensures the confidentiality of *x*-values as the *x*-values are calculated independently by both dealer and participant before distribution begins. Even if intruders are able to collect information about k or more $\leq n$ shares (only the values of $F(x)$), they must also have to accrue the *x*-values to apply Lagrange's interpolation to recover the secret. Thus, even an insecure channel is sufficient to keep confidentiality of the secret. The process of initialization and sharing phase of the scheme are as follows:

1. Dealer chooses two large integers p and q and computes $N = pq$, where $p \equiv 3 \, mod \, 4$ and $q \equiv 3 \, mod \, 4$.
2. Dealer selects an integer $g \in \left[N^{\frac{1}{2}}, N \right]$ such that g is relatively prime to p, q and publishes $\{g, N\}$.
3. Each participant chooses his/her secret shadow as $s_i \in [2, N]$ and computes $R_i = g^{s_i} \, mod \, N$.
4. Each participant sends his/her R_i to the dealer. Dealer ensures that each R_i is unique, otherwise demands a new secret shadow from that participant.
5. Then, dealer randomly chooses $s_0 \in [2, N]$ such that it is relatively prime to $(p-1)$ and $(q-1)$ Dealer also computes $R_0 = g^{s_0} \, mod \, N$ and publishes R_0.
6. Dealer computes $x_i = R_i^{s_0} \, mod \, N$ for each participant.

The dealer assigns an x_i-value for each participant. The participant can calculate his/her own x_i independently from secret shadow and publicly known value R_0 as:

$$x_i = R_0^{s_i} \, mod \, N.$$

In our scheme, RSA (Rivest et al. 1978) algorithm is used such that dealer encrypts each *x*-value by the public key of the participant and only the *x*-value can be decoded by the corresponding participant (private key holder). The plain-text message (here the *x*-value) needs to be pre-processed, so that security can be ensured by random oracle model (Bellare & Rogaway, 1993).

Random oracle model was established in 1993 by M. Bellare and P. Rogaway. In this paradigm, authors stated that a practical protocol is produced by first devising and providing a protocol for random oracle model and then replacing the random oracle by appropriate hash function. The authors argued that this paradigm yields protocols that are more efficient than the standard ones and is applicable for encryption, signatures, and zero-knowledge proof.

Bellare & Rogaway (1995) proposed Optimal Asymmetric Encryption Protocol (OAEP) for RSA, which is proven in the random oracle model. The OAEP algorithm uses a pair of random oracle G and H to process the pain text before the application of asymmetric encryption. The pair of oracle combined with trapdoor one-way permutation function, and the combined scheme is semantically secure under plain-text attack (IND-CPA). Further combination of the scheme with trapdoor function like RSA is also proven secure against chosen cipher attack (NM-CPA). The instantiation of G-Oracle for IND-CCA attack and instantiation of H-Oracle and the use of trapdoor permutation (trapdoor function) for NM-CPA attack are discussed by Boldyreva and Fischlin (2006). The authors also provided a detailed security analysis of OAEP with random oracle model.

OAEP Algorithm for pre-processing the pain text:

Consider the following:

- f is a *k*-bits to *k*-bits trapdoor function (in our scheme, RSA).
- k_0 is chosen such that adversary running time is significantly smaller than s^{k_0}.
- Length of the message x is $k - k_0$ (if the message is smaller, then it has to be padded with zeros).
- G is a "generator" such that $G : \{0,1\}^{k_0} \rightarrow \{0,1\}^n$ and H is the hash function $H : \{0,1\}^n \rightarrow \{0,1\}^{k_0}$.

Encoding:

1. Chose a random $r = \{0,1\}^{k_0}$
2. Generate the encoded stream $E = (x \oplus G(r)) \| (r \oplus H(x \oplus G(r)))$ where $\|$ is the concatenation operator. We can also represent the equation as $E = X \| Y$, where $X = x \oplus G(r)$ and $(Y = r \oplus H(x \oplus G(r)))$.

Decoding:

1. Extract X and Y parts from encoded message E and recover r as $r = Y \oplus H(X)$.
2. Recover the padded message $x = X \oplus G(r)$.

In the proposed model, we use RSA as trapdoor function and MD5 as hash algorithm.

4 PROPOSED MODEL

In this section, we present an improved version of Shamir's (k, n)-threshold secret sharing, which can be considered much secure and verifiable.

Its objectives are as follows:

1. In the proposed version, x-values could not be predicted, as it was shown in the case of normal Shamir's scheme.
2. The shares are distributed by public channels, but they can only be used by authorized user, and hence must provide authentication to the participants.
3. The shares submitted by the participants must be verified, such that any faked share can be easily identified by the combiner.
4. The scheme should also identify whether the dealer is dishonest and he/she supplied fake shares to one or more participants.

The different phases of the model are as follows:

Initialization Phase:

1. Each participant $P_i, i = 1, 2, ..., n$ considers two large primes, p_i and q_i, and computes the following:

$$N_i = p_i q_i \text{ and } \phi_i = (p_i - 1)(q_i - 1).$$

2. Participant $P_i, i = 1, 2, ..., n$ chooses an integer $e_i, 1 \le e_i \le \phi$, such that $gcd(e_i, \phi_i) = 1$.
3. Each participant also computes the secret exponent $d_i, 1 \le d_i \le \phi_i$, such that $e_i d_i \equiv 1 \bmod \phi$.
   The public keys are $Kpu_i = (e, N_i)$ and private keys are $Kpr_i = (d_i, p_i, q_i)$. Participants keep $d_i, p_i, q_i,$ and ϕ_i secret.
4. Participants share the public keys, $Kpu_1, Kpu_2, ..., Kpu_n$ with the dealer and combiner.

Let H be a suitable collision-resistant one-way hash function, which takes an input as binary string of variable length and outputs a hash code, which is a binary string with length q.

Construction of shares:

1. Dealer uses a suitable random number generation function to generate n distinct random numbers with l number of bits – $r_1, r_2, ..., r_n$.
2. Consider the polynomial function used for Shamir's (k, n)-threshold secret sharing as:

$$F(x) = (S + a_1 x + a_2 x^2 + ... + a_{k-1} x^{k-1}) \bmod p$$

then the intermediate shares are computed as (by ensuring that x-values are random):

$$s_1 = (r_1, F(r_1)), s_2 = (r_2, F(r_2)), ..., s_n = (r_n, F(r_n))$$

3. Dealer encrypts the random numbers r_i with the public keys Kpu_i as follows:

$$Er_i = r_i^{e_i} \bmod N \text{ for } i = 1, 2, ..., n$$

4. Use hash function H to generate hash code h_i for each $(Er_i, F(r_i))$:

$$h_i = H\left(Er_i, F(r_i)\right) for\ i = 1, 2, ..., n$$

5. Distribute the shares as follows:

$$S_i = \left(Er_i, F(r_i)\right) for\ i = 1, 2, ..., n$$

6. Dealer also applies the hash function H on the secret S as $h_s = H(S)$. Dealer publishes h_s.
7. Dealer publishes $h_1, h_2, ..., h_n$.

By receiving the shares, the participant can verify if the share is modified by any intruder (by impersonating the dealer) at the public channel. Participant P_i needs to apply hash function to his/her share S_i to obtain the hash code h_i^*. If $h_i^* \neq h_i$, then the S_i share is modified.

1. Participant extracts Er_i and decrypts it as:

$$r_i = Er_i^{d_i} \bmod N.$$

2. Submits $(r_i, F(r_i))$ to the combiner.

Reconstruction of secret:

1. If combiner obtains all k or more $\le n$ shares in the form of $(r_i, F(r_i))$, Shamir's secret sharing can interpolate the value of secret $S = F(0)$.
2. The combiner or participants can regenerate the hash code for the secret obtained and verify with h_s to confirm that it is the actual secret. Any mismatch denotes that one or more participants have supplied faked share or the dealer is dishonest (has supplied false share(s)). To verify the shares, combiner can regenerate $S_i = \left(Er_i, F(r_i)\right)$ from $(r_i, F(r_i))$ using the corresponding public key. The hash function H is applied to obtain the hash code h_i^*. If $h_i^* \neq h_i$, then the S_i share is faked and P_i is the identified cheater. If none of the S_i is faked, then the dealer is identified as dishonest.

5 PERFORMANCE

The obvious features of this scheme are as follows:

Robustness—A (k, n)-threshold secret sharing scheme is called robust only if any k or more $(\le n)$ participants can reconstruct the full secret information. Shamir's secret sharing already holds this property.

Confidentiality—A (k, n)-threshold secret sharing scheme holds confidentiality only if no information about the secret can be obtained if less than k shares are pulled.

Shamir's secret sharing already holds the property, along with the fact that our proposed scheme also promises that malicious users who learned the

Table 1. Comparison of secret sharing schemes.

Scheme by	Secure distribution	Verification of shares	Verification of secret
Zhao et al. (2009)	Yes	No	No
Harn et al. (2009)	No	Yes	No
Ulutas et al. (2011)	Yes	Yes	No
Harn et al. (2013)	No	Yes	Yes
Liu et al. (2015)	No	Yes	Yes
Liu et al. (2016)	No	Yes	Yes
Our proposed scheme	Yes	Yes	Yes

shares on the public channels had no clue about the x values. Hence, malicious users or intruders cannot obtain a useful share.

Traceability—A (k,n)-threshold secret sharing holds traceability, if it is able to detect any participant P_i who sends a fake share $S_i^* = S_i$ to the combiner.

Our proposed method uses hash function to make the shares verifiable.

Participant authentication and verification of shares—In our proposed method, only a legal participant holds the legal private key to access his/her share (as the share is encrypted by corresponding public key by dealer).

No need to have secure channel for distribution—As the x values are encrypted by the pubic keys before distribution, only the legal participants (the one possessing the private key) can receive the x values.

Verifiable secret—Hash function ensures the accuracy of the secret obtained from the shares.

Comparisons of different secret sharing schemes based on verifiability of shares and secrets are shown in Table 1.

6 CONCLUSION

Despite being a perfect secret sharing scheme, Shamir's secret sharing cannot resist attack on the distribution channel. If the shares are communicated through insecure channel, then malicious user may learn the shares from the public channel. In our proposed scheme, we proposed a method where shares can be communicated in a secure way over a public channel. The shares can be decrypted only by authorized participants. The proposed scheme also exhibits properties to verify whether the shares received by shareholders are consistent under the condition that both the secrecy of shares and secret are maintained.

REFERENCES

Asmuth, C. & J. Bloom (1983). A Modular Approach to Key Safeguarding. *IEEE Transactions on Information Theory* 29(2), 208–210.

Bellare, M. & P. Rogaway (1993). Random Oracles are Practical: A Paradigm for Designing Efficient Protocols. In *Proc. of the 1st ACM conference on Computer and communications security*, New York, NY, USA, 62–73.

Bellare, M. & P. Rogaway (1995). Optimal Asymmetric Encryption—How to Encrypt with RSA. In *Proc. Eurocrypt '94 (Springer-Verlag)*, 950, 92–111.

Blakley, G.R. (1979). Safeguarding cryptographic keys. In *Proc. of AFIPS, 1979, Managing Requirements Knowledge, International Workshop on, Managing Requirements Knowledge*, New York, June, 313–317.

Boldyreva, A. & M. Fischlin (2006). On the Security of OAEP. *Advances in Cryptology – ASIACRYPT 2006*, 4284, 210–225.

Harn, L. & C. Lin (2009). Detection and identification of cheaters in (t, n) secret sharing scheme. *J. Designs, Codes and Cryptography* 52(1), 15–24.

Harn, L., M. Fuyou, & C.-C. Chang (2013). Verifiable secret sharing based on the Chinese remainder theorem. *J. Security and Communication Networks* 7(6), 950–957.

Liu, Y., L. Harn, & C.-C. Chang (2015). A novel verifiable secret sharing mechanism using theory of numbers and a method for sharing secrets. *International Journal of Communication Systems* 28(7), 1282–1292.

Liu, Y. & C.C. Chang. (2016). An Integratable Verifiable Secret Sharing Mechanism. *International Journal of Network Security* 18(4), 617–624.

Mignotte, M. (1982). How to share a secret. In *Proc. of the Workshop on Cryptography*, Burg Feuerstein, Germany, March–April, 371–375.

Rivest, R.L., A. Shamir, & L. Adleman (1978). A method for obtaining digital signatures and public key cryptosystems. *Communications of the ACM* 21(2), 120–126.

Shamir, A. (1979). How to Share a Secret. *Communications of the ACM* 22(11), 612–613.

Thien, C.-C. & J.-C. Lin (2002). Secret image sharing. *Computers and Graphics* 26, 765–770.

Ulutas, M., G. Ulutas, & V. Nabiyev (2011). Medical image security and EPR hiding using Shamir's secret sharing scheme. *Journal of Systems and Software* 84(3), 341–353.

Zhao, R., J.-J. Zhao, F. Dai, & F.-Q. Zhao (2009). A new image secret sharing scheme to identify cheaters. *Computer Standards & Interfaces* 31(1), 252–257.

Communication technology

Computer, Communication and Electrical Technology – Guha, Chakraborty & Dutta (Eds)
© 2017 Taylor & Francis Group, ISBN 978-1-138-03157-9

A new handoff management scheme for reducing call dropping probability with efficient bandwidth utilization

D. Verma, A. Agarwal & P.K. Guha Thakurta
Department of CSE, National Institute of Technology, Durgapur, West Bengal, India

ABSTRACT: A new handoff management scheme is proposed in this paper to reduce call dropping probability for handoff requests without significant increase in call blocking probability for new calls originated under the base station. The proposed approach provides efficient bandwidth utilization. Two sets of users are used in the proposed approach. The call requests are prioritized depending on several constraints. The call patterns under a base station have been analyzed over a period of time. Three types of databases are used to store and predict the requirement of channel reservation. The performance of the system is analyzed through experimental results.

1 INTRODUCTION

In wireless networks the number of Base Stations (BS) work in conjunction to provide service from requesting Mobile Terminals (MTs). Each BS has channels of fixed bandwidth which are allocated to MTs upon service request depending on availability. A MT placed within the transmission range of a BS is serviced by that BS. Each MT within this area is serviced by the above BS (Tripathi 1998). When a MT reaches the borders of its BS's coverage area, the signal strength weakens and an ongoing call on the MT needs to be transferred to another BS through a process known as handoff. Thus each BS must have an efficient algorithm to handle both new and handoff calls effectively while ensuring that call drop within the transmission range of a BS due to handoff failure is kept at a minimum.

Based on network interface, handoff is divided in two parts such as horizontal handoff and vertical handoff. When handoff occurs in same type of network interface, it is known as horizontal handoff and handoff in different network interface is known as vertical handoff.

A voice call can be either complete or incomplete depending on whether it is terminated by the user or forcibly by the network. An incomplete call is of two types: a new call that is blocked or an ongoing call that is terminated due to a failed handoff. The dropping of an ongoing call is more annoying than blocking of a new call from user's point of view and therefore handoff algorithms try to minimize the Call Dropping Probability (CDP)

at the expense of an increase in the rate of new calls being blocked (Call Blocking Probability CBP). In this paper, we try to minimize the CDP for horizontal handoffs.

Non-Prioritized and Prioritized are two basic classification of handoff schemes. In Non-prioritized scheme, calls originating in same cell and handoff calls are treated in same way while in Prioritized scheme, handoff calls are given more priority. Due to fixed bandwidth of Wireless channel, trade-offs among bandwidth allocated to each cell, call blocking probability and call dropping probability are carried out. Handoff prioritization is done by several different ways like reserving channels for handoff calls, queueing of handoff calls, transferring the channels, Sub-Rating scheme (Sgora 2009). To fulfil requirements of handoff call, Channel reservation scheme is implemented in two ways, static channel reservation scheme and dynamic channel reservation scheme. In Static Channel Reservation Scheme (SCRS) a threshold is fixed to provide sufficient bandwidth. Guard Channel Scheme (GCS) and threshold priority scheme are two different types of SCRS (Fang 2002, Bartolini 2001). In Dynamic Channel Reservation Scheme (DCRS) number of channels reserved for handoff calls are dynamically adjusted according to different system parameters or traffic load (Ramanathan 1999, Hou 2001). Dynamic Guard Channel Scheme (DGCS) is used to implement DCRS (Wei 2004). Dynamic adjustment of resources are done on two levels, Local and Collaborative. In local channel reservation scheme, decision is taken by Base Station (BS) considering the value of parameters of its own (Kim 1999). While in Collaborative Channel Reservation Schemes (CCRS), decision is based on the system parameters of neighbouring BS also (Diederich 2005). Handoff queueing scheme refers to queueing of handoff call requests or both originating calls and handoff requests. Any MT present in overlapping area of two BSs sends handoff request to destination BS. If any channel is not available at destination BS then this handoff request is queued at destination BS and source BS continues its service to MT until MT is present in common area or handoff request is not accepted (Marichamy 1999, Hong 1986). Static handoff queueing and dynamic handoff queueing scheme are two approaches to implement handoff queueing. Static handoff queueing consists of two methods, pre-emptive and non-pre-emptive. If any handoff request with lower priority is in queue and any higher priority request arrives, lower priority request will be pre-empted from queue (Chia 2006). In dynamic handoff queueing, handoff requests present in queue are reordered based on motion of user or different system parameters like Received Signal Strength (RSS) and Packet Success Rate (PSR) (Xhafa 2004, Lin 2005).

The work proposed in this paper is mainly concerned on reduction in handoff call dropping probability without affecting call blocking probability. This approach provides higher priority to new calls generated by locals residing in the area of a BS. To achieve this, all users are divided in two sets X and Y, where set X consists of users residing in the area of BS and all other users are put in set Y. Eligibility of any user to be a part of set X is determined depending on the number of hours, that user is connected to current BS. Now highest priority in priority queue is given to handoff requests. In case of new originating call, priority is given to user present in X and any other user has lowest priority. A database containing record of number of handoff requests for each time-stamp of whole day is present and database is updated also at each time interval. A threshold of number of days is used and data present in database before that threshold day is removed to keep the size of database constant in order to improve query time. Number of channels at BS for handoff request are dynamically adjusted according to average number of handoff requests in database of that time interval.

The rest of this paper is organized as follows. Section 2 presents the system model. Section 3 describes proposed approach in detail. Section 4 reports the result which contain two sub-headings 1) simulation 2) graph and result description results. Finally, conclusions and future work are discussed in Section 5.

2 SYSTEM MODEL

In the proposed work, we combine two popular handoff models: Channel Reservation Scheme and Priority Queueing of handoff calls. Non pre-emptive handoff queueing scheme and dynamic guard channel reservation scheme is used. Channel Reservation Scheme sets aside a fraction of the total channels in BS to be allocated to handoff calls only. These reserved channels are known as Guard Channels and cannot be allocated to new calls. Handoff queueing scheme allows handoff and originating call attempts to wait instead of getting terminated in case their respective bandwidth in BS is busy. During allocation of calls, a handoff call is given priority among all the calls for a channel to be allocated.

The presented model divides all MTs in two groups X and Y. X consists of MTs which are connected to the BS in consideration for a longer period of time i.e. the MTs which are local to the area the BS covers. Y consists of the remaining MTs. The BS maintains a record of these MTs for a time period of 30 days such that whenever there is a connection request a simple query will tell if the requesting MT is in group X or Y as shown in Figure 1.

The entire bandwidth is divided into three parts as follows:

a. BW_H: This part is reserved exclusively for handoff calls.
b. BW_X: If BW_H is full, handoff calls will be allocated in this part. After allocating handoff calls, new calls from user in X will be allocated.
c. BW_Y: If BW_H and BW_X both are full, handoff calls will be allocated in this region. If BW_X is full, new calls from user in X will be allocated in this region. After these, new calls from user in Y will be allocated.

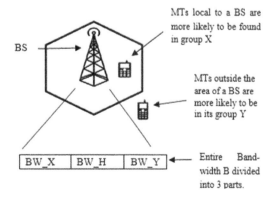

Figure 1. System model for the presented approach.

2.1 Model assumptions

a. Each BS has fixed number of channels.
b. The entire set of MTs is divided into two groups: Group X and Group Y.
c. Group X is the set of MTs which are local with respect to the BS i.e. their connection time to the BS is more than a pre-defined threshold T_0 for a given period of time (30 days). Group Y is the set of MTs which are not in Group X.
d. The call duration for each call has been assigned using a weighted distribution from the analysis in (Holub 2004).
e. The wait time for a call in priority queue is τ seconds.
f. Each BS maintains three databases DB_1, DB_2 and DB_3.

2.2 Initial requirements

DB_1: This database keeps track of all the MTs in Group X. This can be done by keeping track of the time (T) a MT spends in the coverage area of BS over a period of time (say 30 days). If T > T_0 then a row is added to DB_1 which contains the mobile number along with 30 entries each corresponding to a day and stores the total connection time to the BS for that day. DB_1 needs to be updated on a daily basis to continuously add or delete to Group X as the sum of all row entries cross or fall below the threshold T_0. The MTs which are not part of DB_1 are considered to be a part of Group Y.

DB_2: This database keeps track of the average number of channels used at any point of time by the MTs in Group X. For example for every 5 minute interval starting from 00:00 hours till 23:59 hours DB_2 will have the record of the average number of channels that are utilized in that interval by the users in Group X. This database is updated at every 5 minute interval for the above example. The number of channels utilised as such at any interval of time is represented reserve_X.

DB_3: This database keeps track of the number of channels utilized by handoff calls received by a BS at a particular interval of time. As such it also maintains a record similar to DB_2 but stores the total handoff requests for each time interval instead. This number of channels utilized as such at any interval of time is represented by reserve_H.

2.3 Metrics

In order to evaluate the system performance the following metrics are introduced here:

a. Call Dropping Probability (CDP): It denotes the ratio of handoff calls dropped to the handoff requests received at the BS. The CDP should be low such that minimum handoff calls are dropped.
b. Call Blocking Probability (CBP): It is the ratio of the new calls that get blocked to the total new call requests received by the BS. A lower CBP denotes a better chance of a new call getting connected.
c. Channel utilization efficiency: It is the ratio of the number of channels that are allocated (utilized) to the total number of channels available.

3 PROPOSED SCHEME

It is known that the total number of channels of base station have been partitioned into three parts: BW_H, BW_X and BW_Y for handoff calls, calls from users residing in same area as base station and calls from users residing in area of another base station respectively. The partition BW_Y can accommodate handoff calls, calls from user in X as well as calls from user in Y. The partition BW_X can accommodate handoff calls and calls from user in X. The partition BW_H is reserved for handoff calls completely. Size of BW_X and BW_H is determined using the data present in database DB2 and DB3 respectively. Size of BW_H is fixed for one interval according to the expected number of handoff calls in that upcoming interval, where expected number of handoff calls is determined by the maximum of mean value of handoff calls arriving in this upcoming time interval and the succeeding interval in past 30 days. For example if current time is 12:00, then size of BW_H will be the maximum of average number of handoff calls (for 30 days) at 12:05 and 12:10. As we are giving highest priority to handoff calls so size of is the minimum of the channels which are left after reserving for handoff calls and average number of new calls expected (average of number of new calls in past 30 days) to arrive in upcoming interval according to DB2. pq3 is a priority queue in which unallocated handoff requests will be put to wait for finite time. In same way, pq2 and pq1 are priority queues for new call requests from X and new call requests from Y which can't be allocated. Databases are updated continuously according to the number of request arriving in each intervals. DB2 will be updated according to the number of new call requests from user X. DB3 will be updated according to the number of handoff requests that arrived in this interval. When any request arrives; if it is handoff request and if any channel is available in BW_H or BW_X or BW_Y, request will be allocated. If no channel is available, request will be pushed in pq3. If it is new call request from user in X, it will be allocated in BW_X or BW_Y, if no channel is available to serve the request, it will be put into pq2. If it is new call request from Y and channel is available in BW_Y, it will be allocated. If no channel is available request will be put into pq1. In priority queue, call will wait for finite time and after that it will be dropped.

The time required to execute the proposed work is $O(C) + O(\lambda) + O(\alpha)$ where C, λ, α denote total channels, number of calls per second and size of priority queue respectively.

4 SIMULATION AND RESULTS

4.1 Simulation setup

The experimental results for the proposed approach have been carried out using the following setup and parameters with their corresponding values are introduced by Table 1.

4.2 Simulation results

The CDP varies with respect to the number of handoff requests. When more channels are available CDP will be less. Figure 2 denotes the CDP for an incoming handoff call at a BS for the handoff traffic varying from 0 to 5 calls per second. The CDP is higher for a lower bandwidth as there are lesser number of channels reserved (for handoff calls) for a BS with lower bandwidth. For high Bandwidths call dropping is almost 0 till 3 calls/second. As the traffic load increases, the CDP also increases because of the lesser availability of free channels.

The CBP generally increases with the frequency of incoming new call requests. Figure 3 denotes how CBP varies with the frequency of new call requests for the presented model. For a lower bandwidth, the CBP rises faster with call requests as there are lesser number of channels available when more calls arrive and as such chances of a call getting blocked are higher. For higher bandwidths CDP almost remains 0 for up to 3 calls/sec due to the availability of large number of channels to handle the traffic load.

Table 1. Simulation setup and associated parameters.

Item	Value	Unit
Simulation environment	Python, Matlab	
Operating system	Windows 10	
Total bandwidth B	1500, 3000, 4500	kB
Bandwidth required per voice call	10	kB
Range of phone numbers	1–10000	
τ	2	seconds

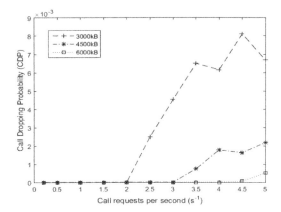

Figure 2. CDP v/s call arrival rate.

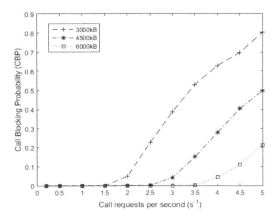

Figure 3. CBP v/s call arrival rate.

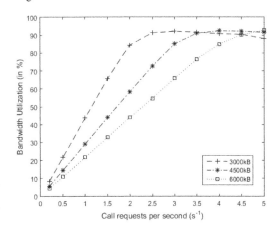

Figure 4. Bandwidth utilization v/s call arrival rate.

Bandwidth Utilization is a good representative of an efficient handoff algorithm. A higher Bandwidth Utilization when call dropping is high and vice versa denotes that the handoff calls are being efficiently handled by the BS. Figure 4 denotes the bandwidth utilization for varying frequency of call requests. A BS with a lower bandwidth almost reaches its full potential (>90%) at a lower traffic load as compared to BS with a higher bandwidth. When compared to Figure 2 it can be seen that CBP is high when bandwidth is almost fully utilized leaving fewer or none channels to be allocated to new calls. Also CDP is correspondingly not so high in Figure 1 as there is a certain amount of bandwidth still kept in reserve for the handoff calls that may arrive in the next interval.

5 CONCLUSION

This is a data analysis based approach implemented by real time simulation. Instead of communicating with other neighboring towers about estimation of handoff or communicating with MTs about their location, SIR, etc. this approach does behavioral analysis of users by tracking call patterns of last 30 days by dividing each day in constant time intervals. This approach reduces the communication overhead as well as computation overhead present in some approaches which use velocity calculation of MT and decision is then taken according to results based on calculation of MTs location, velocity, SIR, etc. Call dropping probability is very low to guarantee Quality of Service (QoS). Also, better quality service was provided to users residing in same area as the BS compared to users which are not local to the concerned BS. Bandwidth Utilization in this approach is also high which denotes the efficiency of the handoff algorithm presented above.

REFERENCES

Bartolini, N. (2001). Handoff and Optimal Channel Assignment in Wireless Networks. *Mobile Networks & Applications. 6(6)*, 511–524.

Chia, C. Y. & M. F. Chang (2006). Channel Allocation for Priority Packets in the GPRS Network. *IEEE Commun. Lett. 10(8)*, 602–604.

Diederich, J. & M. Zitterbart (2005). Handoff Prioritization Schemes Using Early Blocking. *IEEE Communications Surveys & Tutorials. 7(2)*, 26–45.

Fang, Y. & Y. Zhang (2002). Call Admission Control Schemes and Performance Analysis in Wireless Mobile Networks. *IEEE Trans. Veh. Technol. 51(2)*, 371–382.

Holub, J., J. G. Beerends, & R. Smid (2004). A Dependence Between Average Call Duration and Voice Transmission Quality: Measurement and Applications. *Wireless Telecommunications Symposium.* 75–81.

Hong, D. & S. S. Rappaport (1986). Traffic Model and Performance Analysis for Cellular Mobile Radio Telephone Systems with Prioritized and Non-Prioritized Handoff Procedures. *IEEE Trans. Veh. Technol. 35(3)*, 77–92.

Hou, J. & Y. Fang (2001). Mobility-Based Call Admission Control Schemes for Wireless Mobile Networks. *Wireless Communications & Mobile Computing. 1(3)*, 269–282.

Kim, Y. C., D. E. Lee, B. J. Lee, Y. S. Kim, & B. Mukherjee (1999). Dynamic Channel Reservation Based on Mobility in Wireless ATM Networks. *IEEE Commun. Mag. 37(11)*, 47–51.

Lin, T. N. & P. C. Lin (2005). Handoff Ordering Using Link Quality Estimator for Multimedia Communications in Wireless Networks. *Proc. 2005 IEEE Global Telecommunications Conference GLOBECOM 05, Vol. 2*, St. Louis, MO, pp. 1125–1130.

Marichamy, P., S. Chakrabati, & S. L. Maskara (1999). Overview of Handoff Schemes in Cellular Mobile Networks and Their Comparative Performance Evaluation. *In the Proc. IEEE Vehicular Technology Conference (VTC-99), Vol. 3*, Houston, TX, pp. 1486–1490.

Ramanathan, P., K. M. Sivalingam, P. Agrawal, & S. Kishore (1999). Dynamic Resource Allocation Schemes *During Handoff for Mobile Multimedia Wireless Networks. IEEE J. Select. Areas Commun, 17(7)*, pp. 1270–1283.

Sgora, Aggeliki, & Dimitrios D. Vergados (2009). Handoff Prioritization and Decision Schemes in Wireless Cellular Networks: A Survey. *IEEE Communications Surveys & Tutorials. 11(4)*.

Tripathi, Nishith D., Jeffrey H. Reed, & Hugh F. Van Landingham (1998). *Handoff in Cellular Systems*, 26–37.

Wei, Y., C. Lin, F. Ren, R. Raad, & E. Dutkiewicz (2004). Dynamic Handoff Scheme in Differentiated QoS Wireless Multimedia Networks. *Computer Communications. 27(10)*, 1001–1011.

Xhafa, A. E. & O. K. Tonguz (2004). Dynamic Priority Queueing of Handover Calls in Wireless Networks: An Analytical Framework. *IEEE J. Select. Areas Commun. 22(5)*, 904–916.

Computer, Communication and Electrical Technology – Guha, Chakraborty & Dutta (Eds)
© 2017 Taylor & Francis Group, ISBN 978-1-138-03157-9

Graph theory based optimum routing path selection method for wireless sensor network

Bobby Sharma, Nitunjit Brahma & Himangshu Choudhury
Department of Computer Science and Engineering and Information Technology, School of Technology,
Assam Don Bosco University, Guwahati, Assam, India

ABSTRACT: A Wireless Sensor Network (WSN) is a kind of network which consists of specially designed sensors deployed in some area for sensing the environment for various data like temperature, sound, pressure, etc. A wireless sensor node may consist of one or multiple numbers of sensors. WSNs are of high demand because of their tremendous capability to sense the environment, collect data, process data and send data for various purpose. At the same time, these are suffering from various limitations including one major limitation such as energy efficiency. Each node is equipped with limited power backup. A major breakdown may happen in the network due to lack of power in the nodes. It may lead to major discrepancy in the entire network. By keeping energy efficiency as one of the major issue, many routing, power management protocols have been specially designed for WSNs. Routing is one of the major issues in multihop network that has a significant impact on the network's performance. An ideal routing algorithm must be able to find an optimum path for packet transmission within a specified time so as to satisfy the Quality of Service (QoS). In this paper a new graph theory based methodology has been proposed by which it identifies the optimum routing path from a source to destination so that there will be a minimum number of packet drop while it gains maximum throughput.

1 INTRODUCTION

WSNs are basically comprised of small, light weight wireless sensor nodes whose main function is to monitor and collect data from physically challenging environment and pass data to specified destination for processing and analysing (Caytiles 2015, Sharma 2010, Jatav 2012, Chavhan 2012). Distribution of nodes in WSN may be of two different types, either nodes are distributed randomly or these may be in predetermined order. When nodes are distributed in random order then it becomes more challenging for various attacks as well as other errors. In most cases nodes are autonomous. Nodes cooperation is another factor for smooth running of the network. Sensors are main devices that cooperatively working together to collect various data such as pressure, temperature, humidity, intensity, frequency, motion, vibration etc. from the environment. Every sensor nodes consists of sensing device, communication device, controller, memory unit and energy supply unit (Antony 2014).

Various routing algorithms are standardised for WSN. The main goal of traditional routing algorithm is to find the least-cost path from sender to receiver. Nodes in WSN are suffered from resource constraints such as energy, processing capacity, storage, communication range, bandwidth etc.

Because of these, it raises several problems in WSN in spite of having its high potential for implementation. Nodes in WSN contains low powered micro sensors. Routing is the main issue in multi hop, multichip networks. Many network performance parameters depend on routing. Objective of ideal routing algorithm is to find an optimum path for packet transmission without delay (Cheng L. 2014).

WSNs are enormously used by the industry and academia in spite of having its limitations (Dai 2009). These are mainly considered when data collection is done from physical entities and sending data to external user (Buratti 2011). One of the major challenge in the design of wireless sensor network is the conservation of energy for longer live of the network. Traditional routing protocols are not equipped with Energy Harvesting-Wireless Sensor Networks (EH-WSN), that is not powered by energy harvested from environment instead of batteries (Liu 2013).

Due to lack of energy, nodes in WSN may behave abnormally which finally effects the objective of WSN. Energy Efficiency is important role of the Wireless Sensor Networks Researchers (Kannadhasan 2013, Lai 2009). The shortest path problem for routing is one of the classical problem. By identifying shortest path, it tries to minimise the cost. Out of several traditional shortest path algorithms, BFS, Dijkstra's, Bellman Ford algorithms etc. are popular. For a centralised as well as client-server based architecture, such algorithms are very much useful. But for autonomous network, such algorithm requires some additional processing statements by which it can understand the shortest path of autonomous network. In (Buratti 2011), authors mentioned that vertex covering plays an important role in link failure monitoring, location identification, clustering and data aggregation etc.

So, in this paper, we are proposing a graph theory based optimized routing algorithm for wireless sensor network. Section 2 of this paper contains literature review, section 3 contains proposed methodology, section 4 bears results and analysis, section 5 contains conclusion, section 6 contains references.

2 LITERATURE REVIEW

In (Chen 2015), authors proposed lifetime optimization algorithm with mobile sink nodes for wireless sensor networks to minimize the use of battery power. In this cases, Network optimization model is established and it is divided into two parts including movement path selection model and lifetime optimization mode. In (Kaur 2014), authors showed the importance of energy in nodes of WSN. They propose a new algorithm consisting of three phases including formation of cluster and selection of cluster head, in second phase it finds the distance for routing using coordinates of the nodes. Final phase includes data transmission. In (Dagdeviren 2014), authors proposed three algorithm for constructing vertex cover in WSN, first one is a greedy approach to identify vertex by using degree of nodes. That is an adaptation of Parnas & Ron's algorithm. The second approach uses Hoepman's weighted matching algorithm for finding vertex. But in third approach, breadth-first search algorithm is used. In (Mahajan 2015), authors proposed a new methodology based on graph theory for optimum path based on QoS parameters. Fault tolerant mechanism is also adopted for longer live span. In (Cheng L. 2014), the authors mainly emphasised on potential of using genetic algorithm to solve the problem of optimum path by choosing the shortest path. It tries to maintain the Quality of Service (QoS). The energy efficient genetic routing algorithm prolongs the network lifetime. It is a problem solving method based on the concept of natural selection and genetic. Along with that it also discussed about various shortest path algorithms. In (Ahmat 2009), author analyses the importance of graph theory in various type of network including

Internet. It was a primary tool for indentifying the routing path for network. Networking protocols such as border Gateway Protocol, Open shortest Path Protocol etc. The author presents some key graph theory concepts used to represent different types of networks. Some of key graph theory concepts have been discussed by authors to represent network. Some tools are also mentioned for generating graph theory based computer network. In (Nakano 2011), authors mentioned that graph theory results are applicable to communication problems. In this paper, heuristic algorithms are applied for channel assignment problem. In this regards, various results showed that graph theory can be implemented to communication problems. In this case, main focus is on multi-hop wireless network and their relative problems. In (Baranidharan 2011), It is mentioned that energy efficiency in Wireless Sensor Network (WSN) is one of the most important area of research. Most of the research on energy efficiency is found in two broad categories like tree based as well as clustering based protocol (Dai 2009). This paper mainly discussed the different factors and their importance for effecting the clustering. Proposed algorithm is based on Minimum Spanning Tree (MST) and shortest path concept with its strength and limitations. As per author, existing clustering algorithms can be categorised as partitioning, hierarchical, graph theoric, density based and grid based. Authors of this paper also considered the scalability of the network. Multi hop communication between cluster head node to the sink node and having super cluster head node is considered. It has three phases including cluster formation, cluster head selection and data transmission using shortest path. In paper (Fu 2010), authors proposed a new algorithm known as AFLAR. Objective of this algorithm is to meet the QoS requirements like data delay reduction and energy conservation. A novel method for dividing routing request zone and constructs a select equation is also proposed. Select equation is constructed in such a way that which can increase energy awareness. Simulation results showed that proposed algorithm perform well in terms of performance of network lifetime, utilization and consumption balancing of energy and data delay.

3 PROPOSED METHODOLOGY

In the proposed methodology it is assumed that WSN nodes are deployed randomly in wireless media. Nodes will collect data as per specification of sensors from the environment and processed data and send these on hop-by-hop fashion. It is assumed that every node is equipped with location identification system to understand their position. All nodes are assumed to be homogenous. It is assumed that every node understand their neighbour along with their geographical position. Source node will select a neighbour node for communication based on shortest distance as well as their residual energy. In spite of having shortest distance if the residual energy of the node is lesser than predefined threshold energy than such neighbour node will not be selected as next hop for communication. Selection of neighbour node for communication will continue till it reaches destination.

Algorithm

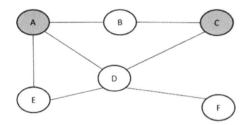

Figure 1.   Wireless Sensor Network.

Input: Assume that A is Source Node and C is Destination Node as shown in Figure 1.
Output: Optimum path
Declared parameter: s_n = source node, d_n = destination node, n_h = next hope, E_{th} = threshold energy

1. Input s_n and d_n
2. If $s_n = d_n$ then
3.   stop communication. // As source and destination nodes are same
4. else
5.   find the neighbour node n_h which is nearest to source node (using shortest path algorithm i.e. dijkstra's algorithm) and keep all other neighbours in list.
6.   check energy of the node n_h as E_r
7.   if $E_r > E_{th}$ then // Comparing energy of the node with threshold energy
8.   consider node as next hop (n_h) for communication
9.   for n_h find shortest path for its neighbour using the shortest path algorithm
10.   if $n_h = d_n$ then
11.   stop communication
12.   else
13.   repeat steps 6-12
14.   end if
15.   else
16.   select another neighbour node from rest of neighbour nodes with shortest distance
17.   repeat steps 6-12
18   end if
19.   end if
20. end if

Table 1.   Simulation parameters.

Routing Protocol	DSDV
No. of Nodes	30, 50, 100
Mac	802.11
Simulation Time	100 sec
Packet Size	512 bytes
CBR Rate	400 Kbits/sec
Initial energy	1000 J
X-dimension of the topography	1000
Y-dimension of the topography	1000

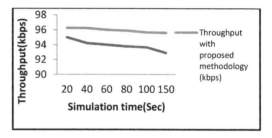

Figure 2.   Throughput vs. simulation time.

4   IMPLEMENTATION, RESULTS AND ANALYSIS

For implementation purpose, simulator NS 2 is used in which a set of WSN nodes are deployed with equal initial residual energy. Simulation parameters are shown in Table 1. The performance of proposed methodology is evaluated in terms of network throughput as well as packet dropping ratio.

Both network throughput and Packet Dropping Ratio (PDR) has been compared for same network environment before and after implementation of the proposed method. From the various results, it is concluded that after implementation of proposed algorithm, network throughput is increasing while PDR is decreasing. For understanding, only throughput analysis has been shown in Figure 2.

5   CONCLUSION

In this paper, importance of WSN has been discussed. Inspite of having wide use of WSN, it has several limitations. One of the major drawbacks of WSN is the limited power. Optimum use of limited power is most essential in WSN. In this paper, graph theory based optimum routing path selection methodology has been proposed. Proposed methodology has been implemented using simulator NS 2. Various results have been compared for throughput and PDR. It is found that after implementation of proposed methodology, throughput of WSN becomes more in comparison to WSN without proposed methodology. Similarly, PDR is getting low after implantation of proposed methodology.

REFERENCES

Ahmat, K. A. (2009). Graph Theory and Optimization Problems for Very Large Networks. *Report number: TR-20091970, cite as: arXiv:0907.3099 [cs.NI] (or arXiv:0907.3099v1 [cs.NI] for this version)*.

Antony D., Raj, A. & Sumathi, P. (2014). Enhanced Energy Efficient Multipath Routing Protocol for Wireless Sensor Communication Networks Using Cuckoo Search Algorithm. *Wireless Sensor Network*, Vol. 6, pages: 49–55.

Baranidharan, B. & Shanthi, B. (2011). A New Graph Theory based Routing Protocol for Wireless Sensor Networks. International journal on applications of graph theory in wireless ad hoc networks and sensor networks (GRAPH-HOC) Vol. 3, No. 4.

Buratti, C. & Fabbri, F. (2011). Throughput Analysis of Wireless Sensor Networks via Evaluation of Connectivity and MAC performance. *Emerging Communications for Wireless Sensor Networks*, ISBN 978-953-307-082-7.

Caytiles, R. D., Kim, J. Y. & Kim K. J. (2015). A Review of the Vulnerabilities and Attacks for Wireless Sensor Networks. *Journal of Security Engineering*.

Chavhan, N. & Maidamwar, P. (2012). A Survey on Security Issues to Detect Wormhole Attack in Wireless Sensor Network. *International Journal on AdHoc Networking Systems (IJANS)* Vol. 2, No. 4.

Chen, Y., Lv, H., Ren, T. & Wang, Z. (2015). Lifetime Optimization Algorithm with Mobile Sink Nodes for Wireless Sensor Networks Based on Location Information. *International Journal of Distributed Sensor Networks*, Article ID 857673, 11 pages, http://dx.doi.org/10.1155/2015/857673.

Cheng L., Yu S. & Zhang B. (2014). Routing protocols for wireless sensor networks with mobile sinks: a survey. *IEEE Communications Magazine*, Vol: 52, Issue: 7, Page(s): 150–157, Print ISSN: 0163-6804, INSPEC Accession Number: 14450731, DOI: 10.1109/MCOM.2014.6852097.

Dagdeviren, O., Kavalci, V. & Ural, A. (2014). Distributed Vertex Cover Algorithms for Wireless Sensor Networks. *International Journal of Computer Networks & Communications (IJCNC)*, Vol. 6, No. 1.

Dai, H.N. (2009). Throughput and Delay in Wireless Sensor Networks using Directional Antennas. *5th International Conference on Intelligent Sensors, Sensor Networks and Information Processing (ISSNIP)*, pages: 421–426, E-ISBN: 978-1-4244-3518-0, Print ISBN: 978-1-4244-3517-3, INSPEC Accession Number: 11141178, DOI: 10.1109/ISSNIP.2009.5416826, IEEE.

Fu, H., & Wang, X. (2010). Adaptive Energy and Location Aware Routing in Wireless Sensor Network. *Wireless Algorithms, Systems, and Applications*, Vol 6221 of the series Lecture Notes in Computer Science. pp. 105–109.

Jatav, V. K. & Tripathi, M., Gaur M. S. & Laxmi V. (2012). Wireless Sensor Networks: Attack Models and Detection. *IACSIT Hong Kong Conferences IPCSIT*, vol. 30, IACSIT Press, Singapore.

Kannadhasan, S., KarthiKeyan, G. & Sethupathi, V. (2013). A graph theory based energy efficient clustering techniques in wireless sensor networks. *Information & Communication Technologies (ICT)*, IEEE, Page: 151–155, ISBN: 978-1-4673-5759-3, INSPEC Accession Number: 13653443.

Kaur, R. & Kaur, R. (2014). Improving Energy Efficiency of Wireless Sensor Network with the Fusion of Graph Theory and Genetic Algorithm. *International Journal of Advance Research in Computer Science and Management Studies*, Volume 2, Issue 7.

Lai, Y.-L. & Jiang J.-R. (2009). Optimal Path Planning for Fault-Tolerant and Energy-Efficient Target Surveillance in Wireless Sensor and Actor Networks. *Tenth International Conference on Mobile Data Management: Systems, Services and Middleware*, Page(s): 531–535, ISSN: 1551-6245, E-ISBN: 978-0-7695-3650-7, Print ISBN: 978-1-4244-4153-2, INSPEC Accession Number: 10720499.

Liu, W. & Wu, Y. (2013). Routing protocol based on genetic algorithm for energy harvesting-wireless sensor networks. *IET Wireless Sensor Systems*, Vol: 3, Issue: 2, Page: 112–118, ISSN: 2043-6386, INSPEC Accession Number: 13886244, DOI: 10.1049/iet-wss.2012.0117, IEEE Xplore DL.

Mahajan, S., Malhotra, J. & Sharma, S. (2015). Energy balanced optimum path determination based on graph theory for wireless sensor network. Vol 5, Issue 6, p. 290–298, Print ISSN 2043 6386, Online ISSN 2043-6394, DOI: 10.1049/iet-wss.2014.0061.

Nakano, K., Sengoku, M., Shinoda, S. & Tamura, H. (2011). An Applications of Graph/Network Theory to Problems in Communication Systems. *ECTI Transactions on Computer and Information Technology*, Vol. 5, No. 1.

Sharma, K. (2010). Wireless Sensor Networks: An Overview on its Security Threats. *IJCA Special Issue on "Mobile Ad-hoc Networks", MANETs*.

Computer, Communication and Electrical Technology – Guha, Chakraborty & Dutta (Eds)
© 2017 Taylor & Francis Group, ISBN 978-1-138-03157-9

Real time motion detection system using low power ZigBee based wireless sensor network

R. Tribedi & A. Sur
Department of Electronics and Communication Engineering, RCC Institute of Information Technology, Kolkata, India

S. Bose
Department of Electronics and Telecommunication Engineering, Jadavpur University, Kolkata, India

ABSTRACT: Real time motion detection is of utmost importance in security and surveillance systems. This paper proposes to build such a system using ZigBee in the field of wireless sensing. A wireless sensor network is an ad-hoc network of small sensor nodes which communicate with radio frequency links, and are deployed in numbers in an environment to sense physical conditions and variables like temperature, light, pressure, humidity, sound etc. Nodes are used for collecting, storing and sharing sensed data. ZigBee is a rather upcoming technology in the field of wireless communication which works on the IEEE 802.15.4 protocol. The low cost and extremely low power consumption of the ZigBee provide a simple system that lasts longer in surveillance. The basic wireless sensor network architecture is composed of ZigBee coordinator, ZigBee router and ZigBee end device. The information from the sensor nodes in the network is sent to the coordinator, the coordinator collects sensor data, stores the data in memory, processes the data, and routes the data to appropriate node via the router. The transmission of data takes place by multi hop routing at the nodes. The Arduino platform when interfaced with ZigBee gives a robust and compact physical structure to each node.

1 INTRODUCTION

The ever advancing field of surveillance and security systems when integrated with the technology of wireless sensing and transmission has the potential to build very powerful and robust hardware frameworks. Computer vision has already taken security systems to a whole new level in recent years. The use of wireless sensor network or WSN provides a rather simple but very effective solution to this dimension. The wireless sensor network comprises of small power devices, called "nodes" where a node can communicate with other nodes either directly or via other nodes. They are small, low power single board computers with a radio for wireless communication which forms the basic building block of the wireless sensor network. A single sensor node has mainly three key components- a microcontroller, sensors and communication subsystem (Hill 2003). The communication subsystem is an important energy intensive subsystem as power consumption must be minimized as far as possible. This is where ZigBee proves its supremacy over other technologies. Sensors and controls do not typically require high bandwidth but they need low latency and

very low power consumption for longer battery lives. The microcontroller brings all other subsystems together. These power devices or nodes can vary greatly in numbers depending on the type of application and other constraints like cost, size of node, energy constraints, communication's bandwidth, memory and computational speed (Healy et al. (2008)). Nodes collect and transfer data in four stages: collecting the data, processing the data, packaging the data and communicating the data. Each node collects data via the sensors, which is then processed and packed into an easily handled form by the microcontroller and communicated to the other nodes using low power radio waves (Yusuf 2014). This investigation uses the Arduino as the microcontroller which provides a compact structure to each node. Usage of XBee module (IEEE 802.15.4) makes it simpler and more power efficient than other Wireless Personal Area Networks (WPAN). ZigBee is ideally suited for simple applications within a home that transmit small amounts of data. These devices are low-rate WPAN (LR-WPAN) specifications, and use minimal power within an operating space of 20–50 meters. ZigBee operates in the Industrial, Scientific and Medical (ISM) radio bands and uses 900 MHz band in the US, 868 MHz band in Europe, and 2.4 GHz worldwide (Farahani 2008). Data transmission can vary be 20 kb/s, 40 kb/s or 250 kb/s. Both Bluetooth and ZigBee are intended for products with limited battery power, but Bluetooth operates best within one room only, but ZigBee operates well across multiple rooms. As opposed to Wi-Fi, ZigBee is a mesh networking standard, meaning each node in the network is connected to each other forming a mesh. ZigBee protocol was designed as "assemble and forget" meaning once it is set up, it can last for a long time. ZigBee-based networks generally consume one fourth of the power required for Wi-Fi networks. ZigBee's battery life is a major plus over Wi-Fi, and needs to be strongly considered if the endpoints will run on batteries. ZigBee networks are primarily intended for very low duty cycle sensor networks. A new network node may be recognized and added in approximately 30 msec. Waking up a sleeping node takes half that time, as does getting access to a channel and transmitting data. Therefore ZigBee applications benefit from the ability to quickly attach information, detach, and go to sleep state, thereby saving a lot of power.

2 ARCHITECTURE

One of the most common ways to establish any communication network (wired or wireless) is to use the concept of networking layers. ZigBee protocol layers are based on the standard Open System Interconnect (OSI) basic reference mode (Farahani 2008). Most networking systems also use at least the first four layers, but many do not use all seven layers. The ZigBee stack architecture includes the MAC, PHY, Network and the Application Support sub-layer (APS) layers. Each of the layers communicates with the layer above it. The PHY defines frequency, power, modulation, and other wireless conditions of the link. The MAC defines the format of the data handling. The remaining layers define other measures for handing the data and related protocol enhancements including the final application (Obaid et al (2014)).

The 802.15.4 standard uses only the first two layers plus the Logical Link Control (LLC) and Service Specific Convergence Sub-layer (SSCS) additions to communicate with all upper layers as defined by

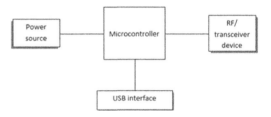

Figure 3.   Block diagram of coordinator.

Characteristics	IEEE 802.11	Bluetooth	ZigBee
Power Profile	Hours	Days	Years
Distance	Upto 100 m	10m	10 to 100m
Network Topology	Point to hub	ad-hoc, small networks	ad-hoc, peer to peer, star or mesh
Complexity	Very complicated	Complicated	Simple
Frequency Range	2.4GHz	2.4GHz	868MHz , 916MHz , 2.4GHz
Cost of terminal unit	High	Low	Low
Integration level & Reliability	Normal	High	High
Ease of use	Hard	Normal	Easy
Application focus	web,email,video	cable replacement	monitoring & control

Figure 1.   Comparison between IEEE 802.11, Bluetooth, ZigBee.

Figure 4.   Hardware of coordinator.

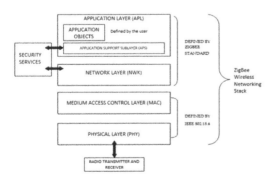

Figure 2.   ZigBee protocol stack.

Figure 5.   Block digram of router.

additional standards. The ZigBee standard defines only the networking, application, and security layers of the OSI and adopts IEEE 802.15.4 PHY and MAC layers as part of the ZigBee networking protocol. Therefore, any ZigBee-compliant device will naturally conform to IEEE 802.15.4 as well. However the IEEE 802.15.4 protocol is independent of the ZigBee standard i.e. a short range wireless network can be based on IEEE 802.15.4 without conforming to ZigBee specific layers. Also it must be noted that the decision to implement the full ZigBee protocol stack or just the MAC and PHY layers depends on the application and future use.

3 EXPERIMENTAL SET UP AND NETWORK IMPLEMENTATION

3.1 *Hardware design*

3.1.1 *Coordinator*
The coordinator consists of the power source, microcontroller, RF/transceiver device and USB interface. The transceiver device is the XBee. A single coordinator is needed for the network. It effectively forms and manages the network as a whole.

3.1.2 *Router*
The router comprises of power source, microcontroller and RF/transceiver device.

Figure 6.   Hardware of router.

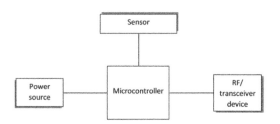

Figure 7.   Block diagram of end-point.

It passes on (route) information to other radios in the network.

3.1.3 *End-point*
The end points/tags are made up of the power source, microcontroller, RF/transceiver device and the sensor. Here the PIR motion detection sensor is used in the end point, which measures infrared radiation in its field and the collected data is transferred back to the coordinator.

3.2 *Software design*

3.2.1 *Working of coordinator node*
The program loaded into the coordinator is written on the IDE, the flowchart for which is given

Figure 8.   Hardware of end point.

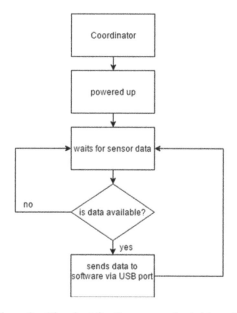

Figure 9.   Flowchart for the program loaded into the coordinator.

below in the figure. At first, the coordinator is powered up and commands are written which makes the coordinator wait for the sensor data from the router nodes. If any data is available from the router nodes then it sends the data to the serial monitor of the Arduino IDE software via the USB COM port through the Arduino board. If the coordinator does not receive any data then it goes back to the cycle of waiting and checking of data availability.

3.2.2 *Working of router node*

The main job of the router is to direct the correct packet of data to the coordinator or other parent routers. The router is powered up and then it is made to wait for data from the child router or sensor. When it receives the data, it checks whether the packet is its own packet or not. If the correct packet is received, then it forwards the data to the parent 'node and continues the cycle of waiting for data from other nodes.' If it does not receive the desired packet, then the packet is discarded and the cycle is continued.

3.2.3 *Working of end-point node*

The end point node has the PIR sensor which works as a standalone system and converts physical data values to electrical voltage values. The sensor device is powered up and sensor data is collected and transmitted to the nodes like routers. The end point transfers the data to the routers.

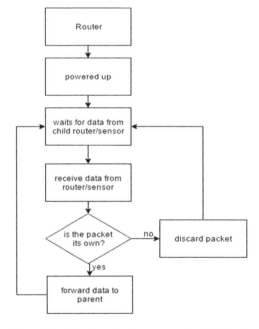

Figure 10.   Flowchart for the program loaded into the router.

4   RESULTS AND DISCUSSION

In this investigation we have been successful in creating a wireless sensor network of multiple nodes. The end device (node) uses the PIR sensor to collect data and route it to the terminal computer i.e. coordinator. Below is presented the results we have obtained in the monitor of the coordinator computer from the end point node. To obtain the results different warm body obstructions were moved in the sensor range of the PIR sensor. The PIR sensor detected the changes in movement between its 2 slots in the form of voltage changes.

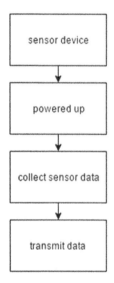

Figure 11.   Flowchart for the program loaded into the end point.

Table 1.   Raw data values shown on the end point LED display.

Sl. No.	End device name	Via router name	Sensor value (raw data)	Remarks/ Inference
1	53	1	980	No obstruction so highest sensor value is obtained
2	53	1	873	An obstruction is introduced in the path of the sensor
3	53	1	450	The obstruction is brought closer
4	53	1	220	The sensor is covered fully

118

Table 2. The various experimental data values from the coordinator monitor are collected in this table.

Sl. No	End device Name	Via router name	API ID	RSSI value (in dBm)	Voltage in photodiode (in V)
1	53	1	129	−63	1.44
2	53	1	129	−55	1.40
3	53	1	129	−50	1.37
4	53	1	129	−51	1.34
5	53	1	129	−54	1.30
6	53	1	129	−52	1.28
7	53	1	129	−65	1.27
8	53	1	129	−54	1.26
9	53	1	129	−59	1.25
10	53	1	129	−57	1.24
11	53	1	129	−54	1.25
12	53	1	129	−59	1.26
13	53	1	129	−58	1.30
14	53	1	129	−54	1.32
15	53	1	129	−50	1.29

Each reading was obtained after a time delay of 1 second as programmed in the microcontroller. The Received Signal Strength Indication (RSSI) values are obtained in dBm ranging from −63 dBm to −50 dBm. As the PIR sensor sends the output via the 10 bit ADC to the Arduino microcontroller, the sensor values range from 0 to 2^{10} i.e. 1024 theoretically. Practically, 0 and 1024 values are not obtained ideally as there will always be some warmness in the atmospheric air or particles in the air which will never lead to 1024 value of the sensor. Also it is not possible to obtain the lowest value of 0 as the PIR sensor will never be fully obstructed by the warmest body in normal conditions. So after calibration, the voltage values vary in the range of 1.2 V–1.44 V.

5 CONCLUSION

wireless sensor networks have seen an immense expansion of applications in both research and industry. It utilizes an efficient form of technology that has the potential to build and integrate vast number of multi-hop wireless sensing and control systems (Sohraby et al. (2007)).This paper tries to investigate this fundamental aspect of WSN with the usage of ZigBee technology. Over the coming years it is likely that ZigBee will increasingly play an important role in the field of computer and communication technology. The various potential applications of WSN can be broadly categorized into different classes such as Environmental data collection, Military applications, Security monitoring, Sensor node tracking, Health application, Home application, and Agriculture and farm application (Pan & Tseng 2007). This study has been successful in building a motion detection system which falls under the category of Environment data collection. Although existing surveillance systems have a much faster data rate but they consume excess power in the sensing and routing process. In order to address the same the ZigBee based system is established which resolves this issue of power consumption and cost.

REFERENCES

Farahani, S. (2008). *ZigBee Wireless Networks and Transceiver.* Burlington, Massachusetts, USA: Newnes.

Healy, M., Newe, T. & Lewis, E. (2008). *Wireless Sensor Node Hardware: A Review.* The Seventh IEEE Conference on Sensors (IEEE SENSORS 2008) Lecce, Italy.

Hill, J.L. (2003). System Architecture for wireless sensor network. Doctoral Dissertation. University of California, Berkeley. 10.

Obaid, T., Rashed, H., Abou-Elnour, A., Rehan, Md., Saleh, M., & Tarique, Md. (2014). ZigBee Technology and its Application in Wireless Home Automation Systems: A survey. *International Journal of Computer Networks &Communication* 6(4): 126.

Pan, M.S & Tseng, Y. (2007). *ZigBee and Their Applications.* Hsinchu, Taiwan: Springer.

Sohraby, K., Minoli, D. & Znati, T. (2007). *wireless sensor networks: technology, protocols, and applications.* New York, USA: John Wiley and Sons Ltd.

Yusuf, O. (2014). Determination of optimal power for ZigBee based wireless sensor networks [Electronic Theses and Dissertations. Paper 5135]. University of Windsor. http://scholar.uwindsor.ca/cgi/viewcontent. cgi?article=6134&context=etd. pp. 1–7.

Computer, Communication and Electrical Technology – Guha, Chakraborty & Dutta (Eds)
© 2017 Taylor & Francis Group, ISBN 978-1-138-03157-9

Optimal path selection for AODV routing protocol in MANET

Soma Manna & Arun Kumar Mondal
Department of ECE, Guru Nanak Institute of Technology, Sodepur, Kolkata, India

Piyu Sarcar
Department of ECE, Narula Institute of Technology, Agarpara, Kolkata, India

ABSTRACT: Routing in mobile Ad Hoc network described by multi-hop wireless link is a delicate task. The lack of permanent infrastructure and the instability of the topological changes due to nodal mobility and changes in wireless propagation conditions during the transmission of time-sensitive information between source and destination pair are challenging in ad hoc network routing. Furthermore, route discovery and route maintenance are the two major tasks in implementing the routing protocol in ad hoc network. The use of multipath routing in ad hoc network with a set of reliable data paths can solve these problems. In the algorithm proposed in this study, we use the AODV multipath routing, where the best route having high data rate and high security level is selected first and another alternate node-disjoined path is selected next as a backup route for instant use during route failure, thereby saving the time and cost of repairing/discovering new routes. Finally, we simulated the networks to depict the proposed algorithm and presented the results obtained. A comparison of the routes and data rates is also made.

1 INTRODUCTION

Advancement of wireless communication emerges a new communication paradigm: self-organized information and communication system having no network infrastructure such as router, switch, and server (Cidon 1999), (Nasipuri 2001). Such a network is referred to as Mobile Ad Hoc Network (MANET) representing a system of wireless nodes that can freely and dynamically move and create arbitrary and temporary network topologies (Cidon 1999) (Nasipuri 1999). The intrinsic attribute of such a network is the dynamic network topology, limited battery power, constrained wireless bandwidth, quality of service, and a large number of heterogeneous nodes that make network management significantly more challenging than stationary and wire network. Each node in MANET acts as a host and a router and communicates with each other via multi-hop wireless links (Manna 2016). As mobile nodes join and link the network, dynamically, sometimes even without a notice, and move randomly, network topology and administration domain membership can change frequently. To combat the inherent instability of these networks, we propose a routing scheme that uses multiple paths based on AODV routing protocol and simulate the networks to find the most secure path among these multiple paths.

Traditional routing protocol for MANETs uses a single path between the source and destination.

When this path fails, a potentially expensive operation was usually performed to locate an alternate route to the destination (Perkins 2000) (www.ietf. org/html.charters/manetcharter.html). This may cause excessive delay, call blocking, and extra overhead on the routing layer of the network. Multipath routing is an alternative to single-path routing, and aims to establish multiple paths between the source and destination pair and has several benefits. In this paper, we focus on AODV reactive routing protocol, which sets up a route between two nodes only when there is a need to send actual traffic between two nodes. AODV routing protocol is accomplished by flooding the network with route request messages, requesting information on the route from the source to the destination. In due course, destination receives the route request message and responds to it with necessary path information. In this paper, we use the AODV routing protocol to find the most secure and reliable path (Lou 2005).

AODV routing protocol is the most secure, reactive routing protocol developed specifically for MANET. In AODV, if a node wants to send data to another node without a route, then it tries to find a new route. Route finding is possible using a route request (RREQ) message. Thus, a node that tries to discover another node sends out an RREQ message, which contains the IP addresses of the transmitting node as well as the destination node. The RREQ message is broadcast over the network with a source node containing a new route to the destination, which in turn responds to the request via route reply (RREP) message. The RREP message is sent to the originator and when the message traverses toward the originator, it establishes the end-to-end path between the source and destination nodes. As wireless channel is a broadcast channel, any node (may or may not be of the same network) within the transmission range of a transmitting node can overhear the other node's transmission. Hence, individual attack is possible in the network. To calculate the probability of finding the attack to every node of a network, we formulate the optimization problem that minimizes the probability of path failures or in other words maximizes the security of the path. This is a combinatorial optimization problem and is formulated as a least-hop count problem of AODV protocol to find the optimal path. In the subsequent section, we will first formulate the problem and then propose an algorithm suitable to solve for optimum path and finally simulate some network to find its performance.

2 PROBLEM FORMULATION

A network can be modeled as an acyclic directed graph $G(V, E)$, where $V = \{1, 2, \ldots n\}$ is the set of nodes and E is the set of trunk (edges) that connects the nodes with $|E| = m$, where m is the edge metric such that $m \in M$, with M being a set of non-negative edge metrics. For the purpose of modeling, we consider that every edge $E(i,j) \in V$ connects the two nodes n_i and n_j when they are in each other's transmission range and are associated with an edge metric m having cost $C_m(i, j)$ with respect to metric $m \in M$. The different link metrics under our considerations are: link cost c, bandwidth b, delay d, delay jitter j, and path/packet loss probability p. These parameters can be used to formulate the optimization problem. A path P(s,t) is a sequence of vertices $V_s, V_1 \ldots V_i, V_{i+1} \ldots V_{t-1}, V_t$ such that $\forall\ s \leq i \leq t$, edge $(V_i, V_{i+1}) \in E$. Depending on each metric $m \in M$, the cost $C_m(s, t)$ of a path p(s, t) is the generic function of the path edge metric, that is:

$$C_m(s,t)\ for\ p(s,t) = \\ f\left((V_s, V_{s+1}) \ldots (V_i, V_{i+1}) \ldots (V_{t-1}, V_t)\right) \quad (1)$$

Examples of other cost functions are given below:

$$C_{bf}\ for\ p(s,t) = \min bf(i, j) \quad \langle buffer \rangle \\ (i,j) \in p(s,t) \quad (2)$$

$$C_h\ for\ p(s,t) = \#edges \quad for\ p(s,t) \quad \langle hop\ count \rangle \quad (3)$$

$$C_{i,j} = \sum \frac{1}{R_{i,j}} \quad (4)$$

where $l_{i,j} \in P(s,t)$ and $R_{i,j}$ is the data rate.

We proposed a node-disjoined shortest pair algorithm considering the minimum cost of network flow and optimum data rate through the paths. Each path is assigned the link cost $C_m(i, j)$ and during each iteration of the algorithm, we obtain a secure path with the minimum cost $C_p p(s, t)$ and the corresponding data rate. The algorithm for finding the optimal path and backup routes may be stated as:

Step 1. Find the first most secure path by minimizing $C_p p(s, t)$, then find the data rate for the path.

Step 2. Modify the routing table deleting data of the intermediate nodes and interlacing arcs of the shortest path. Rearrange the network nodes and transform back to the original network. Then, use the new step for finding the second secured path from the modified routing table having a lower security level than the previous path.

Step 3. Repeat step 2 to obtain one or two backup routes with satisfactory data rates.

Step 4. No further iteration is required as we obtain optimal route and the backup routes.

In this way, we can find multiple secure paths with high data rates. The most secure path is obtained first, which we used for the initial routing; the other routes are kept as the multipath backup route. If the initial route fails, there is no need to search a new route, because the routes with second-level security will be used and so on. We simulated several networks with different number of nodes. In the subsequence section, we showed the two simulated results for the MANETs.

3 SIMULATED RESULT AND CONCLUSIONS

We created a network with a campus area of 500 × 500 m² taking 20 ad hoc nodes forming a network in the transmission range. 200 m. The network is simulated using a high-capacity processor and the OPNET and MATLAB7 platforms. An example of the simulated result is shown in Figure 1.

In the simulated network, the source node designated as 1 initiates the routing procedure by sending an RREQ to its surrounding nodes. This RREQ message sent by the source node is shown in green. The other RREQ messages are shown in cyan, yellow, black, etc. The source node 1 sends the RREQ message to its neighbor nodes 5, 6, 9,

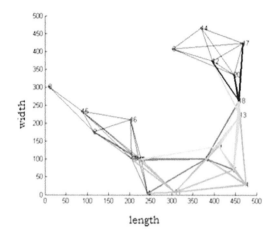

Figure 1. Simulated network with 20 ad hoc nodes.

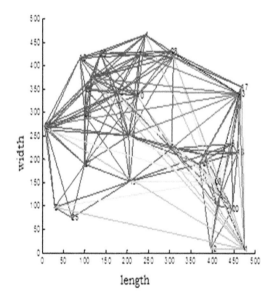

Figure 2. Simulated network with 30 ad hoc nodes.

Table 1. Comparative study of path and data rate.

Network name	Path name	Link	Data rate (kbps)
Network 1	Optimal path	1-13-20	0.45
	Backup path	1-6-18-20	0.33
Network 2	Optimal path	1-23-13-28	0.47
	Backup path	1-17-28	0.35

11, and 13 and the links formed are shown by the green line. Links are formed whenever the nodes 5, 6, 9, 11, and 13 send the RREQ message to their neighbor nodes.

In this simulation, the source will find the optimal path 1-13-20 having the data rate of 0.45 kbps. The second optimal path is 1-6-18-20 having the data rate of 0.33 kbps. We have simulated another network having 30 nodes as shown in Figure 2. The simulation is made using the same platform used for network 1. This simulation is done taking node 1 as source and node 28 as destination. The optimal path obtained in this simulation is 1-23-13-28, the data rate of which is 0.47 kbps. The second optimal path is 1-17-28 with data rate of 0.35 kbps. The simulated results are summarized in Table 1.

4 CONCLUSIONS

In this study, we simulated two networks within an area of 500 m² by taking different number of nodes, and obtained multiple paths for routing. Initially, the most optimal path is used for data transmission, and backup route is kept for future use. In designing suitable multipath routing protocol, the proposed algorithm may be efficiently used for a larger area having larger number of nodes and a suitable multipath with less computational complexity will be obtained.

REFERENCES

Cidon, I., R. Rom & Y. Shavitt (1999). Analysis of multipath routing, *IEEE/ACM Transaction on Networking*, 7(6), 885–896, doi: 10.1109/90.811453.
IETF, Mobile Ad Hoc Networks (MANET) Working Group, [Online] Available: *http://www.ietf.org/html. charters/manetcharter.html.*
Lou, W. (2005). *An efficient N-to-1 multipath routing protocol in wireless sensor networks. In Proc. 2nd IEEE Int. Conference on Mobile Ad-hoc Networks and Sensor system (MAHSS 2005) Washington* DC, 665–672, doi: 10.1109/MAHSS.2005.1542857.
Manna, S., A.K. Mondal & P. Roy (2016). *Enhancement of TCP performance over MANET, Int. Research Journal of Engineering and Technology (IRJET)*, 3(5), 714–716.
Nasipuri, A., R. Castaneda & S.R. Das (2001). *Performance of multipath routing for on-demand protocol in mobile ad hoc networks, Mobile Networks and Applications*, 6(4): 339–349, 339–349, doi:10.1023/A:1011426611520.
Nasipuri, A. & S.R. Das (1999). *On-Demand Multipath Routing for Mobile Ad Hoc Networks, In Proc. 8th Int. Conference on Computer Communication and Networks (IC3N)*, Boston, MA, 64–70.
Perkins, C.E., E.M. Royer & S.R. Das (2000). *Ad-hoc On Demand Distance Vector (AODV) routing. IETF Internet Draft, draft-ietf-manet-aodv-05.txt.*

Computer, Communication and Electrical Technology – Guha, Chakraborty & Dutta (Eds)
© 2017 Taylor & Francis Group, ISBN 978-1-138-03157-9

A study on outage-minimizing routing in a multi-hop wireless network

Vishal Kr. Karwa, Sudipta Ghosh & Surajit Basak
Department of Electronics and Communication Engineering, Guru Nanak Institute of Technology, Kolkata, India

ABSTRACT: In this paper, we investigate the problem of joint power allocation and routing optimization for the minimization of end-to-end outage probability in Decode-and-Forward (DF) multi-hop wireless networks. To solve the problem, we developed a new joint power allocation and routing strategy. First, a closed-form expression for power allocation is derived, and from this solution, routing metric is obtained, which solves the routing problem. We also present distributed implementation of joint power allocation and routing strategy. Simulation results show that the proposed scheme significantly controls end-to-end outage probability.

1 INTRODUCTION

Energy-constraint wireless networks consist of wireless nodes that operate on limited battery energy supply. As replacement or recharging can be infeasible, power minimization is one of the most critical design issues in wireless networks. A conventional multi-hop transmission scheme is an efficient approach for energy saving by relaying the information through several hops and information through several hops and thus reducing the transmit power level. Choosing appropriate relays to forward the source data as well as the transmit power allocation among the nodes can significantly increase power efficiency in wireless networks.

The key objective of any routing session is to maintain the QoS performance. Considering Rayleigh fading channels, we use outage probability to reflect the quality of communication. Babaee (2010) proposed a distributer algorithm exploiting channel state information that seeks to minimize the end-to-end outage probability in Amplify-and-Forward (AF) relay-assisted multi-hop networks, which can be executed in polynomial time. Lang (2011) investigated the minimum power routing problem in end-to-end outage probability-restricted cooperative multi-hop networks. The relays in the networks are suitably grouped to leverage the energy efficiency of MIMO transmission. Gupta (2013a) and Gupta (2013b) distributed joint routing and power allocation strategies to propose Decode-and-Forward (DF) relay-based multi-hop wireless networks for two energy-aware optimization objectives: (i) minimizing total transmission power and (ii) maximizing network lifetime. Ahmadi (2013) also studied outage probability-constrained routing problem in cooperative wireless networks to minimize the transmission power. Mahboobi and Ardebilipour also discussed energy-efficient joint routing and power allocation problem in outage-restricted relay networks with Nakagami-m fading for wireless channels (Mahboobi 2013).

The aim of this paper is to find an optimal route that minimizes end-to-end outage probability in a multi-hop wireless network. It may be noted that an increase in transmit power at each transmitting node reduces both the end-to-end outage and energy efficiency. To the best of the authors' knowledge, the issue of joint power allocation and routing in a total transmitting power-restricted multi-hop wireless networks has not been reported in the literature to date. Therefore, in this paper, we study the joint power allocation and routing problem with an aim to minimize the end-to-end outage probability under sum power constraint. Closed-form expressions of transmission power in each mode for a given route are obtained. From power allocation solution, routing metric is obtained, which satisfies the essential properties of routing protocol such as convergence, optimality, and loop-freeness. Distributed implementation of our proposed solution is also presented.

The rest of the paper is organized as follows. Section 2 describes the system model. Joint routing and power allocation strategy for end-to-end outage minimization is presented in section 3. Distributed routing and power allocation is also implemented in section 4. Section 5 presents the simulation results and finally the paper concludes with section 6.

2 SYSTEM MODEL

2.1 *Network model*

Consider a wireless multi-hop network modeled by an undirected graph $G = (V, E)$, where V is the vertex set indicating the set of $|V|$ nodes of the network and E is the edge set representing the wireless links of the network. All nodes are randomly distributed in a specific area. Each node works on Time-Division-Duplex (TDD) mode, that is, a node cannot transmit and receive simultaneously.

A sequence of data packets are continuously transmitted by the source node V_1 (say) to the destination node V_N (say) through a path $\varphi = (V_1, V_2, .., V_{N-1}, V_N)$ with $N(N \leq |V|)$ nodes, where V_2 to V_{N-1} are the nodes performing the role of intermediate Decode and Forward (DF) relays. The path φ consists of $(N-1)$ hops, where the link between the nodes V_i and V_{i+1} is denoted as i-th hop for $i \in \{1,2,3,...,N\}$. Nodes that do not belong to the path are kept in sleep mode to save battery energy source. Each relay node in the path φ receives the signal transmitted from the preceding node, processes it, and then transmits the signal to the next node in the next time slot. To process the signal, the relay nodes use DF strategy.

2.2 Wireless link model

We assume that the fading amplitudes of the links between any pair of nodes are Rayleigh distributed, which are assumed statistically independent to each other. The additive noises of all receiving nodes are assumed zero-mean complex Gaussian random variables with variance σ^2. For any pair of wireless link, an outage occurs if Signal-to-Noise Ratio (SNR) at the receiver fails to achieve a value higher than or equal to a predetermined threshold γ_{th}. In a multi-hop wireless network with DF relays, end-to-end outage occurs when SNR in any one of the links between V_1 and V_N falls below γ_{th}. Following this, outage probability of a path φ in a multi-hop network (Hasna 2004) is calculated as:

$$P_{out}(\varphi) = 1 - \prod_{i=1}^{N-1} e^{-(\gamma_{th}/\overline{\gamma_i})} \tag{1}$$

The average SNR $\overline{\gamma_i}$ of the i-th link between V_i and V_{i+1} can be written as $\overline{\gamma_i} = G_i P_i$, where G_i is a parameter independent of P_i and contains parameters such as the antenna gains, path loss, shadowing, and noise power and P_i is the transmitted power of V_i. Using Friis propagation formula (Rappaport 2002), G_i can be written as:

$$G_i = \varepsilon_i G_t G_r l^2 / (4\pi)^2 d^\alpha L N_0 \tag{2}$$

where G_t is the transmitter antenna gain, G_r is the receiver antenna gain, l is the wavelength, d is the separation distance between two nodes, L is the system loss factor not related to the propagation,

path loss exponent $\alpha = 2$ for free space, $3 < \alpha < 4$ in urban environment, N_0 denotes noise power at the n-th hop, and ε_i is a parameter that is added to the model to encapsulate the effect of lognormal shadowing.

Hence, the design problem is to find a route that minimizes end-to-end outage probability for a multi-hop wireless network such that sum power does not exceed a predetermined minimum value P_{max}. Mathematically, it can be expressed as:

$$\min_{\varphi \in \varphi(V_1, V_N)} P_{out}(\varphi) \tag{3a}$$

$$\text{Subject to } \sum_{i=1}^{N-1} P_i \leq P_{max} \tag{3b}$$

$$P_i > 0, \quad i = 1,2,...,N-1 \tag{3c}$$

where $\varphi(V_1, V_N) V_1$ and V_N represents set of all possible path between V_1 and V_N. P_i is the non-negative transmission power of each node along the path.

3 JOINT ROUTING AND POWER ALLOCATION FOR END-TO-END OUTAGE MINIMIZATION

The term joint power allocation and routing shows that the proposed routing strategy is based on the power allocation solution. Further, power allocation of each node in the route is required to minimize the end-to-end outage probability. Therefore, first, we determine the power allocation for a given path φ in a multi-hop network that takes into account the goal of routing scheme to minimize the end-to-end outage probability meeting sum power constraint P_{max}.

From (3), the power allocation problem for end-to-end outage minimization may be rewritten as:

$$\min_{P_i} \sum_{i=1}^{N-1} \gamma_{th} / G_i P_i \tag{4a}$$

$$\text{Subject to } \sum_{i=1}^{N-1} P_i \leq P_{max} \tag{4b}$$

$$P_i > 0, \quad i = 1,2,...,N-1 \tag{4c}$$

The aforementioned objective function (4a) is convex and the constraint (4b) is a linear function. The optimization problem is convex and thus has unique solution (Boyd 2004). Therefore, Karush–Kuhn–Tucker (KKT) conditions are necessary and sufficient for optimality. The Lagrangian L with Lagrange multiplier λ associated with the problem (4) is:

$$L(P_1, P_2, ..., P_{N-1}) = \sum_{i=1}^{N-1} \frac{\gamma_{th}}{G_i P_i} + \lambda \left[\sum_{i=1}^{N-1} P_i - P_{max} \right] \qquad (5)$$

Taking the derivative of L with respect to P_i and solving the set of equations, we get optimal transmission power of each node $P_i (i \in \{1, 2, .., N-1\})$, which is given by:

$$P_i = P_{max} \sqrt{\frac{\gamma_{th}}{G_i}} \left[\sum_{i=1}^{N-1} \sqrt{\frac{\gamma_{th}}{G_i}} \right]^{-1} \qquad (6)$$

And the minimum end-to-end outage probability is given by:

$$P_{out}(\varphi) = 1 - \exp \left[-\frac{1}{P_{max}} \sum_{i=1}^{N-1} \sqrt{\frac{\gamma_{th}}{G_i}} \right]^2 \qquad (7)$$

In optimal joint power allocation and routing strategy, the path that requires minimum end-to-end outage probability while maintaining the sum power constraint is to be selected. Hence, the path weight function may be derived from (7) as follows:

$$\omega_{opt}(\varphi) = \sum_{i \in \varphi} \sqrt{\frac{\gamma_{th}}{G_i}} \qquad (8)$$

To minimize the end-to-end outage probability, we need to search for the path with minimum path weight given in (8). Therefore, the routing problem for optimal strategy is:

$$\omega_{opt}(\varphi^*) = arg \mathop{min}\limits_{\varphi \in \varphi(V_1, V_N)} \omega_{opt}^{(\varphi)} \qquad (9)$$

where $\varphi(V_1, V_N)$ represents the set of all possible paths between the source node V_1 and the destination node V_N.

4 DISTRIBUTED ROUTING AND POWER ALLOCATION

4.1 Routing implementation

A routing metric can be represented as a set of elements $(S, \oplus, \omega, \leq)$, where S is the set of all paths in the network, ω is a function that maps a path to a metric value, $\leq$ is the order relation, and $\oplus$ is the path concatenation operation from a link. Before discussing the routing technique, it is important to check monotonicity and isotonicity properties of the routing metric. For paths $\varphi_1, \varphi_2, \varphi_3$ in the network, if $\omega(\varphi_1) < \omega(\varphi_2)$ implies that $\omega(\varphi_1 \oplus \varphi_3) < \omega(\varphi_2 \oplus \varphi_3)$

and $\omega(\varphi_3 \oplus \varphi_1) < \omega(\varphi_3 \oplus \varphi_2)$, then the routing metric is called isotonic (Yang 2008). The routing metric is called monotonic (Yang 2008) if $\omega(\varphi_1) < \omega(\varphi_1 \oplus \varphi_2)$ and $\omega(\varphi_1) < \omega(\varphi_3 \oplus \varphi_1)$, for any paths $\varphi_1, \varphi_2, \varphi_3 \in S$. It can be easily verified that the proposed routing metric following (8) is strictly isotonic and monotonic and the routing metric is additive. Therefore, link-state path vector or distance vector routing protocols such as DSDV (Perkins 1994) or AODV (Perkins 1999) can be easily modified to implement the routing strategy for the proposed optimal routing metric (8).

4.2 Distributed power allocation

For distributed power allocation, each transmitting node along the path φ should be able to calculate its own transmit power without any global knowledge of the network. Without loss of generality, we can assume that each node knows distances to its neighbors. Then, it can be observed from (6) and that each node can calculate its own transmit power according to optimal scheme, if it can obtain the value of path weight vector $\omega_{opt}(\varphi)$. For both DSDV and AODV routing protocols, the path weight vector is known to the source node. For DSDV routing protocol, the path weight vector is stored in the routing table of the source node. For AODV routing protocol, the path weight vector is available to the source node from route reply (RREP) message. Therefore, the source node is required to put path weight vector in the header of the data packet and transmit through the path φ. Each node in the path φ can extract this information and calculate its own transmit power using (6).

5 PERFORMANCE EVALUATION

For performance evaluation of our joint power allocation and routing strategy, we have considered a fully connected wireless network with M nodes, in which all nodes can directly communicate with each other. The nodes are randomly located within a square area of size 50×50 m. The source and sink nodes are located at (0, 0) and (50, 50), respectively. BPSK modulation is adopted and the power consumed in the communication of routing control packets and in the shortest cost path computation is ignored in the simulation. For each network realization, route is obtained by a routing metric and then transmit power is allocated at each node according to the power allocation strategy. We consider optimal routing following (9) and optimal power allocation following (6) strategy. This is referred to as *Optimal-R and PA* in the subsequent discussion. Further comparison is made between our proposed scheme, with the shortest path routing with equal power allocation scheme. This is referred to as *SPR-EPA* scheme. In the SPR-EPA strategy, route is obtained from the shortest path routing, then power allocation is achieved by considering equality of constraint (4b). All the simulations are performed in MATLAB. The results are averaged over 10^4 randomly generated independent network topologies. Values of necessary simulation parameters are found in Table 1.

Figure 1 represents end-to-end outage probability comparison with increasing P_{max} for a fixed network size $M = 20$. It can be observed that as P_{max} increases, end-to-end outage probability decreases. The reason is quite straightforward. For a fixed network size, the required transmission power increases as P_{max} increases, following (6). The best (worst) performance of optimal-R and PA strategy over SPR-EPA strategy is 57% (34%) when $P_{max} = 9(2)$.

Figure 2 depicts the end-to-end outage probability comparison for different network size when $P_{max} = 5$ W. It can be observed that as the network size, that is, the number of nodes in the network,

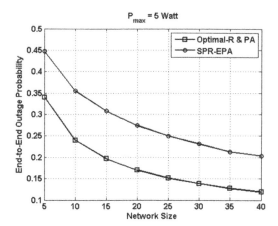

Figure 2. Comparison of end-to-end outage probability with network size.

increases as the end-to-end outage probability decreases. The reason may be explained as follows. Adding more nodes in the network increases the availability of shorter-link distance routes. Thus, transmission power requirement of individual nodes along the route will decrease, which results in the improvement of outage probability. The best (worst) performance of optimal-R and PA strategy over SPR-EPA strategy is 41.3% (23.9%) when the network size is 40 (5).

Table 1. Simulation parameters.

Description with symbols	Value
Transmitter Antenna Gain, G_t	80 dB
Receiver Antenna Gain, G_r	50 dB
Wavelength, l	100 m
Noise power, N_o	–40 dBm
Predetermined Threshold, γ_{th}	3
Path Loss, α	3.5 (Gupta 2013-NCC)
Predetermined sum power threshold, P_{max}	5 W

6 CONCLUSION

In this paper, we studied the problem of joint power allocation and routing optimization for end-to-end outage minimization. From the optimization framework, we obtained closed-form expression for power allocation. In addition, implementation issues of our proposed scheme are discussed, which suggests its practical importance for application in large-scale wireless networks. The simulation results show that our proposed optimal joint routing and power allocation scheme reduces the end-to-end outage probability significantly.

REFERENCES

Ahmadi, P. & B. Jabbari (2013). An outage-aware power saving cooperative routing algorithm in wireless networks. *In: Proc. IEEE WTS*, pp. 1–5.
Babaee, R. & N. C. Beaulieu (2010). Optimal outage efficient routing in amplify-and-forward multihop wireless networks. *In: Proc. IEEE GLOBECOM*, pp. 1–6.
Boyd, S. & L. Vandenberghe (2004), Convex Optimization, *Cambridge Univ. Press*.
Gupta, S. & R. Bose (2013a). Joint routing and power allocation optimization in outage constrained mul-

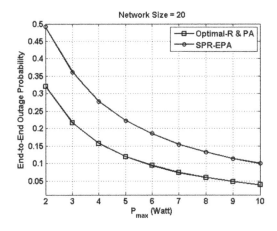

Figure 1. Comparison of end-to-end outage probability with P_{max}.

tihop wireless networks. *In: Proc. IEEE 24th Int. Symposium on Personal Indoor and Mobile Radio Communications (PIMRC)*, pp. 2245–2249.

Gupta, S. & R. Bose (2013b). Joint power allocation and routing optimization in BER constrained multihop wireless networks, *In: Proc. IEEE NCC*, pp. 1–5.

Hasna, O. & M. S. Alouini (2004), Optimal power allocation for relayed transmission over Rayleigh fading channel. *IEEE Transactions on Wireless Communications*, 3(6): 1999–2004.

Lang, Y., D. Wubben & A. Dekorsy (2011). Optimal power routing for end-to-end outage restricted distributed MIMO multi-hop networks. *In: Proc. IEEE ICC*, pp. 1–5.

Mahboobi, B. & M. Ardebilipour (2013). Joint power allocation and routing in full-duplex relay network: an outage probability approach. *IEEE Commu. Letters*, 17(8): 1497–1500.

Perkins, C. E. & P. Bhagwat (1994), Highly dynamic destination-sequenced distance-vector routing (DSDV) for mobile computers, *In: Proc. ACM SIGCOMM*, 24(4): 234–244.

Perkins, C. E. & E. M. Royer (1999), Ad-hoc on-demand distance vector routing, *In: Proc. IEEE Workshop Mobile Comput. Syst. Appl. (WMCSA)*, pp. 90–100.

Rappaport, T. S. (2002), Wireless communications: principles and practice, 2nd edn. *Prentice-Hall*, Englewood Cliffs.

Yang, Y. & J. Wang (2008), Design guidelines for routing metrics in multihop wireless networks, *In: Proc. IEEE INFOCOM*, pp. 1615–1623.

Computer, Communication and Electrical Technology – Guha, Chakraborty & Dutta (Eds)
© 2017 Taylor & Francis Group, ISBN 978-1-138-03157-9

Analyzing image transmission quality using filter and C-QAM

Md. Khaliluzzaman
Department of CSE, Chittagong University of Engineering and Technology (CUET), Chittagong, Bangladesh

Deepak Kumar Chy
Department of EEE, University of Information Technology and Sciences (UITS), Dhaka, Bangladesh

Kaushik Deb
Department of CSE, Chittagong University of Engineering and Technology (CUET), Chittagong, Bangladesh

ABSTRACT: Image Transmission Quality Assessment (ITQA) has become a popular topic of research intense over the last few years. Performance of ITQA over AWGN channel model with and without Raised Cosine filter is shown in this paper. In addition, for better transmission efficiency and achieving the required bandwidth, DCT and 32 cross-constellation QAM (C-QAM) modulation scheme is used. Furthermore, for measuring performance of image quality, PSNR and MSE are taken into account. The simulation results reveal that with lower signal-to-noise ratio, raised cosine filter exhibits better performance than employing no filter. Inter Symbol Interference (ISI), which plays an important role in image transmission, degrades the image quality, which may be eliminated by employing filter.

1 INTRODUCTION

During the past few years, people begin to play more attention on reliable wireless multimedia transmission. For transmitting image, video, and audio based on high-speed data communication, high-quality channel, and high-quality transmission of visual data are in demand. In general, high-quality image transmission acquires high storage capacity and high bandwidth requirements. For that reasons, image compression is performed, so that less storage capacity and required bandwidth is achieved. In addition, ISI degrades the image quality over wireless channel and/or wired channel due to the band-limited characteristic of the transmitted signal. An appropriate filter technique may improved the image quality that is transmitted through the noisy channel.

In this paper, a raised cosine filter is proposed before transmission of compressed image through the noisy communication channel. The typical communication model is tested by transmitting the still image using cross-constellation QAM-32 modulation schemes. The simulation results proved that the proposed communication model for image transmission yields high-quality image when the channel is more vulnerable to noise.

This section provides a descriptive summary of some techniques that have been implemented and tested for image compression and transmission over noisy channels. For example, Mishra et al. (2014) proposed the polar coding for gray-scale image transmission over AWGN channel using M-array QAM. However, they did not suggest the band-limited characteristic of the medium and the influence of inter-symbol interference. On the contrary, Kader et al. (2014) suggested the image transmission using Hierarchical Quadrature Amplitude Modulation (HQAM) for better protection of high-priority data. Nevertheless, no channel model such as AWGN is considered and only salt and pepper noise has been taken into account. Furthermore, Khandokar et al. (2009) used a simple conventional communication model with various M-array modulation techniques with AWGN Channel, but did not show the lower value of E_b/N_o. Only the value of 10 dB has been used. In this paper, we investigate the conventional communication model and compare it with the proposed communication model for image transmission using a filter. In addition, the effects of low and high signal-to-noise ratios have been considered for both communication models for still image transmission.

The remainder of this paper is organized as follows. In section II, theoretical background of DCT compression, 32 C-QAM modulation, raised cosine filter, White Gaussian Noise (AWGN) channel, Signal-to-Noise Ratio (SNR), Bit Error Rate (BER), and E_b/N_0 (Energy per bit-to-Noise power spectral density ratio) have been demonstrated. In section III, the proposed method is described and the experimental results are explained. The paper concludes in section IV.

2 THEORETICAL BACKGROUND

2.1 *DCT compression*

In this paper, the image that will be transmitted through the channel is compressed using the Discrete Cosine Transform (DCT) (Bukhari et al. 2002). The complete image is considered as a single block. Let I be the original image defined as I = f (u,v), {O < u ≤ $M-1$, 0 < v ≤ $N-1$}. M × N is the dimension of the image and f (u, v) denotes the gray-level pixel's at (u, v) coordinates. The M × N DCT coefficients are given by:

$$F(i,j) = \frac{2c(i)c(j)}{\sqrt{MN}} \sum_{u=0}^{M-1}\sum_{v=0}^{N-1} f(u,v)\cos$$
$$\times \left[\frac{(2u+1)i\pi}{2M}\right]\cos\left[\frac{(2v+1)j\pi}{2N}\right] \quad (1)$$

where, c(:) is defined as:

$$c(i) = \begin{cases} \dfrac{1}{\sqrt{2}}, & if\ i = 0 \\ 1, & if\ i \neq 0 \end{cases}$$

Respectively, the inverse DCT coefficients are given by:

$$f(u,v) = \sum_{u=0}^{M-1}\sum_{v=0}^{N-1} \frac{2c(i)c(j)}{\sqrt{MN}} f(i,j)$$
$$\times \cos\left[\frac{(2u+1)i\pi}{2M}\right]\cos\left[\frac{(2v+1)j\pi}{2N}\right] \quad (2)$$

In the following sections, we consider square images (M = N).

2.2 *32 C-QAM modulation technique*

32 C-QAM is a cross-shape constellation. The constellations for odd values of n as Q = 2^n have a cross shape and hence called 32 cross-constellation QAM, where the value of n is 5. The baud rate or symbol rate is T, where T denotes symbol rate or baud rate. In general, the symbol rate is equal to the bit rate divided by the total number of bits transmitted with each symbol. However, the signal bandwidth depends on symbol rate in such situation. Each symbol of 32 cross-constellation QAM carries m = log2 (k) bits of information. Therefore, 32 cross-constellation QAM carries 5 bit of information per symbol. The average energy

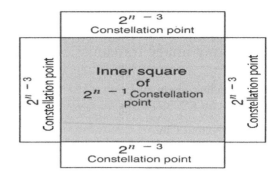

Figure 1. Square 16 QAM expanded to 32 cross-constellation QAM (n = 5).

can be calculated of 32 cross-constellation QAM, which is a revolved version of a 36-QAM square constellation, where the corner points are deleted, as they consume more power than other points in 36-QAM. If a 36-QAM is rotated, the energy is unchangeable. Therefore, the total energy of 36 QAM is first estimated. Then, the energy of the four-corner points is determined. Finally, the estimated energy of 36-QAM will be subtracted by the four-corner points' energy. The resulting energy divided by 32 gives average energy of 32 cross-constellation.

In addition, it is important to note that the four-corner points of 36-QAM possess most power. If these four-corner points are deleted, the amount of the highest power produced by transmitter will be reduced.

2.3 *FIR raised cosine filter*

A pair of raised cosine filter is used in both transmitter and receiver sites, which act as square root-raised cosine filter. In practice, ISI cannot be zero when both filters are implemented due to some numerical decision error in both the design phase and implementation phase.

For ensuring no inter-symbol interference, the following conditions should meet when a pulse-shaping filter is employed (Chy et al. 2015).

1. The pulse shape shows a zero crossing at the sampling point of all pulse intervals other than its own. That is Minimized ISI.
2. The shape of the pulses is such that the magnitude decays readily outside of the pulse interval resulting in high stop band attenuation.

The impulse response of the raised cosine filter and the square root raised cosine filters in time domain are given by (3), (4), (5), (6) (Kumar et al. 2015).

$$h_{RC}(t) = \frac{\sin\left(\dfrac{\pi T}{T}\right)\cos\left(\dfrac{\pi \alpha T}{T}\right)}{\left(\dfrac{\pi T}{T}\right)\ 1-\left(\dfrac{\pi T}{T}\right)^2} \tag{3}$$

This expression can be simplified further by introducing the sinc function $\left(sinc x = \frac{\sin x}{x}\right)$.

$$h_{RC}(t) = \sin c\left(\frac{\pi T}{T}\right)\frac{\cos\left(\dfrac{\pi \alpha T}{T}\right)}{1-\left(\dfrac{\pi T}{T}\right)^2} \tag{4}$$

The sinc function in response of the filter ensures that the signal is band-limited. The time domain or impulse response of the square root-raised cosine filter is given as:

$$h_{RRC}(t) = \frac{\sin\left[\pi(1-\alpha)t\right] + 4\alpha\left(\dfrac{t}{T}\right)\cos\left[\pi(1+\alpha)\dfrac{t}{T}\right]}{\left(\dfrac{\pi T}{T}\right)\left[1-\left(\dfrac{4\alpha t}{T}\right)^2\right]} \tag{5}$$

The overall response of the system is given by (6):

$$h_{RC}(t) = h_{RRC}(t)h_{RRC}(t) \tag{6}$$

The impulse and magnitude responses of Raised Cosine filter are shown in Figures 2 and 3, respectively.

To eliminate ISI, numerous filters have been proposed in the literature. However, Finite Impulse Response (FIR) filter is inevitable among the filters, as they give linear phase, which is a desire for any image or video communication systems for researchers. Moreover, FIR filters ensure stability and fewer finite precision errors.

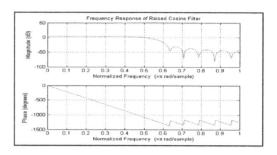

Figure 3. Frequency response of raised cosine filter.

2.4 White Gaussian Noise (AWGN) channel

To produce AWGN noise, standard built-in functions of MATLAB are used. AWGN noise of zero indicates that variance and powers are identical. The image corrupted by channel noise is shown in (7) (John et al. 1998, Samsuzzannan et al. 2010, Bernard et al. 2001):

$$Z = a + n \tag{7}$$

The probability distribution function for this Gaussian noise can be represented as:

$$p(z) = \frac{1}{\sigma\sqrt{2\pi}}\exp\left[-\frac{1}{2}\left(\frac{z-a}{\sigma}\right)^2\right] \tag{8}$$

The model of this noise assumes a power spectral density $G_n(f)$, which is flat for all the frequencies denoted as:

$$Gn(f) = \frac{N_0}{2} \tag{9}$$

The denominator 2 implies that the power spectral density is a two-sided spectrum. For evaluating the performance of the communication system, AWGN model is used in the simulation environment.

2.5 Signal-to-noise ratio

Measuring signal power in comparison with noise power is known as signal-to-noise ratio. Considering AWGN channel model for complex and unpredictable signal as well as channels, Signal-to-Noise Ratio (SNR) can be estimated as (10):

$$SNR = (\text{Signal Power} / \text{Noise Power})$$

$$= \frac{\dfrac{i}{T}\sum_{n=1}^{T}\left[(I_n)^2 + (Q_n)^2\right]}{\dfrac{1}{T}\sum_{n=1}^{T}|N_i,Q|^2 + |N_Q,n|^2} \tag{10}$$

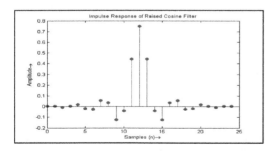

Figure 2. Impulse response of raised cosine filter.

where I_n and Q_n are the in-phase and quadrature components of signals for M-array modulation schemes, whereas N_i,n and N_Q,n are the in-phase and quadrature components of the corresponding complex noise considered in this work.

2.6 Bit Error Rate

Bit Error Rate in digital communication system is the ratio of the number of erroneous bits to the total number of bits given in a specific time when transmission takes place in any communication system:

$$BER = (Bits\ in\ Error)/(Total\ bits\ received) \quad (11)$$

BER is a unit-less performance measure, often expressed as a percentage (Ghosh et al. 2012).

BER can be calculated in terms of the Probability of Error (POE) (Stathaki et al.) and represented by (12):

$$POE = \frac{1}{2}(1 - erf)\sqrt{\frac{E_b}{N_0}} \quad (12)$$

where "erf" is the error function, E_b is the energy in one bit, and N_0 is the noise power spectral density (noise power in a 1 Hz bandwidth). E_b/N_0 (*Energy per bit-to-Noise power spectral density ratio*).

3 PROPOSED METHOD

This section describes the proposed method in detail. The basic steps involved in the proposed method are: (1) converting RGB image into gray-scale image, (2) applying DCT in gray-scale image to be compressed and be quantized applying DCT, (3) encoding the quantized image and finally finding the compressed image, (4) utilizing modulation technique on encode values to obtain the symbols, (5) before transmitting symbol value through AWGN channel, filtering the symbol data values with raised cosine filter and applying the Inverse first Fourier transform to convert the value from frequency domain to time domain, (6) passing the noisy values through the inverse raised cosine filter at the receiver site and then converting to time domain and frequency domain by applying Fourier transform, (7) demodulating these transformed values with QAM 32, (8) decoding the demodulated values accordingly, (9) applying the IDCT to get the original gray-scale values, and (10) retrieving the original image. The workflow of the proposed framework is shown in Fig. 4.

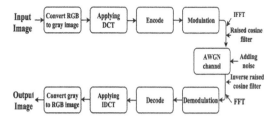

Figure 4. Workflow of the proposed framework.

3.1 Converting RGB image into gray-scale image

In this section, initially the input RGB image is converted into gray-scale image, which contains only one channel value, that is, intensity. The RGB image is converted to gray-scale image as (13):

$$I_{gray} = 0.299 * R + 0.587 * G + 0.114 * B \quad (13)$$

3.2 Applying DCT to compress the gray-scale image

In an image, low frequency values are found in the left upper corner and high frequency values are found in the lower right corner of the compressed image block. As most of the valuable information is stored at low frequencies, discarding certain information from higher frequency has little effect on the overall image quality.

Typically, 8×8 or 16×16 block size is used instead of implementing DCT on the entire image. The DCT transform is usually applied first row-wise and then column-wise. Avoiding the complexity, column-wise DCT operation can be accomplished applying row-wise DCT operator. For this reason, initially, row-wise DCT operation is applied on the image block, then transpose is implemented on the column-wise operation and the row-wise DCT operation is performed again. The procedure is shown in Fig. 5.

The mathematical expression for performing N-point DCT operation is given in (1). For instance, 8-point DCT operation equals 8 in (1). The inputs to the *8-point DCT* operator are eight pixel values $(u(0)......u(7))$. After executing the DCT computation, eight DCT values $(F(0)......F(7))$ are yielded at the operator output. For calculating each DCT value, all input obtained values such as $(u(0)......u(7))$ are used. These values are finally quantized. Quantization is achieved by dividing the transformed image matrix by the quantization matrix. Values of the resultant matrix are then rounded off.

Figure 5. Two-dimensional DCT operations.

3.3 Encode

In this section, the compressed quantization values are encoded to convert into binary code streams. The most commonly used entropy encoders are the Huffman encoder and the arithmetic encoder, although for applications requiring fast execution, simple Run-Length Encoding (RLE) has proven very effective.

3.4 Utilizing 32 C-QAM modulation

In this section, 32 C-QAM, cross-constellation Quadrature Amplitude Modulation is used in this paper, which has five I values and five Q values. This results in 32 possible states for the signal. Here, 32 C-QAM is used, as there is a trade-off between power efficiency and bit error rate.

3.5 Applying raised cosine filter and IFFT

This section uses the root raised cosine filter to eliminate inter-symbol interference, which degrades the image quality severely at low signal-to-noise ratio. Single raised cosine filter at transmitter side cannot reduce the inter-symbol interference. Therefore, a pair of raised cosine filters is used in this paper, which acts as root raised cosine filter. In addition, the original image block is in frequency domain, but raised cosine filter and AWGN model follow real-time operation. For this purpose, Inverse First Fourier Transform (IFFT) is introduced in this proposed method before using raised cosine filter.

3.6 AWGN channel

In practice, according to Shannon's capacity theorem, no channel is noise-free. All transmitted signals are affected by noise, is unpredictable in nature. Precise mathematical operations cannot be done in the presence of noise. Hence, AWGN channel is chosen as channel model for simulation.

3.7 Retrieving original image

In this section, retrieval of original image at the receiver side is demonstrated. First, the noisy image data are passed through the inverse raised cosine filter to obtain the modulated image data in time domain. Then, demodulation process is applied followed by first Fourier transform to obtain the encoded data in frequency domain.

After this, the decoding technique is implemented to obtain quantized image value. Furthermore, Inverse Discrete Cosine Transform (IDCT) operation is performed to obtain the gray-scale value. Finally, the original RGB image is retrieved from the gray-scale image by using appropriate method.

4 SIMULATION AND EXPERIMENTAL RESULTS

In this paper, simulation is carried out for grayscale images. The simulation was performed using MATLAB software.

The Mean Square Error (MSE) and Peak Signal-to-Noise Ratio (PSNR) are typically used to measure the quality of the receiving image with respect to the transmitting image. The MSE and PSNR are measured using (14) and (15):

$$MSE = \frac{1}{MN} \sum_{i=0}^{M-1} \sum_{j=0}^{N-1} \| I(i,j) - R(i,j) \|^2 \tag{14}$$

$$PSNR = 10 log_{10} \frac{\| I(i,j) - R(i,j) \|^2}{MSE} \tag{15}$$

The result of compressed quantized image for the input image with and without filter is shown in Fig. 6(b) and Fig. 6(c), respectively. Fig. 7 shows the processing example of retrieved images with E_b/N_0 (dB) effect, where raised cosine filter is used. From the experimental results, it is seen that the received image quality improved with the increase

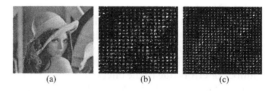

Figure 6. Processing example of quantized image: (a) original image, (b) quantized image using filter, and (c) quantized image without filter.

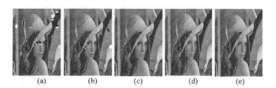

Figure 7. Processing example of output images with different E_b/N_0 (dB) value using raised cosine filter: (a) E_b/N_0 (dB) = 1, (b) E_b/N_0 (dB) = 2, (c) E_b/N_0 (dB) = 3, (d) E_b/N_0 (dB) = 4, and (e) E_b/N_0 (dB) = 5.

Table 1. Experimental results of no. of error, BER, MSE, and PSNR with filter.

E_b/N_0 (dB)	Number of error	Bit Error Rate (BER)	MSE	PSNR
1	1625	$9.9182 \times e^{-4}$	NA	NA
2	467	$2.8503 \times e^{-4}$	NA	NA
3	68	$4.1504 \times e^{-5}$	25.1261	3.8900
4	9	$5.4932 \times e^{-6}$	23.5973	4.1420
5	1	$6.1035 \times e^{-7}$	23.5973	4.1420
6	0	0	23.5973	4.1420
7	0	0	23.5973	4.1420
8	0	0	23.5973	4.1420
9	0	0	23.5973	4.1420
10	0	0	23.5973	4.1420

Table 2. Experimental results of no. of error, BER, MSE, and PSNR without filter.

E_b/N_0 (dB)	Number of error	Bit Error Rate (BER)	MSE	PSNR
1	29675	0.0116	NA	NA
2	14460	0.0056	NA	NA
3	6121	0.0025	NA	NA
4	2148	$8.3906 \times e^{-4}$	NA	NA
5	501	$1.9570 \times e^{-4}$	NA	NA
6	123	$4.8047 \times e^{-5}$	$8.7112 \times e^{36}$	$1.1220 \times e^{-35}$
7	20	$7.8125 \times e^{-6}$	13.7908	7.0874
8	3	$1.1719 \times e^{-6}$	13.7908	7.0874
9	0	0	13.7908	7.0874
10	0	0	13.7908	7.0874

of E_b/N_0 (dB) values. When the value of E_b/N_0 (dB) is higher than five, the bit error rate is zero and the values of MSE and PSNR are constant, that is 23.5973 and 4.1420, respectively. The bit error rate, MSE, and PSNR with respect to E_b/N_0(dB) with raised cosine filter are shown in Table 1.

Figure 8 shows the processing example of retrieving images with E_b/N_0 (dB) effect, where raised cosine filter was not used. The experimental result shows that with the lower E_b/N_0 (dB) value, the received image without filter is more blurred with respect to the images that are received with filter. The bit error rate, MSE, and PSNR without raised cosine filter are shown in Table 2. It is evident that BER is zero when the value of E_b/N_0 (dB) is higher than eight, and the values of MSE and PSNR are constant for those of E_b/N_0 (dB), 13.7908 and 7.0874, respectively. The best-quality images are received by the proposed simulation with $E_b/N_0 = 5$ dB and using the raised cosine filter.

It is seen from Table 2 that for $E_b/N_0 = 5$ dB, the value of MSE is indefinite. It is also observed that by increasing E_b/N_0 by 1 dB, the value of MSE is very large or infinity. This is because of the large number of errors due to the effect of noisy AWGN channel. On the contrary, the interference is much greater than noise caused by ISI.

For instance, assuming $E_b/N_0 = 1, 2, 3, 4,$ and 5 dB, the Bit Error Rate (BER) and PSNR are much better than the transmitting image without filter. At those values, the retrieved images are blurred, thereby unsuitable for human visualization without filter. For higher values of E_b/N_0, where no error is occurred, conventional communication model with QAM-32 reveals good results. For low signal-to-noise ratio, the inter-symbol interference is dominated over noise. As a result, without filter, the image is blurred. This suggests that the proposed method is more applicable than conventional communication model for image transmission. The results of comparison of SNR

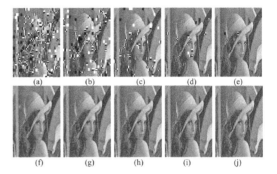

(a) (b) (c) (d) (e)

(f) (g) (h) (i) (j)

Figure 8. Processing example of output images with different E_b/N_0 (dB) values without using raised cosine filter: (a) E_b/N_0 (dB) = 1, (b) E_b/N_0 (dB) = 2, (c) E_b/N_0 (dB) = 3, (d) E_b/N_0 (dB) = 4, (e) E_b/N_0 (dB) = 5, (f) E_b/N_0 (dB) = 6, (g) E_b/N_0 (dB) = 7, (h) E_b/N_0 (dB) = 8, (i) E_b/N_0 (dB) = 9, and (j) E_b/N_0 (dB) = 10.

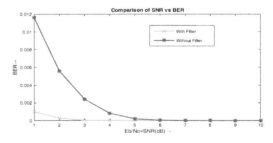

Figure 9. Comparison of SNR and BER with and without filter.

and BER with and without filter are shown in Fig. 9.

Khandokar et al. (2009) obtained a BER of 0.2605 at $E_b/N_0 = 10$ dB using 32-QAM, whereas the proposed method with filter exhibits zero BER at the same value of E_b/N_0. Furthermore, Mishra et al. (2014) obtained a PSNR value of 45 dB for

$E_b/N_0 = 10$ dB, whereas the proposed method with filter shows a value of 4.1420 dB at the same value of E_b/N_0.

5 CONCLUSIONS

For low SNR, when compressed image is affected by noisy channel, it is difficult to retrieve the original image. The fact is that Signal-to-Interference Ratio (SIR) is higher than SNR. Thus, for lower values of E_b/N_0, the images are blurred. However, this problem is mitigated by introducing raised cosine filter in the proposed method. For higher values of SNR, the conventional communication model with QAM-32 shows better performance. In future, the image quality and data protection may be ensured by implementing Hierarchical QAM (HQAM) and sophisticated coding technique using the proposed communication model for image transmission.

REFERENCES

Bernard, S. (2001). Digital communications fundamentals and applications. *Prentice Hall, USA*.

Bukhari, K. Z., Kuzmanov, G. K., & Vassiliadis, S. (2002, November). DCT and IDCT implementations on different FPGA technologies. In *Proceedings of the 13th Annual Workshop on Circuits, Systems and Signal Processing (ProRISC'02).–Veldhoven, The Netherlands* (pp. 232–235).

Chy, D. K., & Khaliluzzaman, M. (2015). Comparative Performance of BER in the Simulation of Digital Communication Systems using Raised Cosine Filter. In *Third Intl. Conf. on Advances in Computing, Electronics and Electrical Technology-CEET*.

Ghosh, A., Majumder, B., Paul, P., Mullick, P., Thakurta, I. G., & Ghosh, S. K. (2012). Comparative BER Performance of M-ary QAM-OFDM System in AWGN & Multipath Fading Channel. *International Journal on Computer Science and Engineering, 4*(6), 1185.

Hanzo, L., Webb, W. T., & Keller, T. (2000). Single-and Multi-carrier Quadrature Amplitude Modulation: Principles and Applications for Personal Communications, WATM and Broadcasting: 2nd.

Kader, M. A., Ghani, F., & Ahmad, R. B. (2014). Image Transmission Over Noisy Wireless Channels Using HQAM and Median Filter. In *The 8th International Conference on Robotic, Vision, Signal Processing & Power Applications* (pp. 145–152). Springer Singapore.

Kumar, D., Khaliluzzaman, M., & Faisal, M. (2015). Performance Analysis of Digital Communication Systems in Presence of AWGN Implementing Filter Technique. *International Journal of Computer Applications, 131*(18), 5–10.

Maiya, S. V., Costello, D. J., & Fuja, T. E. (2012). Low latency coding: Convolutional codes vs. LDPC codes. *IEEE Transactions on Communications, 60*(5), 1215–1225.

Masud, M. A., Samsuzzaman, M., & Rahman, M. A. (2010). Bit Error Rate Performance Analysis on Modulation Techniques of Wideband Code Division Multiple Access. *arXiv preprint arXiv:1003.5629*.

Mishra, A. et al. (2014). Quality Image Transmission through AWGN Channel using Polar Codes. *International Journal of Computer Science and Telecommunications, 5*(1).

Proakis, J. G., Salehi, M., & Bauch, G. (2012). *Contemporary communication systems using MATLAB*. Nelson Education.

Sanyal, A., & Samaddar, S. K. (2012). The Performance Analysis of Fast DCT Algorithms on Parallel Cluster Architecture. *International Journal of Information and Electronics Engineering, 2*(3), 369.

Stathaki, T., Constantinides, A., & Stathakis, G. Equiripple minimum phase piecewise flat FIR filter design from linear phase systems using root moments. In *Proc. of the IEEE ICASSP* (Vol. 98, pp. 15–19).

Tao, L., & Kwan, H. K. (2012). Multirate-based fast parallel algorithms for DCT-kernel-based real-valued discrete Gabor transform. *Signal Processing, 92*(3), 679–684.

Computer, Communication and Electrical Technology – Guha, Chakraborty & Dutta (Eds)
© 2017 Taylor & Francis Group, ISBN 978-1-138-03157-9

Low-cost and secured biometric system-based EVM with instant counting by wireless transmission of data to a central place

Anirban Ghosal, Sudip Chakraborty, Sucharita Maity & Ankita Paul
Department of ECE, JIS College of Engineering, Kalyani, Nadia, India

Partha Roy
Department of Electrical Engineering, Jadavpur University, Kolkata, India

ABSTRACT: This paper focuses on a simplified, low-cost and secured biometric system-based electronic voting machine from which voting data can be transmitted to a central place where counting can be done in every one hour during a voting session. Today, we all know that fake voting is still a major drawback in the elections. Although we have moved from the traditional ballot paper to Electronic Voting Machines (EVMs), the problem of fake voting cannot be avoided completely. In the electronic voting system, the opinion of voters is recorded, stored, and processed digitally. However, the EVM is not capable of detecting fake voters. Even the recorded voting data can be tampered later on when the voting process is over. In our project, we tried to secure the voting process by transmitting the voting data from the EVM within one hour of casting a vote. Here, in the EVM, low-cost 12-bit counter ICs (MC14040B) are used to count the vote and parallel-in/serial-out shift registers (54F/74F676) are used to store the votes. RF multiplexers (ADG901 series) are used to multiplex the results of various candidates for wireless transmission.

1 INTRODUCTION

In a democratic country, it is a fundamental right to express one's opinion by voting in elections. Elections (Sudhakar & Sai 2010) allow the people to choose their representatives and express their preferences for how they are governed. In all earlier elections of India, such as state or central elections, a voter casts his/her vote by marking with stamp against their chosen candidate and then folding the ballot paper as per a prescribed method, before dropping it into the ballot box. This is time-consuming and very much prone to errors. The same method was continued until electronic voting machines were introduced in the election process. In the case of EVMs, materials such as ballot papers, ballot boxes and stamping were completely condensed into a simple box called the ballot unit. However, EVMs retain all the characteristics of voting by ballot papers, while making polling a lot more expedient.

EVMs with in-built memory are compared with those with a Detachable Memory Module (DMM) that can be easily tampered. How difficult is it to tamper votes stored in EVMs? Not very much; this information is based on the response to an application filed under the Right to Information Act (RTI) by a social activist based in Pune, India. In fact, if one were to tamper with votes, one of his/her best options would be an EVM. A scandalous fact that was brought to the fore by Rekha Gore's RTI is that Maharashtra is one of the few states in the country that uses electronic voting machines with Detachable Memory Modules (DMM) for municipal elections. The DMM is a small part of the EVM that contains election results. However, it is detachable from the machine. Therefore, there is a high risk of destroying, tampering, or even replacing the election results. Therefore, the results can be very easily manipulated by replacing the DMM. Japan and many other European countries across the world have discontinued the use of EVMs. In America, EVMs have been heavily criticized. The EVM can also be hacked.

For example, a small display component of the machine can be replaced with an identical component. This component can be programmed to steal a percentage of the votes in favor of a chosen candidate. Signals to activate the program can even be transmitted via a mobile phone. The voting result is displayed on LED segments. There is a 3 mm gap between the segments where a chip can be loaded. This chip can be connected wireless to a mobile phone. If a miscreant manages to place a chip, a signal from a micro-controller will control what is displayed on the screen. Before the commencement of the public counting session, the votes stored in the EVM can be changed by using a specially made pocket-sized device. Inside the machine, there are micro-controllers under which there are electrically enabled programs with a 'read-only' memory. It is used only for storage purposes. However, we can read and write the memory from an external interface. A small clip can be developed with a chip on the top to read votes inside the memory and manipulate the data by swapping the vote from one candidate to another.

2 SYSTEM DESCRIPTION AND DYNAMIC MODEL

A few years back, the Government of India collected the data of Indian citizens for making Aadhar Card. In this system, fingerprints of all valid Indian citizens can be stored. It is a biometric system similar to the KY-M6 Fingerprint Sensor Module that can be interfaced with a computer in which the fingerprint impression of a voter is stored. Before casting his/her vote, the voter should first appear for the finger impression test. Only if the finger impression is recognized by the system will a control system be given to the EVM to activate the machine. The EVM is controlled by a timer circuit set for one hour, after which the polling switch of the EVM will be automatically deactivated. In the EVM, there are several switches for several candidates by means of which voters can cast their votes. Each switch is connected to a different pulse generator and each pulse generator is connected to a different 12-bit up-counter. When a switch is pressed, a 5-volt power supply is provided at the pulse generator, and thus a pulse is generated and fed as a clock into a 12-bit up-counter. A 12-bit up-counter is capable of counting 4096 votes. IC MC14040B can be used as a 12-bit up-counter. Outputs of these up-counters are connected to the parallel-in/serial-out registers. IC 54F/74F676 can be used as a parallel-in/serial-out register. Output of each register is connected to a different ASK modulator and to a different permanent memory. IC LF398 can be used as a ASK modulator. All ASK modulators have the same carrier frequency generated by a local crystal oscillator. All outputs of the ASK modulators are connected to multiplexers for multiplexing the data required for wireless transmission. After one hour, the polling switch of the EVM will be deactivated, as it is controlled by a timer circuit designed to keep the polling switch of the EVM active for one hour, but the power supply to the counter and the register will remain ON. As a next step, the election officer present in the booth will press a button in the EVM to supply power to the ASK modulator and the multiplexer. In doing so, the stored results of several candidates are modulated, multiplexed, and transmitted. The modulated data is encrypted for security before transmission.

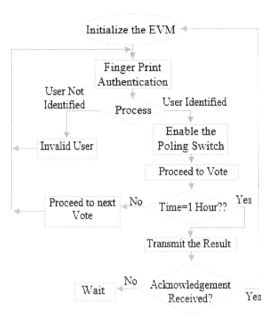

Figure 1. Flowchart of the proposed system.

The receivers of those transmitters are placed in a central place, and one member from each party and one member of the Election Commission will be present in that place. The transmitted data will be coherently detected and all the data of the same candidate will be added by a binary adder and the system is interfaced with a computer in which the results of different parties will be stored. After receiving the data, the member of the Election Commission will send a message to the corresponding booth from where the data were transmitted. Next, the election officer in the booth will reset the timer circuit, counters, and registers. Thus, a stoppage time of 5-minutes for polling will again start.

3 FEATURES OF THE COMPONENTS

Fingerprint biometric: Human fingerprints are unique to each person and can be considered as a sort of signature, certifying the person's identity. Fingerprints are the oldest and most widely used form of biometric identification.

KY-M6 fingerprint sensor: KY-M6 Fingerprint Sensor Module (Sudhakar & Sai 2010) is able to conduct fingerprint image processing, template generation, template matching, fingerprint searching, storage, etc. Compared with similar products from other suppliers, KY-M6 proudly boasts of its following features:

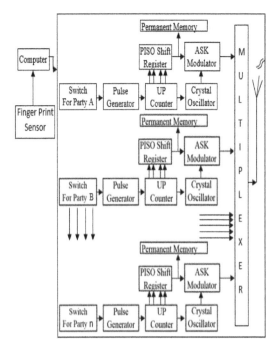

Figure 2. Block diagram of the proposed EVM system with data transmission.

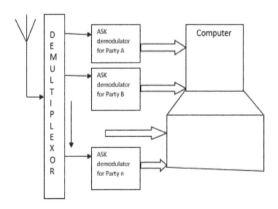

Figure 3. Receiving section of the transmitted result.

- proprietary intellectual property;
- wide range of fingerprints with different qualities;
- highly improved algorithm;
- flexible applications;
- easy to use and expand;
- low power consumption;
- different security levels.

12-Bit binary counter: The MC14040B 12-stage binary counter is constructed with MOS P-channel and N-channel enhancement mode devices in a

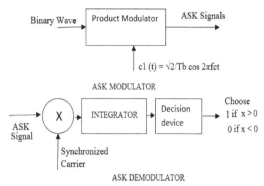

Figure 4. ASK modulation and demodulation scheme.

single monolithic structure. This part is designed with an input wave shaping circuit and 12 stages of ripple-carry binary counter. The device advances the count on the negative-going edge of the clock pulse. Its applications include time delay circuits, counter controls, and frequency-driving circuits.

General description of 54F/74F676 16-bit serial/parallel-in, serial-out shift register: It contains 16 flip-flops with provision for synchronous parallel or serial entry and serial output. When the Mode (M) input is HIGH, information present on the parallel data (P0–P15) inputs is entered on the falling edge of the Clock Pulse (CP) input signal. When M is LOW, data is shifted out of the most significant bit position while the information present on the Serial (SI) input shifts into the least significant bit position. A HIGH signal on the Chip Select (CS) input prevents both parallel and serial operations.

ASK modulation: Amplitude Shift Keying (ASK) is a modulation process, which imparts to a sinusoid two or more discrete amplitude levels. These are related to the number of levels adopted by the digital message. For a binary message sequence, there are two levels, one of which is typically zero. The data rate is a sub-multiple of the carrier frequency.

The formal transmission system can be modified by providing the chaotic scheme. In the chaotic scheme, reliable masking can be established to ensure secure communication. The synchronization of chaotic models in the transmitter end and the receiver end can be accomplished by several techniques. Among them, a very robust and modern controller is the Sliding Mode Controller (SMC). Fig. 5 shows the communication system using chaos masking based on the SMC.

In communication engineering, the chaos theory has been extensively used in recent years due to its noise immunity and straightforwardness (Argyris, et al. 2010, Cuomo et al. 1993, Harb et al. 2004).

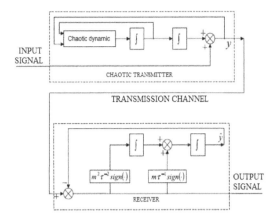

Figure 5. SMC-based chaotic communication scheme.

4 CONCLUSION

This paper aims at enhancing security by eliminating fake and duplicate votes using fingerprint-based authentication. Tampering or hacking of the EVM can be eliminated by transmitting the voting results instantly. In future, measures will be taken to count the votes immediately after a vote has been cast.

REFERENCES

Argyris, A., A. Grivas, M. Hamacher, A. Bogris and D. Syvridis (2010), "*Chaos-on-a-chip Secures Data Transmission in Optical Fibre Links*", Opt. Express 18(5), 5188–8198, DOI:10.1364/OE.18.005188.

Cuomo, K.M. and A.V. Oppenheim, (1993), "*Circuit Implementation of Synchronised Chaos with Application to Communication*," Phys. Rev. Lett 71(1), Vol. 71, no. 65, pp. 65–68.

Harb, B.A. and A.M. Harb, (2004), "*Chaos and Bifurcation in A Third Order Phase Locked Loop*", Chaos, Solitons Fractals 19(3), 667–672, DOI: 10.1016/S0960-0779(03)00197-8.

Sudhakar, M. and B.D.S. Sai (2015), "*Biometric System Based Electronic Voting Machine Using Arm9 Micro-controller*", IOSR – Journal of Electronics and Communication Engineering (IOSR – JECE), e-ISSN:2278–2834, p-ISSN:2278–8735, Vol. 10, Issue 1, Ver. II (Jan–Feb), pp. 57–65.

Computer, Communication and Electrical Technology – Guha, Chakraborty & Dutta (Eds)
© 2017 Taylor & Francis Group, ISBN 978-1-138-03157-9

Performance analysis of the CO-OFDM system in a CR network

S. Nandi
Bengal Institute of Technology, Kolkata, India

M. Sarkar
B.P.P.I.M.T, Kolkata, India

A. Nandi
N.I.T, Silchar, India

N.N. Pathak
BCREC, Durgapur, India

ABSTRACT: In modern communication networks, Orthogonal Frequency Division Multiplexing (OFDM) combined with Cognitive Radio (CR) has the potential to combat with the spectrum scarcity problem by efficiently utilizing the frequency bandwidth of the network. OFDM is a subset of frequency division multiplexing that realizes the Multi-Carrier Modulation (MCM) technique through a single channel, and has many advantages that are critical for high-speed CR networks. This paper carries out a performance analysis based on the Bit Error Rate (BER), Q-factor, input power, and link distances of the Coherent Optical OFDM (CO-OFDM) system using a 512-subcarrier 16 Quadrature Amplitude Modulation (16QAM) technique for the transmission of optical signals at a data rate of 10 Gbps. The results indicate that the possible reliable transmission distance is up to 150 Km without using any compensation scheme in relation to the link distance and input power.

Keywords: CO-OFDM, CR networks, Spectrum utilization, QAM, BER, Q-factor

1 INTRODUCTION

In modern age, the demand for high data rate has been increasing sharply, especially in applications such as multimedia, voice, and data communication including wired and wireless communications. On the one hand, the implementation of various techniques for efficient spectrum utilization with fast communication is not so easy. On the other hand, measurements show that the wide ranges of radio spectrum are hardly used while the bands are used largely. The unused portions of the spectrum can be utilized by using a new technology, i.e. CR which is

a novel concept in wireless communication (Mitola, J. & G. Q. Maguire Jr. 1999). The evolutionary future of the wireless communication will be based on the usage use of computer-based devices. Fig. 1 shows the evolutionary progress made in the usage of communication technology. The traditional radio is basically the hardware-driven generic radio transceiver (Mitola III, J. 2000a). This was followed by the programming concept that was introduced in Programmable Digital Radio (PDR), which is the simple Software-Defined Radio (SDR) progenitor. In the SDR, many signal processing functions such as modulation and coding are performed in soft-

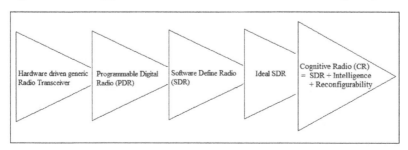

Figure 1. Evolution of radio (hardware-driven generic radio transceiver, PDR, SDR, ideal SDR, and CR).

ware (Mitola III, Joseph. 2004b). The ideal SDR connects the SDR and CR. The CR adds some extra features to the SDR framework. The cognition function of CR makes it possible to change its transmission parameters based on the interaction with the environment in which it operates (Federal Communications. 2003). The CR offers a solution to the spectrum crowding problem using dynamic spectrum management (Xiao, Lu. et al. 2014). A special type of a Multi-Carrier Modulation (MCM) technique is Orthogonal Frequency Division Multiplexing (OFDM) in wireless communication. OFDM is a widely used technology having the capacity to fulfill the requirement of CR inherently or with slight modifications. OFDM is a reliable and effective method. Here, each subcarrier is orthogonal to every other subcarrier and performs individual channel responsibility; thus, resource allocation can effectively enhance the performance (Wong, C. Y. 1999, Rhee, W. & J. M. Cioffi 2000). In the optical domain, OFDM has been classified into two main detection techniques: (a) Direct Detection OFDM (DD-OFDM) and (b) Coherent detection OFDM (CO-OFDM). CO-OFDM has the combined advantages of both the coherent system and the OFDM system (Jang, J. & K. B. Lee 2003, Shen, Z. et al. 2005). The main features of CO-OFDM are linear detection and high spectral efficiency. It has the ability to overcome many optical fiber restrictions such as Polarization Mode Dispersion (PMD) and Chromatic Dispersion (CD), as proposed by Jansen (2008). The OFDM system can efficiently operate in broadband communication using one-tap equalizers in the frequency.

Domain. It also has the ability to adaptively change the modulation order, coding, and signal processing scheme, and transmit the power of each subcarrier present in the modulation as per the requirement and channel conditions (Keller, T. & L. Hanzo 2000). By this adaptation feature, optimization can be achieved in different spheres such as Bit Error Rate (BER) reduction, enhancement of throughput and coverage area, limiting interference to License Users (LU), and elongated battery lifetime.

This paper is organized as follows: Section 2 introduces the system model, i.e. A. CR network architecture and B. OFDM system model. Section III presents the simulation result and discussion. Finally, Section IV draws the conclusions.

2 SYSTEM MODEL

2.1 *Cognitive radio network architecture*

The description of the CR network architecture provided by Mitola & Maguire Jr. (1999) in their paper gives the impression on how a radio-based complex device can help overcome the spectrum scarcity problem. The CR acquires knowledge from the environment through various sensors in the network architecture. The network architecture of the CR, shown in Figure 2, has two types of components: (i) primary network and (ii) CR network (Hong, X. et al. 2009). The existing network is the primary network, also known as the licensed network, where the Primary Users (PUs) have a license to operate in a certain spectrum band (Jondral, Friedrich. K. 2005). The implementation of the Secondary User's (SU) or Cognitive Radio (CR) user's environment (unlicensed network) by using the licensed network is carried out without causing any disturbance to the primary network's infrastructure. Primary base stations control the primary user activities that are not affected by the unlicensed users. The CR network is the dynamic spectrum access network that does not have the license to operate in the desired spectrum band. This necessitates the CR users to equip with some additional features for sharing the licensed spectrum band. The CR base station in the CR network provides a single-hop connection to its users. The spectrum broker in the CR network is responsible for distributing the spectrum resources among different CR networks (Srinivasa, S. & S. A. Jafar 2007). Here, we assume that the transmission scheme within the CR network is OFDM. The system model of OFDM is briefly described in Section 2.2.

2.2 *OFDM system*

The OFDM concept was first proposed by Cimini, L. J. 1985 for the performance analysis of a digital mobile communication channel. Later on, this concept was incorporated widely into various types of communication such as ultra-wideband (WiMedia 2005), WLAN (e.g., IEEE802.11a/g and HiperLAN2), and WiMax (Ghosh, A. et al. 2005). The performance of basic OFDM systems has been extensively researched in the literature (Cho, Y. et al. 2010, Bingham, J. A. C. 1990, Tang, H. 2005, Wang, Z. & G. B. Giannakis 2000, Van Nee, R. & R. Prasad 2000). The first real-time optical OFDM systems were demonstrated by Jin, X.Q. (2009). However, extensive research is still required for the improvement of data rates for reliable data transmission.

In this paper, we evaluate the transmission of optical signals in the CO-OFDM system at 10 Gbps using the Quadrature Amplitude Modulation (QAM) technique. The investigation of the CO-OFDM system in the optical and electrical domains is carried out by the OptiSystem™ simulator and MATLAB, respectively. A block diagram of the CO-OFDM structure is shown in Figure 3. Five main functional blocks are required to describe this CO-OFDM system, which are as follows: *a.* OFDM transmitter, *b.* RF to Optical

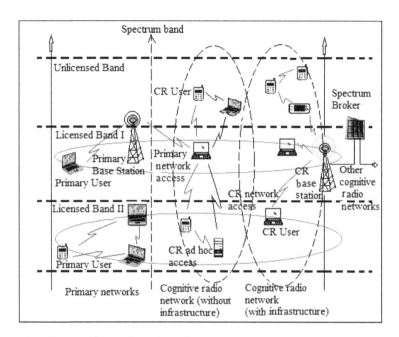

Figure 2. Network architecture of cognitive radio environment.

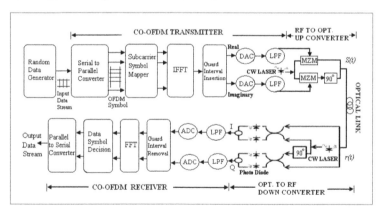

Figure 3. Block diagram of the CO-OFDM system.

(RTO) up-converter, *c.* optical link, *d.* Optical-To-RF (OTR) down converter, and *e.* OFDM receiver (Shieh, W. & C. Athaudage 2006).

In the CO-OFDM transmitter section, a Pseudo-Random Binary Sequence (PRBS) generator generates the random data stream. The generated serial data stream is then converted into a parallel data stream using a serial-to-parallel converter. The data are then mapped onto any constellation diagram using some modulation techniques such as QAM and PSK and then passed onto an IFFT modulator to obtain the OFDM spectrum. At the transmission block, both modulation and multiplexing are achieved digitally using an Inverse Fast Fourier Transform (IFFT). Guard Interval (GI) is inserted into the OFDM spectrum before transmission. The data are then directly up-converted to the optical frequency. A CW laser and two Mach–Zehnder modulators are used in this up-converter block. The laser line width is set at 0.15 MHz, with adjustable launch power, and the frequency of the carrier wave is set at 193.1 THz. The signal $S(t)$ is then propagated through the optical link and degraded due to fiber impairments. A coherent receiver with a local oscillator is used to down-convert the received data $r(t)$ to the RF domain. The receiver section uses two balanced receivers for photo-detection. It is followed by the OFDM demodulator section. The IFFT block is paired with a FFT block at the receiver section. The FFT converts the time domain signal into the frequency domain signal. Thus, all the points in the time–frequency grid of the OFDM system's operating band can be scanned without any additional hardware or computational process. The subcarrier frequencies are mathematically orthogonal over one OFDM symbol period, which is shown in the below mathematical expression; finally, the data are demodulated and sent to the detector and decoder for BER measurements. The above schematic diagram demonstrates a 10 Gbps coherent 512-subcarrier 16-QAM CO-OFDM system. Here, the transmitted signal $S(t)$ is represented as:

$$S(t) = \sum_{i=-\alpha}^{+\alpha} \sum_{k=1}^{N_{SC}} C_{ki} S_k (t - iT_s) \qquad (1)$$

$$S_k(t) = \Pi(t)\exp(j2\Pi f_k t) \qquad (2)$$

$$\Pi(t) = \begin{cases} 1, & (0 < t \leq T_s) \\ 0, & (t \leq 0, t > T_s) \end{cases} \qquad (3)$$

where C_{ki} is the ith information symbol at the kth subcarrier; S_k is the waveform of the kth subcarrier; N_{SC} is the number of subcarriers; f_k is the frequency of the subcarrier; and T_s is the symbol period. Each subcarrier can be matched or correlated by the filters used in the receiving end. The detected information symbol at the receiver $\acute{C}_{ki}$ is represented as:

$$\acute{C}_{ki} = \int_0^{T_s} r(t - iT_s) S^*_k dt = \int_0^{T_s} r(t - iT_s) exp(-j2\Pi f_k t) dt \qquad (4)$$

where $r(t)$ is the received time domain signal, which is non-overlapping and band limited. The orthogonality in OFDM derived from the correlation between any two subcarriers is represented by:

$$\delta_{ki} = \frac{1}{T_s} \int_0^{T_s} S_k S^*_l dt = \frac{1}{T_s} \int_0^{T_s} \exp\left(j2\Pi(f_k - f_l)t\right)dt$$
$$= \exp\left(j\Pi(f_k - f_l)T_s\right)\frac{\sin\left(\Pi(f_k - f_l)T_s\right)}{\Pi(f_k - f_l)T_s} \qquad (5)$$

If the relation $(f_k - f_l) = m\frac{1}{T_s}$ (6) is satisfied, then it can be proved that the two subcarriers are orthogonal to each other, and the frequencies spaced at multiple of inverse of the symbol rate can be restored by the matched filter from equation (4) without Inter-Carrier Interference (ICI) of the overlapped signal. Thus, the spectrum is utilized in an efficient manner (Shieh, W. 2008).

3 RESULTS AND DISCUSSION

The performance of the CO-OFDM system was analyzed for the channel length ranging from 50 Km to 450 Km using the 16QAM modulation technique. The signal quality was assessed based on the constellation diagram before reaching the threshold limit in the OFDM receiver.

Figure 4 shows the variation of Q-factor with input CW laser power. Initially, with the increase in the input power, the value of Q-factor increases, thus yielding a better signal strength at the receiver. However, after reaching a peak value, the signal performance starts to degrade due to the increased nonlinear interference effects when the input power exceeds 5dBm. Therefore, a moderate value of input laser power is favorable for a reliable transmission.

Figure 5 illustrates the variation of Q-factor with link distance. It is evident that as the transmission length increases, the value of Q factor decreases, and the signal quality degrades sharply from 150 Km. However, the distance can be

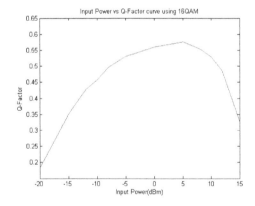

Figure 4. Variation of Q-factor with input power.

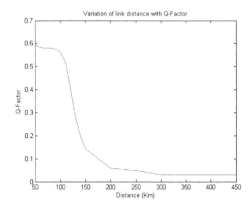

Figure 5. Variation of Q-factor with link distance.

significantly increased by increasing the transmitted power. Moreover, a minimum BER is obtained at the maximum Q-factor level from the simulation of this system model, which proves that Q-factor and BER are inversely proportional to each other.

4 CONCLUSION

In this paper, we demonstrate the transmission of the OFDM spectrum in a multi-user environment of the CR network. The 16QAM modulation technique is used for the simulation of a coherent optical OFDM system. The experimental results indicate that, in our system, the Q-factor varies inversely with the expected values of BER. Until a certain input power level, the output signal reception is better at the receiver, but further increase in the input power level leads to the deterioration of the received signal due to the nonlinear interference effects. Therefore, it is very important to identify the threshold limit of the input power before transmission. It can also be concluded that up to a link distance of 150 Km, the channel performance is better maintained.

Furthermore, the simulated results obtained in this paper agree quite satisfactorily with the available experimental data and with other theoretical works on the CO-OFDM system in the CR network.

REFERENCES

Bingham, J.A.C. (1990). Multicarrier modulation for data transmission: an idea whose time has come. *IEEE Communications Magazine, 28, 5,* 5–14.

Cho, Y., J. Kim, W. Wang, & C. Kang (2010). MIMO-OFDM Wireless Communications with MATLAB®. *Wiley-IEEE Pres, Singapore.* 1–431.

Cimini, L.J. (1985). Analysis and Simulation of a Digital Mobile Channel Using Orthogonal Frequency Division Multiplexing. *IEEE Transactions on Communications, 33, 7,* 665–675.

Federal Communications Commission. (2003). Facilitating Opportunities for Flexible, Efficient and Reliable Spectrum Use Employing Cognitive Radio Technologies, *03-322,* 1–53.

Ghosh, A., D.R. Wolter, J.G. Andrews, & R. Chen (2005). Broadband wireless access with WiMax/802.16: current performance benchmarks and future potential. *IEEE Communications Magazine. 43, 2,* 129–136.

Hong, X., Z. Chen, C.X. Wang, S.A. Vorobyov, & J.S. Thompson (2009). Cognitive radio networks. *IEEE Veh. Technol. Mag. 4, 4,* 76–84.

Jang, J. & K.B. Lee (2003). Transmit power adaptation for multiuser OFDM systems. *IEEE J. Select. Areas Commun. 21, 2,* 171–178.

Jansen, S.L., I. Morita, T.C.W. Schenk, N. Takeda, & H. Tanaka (2008). Coherent Optical 25.8-Gb/s OFDM transmission over 4160 Km SSMF. *J. Lightw. Technol. 26, 1,* 6–15.

Jin, X.Q., R.P. Giddings, & J.M. Tang (2009). Real-time transmission of 3 Gb/s 16-QAM encoded optical OFDM signals over 75 km SMFs with negative power penalties. *Opt. Exp. 17,* 14574–14585.

Jondral, Friedrich. K. (2005). Software Defined Radio—Basics and Evolution to Cognitive Radio. *EURASIP Journal on Wireless Communications and Networking.* 275–283.

Keller, T. & L. Hanzo (2000). Adaptive modulation techniques for duplex OFDM transmission. *IEEE Trans. Veh. Technol. 49, 5,* 1893–1906.

Mitola III, J. (2000a). Cognitive radio: an integrated agent architecture for software defined radio. Ph.D. dissertation. *Computer Communication System Laboratory, Department of Teleinformatics, Royal Institute of Technology (KTH), Stockholm, Sweden.* 1–313.

Mitola III, Joseph. (2004b). Software Radio Architecture: Object-Oriented Approaches to Wireless Systems Engineering. *John Wiley & Sons.*

Mitola, J. & G.Q. Maguire Jr. (1999). Cognitive radio: Making software radios more personal. *IEEE Personal Communications.6, 4,* 13–18.

Rhee, W. & J.M. Cioffi (2000). Increase in capacity of multiuser OFDM system using dynamic subchannel allocation. *In VTC Spring. 2,* 1085–1089.

Shen, Z., J.G. Andrews, & B.L. Evans (2005). Adaptive resource allocation in multiuser OFDM systems with proportional rate constraints. *IEEE Trans. Wireless Commun. 4, 8,* 2726–2737.

Shieh, W. & C. Athaudage (2006). Coherent optical orthogonal frequency division multiplexing. *Electron. Lett. 42,* 587–589.

Shieh, W., H. Bao, & Y. Tang (2008). Coherent Optical OFDM: theory and design. *Optics Express, OSA. 16, 2.* 841–859.

Srinivasa, S. & S.A. Jafar (2007). Soft sensing and optimal power control doe cognitive radio. *In Proc. IEEE Global Telecommun. Conf.* 1380–1384.

Tang, H. (2005). Some Physical layer issues of wide-band cognitive radio system. *Proceedings, IEEE DySPAN,* 151–159.

Van Nee, R. & R. Prasad (2000). OFDM for wireless multimedia communications. *Artech House, Massachusetts, USA.* 1–280.

Wang, Z. & G.B. Giannakis (2000). Wireless multicarrier communications. *IEEE Signal Processing Magazine, 17, 3,* 29–48.

WiMedia (2005). Multiband OFDM physical layer specification. *WiMedia Alliance.* 1–204.

Wong, C.Y., R.S. Cheng, K.B. Lataief, & R.D. Murch. (1999). Multiuser OFDM with adaptive subcarrier, bit, and power allocation. *IEEE J. Select. Areas Commun. 17, 10,* 1747–1758.

Xiao, Lu., Ping. Wang, Dusit. Niyato, & Ekram. Hossain (2014). Dynamic Spectrum Access in Cognitive Radio Networks with RF Energy Harvesting. *IEEE Wireless Communication. 21, 3,* 102–110.

Computer, Communication and Electrical Technology – Guha, Chakraborty & Dutta (Eds)
© 2017 Taylor & Francis Group, ISBN 978-1-138-03157-9

High temperature electrical properties of thin Zirconium di-oxide films on ZnO/Si layers

S.K. Nandi

Department of Physics, Rishi Bankim Chandra College, Naihati, 24-Parganas (North), West Bengal, India

ABSTRACT: Zirconium di-oxide (ZrO_2) films have been deposited on ZnO/Si substrate by microwave plasma cavity discharge system. Using Metal Insulator Semiconductor (MIS) capacitor structures, the reliability and the leakage current characteristics of ZrO_2 films have been studied at room and also high temperature. Poole-Frenkel current conduction mechanism is found to dominate at high temperature.

1 INTRODUCTION

Zinc Oxide (ZnO) is an n-type semiconductor with 3.3 eV direct band gap at room temperature and is a potential candidate for short-wavelength (UV/violet/blue) optical devices such as LED or laser diodes etc. ZnO crystallites show wurtzite structure, i.e. a complete hcp-lattice with oxygen atoms can be inserted into the zinc hcp-lattice and leads to the formation of a structure of close-packed atom planes stacked on each other in the c-direction, worked by Jiwei et al. 2000. ZnO have been used in many applications such as piezo-electric transducers, optoelectronics and thin-film transistors shown by Srikant & Clarke 1998. As a consequence, there is a renewed interest in the study of the properties of ZnO relevant for electronic device applications. In addition, to continue the scaling trend of MOS technology and the development of MOSFETs, requires new materials having a high dielectric constant (high-ε) with a low leakage current than SiO_2. Among many high dielectric constant insulators ZrO_2 has high dielectric constant (15–22), wide band gap (4.6–7.8 eV) worked by Balog et al. 1977, high breakdown field (15–20 MV cm^{-1}) and superior thermal stability.

2 DEVICE FABRICATION

The Al-ZrO_2-ZnO/Si Metal-Insulator-Semiconductor (MIS) capacitors used in this study were deposited on undoped ZnO (100 nm) thin film deposited on n-Si (100) at 450°C using rf magnetron sputtering of sintered commercial 2-inch ZnO target (Purity~99.999%). Zirconium tetra-tert butoxide [Zr(OC(CH$_3$)$_3$)$_4$] was used as source material for the deposition of ZrO_2 thin films.

[Zr(OC(CH$_3$)$_3$)$_4$] was vaporized from a bubbler kept at room temperature and was carried to the quartz process/deposition chamber of the microwave (700 W, 2.45 GHz) cavity discharge system through a gas line. The process chamber was maintained at a pressure of 500 mTorr and temperature of 150°C during deposition.

O_2 as a carrier gas. For metal contact, Al was evaporated on an area 1.96×10^{-3} cm^2 through a shadow mask.

The plot of deposition rate of ZrO_2 film is shown in Figure 1. The deposition behavior under the present experimental condition may be described by a parabolic rate equation

$$t_{ox}^2 = C \cdot t \qquad (1)$$

where, t_{ox} is the thickness of oxide grown in time t and C is the parabolic rate constant. The calculated value of C is 8.5×10^3 Å/min from the solid line as shown in Figure 1.

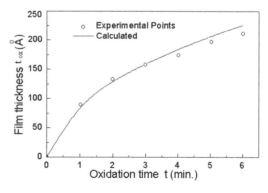

Figure 1. ZrO_2 film thickness as a function of time for microwave plasma oxidation.

3 RESULT AND DISCUSSIONS

The surface structure of the ZnO/Si films can be observed in Figure 2. By Scanning electron microscopy, is shown that structure is columnar. The chemical compositions of ZrO_2 (9 nm) films on ZnO/Si films were investigated by X-ray Photoelectron Spectroscopy (XPS), in Figure 3, the broad energy XPS spectra of deposited ZrO_2 films on ZnO/Si layers. High-resolution recording of spectrum shows the XPS peak positions of Si 2p (100 eV), Si 2 s (149 eV), Zr 3d (183 eV), C 1 s (284.5 eV), O 1 s (532), and Zn 2p (1022 eV and 1045 eV) for deposited ZrO_2 on ZnO/Si. Here all the binding energies were corrected for sample with reference to the C 1 s line around at 284.5 eV. By the high-resolution XPS spectra of Zr 3d it is also observed that the core-level spectrum of Zr 3d at binding energies 182.3 eV for Zr $3d_{5/2}$ and 184.6 eV for Zr $3d_{3/2}$. The peak of Zr $3d_{5/2}$ at 182.3 eV is a typical characteristic of the Zr^{4+} in ZrO_2 observed by Sun et al. 2000.

High frequency (1 MHz) current-voltage (J-V) in Figure 4 and conductance-voltage (G-V) Figure 5 characteristics of the deposited ZrO_2 films were measured. The values of interface trap density (D_{it}) and fixed oxide charge density (Q_f/q) were found to be 1.3×10^{12} cm^{-2} eV^{-1} and 2.4×10^{12} cm^{-2}, respectively. Leakage current density through MIS capacitors, as a function of applied gate bias at different high temperature also seen in Figure 5, which at high temperature current conduction follows the Poole-Frenkel mechanism in the range 1.1 MV/cm to 1.6 MV/cm.

It is also observed that in the G-V curves shift towards the left as the device temperature is increased from 27°C (RT) to 200°C. The flat band voltage (V_{fb}) is found to increase (more negative) with increasing device temperature as shown in Figure 5 (study temperature range was 27–200°C), which indicates that generation of positive charges

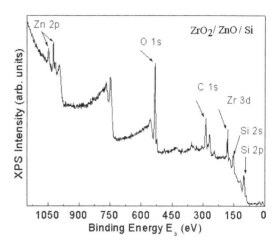

Figure 3. Wide scan XPS spectrum of ZrO_2/ZnO/Si film.

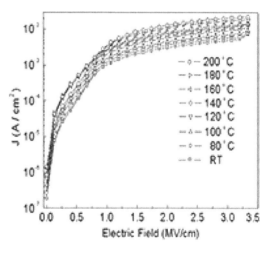

Figure 4. Current density J as a function of gate voltage across the gate dielectric.

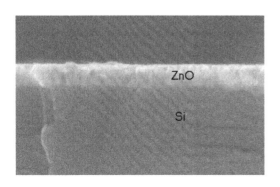

Figure 2. SEM view of the rf sputtered ZnO film deposited on Si.

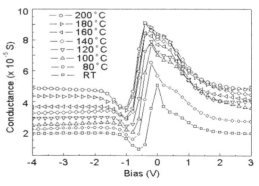

Figure 5. Normalized high frequency (1 MHz.) G-V characteristics at different temperature.

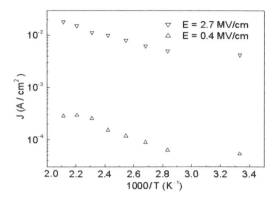

Figure 6.   Temperature dependence current.

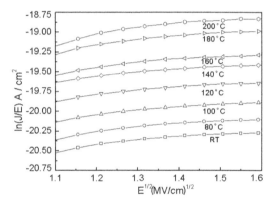

Figure 7.   The Poole-Frenkel (PF) plot.

in the gate oxide. This may be due to the presence of hole trapping centers in the films. The gate current density (J) as a function of the voltage across the gate (V) for the ZrO_2/ZnO/Si layer is shown in Figure 4, from room temperature to 200°C. It is observed that the current density of the films is strongly temperature dependent at low field, i.e., for E < 1 MV/cm, while its temperature dependence is much weaker at higher voltage. It is shown that the temperature dependence of the conductance current at two electric fields, 0.4 MV/cm and 2.7 MV/cm, respectively in Figure 6. This is also known as Arrhenius plot. The Poole-Frenkel (PF) emission, mechanism is used to explain the basic conduction process in this insulators. When the top electrode (gate) is positively biased, i.e. the holes are injected from the gate, the leakage current in the films can be well fitted by ln(J/E) vs. $E^{1/2}$ plot, indicating that the conduction is PF mechanism

$$\ln[J/E] = \frac{\beta_{PF}}{\xi \cdot k \cdot T} \cdot \sqrt{E} + \left[\ln C_t - \frac{q\Phi_b}{\xi \cdot k.T} \right] \quad (2)$$

where $\beta_{PF} = (q^3 / 4 \pi \varepsilon_0 \varepsilon_i)^{1/2}$

A PF plot of the as-deposited film is shown in Figure 7. It is observed that the experimental plot of $\ln(J_{PF}/E)$ vs. $\sqrt{E}$ is a straight line with a positive slope in the range 1.1 MV/cm to 1.6 MV/cm, which follows the P-F mechanism shown by Yassine et al. 1999.

4   CONCLUSIONS

In conclusion, it is observed that leakage current density through MIS capacitors, as a function of applied gate bias at different high temperature, high temperature current conduction follows the Poole-Frenkel mechanism in the range 1.1 MV/cm to 1.6 MV/cm. It is observed that the electrical properties of ZrO_2 films on ZnO/Si substrates are good and the films may be used as gate dielectric for future electronic applications.

REFERENCES

Balog, M., Schieber, M., Michman, M. & Patai S. (1977). The chemical vapour deposition and characterization of ZrO_2 films from organometallic compounds. *Thin Solid Films.* 47: 109–120.

Jiwei, Z., Liangying, Z. & Xi, Y. (2000). The dielectric properties and optical propagation loss of c-axis oriented ZnO thin films deposited by sol-gel process. *Ceramics Int.* 26: 883–885.

Srikant, V. & Clarke, D.R. (1998). On the optical band gap of zinc oxide. *J. Appl. Phys.* 83: 5447–5451.

Sun, Y. M., Lozano, J., Ho, H., Park, H.J. & Veldman S. (2000). Interfacial silicon oxide formation during synthesis of ZrO_2 on Si (100). *Appl. Surf. Sci.* 61: 115–122.

Yassine, A., Nariman, H. E. & Olasupo K. (1999). Field and temperature dependence of TDDB of ultrathin gate oxide. *IEEE Electron. Dev. Lett.* 20: 390–392.

Computer, Communication and Electrical Technology – Guha, Chakraborty & Dutta (Eds)
© 2017 Taylor & Francis Group, ISBN 978-1-138-03157-9

Realization of all-optical frequency-encoded dibit-based OR and NOR logic gates with simulated verification

Partha Pratim Sarkar, Smita Hazra & Bitan Ghosh
Department of Electronics and Communication Engineering, University Institute of Technology,
The University of Burdwan, Burdwan, West Bengal, India

Sankar Narayan Patra
Department of Instrumentation Science, Jadavpur University, Kolkata, West Bengal, India

Sourangshu Mukhopadhyay
Department of Physics, The University of Burdwan, Burdwan, West Bengal, India

ABSTRACT: Optics is recognized as a powerful and prospective candidate for the realization of logic devices, digital optical systems for communication, and computation due to its super-fast speed. All-optical frequency-encoded dibit representation techniques conduct real-time operations with ultra high switching speed rather than all conventional opto-electronic and optical switches. So, a high degree of parallelism can be exploited from these proposed systems. Here, the authors have proposed all-optical frequency-encoded dibit-based OR and NOR logic gates using the optical switches like reflected semiconductor optical amplifiers and add/drop multiplexers.

1 INTRODUCTION

For extremely fast information processing, a photon can be used as a more appropriate information transporter than an electron. All-optical logic gates are also very popular for implementing an all-optical computer and data processor (Ghosh & Mukhopadhyay, 2013). Many all-optical logic gates (Sarkar, Satpati & Mukhopadhyay, 2014) have been developed using optics like flip-flop (Dutta & Mukhopadhyay, 2011), bi-stable multivibrators, latches (Sarkar, Ghosh, Patra & Mukhopadhyay, 2016), etc. A Semiconductor Optical Amplifier (SOA) has been established as a very promising optical device for conducting many all-optical logical operations. Different types of encoding principles such as spatial encoding, polarization encoding (Guo & Connelly, 2007), frequency encoding (Sarkar & Mukhopadhyay, 2014), intensity encoding and phase encoding (Chandra et al. 2014), etc. are needed for implementing the all-optical logic and arithmetic devices. The frequency encoding principle is the most reliable one among all other encoding principles as frequency is the basic characteristic of light and it remains unaltered and unchanged under reflection, refraction, absorption, etc.

In case of data communication, two different states of information can be represented by two different frequencies. Here, the presence of a specific frequency of light is treated as "1" logic state and other specific frequency is treated as "0" logic state. Here, the authors have used two different switches. One of them is a Reflected Semiconductor Optical Amplifier (RSOA) (Wang & Foster, 2012). This is a switch where if a weak probe beam light of wavelength, say $\lambda_1 = 1540$ nm and a strong pump beam of wave length $\lambda_2 = 1550$ nm are injected to the input terminals. The strong pump beam transfers its total power to the weak probe beam and then the weak probe beam becomes stronger and comes out to the output terminal; it thus acts as a perfect wavelength converter. On the other side, another switch, i.e., an optical add/drop multiplexer (Sarkar, Satpati & Mukhopadhyay, 2013) is a frequency-selecting network. It is tuned with a particular biasing current and it reflects a particular frequency of light through it and passes all the other frequencies of light.

In this paper, the authors have proposed the dibit representation technique. In dibit representation, to represent a digit, two consecutive bit positions are considered. These two positions may be considered as two different frequencies, say $\upsilon_1 = 195$ THz and $\upsilon_2 = 193$ THz. Now, let us consider an optical frequency of light beam $\upsilon_1 =$ digital logic state "0" and $\upsilon_2 =$ digital logic state "1". Then, in dibit representation, this digital value "0" is represented as [0] [1] and digital value "1" is represented as [1] [0]. This means that the presence of the two frequencies side by side, $\upsilon_1\upsilon_2$, represents the logic state "0" and $\upsilon_2\upsilon_1$ does the same as the logic state "1". The dibit representation (Mukhopadhyay, 1992) in optics for logical operation was first proposed by Prof. S. Mukhopadhyay. Here, the authors have proposed an alternative approach to realize the optical OR logic gate with the help of RSOA and ADM. Exploiting the dibit representation technique and just interchanging the output bit positions, we may realize the all-optical frequency-encoded dibit NOR logic gate, which is an important member of universal logic families.

2 SCHEME OF REALIZATION OF FREQUENCY-ENCODED DIBIT-BASED OR LOGIC GATE

Here, the input channels are considered as "A" and "B" for a two input OR logic gate. These two channels, i.e., "A" and "B", are subdivided into "A'" and "A''" and "B'" and "B''", respectively, for the representing dibit inputs. Here, one optical light beam of frequency υ_1 is considered as the digital logic state "0," i.e., dibit $[\upsilon_1][\upsilon_2]$ or [0][1] and υ_2 is considered as the digital logic state "1," i.e., dibit $[\upsilon_2][\upsilon_1]$ or [1][0]. Now, the dibit-based two inputs OR logic gate is shown in Figure 1.

For realizing the dibit-based OR gate, at first, υ_1 and υ_2 frequency are given to the input channels of A', A'' and B', B'' simultaneously. So, ADM_1 passes υ_1 frequency to $RSOA_1$ and $RSOA_2$ as a pump beam, and due to presence of a fixed probe beam of υ_2 frequency at $RSOA_2$, υ_2 frequency is the output of $RSOA_2$. This is used as a pump beam of $RSOA_3$. Also, from the B' terminal, υ_1 frequency is passed by ADM_2, and it goes to the $RSOA_3$ as a probe beam. Now, because of the absence of the probe

beam at $RSOA_1$, it may come from ADM_2 as the dropped frequency is absent. Hence, $RSOA_1$ doesn't work. So, υ_1 frequency is transmitted from $RSOA_3$ and comes to the output Y'. After getting the input from the A'' terminal, ADM_3 reflects υ_2 frequency to $RSOA_6$ as a probe beam. So, ADM_4 and $RSOA_4$ do not work. Next, from the B'' terminal, υ_2 frequency is reflected by ADM_5 and acts as a pump beam of $RSOA_5$ as there is a fixed probe beam of υ_1 frequency. Hence, υ_1 frequency is produced at the output and comes to $RSOA_6$ as a pump beam. So, υ_2 frequency comes from $RSOA_6$ to ADM_6 and is passed by ADM_6 because there is no biasing current in it. Finally, we get υ_2 frequency at the Y'' terminal. With the dibit inputs of $[\upsilon_1][\upsilon_2]$ and $[\upsilon_1][\upsilon_2]$, we get $[\upsilon_1][\upsilon_2]$ from the output terminals Y' and Y''.

Now, we apply $[\upsilon_1][\upsilon_2]$ at the A terminal but $[\upsilon_2][\upsilon_1]$ at the B terminal. So, υ_2 frequency is reflected by ADM_2 and hence, there is no probe beam at $RSOA_3$. So, $RSOA_3$ does not work, but due to the presence of the probe beam of υ_2 frequency at $RSOA_1$, υ_2 frequency is created by $RSOA_1$ and is transferred to the output Y'. Next, since υ_1 frequency is applied at the B'' terminal, ADM_5 passes it to ADM_4. Due to the absence of a biasing current at ADM_4, υ_1 frequency comes from ADM_4 and goes to $RSOA_4$. Now, due to presence of a fixed υ_2 frequency as a pump beam at $RSOA_4$, υ_1 frequency comes out from $RSOA_4$. This υ_1 frequency of light beam is produced at the Y'' terminal. So, we get $[\upsilon_2][\upsilon_1]$ from the output terminals Y' and Y'', respectively. This way, we can also get $[\upsilon_2][\upsilon_1]$ from the output terminals Y' and Y'' for the dibit input combinations $[\upsilon_2][\upsilon_1]$, $[\upsilon_1][\upsilon_2]$ and $[\upsilon_2][\upsilon_1]$, and $[\upsilon_2][\upsilon_1]$ at the A and B terminals, respectively.

Now, if we apply $[\upsilon_1][\upsilon_2]$ at the A terminal but $[\upsilon_2][\upsilon_1]$ is applied at the B terminal. So, υ_2 frequency

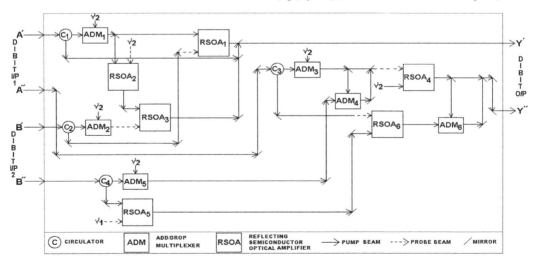

Figure 1.   Schematic diagram of a frequency-encoded optical dibit-based OR logic gate.

is reflected by ADM_2 and hence, there is no probe beam at $RSOA_3$. So, $RSOA_3$ does not work, but due to presence of a probe beam of υ_2 frequency at $RSOA_1$, υ_2 frequency is created by $RSOA_1$ and is transferred to the output Y'. Next, since υ_1 frequency is applied at the B" terminal, ADM_5 passes it to ADM_4. However, due to the absence of a biasing current at ADM_4, υ_1 frequency comes from ADM_4 and goes to $RSOA_4$. Now, due to the presence of a fixed υ_2 frequency as a pump beam at $RSOA_4$, other RSOA blocks are not functioning. Hence, υ_1 frequency is created from $RSOA_4$ and υ_1 frequency is produced at the Y" terminal getting $[\upsilon_2][\upsilon_1]$ from the output terminals Y' and Y", respectively. This way, we can also get $[\upsilon_2][\upsilon_1]$ from the output terminals Y' and Y" for the dibit input combinations $[\upsilon_2][\upsilon_1]$, $[\upsilon_1][\upsilon_2]$ and $[\upsilon_2][\upsilon_1]$, and $[\upsilon_2][\upsilon_1]$ at the A and B terminals, respectively.

2.1 Simulation process of frequency-encoded dibit-based OR gate

Here, for the simulation of the frequency-encoded optical dibit OR gate using Reflected Semiconductor Optical Amplifier (RSOA) and Add/Drop Multiplexer (ADM), Simulink tools of MATLAB (R2008a) software have been used. Two input units (DIBIT_INPUT 1 and DIBIT_INPUT 2) are used for applying dibit inputs. We can get the output result from the "Dibit Output" unit consisting of two consecutive bit positions "1st Bit" and "2nd Bit".

Now, in the case of a mathematical model of RSOA, shown in Figure 2, the RSOA blocks are properly programmed in "C" language for choosing the proper output at the output terminal. These blocks have two inputs: one is "pr" (probe beam) and the other is "pm" (pump beam). The output is termed as "y". When both the inputs in the form of a pump beam and a probe beam are available, the probe beam comes at the output of the RSOA. At the input terminal of the RSOA, if the pump beam and probe beam are considered as "8 peta Hz" = υ_2 = digital logic state "1" and "3 peta Hz" = υ_1 = digital logic state "0," respectively, then at the output terminal of these blocks, we get "3 peta Hz" = υ_1 = digital logic state "0". If the value of the pump beam and probe beam are altered, the value of the output changes accordingly. Similarly, another switch, i.e., an Add/Drop Multiplexer (ADM), has been simulated in such

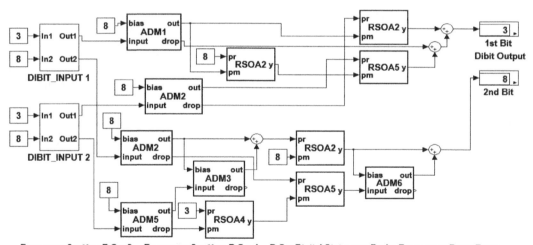

Frequency $3 = V_1 = D.S = 0$, Frequency $8 = V_2 = D.S = 1$, D.S.= Digital State, pr = Probe Beam, pm = Pump Beam
ADM = Add/Drop Multiplexure, RSOA = Reflected Semiconductor Optical Amplifier

Figure 2. Functional model of the all-optical frequency-encoded dibit OR gate.

Table 1. Truth Table of dibit-based optical OR gate.

1st DIBIT i/p		1st Digital i/p	2nd DIBIT i/p		2nd Digital i/p	DIBIT o/p		Digital o/p
A'	A"	A	B'	B"	B	Y'	Y"	Y
$\upsilon_1[0]$	$\upsilon_2[1]$	0	$\upsilon_1[0]$	$\upsilon_2[1]$	0	$\upsilon_1[0]$	$\upsilon_2[1]$	0
$\upsilon_1[0]$	$\upsilon_2[1]$	0	$\upsilon_2[1]$	$\upsilon_1[0]$	1	$\upsilon_2[1]$	$\upsilon_1[0]$	1
$\upsilon_2[1]$	$\upsilon_1[0]$	1	$\upsilon_1[0]$	$\upsilon_2[1]$	0	$\upsilon_2[1]$	$\upsilon_1[0]$	1
$\upsilon_2[1]$	$\upsilon_1[0]$	1	$\upsilon_2[1]$	$\upsilon_1[0]$	1	$\upsilon_2[1]$	$\upsilon_1[0]$	1

155

a way that if a series of frequencies are given at the input of the ADM, it drops a particular frequency depending upon the biasing current and passes all other frequencies. The frequency of the dropped terminal changes if the biasing current changes, Here, if at the input terminal of the ADM, the available frequency is "8 peta Hz" = υ_2 = digital logic state "1" and "3 peta Hz" = υ_1 = digital logic state "0", for a particular biasing current, this block drops one frequency and passes another frequency. The exact opposite appens if the value of the biasing current changes. Now, these two blocks have been connected maintaining the similarity of the all optical dibit-based OR logic gate's block diagram (shown in Fig. 1). Now, at the dibit input terminal of the simulated model, if the digital logic state "1" = [υ_2] [υ_1] = say [8 peta Hz] [3 peta Hz] and the digital logic state "0" = [υ_1] [υ_2] = say [3 peta Hz] [8 peta Hz] are taken, we get the digital logic state "1" = [υ_2] [υ_1] = say [8 peta Hz] [3 peta Hz] at the output terminal of this simulated block. Similarly, the output repeats the same value for the combination of the digital logic state "0" = [υ_1] [υ_2] = say [3 peta Hz] [8 peta Hz] with digital logic state "1" = [υ_2] [υ_1] = say [8 peta Hz] [3 peta Hz] and digital logic state "1" = [υ_2] [υ_1] = say [8 peta Hz] [3 peta Hz] with digital logic state "1" = [υ_2] [υ_1] = say [8 peta Hz] [3 peta Hz]. But, at the output, we get the digital logic state "0" = [υ_1] [υ_2] = say [3 peta Hz] [8 peta Hz] for the combination of dibit inputs, digital logic state "0" = [υ_1] [υ_2] = say [3 peta Hz] [8 peta Hz] and digital logic state "0" = [υ_1] [υ_2] = say [3 peta Hz] [8 peta Hz].

So, finally, it is shown that both the block diagram of the all-optical frequency-encoded dibit-based OR logic gate and the simulative functional model of it fully support the truth table (shown in Table 1).

3 SCHEME OF REALIZATION OF FREQUENCY-ENCODED DIBIT-BASED NOR GATE

Similarly, for the realization of a dibit-based NOR logic gate, we can get the NOR output ("Y'" and "Y''") just by interchanging the bit positions of the OR output ('Y_0'" and 'Y_0'") for the two dibit inputs "A'", "A''" and "B'", "B''" as shown in Figure 3. It provides the advantage of implementing the dibit NOR gate using the dibit OR gate without the use of any extra circuits.

3.1 Simulation process of frequency-encoded dibit-based NOR gate

Similarly, for a frequency-encoded optical dibit NOR gate, Simulink tools of MATLAB (R2008a) software have been used, which is shown in Figure 4. Here, all the model units are the same with the functional model of frequency-encoded optical dibit OR gate, except the "Dibit Output" unit. We can get the NOR output just by interchanging the bit positions of the OR output. Now, if the digital logic state "0" = [υ_1] [υ_2] = say [3 peta Hz] [8 peta Hz] and digital logic state "1" = [υ_2] [υ_1] = say [8 peta Hz] [3 peta Hz] are applied at the dibit input terminal of the simulated model, we get the digital logic state "0" = [υ_1] [υ_2] = say [3 peta Hz] [8 peta Hz] at the output terminal of this simulated block. Similarly, the output repeats the same value for the

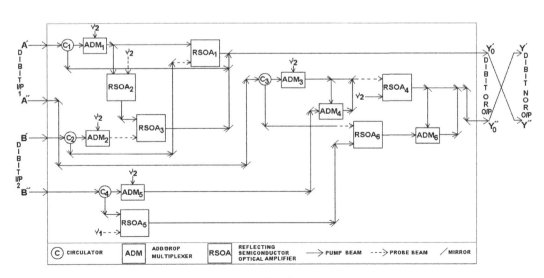

Figure 3. Schematic diagram of frequency-encoded optical dibit-based NOR gate.

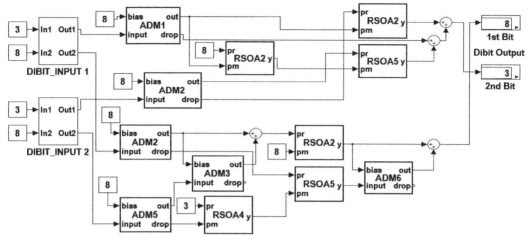

Frequency 3 = ν₁ = D.S = 0,   Frequency 8 = ν₂ = D.S = 1,   D.S.= Digital State, pr = Probe Beam, pm = Pump Beam
ADM = Add/Drop Multiplexure, RSOA = Reflected Semiconductor Optical Amplifier

Figure 4.   Functional model of the all-optical dibit NOR gate.

Table 2.   Truth table of dibit-based optical NOR gate.

1st DIBIT i/p		1st Digital i/p	2nd DIBIT i/p		2nd Digital i/p	DIBIT o/p		Digital o/p
A'	A"	A	B'	B"	B	Y'	Y"	Y
$\upsilon_1[0]$	$\upsilon_2[1]$	0	$\upsilon_1[0]$	$\upsilon_2[1]$	0	$\upsilon_2[1]$	$\upsilon_1[0]$	1
$\upsilon_1[0]$	$\upsilon_2[1]$	0	$\upsilon_2[1]$	$\upsilon_1[0]$	1	$\upsilon_1[0]$	$\upsilon_2[1]$	0
$\upsilon_2[1]$	$\upsilon_1[0]$	1	$\upsilon_1[0]$	$\upsilon_2[1]$	0	$\upsilon_1[0]$	$\upsilon_2[1]$	0
$\upsilon_2[1]$	$\upsilon_1[0]$	1	$\upsilon_2[1]$	$\upsilon_1[0]$	1	$\upsilon_1[0]$	$\upsilon_2[1]$	0

combination of the digital logic state "1" = [υ_2] [υ_1] = say [8 peta Hz] [3 peta Hz] with the digital logic state "0" = [υ_1] [υ_2] = say [3 peta Hz] [8 peta Hz] and digital logic state "1" = [υ_2] [υ_1] = say [8 peta Hz] [3 peta Hz] with digital logic state "1" = [υ_2] [υ_1] = say [8 peta Hz] [3 peta Hz]. But, at the output, we get the digital logic state "1" = [υ_2] [υ_1] = say [8 peta Hz] [3 peta Hz] for the combination of dibit inputs, digital logic state "0" = [υ_1] [υ_2] = say [3 peta Hz] [8 peta Hz] and digital logic state "0" = [υ_1] [υ_2] = say [3 peta Hz] [8 peta Hz].

So, from the above realization, the block diagram of the all-optical frequency-encoded dibit-based NOR logic gate as well as the simulative functional model of it entirely support the truth table of the dibit-based optical NOR gate, which is shown in Table 2.

4   CONCLUSIONS

This dibit representation technique is very much accurate and reliable one because it supports for decline of bit error problem by increasing high signal to noise ratio. It can also expect high degree of parallelism. The truth table satisfies these dibit based OR and NOR logic gate. Therefore its performance can directly be utilized for developing and verifying the performances of different logic devices based on frequency encoding principle. Also by using this dibit representation one can implement for other optical operations like flip-flops, multivibrators, latch etc.

REFERENCES

Chandra, S.K. et al. (2014). All-optical phase encoded 4-to-1 phase multiplexer using four wave mixing in semiconductor optical amplifier. *Optik—International Journal for Light and Electron Optics* 125(23): 6953–6957.

Dutta, S. & Mukhopadhyay, S. (2011). Alternating approach of implementing frequency encoded all-optical logic gates and flipflop using semiconductor optical amplifier. *Optik—Int. J. Light Electron Opt.* 122(2): 125–127.

Ghosh, B. & Mukhopadhyay, S. (2013). A novel realization of all-optical dibit represented frequency encoded Boolean and quaternary inverters without switching device. *Optik—International Journal for Light and Electron Optics* 124(21): 4813–4815.

Guo, L.Q. & Connelly, M.J. (2007). A poincare approach to investigate nonlinear polarization rotation in semiconductor optical amplifiers and its applications to all optical wavelength conversion. *Proc. of SPIE* 6783(678): 1–5.

Mukhopadhyay, S. (1992). Binary optical data subtraction by using a ternary dibit representation technique in optical arithmetic problems. *Applied Optics* 31(23): 4622–4623.

Sarkar, P.P., Ghosh, B., Patra, S.N. & Mukhopadhyay, S. (2016). A simulative study of all optical frequency encoded dibit based universal NAND and NOR logic gates using reflective semiconductor optical amplifier and add/drop multiplexer. *Opticheskii Zhurnal* 83(4), pp. 80–87.

Sarkar, P.P., Satpati, B. & Mukhopadhyay, S. (2013). New simulative studies on performance of semiconductor optical amplifier based optical switches like frequency converter and add-drop multiplexer for optical data processors. *Journal of Optics* 42(4): 360–366.

Sarkar, P.P., Satpati, B. & Mukhopadhyay, S. (2014). Analytical and simulative studies on optical NOR and controlled NOR logic gates with semiconductor optical amplifier. *Optik—International Journal for Light and Electron Optics* 125(3): 1333–1336.

Sarkar, P.P. & Mukhopadhyay, S. (2014). All optical frequency encoded NAND logic operation along with the simulated result. *Journal of Optics* 43(3): 177–182.

Wang, K.Y. & Foster, A.C. (2012). Ultralow power continuous-wave frequency conversion in hydrogenated amorphous silicon waveguides. *Optics letters* 37(8): 1331–1333.

Computer, Communication and Electrical Technology – Guha, Chakraborty & Dutta (Eds)
© 2017 Taylor & Francis Group, ISBN 978-1-138-03157-9

Analysis of optical gain in a Tin-incorporated group IV alloy-based transistor laser for a mid-infrared application

Ravi Ranjan & Mukul Kr. Das

UGC-SAP Research Laboratory, Department of Electronics Engineering, Indian Institute of Technology (Indian School of Mines), Dhanbad, India

ABSTRACT: In this work, we analyze the polarization-dependent TE mode optical gain in Si-$Si_{0.12}Ge_{1-y}Sn_y$—$Si_{0.11}Ge_{0.73}Sn_{0.16}$ based Transistor Laser (TL) with intrinsic i-$Ge_{1-x}Sn_x$ single Quantum Well (QW) in the base for different Sn concentrations in the well and barrier. The electronic band structure and wave function of Γ-conduction band, Heavy Hole (HH) valance band are also shown during analysis. Results show that TL works in the mid-infrared (2–4 μm) region. This analysis is helpful to develop a low-cost optoelectronics system based on group IV photonics.

1 INTRODUCTION

The potential applications of a low-cost mid-infrared device motivate many researchers to work on devices based on group IV materials (Si and Ge) and their alloys (Soref 2010). Yet, the indirect bandgap nature of these materials prevents them to be used as active light-emitting devices. However, an alloy of the semimetal cubic α-Sn and Ge produces a tuneable direct energy gap semiconductor (Kouvetakis et al. 2006). The band engineering of Ge with α-Sn provides an opportunity for all group-IV direct bandgap light-emitter devices in the mid-infrared region (Kouvetakis et al. 2006). Several research papers have been reported on GeSn/SiGeSn-based mid-infrared devices such as lasers (Chang et al. 2010), modulators (Moontragoon et al. 2010), detectors (Werner et al. 2011), etc. in recent years.

In this context, the authors proposed a simple theoretical model for Tin-incorporated group IV material-based mid-infrared TL (2.76 μm) and calculated its optical parameters like the differential optical gain, transparency and threshold carrier density (Ranjan & Das 2016). A TL is one of the optoelectronic devices that is a combination of a transistor and laser, invented by M. Feng and N. Holonyak (Holonyak & Feng 2006). In a TL, a quantum well is inserted in the base region of the heterojunction bipolar transistor, which works as an active layer and causes a laser action, so it can take an electrical input and simultaneously give an optical as well as electrical output (Feng et al. 2005).

The motivation for the present work comes from Soref (2010), where the author illustrated the mid-infrared window of the waveguide core as well as the cladding materials in 2–4 μm wavelength for on-chip CMOS optoelectronics systems based on group IV photonics. So, simultaneously, it is also required to develop an optical source that can be integrated with other circuit components on a single chip for the application in the 2–4 μm range. In this work, we analyzed the optical gain of group IV based TL for the 2–4 μm mid-infrared wavelengths. The analysis is required for a better understanding of the effect of the Sn content on the TL optical properties as well as to realize all of the group IV optoelectronics system.

2 THEORETICAL MODEL AND ANALYSIS

The schematic structure of a TL with variable Sn concentration in the well and barrier considered in our analysis is shown in Figure 1. The n-type Si material forms an emitter, the p-type $Si_{0.12}Ge_{1-y}Sn_y$ a base, and n-type $Si_{0.11}Ge_{0.73}Sn_{0.16}$ a collector. An intrinsic $Ge_{1-x}Sn_x$ QW is inserted in the $Si_{0.12}Ge_{1-y}Sn_y$ base for laser action. The collector layer is lattice matched with a strain-relaxed $Ge_{0.87}Sn_{0.13}$ buffer layer, which is used for subsequent growth of the barrier and well. The width of the QW and barrier is less than the critical thickness. The band structure of Γ and L valley in the conduction band and HH and the Light Hole (LH) valence band in the well and barrier is calculated using the model solid theory (Walle 1989). The calculated band profile is used to find Eigen energies and corresponding wave functions in Γ conduction band and HH band by solving the following Schrödinger equation with effective mass approximation (Chuang, 2009).

$$\left[\frac{-\hbar^2}{2}\frac{\partial}{\partial z}\frac{1}{m_P}\frac{\partial}{\partial z} + \frac{\hbar^2 k_t^2}{2m_P} + V_P(z)\right]\psi = E_P\psi \qquad (1)$$

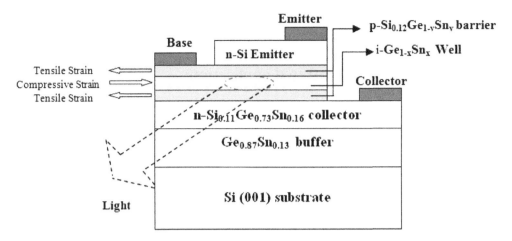

Figure 1. Schematic structure of *npn* Si-Si$_{0.12}$Ge$_{1-y}$Sn$_y$—Si$_{0.11}$Ge$_{0.73}$Sn$_{0.16}$ based Transistor Laser (TL) with strain-balanced *i*-Ge$_{1-x}$Sn$_x$ single Quantum Well (QW) in the base.

where z is the position variable; Ψ is the wave function; E_p and m_p are the Eigen energy and effective mass in different bands, respectively; and k_t is the wave vector perpendicular to k_z (growth axis). V_p is the potential profiles of different bands, which includes the strain effect. The optical gain is estimated at the band edge in the QW with the help of the Fermi golden rule (Chuang 2009). Eigen energies and wave function are taken along the z-direction in the conduction band and HH band. The expression for gain cofficient is given by

$$g(\hbar\omega) = C_0 \left| I_{h1}^{e1} \right|^2 \int_0^\infty dE_t \rho_r^{2D} \left| \hat{e} \cdot p_{cv} \right|^2$$
$$\times \frac{H/2\pi}{\left[E_{h1}^{c1}(0) + E_t - \hbar\omega \right]^2 + (H/2)^2}$$
$$\times \left[f_c^1(\hbar\omega) - f_v^1(\hbar\omega) \right] \qquad (2)$$

where C_0 is the constant, n_r is the refractive index of the well material, ρ_r^{2D} is the reduced density of the state function, L_z is the width of the QW along the z-direction, and m_r is the reduced effective mass. I is the overlap integral of the conduction and valance subband wave function. $\left| \hat{e} \cdot p_{cv} \right|^2$ is the momentum matrix element between the conduction band and the valance band, where $\hat{e}$ is the polarization unit vector. H is the full width at half-maximum of the Lorentzien function. E_{h1}^{c1} is the bandgap between E_{c1} and E_{h1}, which are the bound state Eigen energies in Γ conduction band and HH valance band, respectively. f_c^1 and f_v^1 are the fermi occupation probabilities in the conduction and valance band. The detailed theoretical calculation about the band structure and optical gain and values of the material parameters are available in Ranjan & Das (2016).

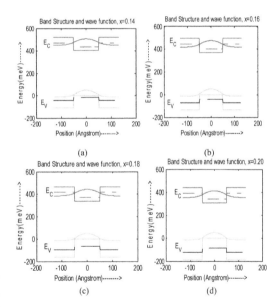

Figure 2. Potential profiles and wave function in the base region of the TL for different Sn concentrations in the well.

3 RESULTS AND DISCUSSIONS

Fig. 2 shows the potential profiles of various bands and the wave function of the Γ-conduction band and HH valance band in the base region of the TL for different.

Sn concentrations in the well region. Since the compressive strain is present in the well, the HH valance band is above the LH valance band, but it is reverse in the barrier region. The bandgap of

Sn is lowest among Si, Ge, and Sn, with increasing Sn concentration in the well, bandgap decreases, which is shown in Fig. 2a–d. The composition of Sn in the barrier is fixed ($Si_{0.12}Ge_{0.76}Sn_{0.12}$), so the bandgap between the Γ and L valley is constant (49.3 meV). The band offset between the well and the barrier is finite, so the wave function is not confined only in the well region.

Fig. 3 shows the TE mode optical gain for different Sn concentrations in the well region. The bandgap of the Γ-conduction band decreases with the Sn concentration, so the peak of the optical gain is red shifted. The figure shows that the entire peak is almost constant; this is because the injected carrier is constant in the well region for different Sn concentrations. This is due to the same amount of carrier transport from the Γ valley of the barrier to the Γ valley of the well and the same is applicable for the L valley of the well and barrier.

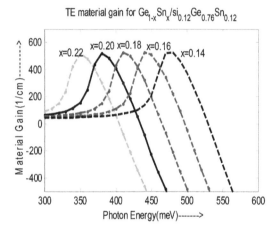

TE material gain for $Ge_{1-x}Sn_x/si_{0.12}Ge_{0.76}Sn_{0.12}$

Figure 3. TE mode optical gain for different Sn concentrations in the well region.

Table 1. Values of some material parameters in the SiGeSn/GeSn-based TL for different compositions of Sn in the well.

QW composition	Eg_Γ, Eg_L (meV) well	Eg_Γ, Eg_L (meV) Barrier	TE Gain (cm⁻¹)	Photon Energy (meV)	Wave length (μm)
$Ge_{0.86}Sn_{0.14}$	404.5, 433.8	520.8, 471.6	524.9	480	2.58
$Ge_{0.84}Sn_{0.16}$	371.2, 401.1	493.8, 444.5	527.6	440	2.82
$Ge_{0.82}Sn_{0.18}$	339.0, 369.5	466.8, 417.6	527.6	410	3.03
$Ge_{0.80}Sn_{0.20}$	308.0, 339.1	440.0, 390.7	521.3	380	3.26
$Ge_{0.78}Sn_{0.22}$	278.1, 309.8	413.3, 364.0	513.0	350	3.54

In Table 1, the different material parameters like bandgap of Γ and L conduction subbands (Eg_Γ and Eg_L) in the well and barrier, optical gain, and wavelength are tabulated. By observing the tabulated data, we can see that the Γ conduction band is the lowest bandgap, which is necessary for the laser action in the TL.

4 CONCLUSION

In present study, we analyzed the polarization-dependent TE mode optical gain in the SiGeSn/GeSn-based TL for the mid-infrared (2–4 μm) wavelength. The change in Sn concentrations (0.14–0.22) in the well region decreased the bandgap of the Γ-valley conduction band with increasing Sn concentration in the well. This phenomenon shifted the peak of the optical gain from 2.58 μm to 3.54 μm wavelength. This analysis is helpful to design all group IV optoelectronics system.

REFERENCES

Chang, G., S. Chang & S. L. Chuang (2010). Strain-Balanced Ge$_z$Sn$_{1-z}$–Si$_x$Ge$_y$Sn$_{1-x-y}$ Multiple-Quantum-Well Lasers. IEEE J. Quantum Electron. 46(12), 1813–1820.

Chuang, S. L. (2009). Physics of photonic Devices. 2nd ed. New York, Wiley.

Feng, M., N. Holonyak, Jr., G. Walter & R. Chan (2005). Room temperature continuous wave operation of a heterojunction bipolar transistor laser. Appl. Phys. Lett. 87, 131103.

Holonyak, N. Jr. & M. Feng (2006). The Transistor Laser. IEEE Spectrum, 43, 50–55.

Kouvetakis, J., J. Menendez & A. V. G. Chizmeshya (2006). Tin-based group IV semiconductors: new platforms for opto- and microelectronics on silicon. Annual Review of Materials Research, vol. 36, pp. 497–554.

Moontragoon, P., N. Vukmirovic, Z. Ikonic & P. Harrison (2010). SnGe Asymmetric Quantum Well Electroabsorption Modulators for LongWave Silicon Photonics. IEEE J. Sel. Top. Quantum Electron. 16, 100–105.

Ranjan, R. & M. K. Das (2016). Theoretical estimation of optical gain in Tin incorporated group IV alloy based transistor laser. Optical and Quantum Electronics, Vol. 48, Article no. 201.

Soref, R. (2010). Mid-infrared photonics in silicon and germanium. Nat. Photonics, 4, 495.

Werner, J., M. Oehme, M. Schmid, M. Kaschel, A. Schirmer, E. Kasper & J. Schulze (2011). Germanium-tin p-i-n photodetectors integrated on silicon grown by molecular beam epitaxy. Appl. Phys. Lett. 98, 061108.

Walle, C.G.V.D. (1989). Band lineups and deformation potentials in the model-solid theory. phys. Rev. B, vol. 39, no. 3, pp. 1871–1883.

Computer, Communication and Electrical Technology – Guha, Chakraborty & Dutta (Eds)
© 2017 Taylor & Francis Group, ISBN 978-1-138-03157-9

An efficient low-power 1-bit full adder using a multi-threshold voltage scheme

Raushan Kumar & Sahadev Roy
Department of ECE, NIT, Arunachal Pradesh, India

Arpan Bhattacharyya
Department of CSE, Government College of Textile Technology, Berhampore, India

ABSTRACT: Power and efficiency are a major concern in any digital system, and their minimization is tough in full-adder circuits. In the previous work on this topic, the energy consumption was in the range of microwatt. This paper proposes an efficient 14-transistor 1-bit full-adder circuit under a multi-threshold voltage scheme. This circuit is designed using a Complementary Metal Oxide Semiconductor (CMOS) in 180-nm technology. The power consumption of this circuit is achieved in nanowatt scale for different frequencies; this is less than other existing adder circuits for the same testing condition. The design and implementation of the proposed model were analyzed and verified through the SPICE simulation platform and the average power consumed by this circuit was 17×10^{-9} W at an operating frequency of 500 MHz, which is less than the values for other conventional methods.

1 INTRODUCTION

Due to the rapid progression of portable electronic devices and personal assistant devices like laptops and cellular devices, investigation of the low-power design of VLSI systems become essential (Roy, Saha, & Bhunia, 2016). With increases in chip density and the number of transistor power consumption and effective fabrication are of the VLSI circuit is also increasing. This further caused many problems, which adversely affected consistency and packaging volume density. Due to these two reasons, the cost of the VLSI-based design also went up with high-power consumption and increasing delay. Currently, low-power consumption with less time delay and small size are the minimum requirements for modern IC designers.

The major sources of power deceptions in VLSI manly CMOS circuit blocks are switching power consumption for switching capacitances charging and discharging; short-circuit current conduction from the power source to the ground for the synchronized working principle of the n-MOS and the p-MOS logic blocks; and the leakage current drown in the transistor, which is known as static power dissipation due to the (Naveen & Thanushkodi, 2013).

Binary addition is essential and most often used for arithmetic operation in processors, embedded systems and different System on a Chip (SoC),

Application-Specific Integrated Circuits (ASIC), etc. Hence, the binary adders are an important constructing part of the VLSI circuits. An efficient design procedure of these adders increases the performances of the overall VLSI circuits. In the recent years, the remarkable development of CAD technology, made the design of various types of adder circuits possible. The primary circuits of a CMOS 28 transistor adder designed using the combination of the pull-up circuit with the pull-down circuit (Kumar M., 2012) and using 14 NMOS along with 14 PMOS transistors are most broadly reported (Haring, 1966). The use of a transmission gate and Pass Transistor Logic (PTL) 16 transistors hybrid full-adder cell has also been reported (Bhattacharyya, Kundu, Ghosh, Kumar, & Dandapat, 2015). A 32 transistors Complementary Pass-Transistor Logic (CPL) adder with a high-power dissipation and better dynamic ability is reported in (Kumar, Roy, & Bhunia, 2016). Transmission gate 20 transistors CMOS adder circuits using transmission gates are proposed in (Shalem, John, & John, 1999). The TFA transmission function-based full adder using 16 transistors is proposed in (Yano, Yamanaka, Nishida, Saito, Shimohigashi, & Shimizu, 1990). A Multiplexer-Based Adder (MBA) consisting of 12T transistors are the elimination of the direct path to the source (Zimmermann & Fichtner, 1997). A static energy recovery full adder using

10T also suffers from higher delay (Tung, Hung, Shieh, & Huang, 2007). The circuit level, a power-optimized design, is desirable with a less number of transistors count, very less power absorption even for no-load condition or high-speed operations, and output voltage swing in a tolerable limit. In this paper, a modified new low-power 1-bit full adder with four transistor XOR module and one multiplexer block with six transistors has been performed.

The paper is prearranged as follows: Section 2 consists of the illustration of the MTVL scheme. Section 3 discusses an illustration of signal bit 14T full adder. Section 4 proposes a multi-threshold full adder. Section 5 consists of the result and comparison discussion and section 6 consists of the conclusions.

2 14T TRANSMISSION GATE FULL ADDER

A one-bit full-adder circuit can be realized by using a different combination 2: 1 multiplexer circuit along with XNOR/XOR logic blocks. Two XOR cells with a multiplexer using four transistors transmission gate one-bit full-adder network is shown in Fig. 1. The resultant sum bit is generated by using two XNOR gates and carry (Cout) is realized from the multiplexer circuits (Sharma, Periasamy, Pattanaik, & Balwinder, 2013). The main advantage of the transmission gate full-adder noise merging is better than that of the existing methodology.

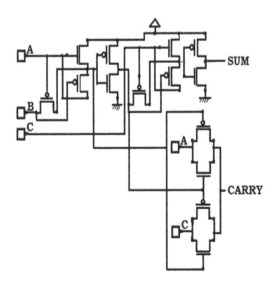

Figure 1. Illustration of 14T based on a transmission gate full adder.

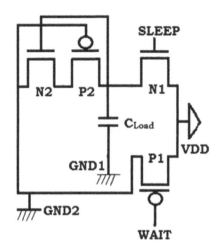

Figure 2. Illustration of a multi-threshold circuit.

3 AN ILLUSTRATION OF THE MTVL SCHEME

Multi-threshold schemes work based on diode basing logic. The main advantage of this technique is that it reduces noise and also the leakage current. In this procedure, high threshold CMOS transistors P1, P2, N1, and N2 have been used for minimized leakage current. Assembling the transistors N1, N2, and P2 (Fig. 2) also reduce the reserve leakage current due to sleep and active state, hence discharging the virtual ground-1 voltage divided into two portions, which will minimize noise effectively. In the case of forward biasing, the effective threshold voltage of the wait transistor can be minimized, but the size of the transistor is constant. The threshold voltage can be minimized by the wait transistor during the mode change, i.e., the sleep-to-wait mode of the CMOS transistor. The load capacitor Cload is inserted at the intermediate node ground-2 and the current may flow through transistor P2 although it is in the sleep mode (Verma, Kumar, & Marwah, 2014). In this way, the switching action is possible, generating minimal activation noise during sleep, wait, and active phase of the CMOS transition.

4 PROPOSED DESIGN METHODOLOGY

Here, we discuss the design methodology of the proposed high-speed, low-power, efficient one-bit full-adder circuit. The basic designed concept is based on the multi-threshold logic. Three blocks are categorized here to perform the addition operation. Here, the full-adder circuit consists of two four-transistor XOR gate and six transistors 2:1 multiplexor. The

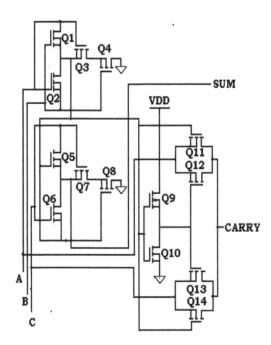

Figure 3. Proposed 14T full adder.

$$D_{ynamic} = \sum \alpha C_L V^2_{dd} f \qquad (2)$$

In the above expression, we can calculate the dynamic power of the full-adder circuit, where α is the switching action of the load, and Cload is a charging-discharging loaded capacitor, and the clock frequency is represented as f.

5 RESULT ANALYSIS AND DISCUSSION

5.1 *Power consumption*

This section discusses the power consumption at different supply voltages. Here, the analysis was done by using the variable supply voltage, ranging from 1V to 1.8V. The comparative analysis is conceded on 180 nm CMOS technology. The experimental results show that the modified 14T full adder consumes less power among the other 14T transmission gate based full adder (Sharma, Periasamy, Pattanaik, & Balwinder, 2013). The analysis of the average power is shown in Table 1 and Fig. 4.

full adder consists of three inputs A, B, and for the previous carry in (Cin) and two output ports for the sum and carry bits. We have modified this style with the concept of the multi-threshold voltage logic scheme, so the circuit is more suitable for low Vdd operations. The main basic idea is to design this circuit's low and high threshold voltage operation. The high Vth transistors reduce the leakage current effectively but with the performance degraded, whereas low Vth transistors give better performances but have high leakage currents. Fig. 3 shows the proposed low-power high-speed 14T one-bit full-adder circuit. In the MTVL scheme, we measured the high Vth value for both the transistors used in the transmission gate. In this proposed design, we achieved less average power, less power delay product, and also less chip size area of this full adder.

4.1 *Component of power consumption*

There are several components used in the minimization of power consumption. The short-circuit power dissipation and statics CMOS power are similar to the switching power. Show that the total power is written in the form of a sum of all power consumption.

$$P_{total} = P_{dynamic} + P_{short circuit} + Ec_{stati} \qquad (1)$$

The dynamic power has a quadratic relationship with the supply voltage (Vdd), and it can be expressed as

Table 1. Comparison table based on power consumptions at different frequencies.

		14T	Proposed 14T Full-adder
Power supply (V)	Frequency (MHz)	Average power (μw)	Average power (nw)
1.0	100	01.94	01.22
	200	02.05	01.53
	250	02.06	01.56
	400	02.79	01.57
	500	02.85	01.69
1.2	100	04.00	01.77
	200	04.15	02.15
	250	04.47	02.18
	400	04.50	02.21
	500	04.62	02.38
1.4	100	07.00	02.43
	200	07.04	02.71
	250	07.48	02.75
	400	07.84	03.14
	500	07.86	03.15
1.6	100	09.81	06.82
	200	10.30	06.98
	250	10.34	07.30
	400	10.35	07.41
	500	11.20	07.95
1.8	100	10.03	10.00
	200	11.10	14.20
	250	11.20	16.00
	400	11.30	16.20
	500	13.70	17.00

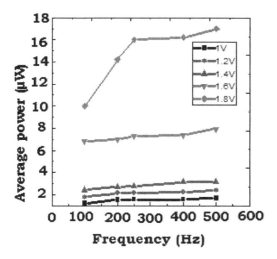

Figure 4. Frequency vs. average power dissipation at different supply voltage.

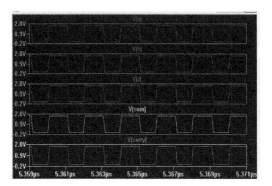

Figure 5. Simulation result of the proposed full adder.

5.2 *Comparison of power delay product*

Power delay product due depends on the supply voltage and transmission delay. We analyzed the proposed circuit in 1V to 1.8V and frequency range from 100 MHz to 500 MHz. The comparative analysis was done in 180 nm CMOS technology. The result of the modified 14T full adder has less PDP compared to the existing 14T transmission gate full adder (Sharma, Periasamy, Pattanaik, & Balwinder, 2013) circuit shown in Figure 5 and Table 2.

There was a reduction of the dynamic power and voltage, but in the case of the supply voltage (less than 1V), the circuit performance was slowly degraded. The proposed circuit was able to overcome this issue. The results show that power consumption of the circuit is in nanowatt range, where the existing 14T power drawn is in milli watt range.

Table 2. Comparison between the existing 14T and proposed adder at 250 MHz frequency.

VDD	Average power	Delay (PS)	PDP (fj)
1.0	2.06 µW	565.34	1.164
	1.56 nW	535.02	0.0008
1.1	3.62 µW	559.77	1.82
	1.87 nW	512.85	0.0009
1.2	4.47 µW	554.20	2.477
	2.18 nW	490.69	0.0010
1.3	5.97 µW	529.60	3.13
	2.46 nW	469.50	0.0011
1.4	7.48 µW	505.11	3.778
	2.75 nW	449.05	0.0012
1.5	8.91 µW	451.38	3.944
	5.02 nW	401.64	0.0018
1.6	10.34 µW	397.66	4.111
	7.30 nW	354.23	0.0025
1.7	10.77 µW	392.27	4.222
	11.65 nW	332.17	0.0037
1.8	11.20 µW	386.88	4.333
	16.00 nW	310.12	0.0049

Simulation result of the existing 14T full adder

Simulation result of the proposed 14T full adder

6 CONCLUSION

There are several techniques for the full-adder circuits design. The main aim was to reduce the leakage current and biasing voltage. But MTVL design techniques are much suitable for controlling the leakage current and also the reducing power consumption. We presented an approach to derive the adder circuits from MTVL that combines the advantages of less power consumption, PDP, and area. The output characteristic analysis was carried out at different power supplies and different frequencies in 180 nm CMOS technology by the SPICE and MICROWIND simulation tool. The proposed low-power 14T one-bit full-adder structure is suitable for low-voltage applications.

REFERENCES

Bhattacharyya, P., Kundu, B., Ghosh, S., Kumar, V., & Dandapat, A. (2015). Performance Analysis of a Low-Power High-Speed Hybrid 1-bit Full Adder Circuit. *IEEE Transactions on Very Large Scale Integration (VLSI) Systems, 23* (10), 2001–20.

Haring, D. R. (1966). *IEEE Trans. on Electronic Computers, EC-15* (1), 45–65.

Kumar, M. (2012). Design of 9-transistor single bit full adder. *Proceedings of the Second International Conference on Computational Science, Engineering and Information Technology* (pp. 337–340). ACM.

Kumar, R., Roy, S., & Bhunia, C. T. (2016). Study of Threshold Gate and CMOS Logic Style Based Full Adder Circuits. *3rd Int. IEEE Conference on Electronics and Communication Systems (ICECS)* (pp. 173–179). Coimbatore: IEEE.

Naveen, R., & Thanushkodi, K. (2013). Low-leakage Full Adder circuit using Current Comparison Based Domino Logic. *International Conference on Current Trends in Engineering and Technology (ICCTET)* (pp. 41–45). IEEE.

Roy, S., Saha, R., & Bhunia, C. T. (2016). On Efficient Minimization Techniques of Logical Constituents and Sequential Data Transmission for Digital IC. *Indian Journal of Science and Technology, 9* (9), 1–9.

Shalem, R., John, E., & John, E. (1999). A novel low power energy recovery full adder cell. *Proceedings of Ninth Great Lakes Symposium on VLSI* (pp. 380–383). IEEE.

Sharma, S., Periasamy, C., Pattanaik, M., & Balwinder, R. (2013). Activation noise aware ultra low power diode based multi-threshold CMOS technique for static CMOS adders. *Annual International Conference on Emerging Research Areas and 2013 International Conference on Microelectronics, Communications and Renewable Energy (AICERA/ICMiCR)* (pp. 1–6). IEEE.

Tung, C. K., Hung, Y. C., Shieh, S. H., & Huang, G. S. (2007). A Low-Power High-Speed Hybrid CMOS Full Adder for Embedded System. *IEEE Conference on Design and Diagnostics of Electronic Circuits and Systems, pp. 1−4.*

Verma, S., Kumar, D., & Marwah, G. K. (2014). New High New High-Performance 1-Bit Full Adder Using Domino Logic. *n Computational Intelligence and Communication Networks (CICN) International Conference on (pp. 961–965). IEEE.*

Yano, K., Yamanaka, T., Nishida, T., Saito, M., Shimohigashi, K., & Shimizu, A. (1990). A 3.8-ns CMOS 16×16-b multiplier using complementary pass-transistor logic. *IEEE Journal of Solid-State Circuits, 25* (2), 388–395.

Zimmermann, R., & Fichtner, W. (1997). Low-power logic styles: CMOS versus pass-transistor logic. *IEEE J. Solid State Circuits, 32* (7), 1079–1090.

Computer, Communication and Electrical Technology – Guha, Chakraborty & Dutta (Eds)
© 2017 Taylor & Francis Group, ISBN 978-1-138-03157-9

Power efficient, high frequency and low noise PLL design for wireless receiver applications

B. Supraja
Department of EIE, VNR Vignana Jyothi Institute of Engineering and Technology, (Autonomous), Hyderabad, India

N. Ravi & T. Jayachandra Prasad
Department of Physics, RGM College of Engineering and Technology, (Autonomous), Nandyal, Andhra Pradesh, India
Department of ECE, RGM College of Engineering and Technology, (Autonomous), Nandyal, Andhra Pradesh, India

ABSTRACT: In this paper, we implemented a Phase Locked Loop (PLL) to operate in higher frequencies and low power for wireless receivers. The PLL is implemented with proposed designs at block level. A new design is proposed for CP to avoid the current mismatch, which cause to produce spurs in the output of the PLL. A hybrid type of FD is proposed to enrich the power efficiency of the PLL. The efficiency is achieved at each block of the PLL. Therefore, Power Efficient PLL (PEPLL) for wireless communication is constructed and implemented by Tanner EDA tool. The parameters such as power and delay of the PEPLL are calculated using H-Spice for 180 nm technology at 2.0 V. The PEPLL consumes only 1.2155 mW of power and requires only 88 MOS devices.

1 INTRODUCTION

At present, modern nanometer CMOS technologies are very efficient, in terms of power consumption and speed, for the design of digital integrated circuits. High speed, low power Phase-Locked Loops (PLL) are extensively used in frequency synthesis, clock generation and recovery circuits for microprocessors and wireless communication systems (Yasuaki 1997). The PLL frequency synthesizer cannot help having a slow lock up time. For the improvement of the lock up time, first, to find objective frequency, we propose a new Phase Frequency Detector (PFD) which detects the phase and frequency difference between the reference frequency Cref and feedback one Cout (Bo Li et al. 2011, Mhd Zaher Al 2007, Chih-Wei et al. 2009, Soares et al. 1999, Qiuting et al. 1996).

Usual requirements while designing PLL is small acquisition time, maximum locking range and minimum phase error variance. To meet these requirements various structures have been proposed for charge pump PLL (Yuan et al. 1989).

Conventional CMOS charge pump circuits have some current mismatching characteristics. The current mismatch of the charge pump in the PLLs generates a phase offset, which increases spurs in the PLL output signals. In particular, it reduces locking range in wide range PLLs with a dual loop

scheme (Jae-Shin et al. 2000, Joung-Wook et al. 2014).

2 PFD ARCHITECTURES

The Conventional PFD is designed using 2 flip flops and AND gate as shown in Fig. 1. However As the technology has a rapid growth, power, area and delay are considered as the major challenges while designing circuits and fabricating the ICs using VLSI technology. The number of devices majorly affects the area, hence the area can be reduced by reducing the number of devices. Thus, by using De Morgan's Law i.e. $A.B = (\overline{A} + \overline{B})$ the AND gate in PFD can be replaced by using the NOR gate with the complement form of UP and Down signals as inputs to the NOR gate. The architecture of PFD using NOR gate is shown in Fig. 2.

Due to the presence of a logic gate (AND in Fig. 1 or NOR in Fig. 2) in reset path, the time desired to charge the gate and reset both flip-flops will be added to the reset delay time in the internal components of the flip flops and produce a large dead zone. The change is to remove the reset path and reduce the delay time that causing the dead zone problem. Thus the PFD with the removal of reset path by directly driving the reset

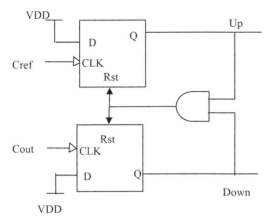

Figure 1. PFD using AND gate.

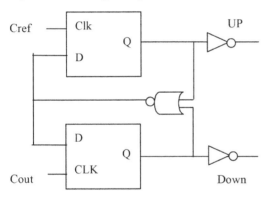

Figure 2. PFD using NOR gate.

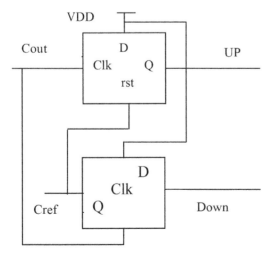

Figure 3. PFD with no gate.

with the CLK is as shown in Fig. 3. The D Flip Flop is modified so that the PFD with No gate functions exactly as Conventional PFD and the delay in reset path is removed resulting in the ideal PFD behavior. The simulation results of different structures are shown in Table 3 where the power and delay of the designs are given to evaluate the suitable design.

3 STRUCTURE AND OPERATION OF CHARGE PUMP

Charge pump is a tri state electronic switch controlled by the three states of the PFD. The charge pump introduces a constant current for a definite period of time equal to the phase difference between during the charging period and it draws the current from the loop filter or a capacitive load for a certain period of phase difference during discharging. Thus PFD drives the Charge Pump (CP) and adjusts the amount of current to be injected into or drawn from the loop filter or load capacitor (Intissar et al. 2013, De-Zhi Wang et al. 2013).

The basic architecture of the charge pump is shown in Fig. 4. The CP consists of 2 switched current sources driving a capacitive load. Whenever the Up signal is high switch S1 closes, resulting in the charging of load capacitor i.e., current is injected into the loop filter. When Down signal is high switch S2 is closed resulting in the discharging of load capacitor i.e., current is drawn from loop filter. The association of the Charge pump circuit with the loop filter converts the logic states generated by PFD into an analog voltage V_{ctrl} for controlling the VCO. The VCO converts the V_{ctrl} into frequency (Amr Elshazly et al. 2013, Youngshin et al. 2002, Onur 2008, Li Zhiqun et al. 2011).

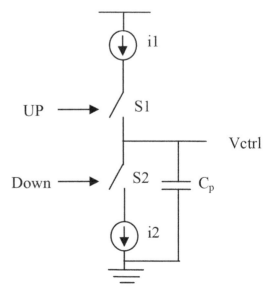

Figure 4. Basic architecture of charge pump.

3.1 *Proposed Charge pump*

The charge pump with active loop filter possess perfect current match, but the number of transistors used in this charge pump is high and also the operational amplifier is used to construct the active loop filter. Hence, more area is required for this type of charge pumps. The charge pump with the passive loop filter has less number of transistors when compared with the charge pump with active loop filter and structure is also very simple and easy to construct. No need of inverters because the complemented form of control signals is not used in this charge pump. This charge pump also avoids the charge injection and clock feed through the proper sizing of transistors. The usage of current sources with the same value is necessary to reduce jitter. But charge sharing cannot be eliminated in this charge pump. Hence, by analyzing the advantages and disadvantages of charge pump with active loop filter and charge pump with passive loop filter a new charge pump is proposed.

The proposed charge pump is very simple to construct and looks very similar to charge pump (Song M et al. 2009). The non-ideal effects of the CP can be reduced by choosing the position of switches carefully. In the proposed charge pump we exchange the positions of switches and current sources. The Fig. 5 represents the proposed CP in which the position of switches and current sources are exchanged. The CP also characterizes that charge injection and clock feed-trough caused by switches can not directly impact the output voltage. In addition to this, as the terminals P, Q and R are at the similar point the charge distribution

phenomenon is also decreased. Thus the proposed charge pump is free from non idealities such as charge injection, clock feed through and charge sharing. Thus, current mismatch is reduced in proposed charge pump.

4 FREQUENCY DIVIDER

The conventional circuit implementation of CMOS TSPC flip flop is as shown in Fig. 7. Both the input and output signals of the TSPC divider are single ended. The Conventional TSPC consists of 10 transistors (Lin J et al. 2004). Later modifications are done to further reduce the number of transistors. Thus the general TSPC consists of 9 transistors. There are various construction rules

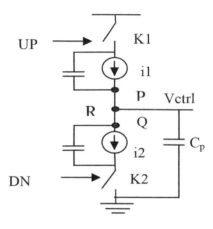

Figure 6.   Position of switches in proposed charge pump.

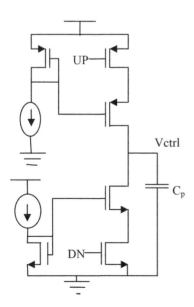

Figure 5.   Proposed charge pump.

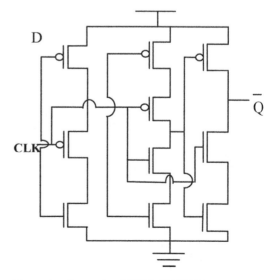

Figure 7.   Conventional TSPC (DIVCON).

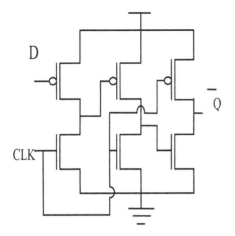

Figure 8. Extended TSPC (E-TSPC).

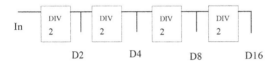

Figure 9. Divide by 16 frequency divider.

Table 1. Divide by 16 using various flip flops.

DESIGN	Div By 2(1)	Div By 2(2)	Div By 2(3)	Div By 2(4)
FD1	DIV	DIV	DIV	DIV
DIVCON	DIVCON	DIVCON	DIVCON	DIVCON
FDs	DIVS	DIVS	DIVS	DIVS
FDetspc	E-TSPC	E-TSPC	E-TSPC	E-TSPC
FDn	DIVN	DIVN	DIVN	DIVN
FDH1	E-TSPC	DIVN	DIVN	DIVN
FDH2	E-TSPC	E-TSPC	DIVN	DIVN

Table 2. Power and delay analysis of Divide by 16 using various flip flops.

DESIGN	POWER (mW)	DELAY (nS)	PDP (pJ)	EDP (J^2) 10^{-20}	No. of Transistors
FD1	0.381	66.9	25.4	170.5	36
DIVCON	0.536	0.878	0.47	0.04	40
FDs	0.407	26.4	10.74	28.38	36
FDetspc	7.29	58.1	423.3	2459	24
FDn	0.479	25.8	12.34	31.83	36
FDH1	0.127	129	16.30	210.0	33
FDH2	3.89	11.8	45.74	538.0	30

to design TSPC (Deyun Cai et al. 2010). Various flip flops are proposed in the recent years to meet the desires of low power, high speed performance (Staszewski et al. 2005, Siliang et al. 2012, Chao et al. 2010). This section deals with some of the basic flip flops that are generally used. All the flip flops discussed in this section are single ended, i.e. single ended input and output.

4.1 Implementation of divide by 16 frequency divider

The frequency divider with division ratio 16 is implemented by cascading 4 divide by two stages. The schematic diagram and waveforms of divide by 16 frequency divider is shown in Fig. 9. The divide by 2 block can be implemented by using any one of the TSPC or E TSPC DFF. The homogeneous frequency divider can be implemented by using all the cascaded divide by 2 circuits implemented using same DFF. The alternative method is to cascade different TSPC or ETSPC blocks to implement a divide by 16 frequency divider. Various frequency dividers are designed by using the different combinations of DFF at divide by 2 blocks. Table 1. Gives the information about the divide by 2 blocks used in various frequency divider designs. The power and delay analysis of various frequency divider designs is shown in Table 2.

5 RESULTS AND DISCUSSIONS

Various PFD designs are simulated using Tanner EDA tool. The Power and Delay analysis of existing and proposed PFD designs are shown in Table 3. Each design has its own advantages and disadvantages. Supported the necessity acceptable PFD style is to be selected. The results of the various PFDs are tabulated in Table 3. The proposed PLL is effective in all the parameters when compared with the existing PLL.

1. Effective PFD design with the smallest amount power delay product reduces the delay in reset path and dead zone downside is avoided.
2. The usage of the proposed charge pump not only reduces the area, but also makes the PLL free from spurs. The charge sharing problem in existing PLL is reduced in proposed charge pump by exchanging the positions of switches and current sources.
3. The ring VCOs with 3 delay cells and differential ring VCO with 4 delay cells does not use the AFC. Thus the area required to build them on IC is very less when compared with the area required to build the differential ring VCO with AFC. The majority of power consumption is due to the VCO. As the VCO designs without AFC uses very less number of components

Table 3. Power and delay analysis of various PFD designs.

DESIGN NAME	Power (mW)	Delay (nS)	PDP (pJ)	Flip flop	Logic Gate	No. of Transistors
DFFPFD (Lin J et al. 2004)	0.07	20.859	1.649	Modified DFF	NO GATE	16
DFFNOR (PROPOSED)	0.5642	0.1103	0.0622	Modified DFF	NOR	20
PFD NOR GATE USING GDI (PROPOSED)	0.5067	0.2318	0.1174	DFF for Dynamic PFD	NOR	20
DYNPFD	0.8121	0.4815	0.3911	DFF for Dynamic PFD	NOR	20
DFFPFDCONAND (PROPOSED)	0.3237	0.0321	0.0106	Modified DFF	AND	22
DYNAND (PROPOSED)	0.2985	0.0842	0.0251	DFF for Dynamic PFD	AND	22
TRDPFD (Chih-Wei et al. 2009)	0.3437	0.3051	0.1049	DFF for conventional PFD	AND	22
TGCMOS PFD (PROPOSED)	0.7065	0.2820	0.1992	DFF for Dynamic PFD	NOR	24
DYNPFDMUX (PROPOSED)	0.8144	0.3018	0.2458	DFF for Dynamic PFD	NOR	28

Table 4. Comparison of Proposed PLL with Conventional PLL (Joung et al. 2002).

Design name	Total power (mW)	Propagation delay (nS)	Peak power (W)	No. of MOS devices	Transition time (S)
Existing PLL	8.0678	7.5462	71.91	561	290.83
Proposed PLL	1.2155	33.312	0.0412	88	159.77

(devices) the power consumption is also very low. But the ring VCO with 3 delay cells generate the low frequency signal when compared with the differential ring VCO with 4 delay cells.
4. Divide by 16 Frequency divider (FDH2) is selected by considering the trade-off between the PDP, EDP and number of devices.

The existing PLL is implemented from Joung et al. The proposed PLL is compared with the existing PLL. The PEPLL has shown more power efficient than that of the existing PLL. The PEPLL consumes only 1.2155 mW of power only, but the conventional PLL consumes 8.06 mW as shown in Table 4. The PEPLL requires only 88 MOS components to implement the design. This is not so for the existing.

The transition time required for PEPLL is less than that of the conventional. Precaution is taken to design charge pump. Current mismatch is avoided. Therefore the PEPLL is a spur reduction design, shows more improvement in the simulation results.

6 CONCLUSION

Different designs for each block of Phase Locked Loop namely phase frequency detector, charge pump with loop filter, voltage controlled oscillator and frequency divider are introduced. Every block is individually simulated. By considering the parameters like power, area delay, loading effect of

individual blocks various phase locked loops are designed. Among all designs the proposed design (PEPLL) possesses a high frequency with spur suppression operating. The proposed design has less number of devices, thus reducing the area and power consumption. Hence the PEPLL is a high frequency spur suppression power efficient design.

REFERENCES

Amr Elshazly, Rajesh Inti, BrianYoung, and Pavan Kumar Hanumolu (2013). Clock Multiplication Techniques Using Digital Multiplying Delay-Locked Loops, *IEEE Journal of Solid-State Circuits*, Vol. 48, No. 6, 1416–1428.

Bo Li, Yiming Zhai, Bo Yang, Thomas Salter, Martin Peckerar, and Neil Goldsman (2011). Ultra low power phase detector and phase-locked loop designs and their application as a receiver, *Microelectronics Journal* 42, 358–364.

Chao Guo, Siheng Zhu, Jun Hu, Jing Diao, Houjun Sun and Xin Lv (2010). Design and Optimization of Dual Modulus Prescaler Using the Extended True Single Phase Clock, *IEEE Conference ICMMT Proceedings*, 636–638.

Chih-Wei Chang and Yi-Jan Emery Chen (2009). A CMOS True Single-Phase-Clock Divider With Differential Outputs, *IEEE Microwave and Wireless Components Letters*, Vol. 19, No. 12, 813–815.

Deyun Cai, Haipeng Fu, Danfeng Chen, Junyan Ren, Wei Li and Ning Li (2010). An improved Phase/Frequency Detector and a glitch-suppression Charge Pump design for PLL Applications, *IEEE Conference Proceedings*.

De-zhi Wang, Ke-feng Zhang and Xue-cheng Zou (2013). High Current Matching over Full-Swing and Low-Glitch Charge Pump Circuit for PLLs, *Radioengineering*, Vol. 22, No. 1, 153–158.

Intissar Toihria, Rim Ayadi and Mohamed Masmoudi (2013). An Effective CMOS Charge Pump-Phase Frequency Detector Circuit for PLLs Applications, *10th International Multi-Conference on Systems, Signals & Devices* Hammamet, Tunisia, March 18–21, 1–7.

Jae-Shin Lee, Min-Sun Keel, Shin-I1 Lim and Suki Kim, (2000). Charge pump with perfect current matching characteristics in phase-locked loops, *Electronics Letters,* 9th November 2000, Vol. 36, No. 23, 1907–1908.

Joung-Wook Moon, Kwang-Chun Choi and Woo-Young Choi (2014). A 0.4-V, 90 ~ 350-MHz PLL With an Active Loop-Filter Charge Pump, *IEEE Transactions on Circuits and Systems—II: Express Briefs*, Vol. 61, No. 5, 319–323.

Li Zhiqun, Zheng Shuangshuang, and Hou Ningbing (2011). Design of a high performance CMOS charge pump for phase-locked loop synthesizers, *Journal of Semiconductors*, Vol. 32, No. 7, 1–5.

Lin. J, B. Haroun, T. Foo, J.-S.Wang, B. Helmick, S. Randall, T. Mayhugh, C. Barr, and J. Kirkpatrick (2004). A PVT tolerant 0.18 MHz to 600 MHz self-calibrated digital PLL in 90 nm CMOS process, in *IEEEISSCC Dig. Tech.* Papers, 488–489.

Mhd Zaher Al Sabbagh (2007). 0.18 µm Phase/Frequency Detector and Charge Pump Design for Digital Video Broadcasting For Handheld's Phase-Locked-Loop Systems, MS thesis.

Onur Kazanc (2008). High Frequency Low-Jitter Phase-Locked Loop Design, Project Report, Yeditepe University.

Qiuting Huang, and Robert Rogenmoser (1996). Speed Optimization of Edge—Triggered CMOS Circuits for Gigahertz Single-Phase Clocks, *IEEE J. Solid-state Circuits,* vol. 31, no. 3, 456–465.

Siliang Hua, Hua Yang, Yan Liu, Quanquan Li and Donghui Wang (2012). A Power and Area Efficient CMOS Charge-Pump Phase-Locked Loop, *IEEE Conference Proceedings.*

Song. M, Y.-H. Kwak, S. Ahn, W. Kim, B. Park, and C. Kim (2009). A 10 MHz to 315 MHz cascaded hybrid PLL with piecewise linear calibrated TDC, in Proc. IEEE CustomInt. Circuits Conf. (CICC), pp: 243–246.

Soares. J. N, Jr. and W. A. M. Van Noije (1999). A 1.6-GHz dual modulus prescaler using the extended true-single-phase-clock CMOS circuit technique (E-TSPC), *IEEE J. Solid-state Circuits*, vol. 34, no. 1, 97–102.

Staszewski. R. B, J. L.Wallberg, S. Rezeq, C.-M. Hung, O. E. Eliezer, S. K. Vemulapalli, C. Fernando, K. Maggio, R. Staszewski, N. Barton,M.-C. Lee, P. Cruise, M. Entezari, K. Muhammad, and D. Leipold (2005). All-digital PLL and transmitter for mobile phones, *IEEE J. Solid-State Circuits,* vol. 40, no. 12, 2469–2482.

Yasuaki Sumi (1997). Speedup of Lock up time in the PLL Frequency Synthesizer using Frequency Detector, 1997 *IEEE International Symposium on Circuits and Systems*, June 9–12, 1917, Hong Kong, 1365–1368.

Youngshin Woo, Young Min Jang and Man Young Sung, (2002). Phase Locked Loop with dual phase frequency detectors for high frequency operation and fast acquisition, *Microelectronics Journal* 33, 245–252.

Yuan. J and C. Svensson (1989). High-speed CMOS circuit technique, *IEEE J. Solid-state Circuits*, vol. 24, no. 1, 62–70.

Computer, Communication and Electrical Technology – Guha, Chakraborty & Dutta (Eds)
© *2017 Taylor & Francis Group, ISBN 978-1-138-03157-9*

Oscillating wave propagation inside DNG material based 1D photonic crystal

Bhaswati Das & Arpan Deyasi
RCC Institute of Information Technology, Kolkata, West Bengal, India

ABSTRACT: Remarkable oscillating wave is observed instead of monotonic decaying feature inside one-dimensional photonic crystal when conventional SiO$_2$-air composition is replaced by DNG material-air combination; and operating wavelength is set just below or above Bragg wavelength (1.55 µm). Results are observed for nano-fishnet with rectangular void, elliptical void and paired nanorod structures. Coupling coefficient between forward and backward propagating waves is varied for all the structures independently. Result speaks in favor of guided transmission in metamaterial based photonic crystal for optical communication for longer length with negligible loss.

1 INTRODUCTION

The Photonic Crystal (PhC) is the periodic variation of refractive indices of constituent materials along the propagation direction, leads to electromagnetic bandgap (Yablonovitch 1987). The structure is already established as photonic bandpass filter (Robinson *et al.*, 2011) and is also used as making photonic crystal fiber (Belhadj *et al.* 2006). This novel microstructure is already used in making optical transmitter (Szczepanski 1988), receiver (Fogel *et al.* 1998), sensor (Shanthi *et al.* 2014), memory (Lima *et al.* 2011), quantum device (Jiang *et al.* 1999) etc. It works as building block of the next generation communication system. But works are not reported about the Double Negative Material (DNG material) based PhC design, though importance of metamaterial is already verified in high-frequency communication application. Design of high-frequency antenna using metamaterial is the subject of research in the preset decade due to the novel property of improved signal-to-noise ratio owing to the negative refractive index material (Xiong *et al.* 2012, Hwang *et al.* 2009). The rejection of noise in the desired frequency spectrum, more precisely, in the region of optical communication (where minimum attenuation is achieved) is the choice of interest for communication engineers. Therefore, study of the property of metamaterial array is one of the key areas in designing photonic integrated circuit.

In last decade, various research works are published in literatures of repute involving 1D photonic crystal (Xu *et al.* 2007, Rudzinski 2007) due to the ease of fabrication and less complex mathematical modeling. Also a few papers are published involving 2D microstructure (Zhang *et al.* 2004) where different numerical methods are involved to solve the eigenvalue equations for calculating band structure. Materials involved in these works are conventional SiO$_2$/air composition (Gao *et al.* 2011), or semiconductor heterostructure (Maity *et al.* 2013). But as far the knowledge of the authors, research work involving DNG materials are published rarely. The present work shows the forward and backward wave nature inside 1D photonic crystal, which exhibits that the wave can travel for a longer grating length compared to that obtained for conventional PhC by setting operating wavelength slightly less or greater than Bragg wavelength. Effect of coupling coefficient on wave propagation is studied for all the three metamaterial structures. Results are important for analysis of guided transmission inside 1D photonic crystal structure.

2 MATHEMATICAL MODELING

In this paper, we have considered three structures of metamaterials namely nano-fishnet with rectangular void (r.i = −1), nano-fishnet with elliptical void (r.i = −4) and paired nanorod (r.i = −0.3). Though fishnet structures are three dimensional in nature, but we have considered propagation of electromagnetic wave in one-direction only. This si due to the fact that we are only interested in wave propagation in oscillating form inside the system, and since in three-dimensional structure, the nature and period of propagating wave is almost same in all three directions, hence it is quite understandable to study the behavior in any one of the direction. Moreover, Bragg grating concept is easily applicable in one-direction, and hence computation becomes less time-consuming; which will provide appropriate result. Since we expect the periodic nature of wave, hence we safely assume one-direction of wave propagation with coupled mode theory.

Inside a photonic crystal, the forward and backward propagating waves may be represented as

$$\frac{da(z)}{dz} = j\beta_a a(z) \tag{1}$$

$$\frac{db(z)}{dz} = -j\beta_b b(z) \tag{2}$$

assuming the waves are coupled. Subject to the appropriate boundary condition, solution of the equations is

$$a(z) = D\exp\left[\frac{jz(\beta_a - \beta_b)}{2}\right]$$
$$\times \exp\left[\frac{\sqrt{4\kappa_{ab}\kappa_{ba} - (\beta_a + \beta_b)^2}}{2}z\right] \tag{3}$$

$$b(z) = b\exp\left[\frac{jz(\beta_a - \beta_b)}{2}\right]$$
$$\times \exp\left[\frac{-\sqrt{4\kappa_{ab}\kappa_{ba} - (\beta_a + \beta_b)^2}}{2}z\right] \tag{4}$$

where we assume that total power is conserved. With the boundary conditions $b(0) = b_0$ and $a(L) = 0$, these equations may be solved analytically. Expressions of forward and backward waves thus can be given as

$$a(z) = b_0 \frac{\kappa\exp[-j\Delta\beta z]\sinh[\alpha(z-L)]}{\Delta\beta\sinh\alpha L - j\alpha\cosh\alpha L} \tag{5}$$

$$b(z) = b_0\kappa\exp[j\Delta\beta z] \times$$
$$\frac{[\Delta\beta\sinh\{\alpha(z-L)\} + j\alpha\cosh\{\alpha(z-L)\}]}{-\Delta\beta\sinh\alpha L + j\alpha\cosh\alpha L} \tag{6}$$

where $\alpha = \sqrt{\kappa^2 - \Delta\beta^2}$

Here we assume that both the propagation constants of forward and backward waves are equal. It may be mentioned that attenuation of transverse waves in 2D plane is not taken into consideration

as it is only possible when operating wavelength becomes equal to Bragg wavelength (Huang *et al.,* 2016). The diagrams of the structures are given below:

3 RESULTS AND DISCUSSIONS

Using Eq. (5) and Eq. (6), forward and backward wave profiles are simulated and plotted as a function of grating length for one-dimensional photonic crystal. Fig. 2a and Fig. 2b exhibit the forward wave propagating characteristics inside

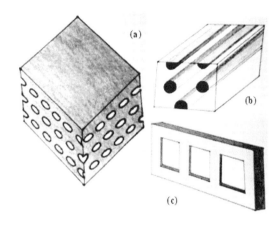

Figure 1. Schematic diagram of (a) nano-fishnet with elliptical void, (b) paired nanorod, (c) nano-fishnet with rectangular void.

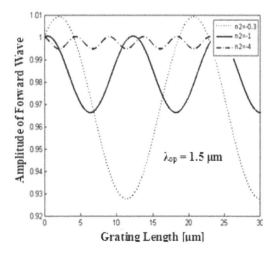

Figure 2a. Forward wave profile for different DNG material/air compositions as a function of grating length for operating wavelength 1.5 μm.

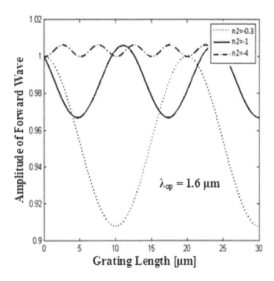

Figure 2b. Forward wave profile for different DNG material/air compositions as a function of grating length for operating wavelength 1.6 μm.

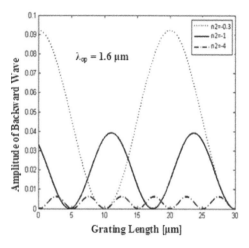

Figure 3b. Backward wave profile for different DNG material/air compositions as a function of grating length for operating wavelength 1.6 μm.

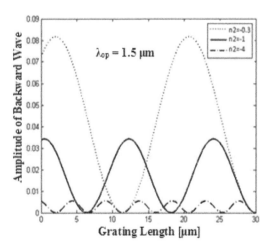

Figure 3a. Backward wave profile for different DNG material/air compositions as a function of grating length for operating wavelength 1.5 μm.

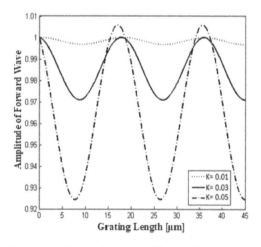

Figure 4a. Forward wave profile at 1.5 μm operating wavelength under different coupling conditions for paired nanorod.

1D PhC. Bragg wavelength is set at 1.55 μm, and profiles are generated for operating wavelength 1.5 μm and 1.6 μm respectively.

From the plots, it is seen as propagation of forward wave becomes almost unaffected w.r.t operating wavelength when nano-fishnet with elliptical void structure (n = −4) is considered, whereas passing through paired nanorod structure (n = −0.3) gives maximum variation. The same is also true for backward wave propagation.

Fig. 4 – Fig. 7 show the comparative analysis of variation of propagating wave for different coupling coefficients for three different DNG material/air compositions, namely paired nanorod (Fig. 4a, Fig. 5a, Fig. 6a, Fig. 7a), nano-fishnet with rectangular void (n = −1) (Fig. 4b, Fig. 5b, Fig. 6b, Fig. 7b) and nano-fishnet with elliptical void (n = −4) (Fig. 4c, Fig. 5c, Fig. 6c).

A comparative study reveals that for higher coupling between forward and backward propagating waves leads maximum oscillation for the wave propagation, whereas loose coupling makes almost

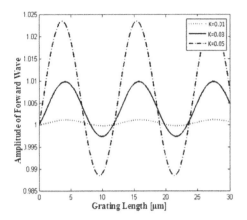

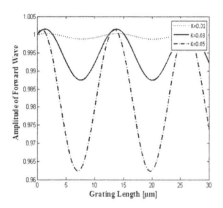

Figure 5b. Forward wave profile at 1.6 μm operating wavelength under different coupling conditions for nanofishnet with rectangular void.

Figure 4b. Forward wave profile at 1.5 μm operating wavelength under different coupling conditions for nanofishnet with rectangular void.

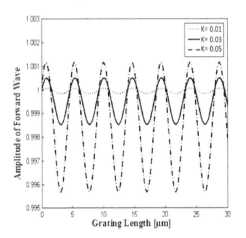

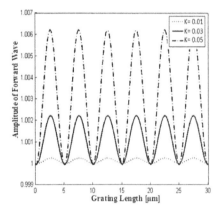

Figure 5c. Forward wave profile at 1.6 μm operating wavelength under different coupling conditions for nanofishnet with elliptical void.

Figure 4c. Forward wave profile at 1.5 μm operating wavelength under different coupling conditions for nanofishnet with elliptical void.

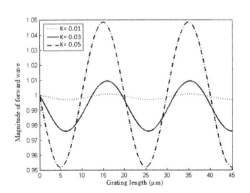

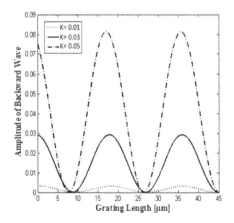

Figure 5a. Forward wave profile at 1.6 μm operating wavelength under different coupling conditions for paired nanorod.

Figure 6a. Backward wave profile at 1.5 μm operating wavelength under different coupling conditions for paired nanorod.

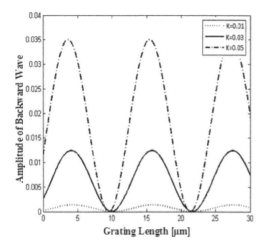

Figure 6b. Backward wave profile at 1.5 μm operating wavelength under different coupling conditions for nano-fishnet with rectangular void.

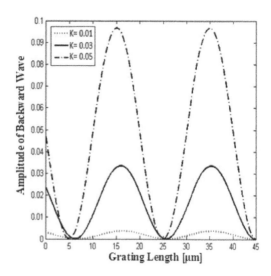

Figure 7a. Backward wave profile at 1.6 μm operating wavelength under different coupling conditions for paired nanorod.

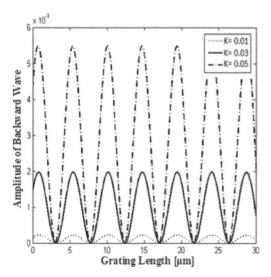

Figure 6c. Backward wave profile at 1.5 μm operating wavelength under different coupling conditions for nano-fishnet with elliptical void.

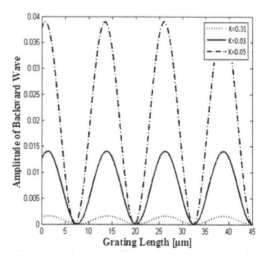

Figure 7b. Backward wave profile at 1.6 μm operating wavelength under different coupling conditions for nano-fishnet with rectangular void.

un-attenuated transmission. The magnitude of periodic oscillation is highest for paired nanorod structure. A sharp contrast is also observed about the magnitude of the oscillatory wave is observed for 1.5 μm and 1.6 μm operating wavelength.

If operating wavelength becomes greater than Bragg wavelength, then magnitude of propagating wave increases irrespective of material compositions for both forward and backward waves.

The magnitude of waves (for both forward and backward) rapidly changes with the change of DNG material. This is evident from the compara-

tive study of Fig. 4, Fig. 5, Fig. 6, Fig. 7. For backward wave, it may be noted that lower coupling coefficient does not make noticeable contribution. This is true for forward wave propagation also.

4 CONCLUSION

For conventional PhC, both forward and backward waves show monotonic decreasing behavior which suggests that electromagnetic wave propagation is not possible for a longer distance. But DNG material with air interface exhibits completely different behavior which suggests that e.m wave may propagate a longer grating length compared to the earlier founding's, if operating wavelength is set at slightly higher or lower than Bragg wavelength. Result suggests that nano-fishnet with elliptical void may be used as a photonic waveguide in the micrometer range for all photonic integrated circuits. Coupling between forward and backward waves plays critical role in determining shape of the propagating wave. Since these structures also exhibit bandpass filter characteristics at the Bragg wavelength (depending on structural parameters), so it can serve the dual purpose.

REFERENCES

Belhadj, W., Malek, A.F., Bouchriha, H. (2006) Characterization and Study of Photonic Crystal Fibres with Bends, Material Science and Engineering: C, 26, 578–579.

Fogel, I.S., Bendickson, J.M., Tocci, M.D., Bloemer, M.J., Scalora, M., Bowden, C.M., Dowling, J.P. (1998) Spontaneous Emission and Nonlinear Effects in Photonic Bandgap Materials, Pure and Applied Optics, 7, 393–408.

Gao, Y., Chen, H., Qiu, H., Lu, Q., Huang, C. (2011) Transmission Spectra Characteristics of 1D Photonic Crystals with Complex Dielectric Constant, Rare Metals, 30, 150–154.

Huang, H.H., Lin, C.K., Tan, K.T. (2016) Attenuation of transverse waves by using a metamaterial beam with lateral local resonators, Smart Materials and Structures, 25(8), 085027.

Hwang, R.B., Liu, H.W., Chin, C.Y. (2009) A metamaterial-based E-plane Horn Antenna, Progress in Electromagnetics Research, 93, 275–289.

Jiang, Y., Niu, C., Lin, D.L. (1999) Resonance Tunneling through Photonic Quantum Wells, Physical Review B, 59, 9981–9986.

Lima, Jr. A.W., Ferreira, A.C., Sombra, A.S.B. (2011) Optical Memory made of Photonic Crystal working over the C-band of ITU, Journal of Optical Fiber Communication Research, 1–16.

Maity, A., Chottopadhyay, B., Banerjee, U., Deyasi, A. (2013) Novel Band-Pass Filter Design using Photonic Multiple Quantum Well Structure with p-Polarized Incident Wave at 1550 nm, Journal of Electron Devices, 17, 1400–1405.

Robinson, S., Nakkeeran, R. (2011) Two Dimensional Photonic Crystal Ring Resonator based Bandpass Filter for C-Band of CWDM Applications, Nat. Conf. Comm, 1–4.

Rudzinski, A. (2007) Analytic Expressions for Electromagnetic Field Envelopes in a 1D Photonic Crystal, ACTA Physics Polonica A, 111, 323–333.

Shanthi, K.V., Robinson, S. (2014) Two-Dimensional Photonic crystal based Sensor for Pressure Sensing, Photonic Sensors, 4, 248–253.

Szczepanski, P. (1988) Semiclassical Theory of Multimode Operation of a Distributed Feedback Laser, IEEE Journal of Quantum Electronics, 24, 1248–1257.

Xiong, H., Hong, J.S., Peng, Y.H. (2012) Impedance Bandwidth and Gain Improvement for Microstrip Antenna using Metamatrials, Radio Engineering, 21, 993–998.

Xu, X., Chen, H., Xiong, Z., Jin, A., Gu, C., Cheng, B., Zhang, D. (2007) Fabrication of Photonic Crystals on Several Kinds of Semiconductor Materials by using Focused-Ion Beam Method, Thin Solid Films, 515, 8297–8300.

Yablonovitch, E. (1987) Inhibited Spontaneous Emission in Solid State Physics and Electronics, Physical Review Letters, 58, 2059–2061.

Zhang, Z., Qiu, M. (2004) Small-volume Waveguide-section High Q Microcavities in 2D Photonic Crystal Slabs, Optics Express, 12, 3988–3995.

Computer, Communication and Electrical Technology – Guha, Chakraborty & Dutta (Eds)
© 2017 Taylor & Francis Group, ISBN 978-1-138-03157-9

Comparative analysis of filter performance in DNG material based photonic crystal structure

Arpan Deyasi, Solanki Ghosh, Ruma Dutta & Varsha Shaw
RCC Institute of Information Technology, Kolkata, India

ABSTRACT: Photonic Filter is designed at 1.55 µm using double negative refractive index materials. Both paired nanorod (r.i is −0.3) and nano-fishnet structure with elliptical void (r.i is −4) are considered for comparative analysis. Electromagnetic wave propagation for both normal and oblique incidences (TE mode only) are considered for analysis purpose, and simulation results reveal that passband at the desired zone may be obtained by suitable choice of structural parameters and external conditions. Results also indicate that suitable choice of defect density enhances the filter performance with less ripple in passband; and smooth nature of stopband slopes, compare to that obtained for defect-free (ideal) structure. Results are important in designing optical filter at the desired spectrum.

1   INTRODUCTION

In the age of photonics, design and physical realization of all-optical-circuit is one of the subjects of research (Li *et al.*, 2011), which may provide as a possible and effective replacement of optoelectronic counterparts. Electromagnetic wave propagation inside the structure and its tailoring for specific application thus becomes one key area of study, and photonic crystal now-a-days becomes one prime candidate for that purpose (Shambat *et al.*, 2009) due to its novel property of restricting e.m wave in certain wavelength ranges, and allowing other spectra (Chen *et al.*, 1996). This is possible with the difference of refractive indices of the layers and the dimension of different layers inside the unit block of the periodic structure. Among the different structures, 1D microstructure is easily realizable from both theoretical (Gao *et al.*, 2011) and experimental (Limpert *et al.*, 2004) stand-point. The structure has already proved its effectiveness in different communication systems (Srivastava *et al.*, 2008, Limpert *et al.*, 2003) with improved efficiency. It is already effectively utilized in optical filter (Mao *et al.*, 2008, Biswas *et al.*, 2015).

Metamaterial is known in the field of antenna engineering as it effectively enhances the SNR of the device (Segal *et al.*, 2015). Superlens biosensor is already designed using LHM material (Dorrani *et al.*, 2012). Various radiating and guiding structures are already reported in literatures using metamaterial (Engheta *et al.*, 2006, Marques *et al.*, 2008, Cui *et al.*, 2010). Reflectance spectrum of 1D DNG material based structure is calculated (Srivastava *et al.*, 2016) in recent past. Bandgap properties are calculated (Aghajamali

et al., 2014) calculated bandgap properties of ternary and binary lossy photonic crystals. Photonic crystal containing magnetized cold plasma defect (Aghajamali *et al.*, 2015) is computed theoretically in presence of defect mode. In this paper, transmittivity property of defected 1D photonic crystal is calculated using transfer matrix method, where both paired nanorod and nano-fishnet structure with elliptical void are separately considered. Results are obtained for TE mode of propagation, and are compared with that obtained in ideal structure. Position of ripple in passband is optimized by suitably choosing dimension of different layers. Findings will play key role in metamaterial based photonic filter design for high frequency antenna.

2   MATHEMATICAL FORMULATION

For simulation purpose, we have considered two types of metamaterials namely, paired nanorod (r.i = −0.3), and nano-fishnet with elliptical void (r.i = −4). Schematic figures of these structures are shown in figure below:

Considering the phase factor of the field propagating through uniform medium, propagation matrix is given as the function of barrier and well widths

$$P_{1,2} = \begin{pmatrix} \exp[jk_{1,2}d_{1,2}] & 0 \\ 0 & -\exp[jk_{1,2}d_{1,2}] \end{pmatrix} \quad (1)$$

where $d_{1,2}$ is the dimension of barrier/well layer, $k_{1,2}$ is the propagation vector. In presence of defect, Eq. (1) will be modified as

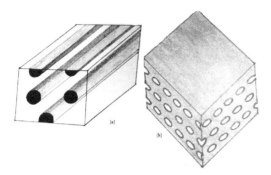

Figure 1. Schematic structures of (a) paired nanorod and (b) nano-fishnet with elliptical void.

$$P_{1f,2f} = \begin{pmatrix} \exp[jk_{1,2}d_{1,2}]f & 0 \\ 0 & -\exp[jk_{1,2}d_{1,2}]f \end{pmatrix} \quad (2)$$

where 'f' is the defect density. Thus, transfer matrix for the elementary cell (constituting of one barrier and one well layer) for ideal and defected structures are respectively given by

$$M = M^T{}_1 P_1 M^T{}_2 P_2 \quad (3)$$

and

$$M = M^T{}_1 P_{1f} M^T{}_2 P_2 \quad (4)$$

where M is the transfer matrix between the adjacent layers, given by

$$M^T{}_{1,2} = \frac{1}{t}\begin{pmatrix} 1 & r_{21,12} \\ r_{21,12} & 1 \end{pmatrix} \quad (5)$$

where '$t12$' is the transmittivity between two adjacent layers. For p-polarized incident wave at angle θ_1, interface reflectivities are given by

$$r_{12} = -r_{21} = \frac{n_1\cos(\theta_2) - n_2\cos(\theta_1)}{n_1\cos(\theta_2) + n_2\cos(\theta_1)} \quad (6)$$

For s-polarized incident wave at angle θ_1, interface reflectivities are given by

$$r_{12} = -r_{21} = \frac{n_1\cos(\theta_1) - n_2\cos(\theta_2)}{n_1\cos(\theta_1) + n_2\cos(\theta_2)} \quad (7)$$

For a perfectly periodic medium composed of N such elementary cells, the total transfer matrix for such a structure is

$$M_{tot} = M_N \quad (8)$$

Transmission coefficient is given by

$$T = \frac{1}{M_{11}{}^2(tot)} \quad (9)$$

3 RESULTS AND DISCUSSIONS

Using the Eq. (8) as mentioned in the previous section, comparative analysis of transmittivity profile is plotted in presence and absence of defect. It is seen from Fig. 2 that with the change in refractive index of metamaterial a large change is occurred in optical bandwidth. The plots are made for normal incidence of electromagnetic wave. It is worthwhile to mention that in all the figures, layer thicknesses of metamaterial and air are nomenclature as 'd_3' and 'd_4'.

It is seen that ripple rejection reduced for material system with refractive index −4 (nano-fishnet with elliptical void). A sharp notch is present in the graph for refractive index −0.3 (paired nanorod). In this case of refractive index −4 we plot the graph for $d_3 = 2.3$ μm and $d_4 = 0.45$ μm and for refractive index = −0.3 we plot the graph for $d_3 = 3.85$ μm and $d_4 = 0.75$ μm.

In Fig. 3, analysis is made for TE mode. It is seen that amount of ripple inside passband is reduced for elliptical nano-fishnet structure (refractive index −4). Presence of sharp notch is again displayed for refractive index −0.3. In the case of refractive index −4, graphs are drawn for $d_3 = 2.3$ μm and $d_4 = 0.45$ μm and for refractive index −0.3; results are plotted for $d_3 = 3.54$ μm and $d_4 = 0.59$ μm.

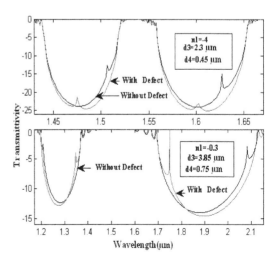

Figure 2. Transmittivity with wavelength for normal incidence of EM wave in presence and absence of defect for different refractive index of metamaterial.

It is also observed that for both the cases, two stopband regions are not symmetric w.r.t wavelength axis; and the effect is more pronounced for the paired nanorod based structure. The lower wavelength edge of passband in this case is very sharp compared to the other end.

Next the effect of material and structural parameters are studied for calculating transmittivity. It is observed from Fig. 4 that fewer ripples are observed elliptical nano-fishnet structure than the paired nanorod. One interesting thing may be noted down that the change of dimension for

metamaterial is more significant in case of nano-fishnet structure than the paired nanorod. Similarly, the effect of air thickness is depicted in Fig. 5. Hence the choice of dimension is very significant in designing optical filter.

Fig. 6 reflects the effect of incidence angle for TE mode propagation. With increase of incidence angle, redshift is observed for both the cases, but the amount of shift is significant for paired nanorod structure. Again, a close measurement of shift reveals that it is slightly larger for stopband at higher wavelength region.

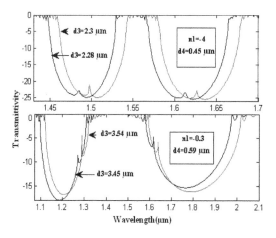

Figure 3. Transmittivity with wavelength for oblique incidence of TE mode in presence and absence of defect for different refractive index of metamaterial.

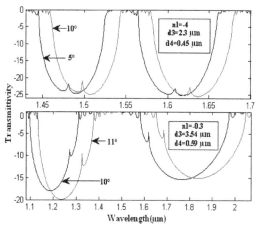

Figure 5. Transmittivity with wavelength for oblique incidence of TE mode in presence of defect for different propagation length of air thickness & for different refractive index of metamaterial.

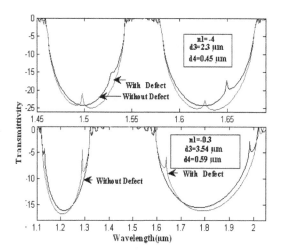

Figure 4. Transmittivity with wavelength for oblique incidence of TE mode in presence of defect for different propagation length of metamaterial & for different refractive index of metamaterial.

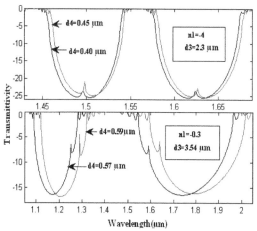

Figure 6. Transmittivity with wavelength for oblique incidence of TM mode in presence of defect for different incident angle & for different refractive index of metamaterial.

4 CONCLUSION

Simulated findings suggest that transmittivity of DNG material based 1D PhC system has improved noise rejection property compared to those made with conventional SiO_2-air system as claimed in earlier literatures. Butterworth property in the proposed optical filter is observed though little amount of ripple is present in passband. Layer dimensions and angle of incidence can significantly affect the device performance. Results are obtained for physically existing metamaterials, so that it can be implemented practically. Simulation is important for metamaterial based optical filter fabrication.

REFERENCES

Aghajamali, A., Akbarimoosavi, M., Barati, M. (2014). Properties of the Band Gaps in 1D Ternary Lossy Photonic Crystal Containing Double-Negative Materials, Advances in Optical Technologies, 2014, 1–7.

Aghajamali, A., Zare, A., Wu C. J. (2015). Analysis of defect mode in a one-dimensional symmetric double-negative photonic crystal containing magnetized cold plasma defect, Applied Optics, 54(29), 8602–8606.

Biswas, P., Deyasi, A. (2015). Computing Transmittivity of One-Dimensional Defected Photonic Crystal under Polarized Incidence for Band-Pass Filter Applications, Journal of Electron Devices, 21, 1816–1822.

Chen, J. C., Haus, H. A., Fan, S., Villeneuve, P. R., Joannopoulos, J. D. (1996). Optical Filters from Photonic Band Gap Air Bridges, Journal of Lightwave Technology, 14, 2575–2580.

Cui, T. J., Liu, R., Smith, D. R. (2010). Metamaterials: Theory, Design, and Applications. New York: Springer Science.

Dorrani, Z., Mansouri-Birjandi, M. A. (2012). Superlens Biosensor with Photonic Crystals in Negative Refraction, International Journal of Computer Science Issues, 9, 57–60.

Engheta, N., Ziolkowski, R. W. (2006). Electromagnetic Metamaterials: Physics and Engineering Explorations, New York: Wiley-IEEE Press.

Gao, Y., Chen, H., Qiu, H., Lu, Q., Huang, C. (2011). Transmission Spectra Characteristics of 1D Photonic Crystals with Complex Dielectric Constant, Rare Metals, 30, 150–154.

Li, Y., Bhardwaj, A., Wang, R., Jin, S., Coldren, L., Bowers, J., Herczfeld, P. (2011). All-optical ACP-OPLL Photonic Integrated Circuit, IEEE MTT-S International Microwave Symposium Digest, 1–4.

Limpert, J., Liem, A., Reich, M., Schreiber, T., Nolte, S., Zellmer, H., Tünnermann, A., Broeng, J., Petersson, A., Jakobsen, C. (2004). Low-Nonlinearity Single-Transverse-Mode Ytterbium-Doped Photonic Crystal Fiber Amplifier, Optic Express, 12, 1313–1319.

Limpert, J., Schreiber, T., Nolte, S., Zellmer, H., Tunnermann, T., Iliew, R., Lederer, F., Broeng, J., Vienne, G., Petersson, A., Jakobsen, C. (2003). High-power air-clad large-mode-area photonic crystal fiber laser, Optic Express, 11, 818–823.

Mao, D., Ouyang, Z., Wang, J. C. (2008). A Photonic-Crystal Polarizer Integrated with the Functions of Narrow Bandpass and Narrow Transmission Angle Filtering, Applied Physics B, 90, 127–131.

Marques, R., Martin, F., Sorolla, M. (2009). Metamaterials with Negative Parameters: Theory, Design, and Microwave Applications, New York: Wiley.

Segal, N., Zur, S. K., Hendler, N., Ellenbogen, T. (2015). Controlling Light with Metamaterial-based Nonlinear Photonic Crystals, Nature Photonics, 9, 180–184.

Shambat, G., Mirotznik, M. S., Euliss, G., Smolski, V. O., Johnson, E. G., Athale, R. A. (2009). Photonic Crystal Filters for Multi-band Optical Filtering on a Monolithic Substrate, Journal of Nanophotonics, 3, 031506.

Srivastava, S. K., Ojha, S. P. (2008). A Novel Design of Nano Layered Optical Filter using Photonic Band Gap Materials, 2nd National Workshop on Advanced Optoelectronic Materials and Devices, 225–230.

Srivastava, S., Aghajamali, A. (2016). Study of optical reflectance properties in 1D annular photonic crystal containing double negative (DNG) metamaterials, Physica B: Condensed Matter, 489, 67–72.

Computer, Communication and Electrical Technology – Guha, Chakraborty & Dutta (Eds)
© *2017 Taylor & Francis Group, ISBN 978-1-138-03157-9*

Sn-concentration dependent absorption in strain balanced GeSn/SiGeSn quantum well

Prakash Pareek & Mukul K. Das

UGC SAP Laboratory, Department of Electronics Engineering, Indian Institute of Technology (Indian School of Mines), Dhanbad, Jharkhand, India

ABSTRACT: This paper presents a study on Sn content dependent absorption in SiGeSn/GeSn strain balanced Quantum well. The motivation for incorporation of Tin (Sn) in Germanium is briefly narrated. Eigen energy states for different Sn concentrations are obtained for Γ valley Conduction Band (ΓCB), Heavy Hole (HH) band and Light Hole (LH) band by solving coupled Schrödinger and Poisson equations simultaneously. Sn concentration dependent absorption spectra for HH-ΓCB transition reveals that significant absorption observed in mid infrared range (3–5 μm). So, $Ge_{1-x}Sn_x$ quantum well can be used for mid infrared sensing applications.

1   INTRODUCTION

Over the last few years, a lot of research has been conducted to realize low cost on chip photosensitive devices. Direct band gap material is preferred in these devices for efficient radiative transition. So photonic technology relies heavily on direct band gap materials of III-V group like GaAs (Bhattacharya 1994). But their high cost and incompatibility to silicon technology restrict them to be used as a low cost monolithic photosensitive device. This fact forces the concerned researchers to work towards a new group of semiconductor materials which can act as a replacement for III-V group.

Recently there has been a great deal of interest among researchers on the design and analysis of photosensitive devices based on Tin (Sn) incorporated Group-IV alloys (Kouvetakis et al. 2006). Alloying Ge with α-Sn can effectively reduce the direct-bandgap of Ge more than its indirect bandgap and, hence, a direct-bandgap GeSn alloy can be realized. Thus GeSn alloy can enable the design of low cost group IV photonic devices which are as efficient as their III-V group counterpart (Gassenq et al. 2012). However, due to large lattice mismatching of Ge and Sn, strain plays a major role in the operation of such devices. The strain role becomes more pronounced in case of multilayer structure like quantum well. Strain balanced structures are suggested to be the best way to tackle excessive strain in multiple layer structure like Quantum well (Daukes et al. 2002).

This paper focuses on the potential of Tin doped group-IV alloys specially GeSn to be used as a active layer in photosensitive devices like detector. Absorption coefficient is a significant performance parameter in these devices. In this study, Sn content dependent absorption coefficient is evaluated in strain balanced SiGeSn/GeSn strain balanced quantum well. Eigen energy states for different Sn contents in GeSn layer are obtained in the well for Γ valley Conduction Band (ΓCB), Heavy Hole (HH) band and Light Hole (LH) band separately by solving coupled Schrödinger and Poisson equations selfconsistently. The absorption characteristics are evaluated for different Sn contents in well. The result revealed that HH-ΓCB transition observes high absorption coefficient within 3–5 μm range of wavelength. Thus this well structure can be used as a infrared sensor in various applications (Roelkens et al. 2014).

2   THEORETICAL FORMULATION

2.1   *Model description*

The Quantum well structure considered in our analysis consists of tensile strained SiGeSn barriers and compressively strained GeSn well which ensures the strain balanced condition for quantum well. The growth axis of the structure is assumed along (001) axis. A 76Å thick Ge_xSn_{1-x} layer is sandwiched between two tensile strained $Si_{0.09}Ge_{0.8}Sn_{0.11}$ layers to form a type-I Single Quantum-Well (SQW) as shown in Fig. 1. The thickness of barrier is computed as 35 Å by using strain balance condition (Daukes et al. 2002). The dimension of the well is chosen in such a way that only one bound state formed in each of the conduction band and valence band. The Sn composition (x) in QW is required to obtain three condition simultaneously,

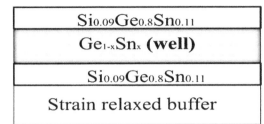

| $Si_{0.09}Ge_{0.8}Sn_{0.11}$ |
| $Ge_{1-x}Sn_x$ **(well)** |
| $Si_{0.09}Ge_{0.8}Sn_{0.11}$ |
| **Strain relaxed buffer** |

Figure 1. Schematic of strain balanced GeSn/SiGeSn quantum well.

(i) Type I SiGeSn/GeSn quantum well for better quantum carrier confinement, (ii) direct band gap in GeSn layer and (iii) a higher value of HH band offset than that of LH band (to be calculated later) for TE mode operation, which features HH-ΓCB dominant transition. A fully relaxed GeSn layer is used as a buffer layer.

The composition of buffer layer should be selected to maintain strain balanced condition in the QW.

2.2 *Band alignment calculations*

Accurate band profile and band discontinuities of the proposed QW structure is the vital requirement for precise modeling of the direct band interband absorption. The strain plays an important role in calculation of band structure and then band discontinuities at the interface of proposed multilayer structure. Our primary concern is to investigate direct interband transition which involves obtaining bandstructure at band edge (Γ-valley conduction band) in GeSn QW by evaluating strain dependent band profile. Moreover, we have also assumed no coupling between the conduction band and valence bands, which is an appropriate estimation for group IV alloys. In this context, model solid theory suggested by Van de Walle, is well suited to our proposed model (Chris & Walle 1989). Model solid theory is said to be one of the most reliable method to calculate band line ups and their alignment at zone center (Γ-valley) in a strained heterostructure. So we followed this theory to calculate band profile in ΓCB, conduction band, HH band and LH band for both well and barrier.

2.3 *Coupled Schrödinger—Poisson self-consistent solution*

After obtaining band profile, Eigen state energy of each band is required in order to evaluate direct absorption characteristics of QW. We obtained quantized energy states for Γ-CB, LH and HH band in QW by solving coupled Schrödinger and Poisson equation self consistently. Self-consistent solution is required to study the quantum confinement effect of carriers in QW more accurately considering variation in charge density of carriers. The Schrödinger equation with effective mass approximation and considering the strain effect is considered in our analysis and is given as (Chuang 1995):

$$\left[\frac{-\hbar^2}{2} \frac{\partial}{\partial z} \frac{1}{m_j} \frac{\partial}{\partial z} + \frac{\hbar^2 K_t^2}{2m_j} + V_j(z) \right] \psi = E_j \psi \quad (1)$$

where, ħ is Planck's constant, z is position variable, K_t is the transverse wave vector, Ψ is wave function, E is Eigen energy, m and V are respectively the effective mass and band profile. Suffix X stands for type of band e.g., j = c for Γ conduction band, j = hh for HH valence band and j = lh for LH band. As the Γ valley is of only interest here, the transverse wave vector, K_t is taken as zero. The equation is solved using Finite Difference Method (FDM) to obtain Eigen energies and wave function in the well (Datta 2005). The whole region of interest is divided into N number of small elements of equal width, and the equation is solved for each of the elements.

After obtaining Eigen energies and their corresponding wave functions, the position dependent charge density of carriers in well is calculated by summing the square of the wave function at each spatial element (Δz) and multiplying this quantity by the number of carriers in each bound state. The expression for electron density (n(z)) and hole density (p(z)) in well is given as (Tan & Snider 1990):

$$n(z) = \sum_n N_n \left| \psi_n^{CB\Gamma}(z) \right|^2 ;$$
$$p(z) = \sum_m N_m \left| \psi_m^{HH/LH}(z) \right|^2 \quad (2)$$

where n and m are the number of subbands in ΓCB, valence band (LH and HH) respectively, N_n is number of electrons in nth sub-band in conduction band and N_m is number of holes in mth sub-band of HH band and LH band.

The obtained position dependent carrier charge densities (n(z) and p(z)) are then used in Poisson equation. Poisson equation relates the potential to the charge density distribution as given in Eqn. 3 (Tan & Snider 1990).

$$\frac{d^2V}{dz^2} = -\frac{q}{\varepsilon} \left(n(z) - p(z) + N_A - N_D \right) \quad (3)$$

where n(z) and p(z) are the electrons and holes charge density distribution as obtained above.

N_a and N_d are acceptor and doping impurities in QW respectively and V is the spatial electrostatic potential in QW. Dirichlet and Neumann boundary conditions are also considered while solving the Poisson equation. Poisson equation was also solved by using finite difference numerical technique. In order to obtain self-consistent Eigen state energies both Poisson equation and Schrödinger equations are solved simultaneously until their solutions are converged (Stern 1970).

2.4 Evaluation of absorption coefficient

After calculating Eigen energies and wavefunctions for Γ-CB, HH band LH band, direct interband transition characteristic of the QW structure is studied. Absorption coefficient is evaluated for QW with the help of Fermi golden rule and using the following mathematical expression, for absorption coefficient α (Chuang 1995).

$$\alpha(\hbar\omega) = \frac{\pi q^2}{n_r c \varepsilon_0 m_0^2 \omega} \sum_{n,m} |I_m^n|^2 \cdot |P_{CV}|^2 \cdot \rho_{r\,2D} \cdot$$

$$\frac{1}{\sigma\sqrt{2\pi}} \exp\left[-\frac{(\hbar\omega - E_0 + E_{cn} - E_{vm})^2}{(2\sigma)^2}\right](f_v - f_c) \quad (4)$$

where, ω is photon frequency, q is electronic charge, c is speed of light, n_r is refractive index of well, ε_0 is static dielectric constant, m_0 is electron rest mass. E_0 is direct band gap of $Ge_{1-x}Sn_x$ which is calculated by linear interpolation of direct band gap of Ge and Sn considering bowing parameter. E_{cn} and E_{vm} are bound state Eigen energies for n_{th} subband in Γ conduction band, and for m_{th} subband in valence band (v = hh for HH band, v = lh for LH band) respectively. I is overlap integral of the Γ conduction subband wave function and valence subband wavefunction given by following equation. ρ_{r2D} is reduced joint density of states in QW. QW is assumed to be undoped in this work (ideal case), only valence band is filled with carriers so carrier probability occupancy of conduction band is taken as zero ($f_c = 0$, $f_v = 1$). Gaussian line shape function is also considered for inhomogeneity in GeSn alloy. P_{cv} is momentum matrix element between conduction band and valence band for Bloch state. It is a key parameter to evaluate absorption in both bulk and nanostructures. Indeed, its magnitude value indicates the strength of the interaction between photon with electron. Moreover under compressive strain TE mode is dominant than TM mode (Chuang 1995). Hence, TE polarization which parallel to the plane of QW layer is assumed in present study. In TE mode momentum matrix element of HH-Γ-CB transition is much greater than that of LH-ΓCB Γ transition at $K_t = 0$.

3 RESULTS AND DISCUSSION

In this study, Sn concentration in $Ge_{1-x}Sn_x$ quantum well layer is varied from 0.15 to 0.18 to obtain quantum mechanical characteristic of strain balanced quantum well and then direct absorption characteristic. As for compressive strained GeSn direct band gap nature induced in GeSn layer for $x \geq 0.15$ (Yahyaoui et al. 2014). Thus the required concentration of Sn should be atleast 15% in $Ge_{1-x}Sn_x$ layer. Moreover, beyond 0.18, the inter layer strain increases and it is intolerable and infeasible in actual situations.

The calculated band profile (band offset of Γ conduction band and heavy hole band) with Eigen state energies (E_{c1} for ΓCB, E_{HH} for HH band and E_{LH} for LH band) for QW at different Sn concentration is shown in Fig. 2. It is clearly observe that HH band (E_{HH}) is up shifted in comparison to light hole band profile. This is attributed due to compressive strain in the well. The band offsets of Γ conduction band and heavy hole band are sufficient large to cause quantum confinement effect of carriers. With increasing in Sn concentration, the shifting of E_{HH} in upward direction increases and bad gap also decreases subsequently. It can be also observed that Eigen wave function of LH band is weakly confined in the well. This is due to a very small negligible band offset for LH band. It also ensures that only HH-ΓCB (E_{HH}-E_{c1}) transition will be dominated in this case.

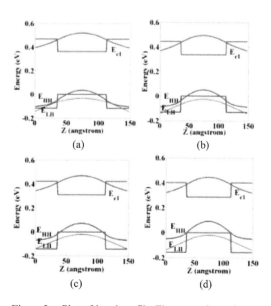

Figure 2. Plot of band profile, Eigen energies and wave functions for different values of x. (a) x = 0.15, (b) x = 0.16, (c) x = 0.17, (d) x = 0.18.

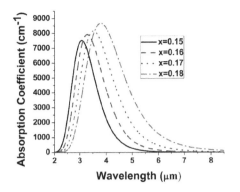

Figure 3. Plot of absorption coefficient (α) for HH-ΓCB direct transition versus wavelength for different values of x.

Absorption coefficient spectra (α) for different x are plotted as a function of wavelength and is shown in Fig. 3. Due to higher optical matrix element of HH to ΓCB transition for TE mode, LH-ΓCB transition is ignored in this study. The figure clearly reveals that significant absorption for HH to ΓCB transition is observed in infrared range of wavelength (3–5 μm). With increasing in Sn concentration peak absorption wavelength shifted to higher wavelength region. This red shifting is due to the lowering of the energy band gap of GeSn with increasing Sn concentration. It is also observed that with increasing Sn concentration, absorption increases.

4 CONCLUSION

This paper explores the potential of GeSn quantum well layer to be used in low cost monolithic photosensitive devices for mid infrared sensing applications. Direct interband absorption is plotted for different Sn content of well for dominant HH-ΓCB transition. The result depicts a significant absorption in 3–5 μm range of wavelength. In addition, increment in absorption with Sn concentration is also observed. Thus this quantum well structure is viable to be used as a low cost mid-infrared sensor.

REFERENCES

Bhattacharya, P. (1994). Semiconductor optoelectronic devices. USA, Prentice Hall.

Chris, G. & V. D. Walle (1989). Band lineups and deformation potentials in the model solid theory. Phys. Rev. B, 39, 1871–1883.

Chuang, S. L. (1995). Physics of optoelctronic devices. New York, John Wiley & Sons inc.

Daukes, N. J., K. Kawaguchi & J. Zhang (2002). Strain balanced criteria for multiple quantum well structure and its signature in x-ray rocking curves. Crystal growth and design, 2(4), 287–292.

Datta, S. (2005). Quantum transport: atom to transisitor. New York, Cambridge University Press.

Gassenq, A., F. Gencarelli, J. V. Compenhout, Y. Shimura, R. Loo, G. Narcy, B. Vincent & G. Roelkens (2012). GeSn/Ge heterostructure short-wave infrared photodetectors on silicon. Optics Express, 20, 27297–27303.

Kouvetakis, J., J. Menedez & A.V.G. Chizmeshya (2006). Tin based group IV semiconductors: new platforms for opto and micro electronics and silicon. Annual review of material research, 36, 497–554.

Roelkens, G., U. D. Dave et. al. (2014). Silicon-Based Photonic Integration Beyond the Telecommunication Wavelength Range. IEEE. Journal of selected topics in quantum electronics, 20(4), 8201511.

Stern, F. (1970). Iteration methods for calculating self-consistent fields in semiconductor inversion layers. Journal of computational physics, 6, 56–67.

Tan, I.-H. & G. L. Snider (1990). A self consistent solution of Schrödinger-Poisson equations using a nonuniform mesh. Journal of applied physics, 68, 4071–4076.

Yahyaoui, N., N. Sfina, J.-L. Lazzari, A. Bournel & M. Said (2014). Wave-function engineering and absorption spectra in $Si_{0.16}Ge_{0.84}/Ge_{0.84}$ $Sn_{0.16}/Si_{0.16}Ge_{0.84}$ strained on relaxed Si0.10Ge0.90 type I quantum well. Journal of applied physics, 115, 033109.

Performance analysis of ZnO/c-Si heterojunction solar cell

Syed Sadique Anwer Askari & Mukul Kumar Das

Department of Electronics Engineering, Indian Institute of Technology (Indian School of Mines), Dhanbad, Jharkhand, India

ABSTRACT: In this paper, performance of ZnO/p-Si solar cell has been analyzed using TCAD tool. Semiconductor-semiconductor, metal-semiconductor heterointerface effects and defects in ZnO have been considered in this analysis. Performance has been studied by varying different physical and device parameters and some best possible designs for enhanced efficiency have been proposed. Thickness of ZnO layer plays an important role on the performance of the device. Short circuit current density of 54 mA/cm^2, open circuit voltage of 0.62767 V and efficiency of 27.88% has been obtained for 80 nm thick ZnO layer of the device.

1 INTRODUCTION

Increasing demand of energy causes enormous emission of CO_2 because energy industry mainly depends on fossil fuel. Solar photovoltaics are intensively studied as an alternative energy source and may replace conventional energy sources based on fossil fuels. The interest in photovoltaic (PV) structures stems from the fact that solar cells are environmentally friendly. Unfortunately, the high cost of the generated electricity from solar photovoltaics restricts its choice to the common people. First practical solar cell having efficiency of ~6% was developed using a diffused Si p-n junction by Bell Laboratories in 1954 (Tsokos 2012). Instead of its various drawbacks, single crystal silicon based solar cell dominates the commercial market due to the low cost and matured technology of Si-based devices and also due to very stable performance of Si-solar cells. The c-Si solar cell approaches theoretical limit of about 30% (Richter et al. 2013) for a 110-µm-thick solar cell made of undoped silicon but the world's highest energy conversion efficiency of 25.6% (Green et al. 2016) at research level achieved by Panasonic in 2014, for non-concentrating silicon solar cells. Therefore price per watt of c-Si based solar cell reduced dramatically. Yet high cost and low efficiency of solar cell are two stumbling blocks for extensive use of solar cells. Several approaches have already been taken by the researchers to reduce its cost i.e. to enhance efficiency of single solar cell. Approaches mainly include design of new structures and use of new materials etc.

Recently, heterostructure solar cell with c-Si as bottom layer has attracted a great deal of interest among researchers to enhance efficiency. This is mainly to reduce the red and blue losses in c-Si based homojunction solar cell. Wide bandgap Transparent Conducting Oxide (TCO) on lower band gap c-Si solar cell gained lots of attention to the researchers. This is due to potential advantages such as simple processing steps, and low processing temperature. Previously Indium Tin Oxide (ITO) films are used as a wide band gap semiconductor on Si substrate. Due to limited source of Indium on earth, abundant material utilization is required. Zinc oxide (ZnO) can be used as a replacement of ITO. ZnO has good electrical and optical properties as well as low cost, non-toxic, natural abundance and relatively low deposition temperature (Kobayashi et al. 1995). ZnO is more resistant to radiation damage as compared to Si, GaAs and GaN which prevents photo-degradation and ensure longer device lifetime (Look 2001, Janotti 2009). Moreover, n-type ZnO can be easily realized by excess of Zn or by doping Al, Ga or In (Peartona et al. 2005).

Performance of ZnO/p-Si solar cell greatly depends upon the properties of ZnO, heterointerface properties, growth condition of ZnO etc. (Kishimoto et al. 2006, Dong et al. 2007). So, it is very important to study the performance of such device with variation of these parameters and also to suggest the optimum design of the device before its fabrication. To the best of Authors knowledge, performance analysis of ZnO/c-Si has not yet been done in detail considering interface defects and defects in ZnO.

In this paper, we have modeled and analyzed the performance of n-ZnO/p-Si heterojunction solar cell using SILVACO-TCAD tool considering the hetero interface effects and other physical phenomenon related to the solar photovoltaics. Some best possible designs for the enhanced efficiency have also been suggested.

2 DEVICE STRUCTURE AND MODEL

Device structure considered in this analysis is based on p-Si substrate and is shown in Figure 1. Zinc Oxide (ZnO) layer is considered to be grown on p-type silicon wafer and it acts as active n-layer as well as anti-reflection coating layer. The reflection from top layer is minimized if refractive index of AR coating (ZnO in our model) is geometrical mean of two surrounding indices. By taking air (refractive index = 1) at one side and Si (refractive index ~4) at another side, at 600 nm, optimum value of refractive index of AR coating is ~2 which is calculated by Equation 1.

$$n_{AR} = \sqrt{n_{air}n_{Si}} \qquad (1)$$

where n_{AR} = refractive index of antireflection coating layer, n_{air} = efractive index of air, and n_{Si} = refractive index of Si.

Thickness required of AR layer to act as perfect AR coating can be calculated by Equation 2.

$$t_{AR} = \frac{\lambda}{4 * n_{AR}}. \qquad (2)$$

where t_{AR} = thickness of antireflection coating layer, λ = wavelength corresponding to refractive index, and n_{AR} = refractive index of antireflection coating layer.

The most favorable thickness of AR layer for 600 nm wavelength is 76 nm. Fortunately, refractive index of ZnO at 600 nm is ~2 and thickness ~80 nm very close to ideal value required for antireflection (AR) coating of Si surface. It also prevents solar cell from radiation damage.

Thicknesses of TCO, emitter and absorber region are optimized by using SILVACO-ATLAS. Aluminum (Al) layer of 100 nm thickness is used as bottom metal contact. The device performance is evaluated by implanting surface recombination model, interface trap model, carrier transport model at semiconductor-semiconductor interface and semiconductor-metal interface. SILVACO simulation is further done based on implementation of Shockley-Read Hall recombination, Auger recombination, bandgap narrowing effects and Fermi-Dirac statistics. The simulation has been performed using Transfer matrix method and under AM1.5 solar spectrum at room temperature. The corresponding power density of input beam is 100 mW/cm² is considered.

Energy band diagram of proposed structure is illustrated in Figure 2. Without considering lattice mismatch between two materials, Anderson model (or so called electron affinity model) is used which directly relates the conduction band discontinuity with electron affinity difference of two semiconductor materials. As shown in energy band diagram, a large valance band discontinuity is present in the ZnO/Si heterojunction. Due to this discontinuity, chance of transport of photogenerated holes from Si to n-ZnO is reduced. This again reduces the availability of holes in n-ZnO which in turn reduces the rate of recombination.

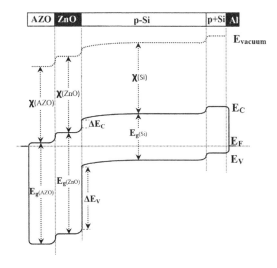

Figure 1. Schematic diagram of Hetero-junction solar cell based on crystalline p-Si and ZnO.

Figure 2. Schematic energy band diagram of ZnO/p-Si Heterojunction Solar cell under equilibrium condition.

3 RESULTS AND DISCUSSION

3.1 *Effects of emitter thickness*

Due to wider band gap of ZnO emitter, it absorbs higher energy photons and lower energy photons are transmitted to the bottom Si layer. Now thickness of emitter layer plays an important role because transmission of photons through this layer greatly depends on it. We have used transfer matrix method for determining this transmission. Larger thickness of ZnO layer may produce high photocurrent in the emitter but that may cause less photocurrent in the bottom layer due to arrival of less number of photons. At the same time, diffusion length of material is closely related to the optimum thickness of a layer in the solar cell. So, it is very important to study the performance with variation of thickness of emitter layer. V-J curve for different thicknesses of ZnO layer is shown in Figure 3. It is clear from figure that the short circuit current density initially increases significantly whereas open circuit voltage increases insignificantly with increasing the thickness of ZnO layer. This is due to the enhanced absorption of photons in the emitter for its large thickness. But after 80 nm thickness of ZnO layer, short circuit current density decreases. This is due to the combined effect of absorption in emitter and Si layers both. So, choice of emitter thickness is crucial to obtain improved performance.

To study the variation of efficiency with emitter thickness, plot of Power Conversion Efficiency (PCE) of solar cell as a function of ZnO layer thickness is shown in Figure 4. Efficiency initially

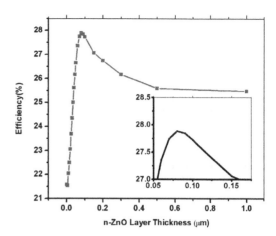

Figure 4. Effect of emitter (ZnO) thickness on power conversion efficiency of ZnO/p-Si heterojunction solar cell. Inset shows the enlarge view of efficiency curve to show maximum power conversion efficiency reached at 80 nm.

increases sharply with the thickness of ZnO layer but decreases slowly after a particular value of emitter thickness (80 nm here). Maximum efficiency of 27.88% is obtained corresponds to the open circuit voltage of 0.62767 V and short circuit current density of 54 mA/cm^2. Decrement of efficiency is due to the combined effects of two phenomenons as explained in the following. Transmission of light into the bottom layer decreases with increase in the emitter thickness. At the same time, though the absorption of light is enhanced as ZnO thickness increases, but the photogenerated carriers, produced outside the diffusion length of holes in ZnO, cannot reach the contacts and hence unable to contribute to the photocurrent. As a result, efficiency decreases. Moreover, more recombination takes place due to presence of more recombination centers in thick ZnO layer.

Series resistance is also increased as ZnO thickness is increased.

3.2 *Effects of absorber layer thickness*

Absorber (p-Si) layer thickness has also important role on the performance of heterojunction solar cell under consideration. Voltage-current (V-J) curve is shown in Figure 5 for different thicknesses of absorber layer. The thickness of absorber layer has been varied from 10 µm to 160 µm in this plot. It is seen from figure that both the short circuit current density and open circuit voltage increase as the thickness of absorber layer increases. Thick absorber layer causes more absorption which in turn causes more generation of carriers and hence enhancement of photocurrent. It is interesting to

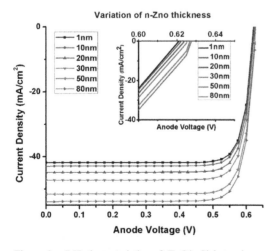

Figure 3. J-V characteristics of ZnO/p-Si heterojunction solar cell for different emitter (ZnO) thickness. Inset shows the enlarge view of J-V curve to show open circuit voltage for different emitter thickness.

observe that photocurrent density is very less for very low value of absorber layer thickness.

When absorber layer thickness becomes very low, the back contact interface becomes very close to the depletion region. Thus, photogenerated electrons are captured easily by back contact due to surface recombination and, so, fewer electrons will contribute to the photo-current. This recombination rate can be decreased by adding a BSF (p^+-Si) layer. The increment in V_{oc} and J_{sc} results in increment of PCE. Plot of PCE as a function of absorber layer thickness is shown in Figure 6.

4 CONCLUSIONS

We have studied the performance of ZnO on c-Si heterojunction solar cell for different values of physical and device parameters. Top ZnO layer and bottom c-Si layer both have important role on the efficiency of the device. Effect of ZnO layer thickness is more significant than that of bottom (Si) layer thickness. Choice of ZnO thickness is very important concerning the absorption of shorter wavelengths, transmission of longer wavelengths for bottom layer and anti-reflection coating. Obtained optimum conversion efficiency is 27.88% with V_{oc} = 0.62767 V, J_{sc} = 54 mA/cm^2 for ZnO layer thickness of 80 nm.

ACKNOWLEDGMENTS

This work is partly supported by MHRD, Govt. of India through the Centre of Excellence in Renewable Energy, Project under FAST (Project No. MHRD (CoE)/RE/2014-2015/402/INST) at Indian School of Mines, Dhanbad.

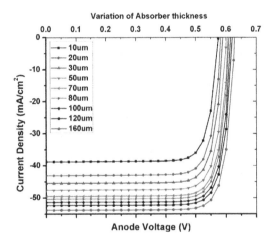

Figure 5. J-V characteristics of ZnO/p-Si heterojunction solar cell for different absorber (p-Si) thickness.

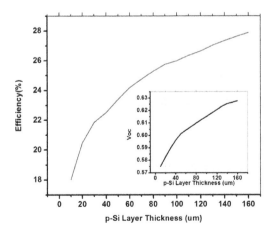

Figure 6. Effect of absorber (p-Si) thickness on power conversion efficiency of ZnO/p-Si heterojunction solar cell. Inset shows the variation of open circuit voltage with change in absorber (p-Si) thickness.

REFERENCES

Dong, B. Z., Fang, G. J., Wang, J. F., Guan, W. J. and Zhao, X. Z. (2007). "Effect of thickness on structural, electrical, and optical properties of ZnO: Al films deposited by pulsed laser deposition". *Journal of Applied Physics.* 101(3): 33713–33717.

Green, M. A., Emery, K., Hishikawa, Y., Warta, W., and Dunlop, E. D. (2016). Solar cell efficiency tables (version 47). *Progress in Photovoltaics: Research and Applications.* 24(1): 3–11.

Janotti, A. and Van de Walle, C. G. (2009). Fundamentals of zinc oxide as a semiconductor. *Reports on Progress in Physics,* 72(12): 126501

Kishimoto, S., Yamamoto, T., Nakagawa, Y., Ikeda, K., Makino, H. and Yamada, T. (2006). Dependence of electrical and structural properties on film thickness of undoped ZnO thin films prepared by plasma-assisted electron beam deposition. *Superlattices and Microstructures.* 39(1–4): 306–313.

Kobayashi, H., Mori, H., Ishida T. and Nakato, Y. (1995). Zinc oxide/*n*-Si junction solar cells produced by spray-pyrolysis method. *Journal of Applied Physics.* 77(3): 1301–1307.

Look, D.C. (2001). Recent advances in ZnO materials and devices. *Materials Science and Engineering: B.* 80(1–3): 383–387

Peartona, S. J., Nortona, D. P., Ipa, K., Heoa, Y. W. and Steinerb T. (2005). Recent progress in processing and properties of ZnO. *Progress in Materials Science.* 50(3): 293–340.

Richter, A., Hermle, M. and Glunz, S. W. (2013). Reassessment of the Limiting Efficiency for Crystalline Silicon Solar Cells. *IEEE Journal of Photovoltaics,* 3(4): 1184–1191.

Tsokos, K. A. (2012), *Physics for the IB diploma, 5th edition.* United Kingdom: Cambridge University Press.

Electrical technology

Computer, Communication and Electrical Technology – Guha, Chakraborty & Dutta (Eds)
© 2017 Taylor & Francis Group, ISBN 978-1-138-03157-9

Line to line fault detection in a multi-bus power system by harmonic analysis

D.K. Ray
MCKV Institute of Engineering, Kolkata, India

S. Chattopadhyay
Ghani Khan Choudhury Institute of Engineering and Technology, Malda, India

K.D. Sharma & S. Sengupta
University of Calcutta, Kolkata, India

ABSTRACT: This paper deals with the identification of a faulty load bus in a standard multi-bus power system for line to line fault, monitoring the system parameters in presence of various frequencies in P-δ plane. Advanced variable slack bus incorporated converged power flow analysis has been used for determining the load angle, real power and feeder operating points for normal and at fault in various load buses of the network in presence of some chosen harmonic frequencies in the system. The shifting of the normal and fault operating points have been monitored on the normal and fault P-δ planes, and depending on the characteristic feature of the operating points in the P-δ plane, respective fault buses have been identified by developing rule sets.

1 INTRODUCTION

Industrial distribution networks are prone to faults, which sometimes cause interruption in supply. These faults have to be predicted far before the damage of a system and this requires the presence of fast fault detectors in a network. A GSM based signaling system (Ugale et al. 2016) has been seen to detect the changes in voltage-current at fault, which is also able to classify various faults depending on the changed parameters of distribution substation connected therewith. An arduino based tripping mechanism has been seen to be developed (Morey et al. 2015) for a small three phase with a single 230/12 V transformer for permanent and temporary faults. The system has been seen to be restored within 5–10 seconds for temporary fault in the system. Discrete Wavelet Transform based power system fault diagnosis (Jose et al. 2014) has been seen in simulation studies. Wavelet Transform and Neural Network based (Abohagar et al. 2013) fault detection has been seen where wavelet transform has been used for determining the healthy and fault coefficients. The fault coefficients obtained from wavelet transform has been seen to train the neural network which has been used for detecting various unsymmetrical faults in a simulated system. The analysis has been seen to be done in PSCAD software. Development of an automatic tripping circuit using 555 timer, voltage regulator (LM7805), relays, comparators, transformer

(230/12V) has been seen for isolating the circuit components for temporary and permanent faults, wherein the circuit resets back to normal condition for temporary fault and remains permanently off for long duration faults (Bakanagari et al. 2013). Unsymmetrical fault assessment has been seen on a 110 kV sub-grid coupling using a superconducting fault current limiter, which depicts highest voltage drop for double line to ground fault during the limitation in the relevant range of impedance (Stemmle et al. 2007). Modeling of a multi-bus network has been seen for multiple distributed generation injection in order to analyze the dynamic performance, steady state and fault conditions (Davidson et al. 2004). Linear graph theory approach based network modeling has been seen wherein generalized model based converged power flow solution presents an accurate estimation of voltages, currents and power flows in faulted three phase unbalanced non-radial distribution system (Halpin et al. 1994). Assessment of harmonic voltages and currents has been seen in a multi-bus power system for symmetrical fault in load bus, which depicts vivid change of system parameters at fault in load bus considering various frequency components present in the system (Kar et al. 2015). Identification of faulty bus in a multi-bus power system has been seen, monitoring the feeder stability zones, for the occurrence of symmetrical short circuit fault in load bus of the system in presence of various inter and sub-synchronous inter-harmonics present in the system (Kar et al. 2016) using P-δ planes. None of the analysis, as before, has been seen to deal with the fault bus identification monitoring the load angle and feeder operating point shifting in harmonic P-δ plane to accurately depict which type of fault has actually occurred in the system. In this paper a technique has been introduced for fault bus identification in a multi-bus power system, monitoring the load angle and feeder operating point shifting in normal and harmonic P-δ planes. Here an advanced variable slack bus incorporate converged load flow solution has been used for determination of the load angle and feeder operating points, whose deviation has been assessed in normal and harmonic P-δ planes to depict the faulty bus in the system.

2 FAULT BUS IDENTIFICATION USING HARMONIC P-DELTA PLANE

The fault bus identification technique has been done on a test standard IEEE 9 bus system. Modeling of the system has been done in ETAP software with conventional line and bus data of the system (IEEE Std. 1459-2010) for acquiring the system data at Line to Line (LL) fault in some chosen load buses of the system, in presence of various harmonic frequencies in the network. The modeled network has been presented in Figure 1. However this analysis technique can also be implemented for other standard transmission and distribution systems, since this offline analysis technique has been implemented on a standard bus system where LL fault has been done in the system modeled in ETAP, wherein the fault restoration time has not been considered. Here the system has been analyzed at fault condition only and the 7th and 11th harmonic frequencies have been used to justify the effect of these frequencies in the detection of remote end faults in the system with the developed P-δ plane analysis technique. These frequencies has been chosen since some benchmark report, IEEE Std. 1459-2010, has been seen to use some of these arbitrary frequency values for determining the performance of multi-bus power system networks.

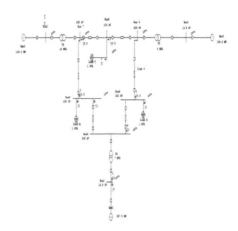

Figure 1.   Standard bus system modeled in ETAP.

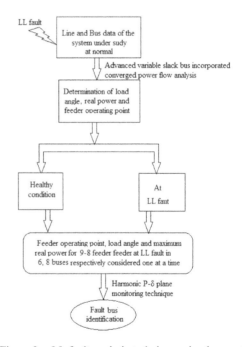

Figure 2.   LL fault analysis technique using harmonic P-δ plane.

Table 1.   Study of maximum real power, load angle and feeder operating points at LL fault in buses 6 and 8.

| Frequency | Changes observed | | | | | | | | |
| | Pmax (p.u.) | | | δ (p.u.) | | | P (p.u.) | | |
	h	f_{LL6}	f_{LL8}	h	f_{LL6}	f_{LL8}	h	f_{LL6}	f_{LL8}
50 Hz	75.79	24	22.26	−0.5	0.7	0.8	−18.31	6.07	5.75
350 Hz	8.15	7.8	23.87	−6.18	−3.1	−3.18	−4.89	−3.41	−2.42
550 Hz	3.96	28.86	27.56	−3.26	−3.11	−3.2	−5.5	−15.12	−14.26

The fault data obtained from the above analysis has been used for building the harmonic P-δ planes at LL fault in buses 6 and 8 of the system in presence of certain harmonic frequencies in the network. The algorithm of the developed analysis technique has been shown in Figure 2.

The real power, load angle and feeder operating points of 9–8 feeder has been analyzed at fault in buses 6 and 8, considered only one at a time, in presence of some chosen harmonic frequencies (7th and 11th). The data of the above analysis has been shown in Table 1. In Table 1, "h" denotes the data at healthy condition of the system, "f_{LL6}" and "f_{LL8}" denotes the data of the system at fault in buses 6 and 8 considered one at a time. From the data obtained in Table 1, harmonic P-δ curves have been developed (Figure 3), wherein the feeder operating

points have been placed. The P-δ curves have been plotted from the concept of occurrence of individual harmonic frequencies in a system (Kar et al. 2015, Kar et al. 2016, IEEE Std. 1459-2010). The deviation of the feeder operating points and the change in direction of load angle of the feeders has been identified for depicting the faulty bus in the system. This analysis has been done in MATLAB.

In Figure 3, the operating points have been seen to vary widely in the harmonic P-δ planes.

3 EXACT FAULT BUS IDENTIFICATION RULE SETS

The data obtained from Table 1 has been analyzed in addition to the Figure 3, from which some

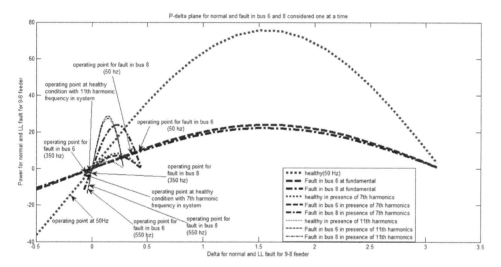

Figure 3. Harmonic P-δ plane for 9–8 feeder at LL fault in Bus 6, 8 considered one at a time.

Table 2. Feature extraction for identifying operating zone of feeders in the system.

Frequency	Location of operating points		
	At healthy condition	At fault in bus 6	At fault in bus 8
50 Hz	Stable motoring zone	Stable generating zone	Stable generating zone
350 Hz	Stable motoring zone	Stable motoring zone	Stable motoring zone
550 Hz	Stable motoring zone	Stable motoring zone	Stable motoring zone

Table 3. Feature extraction for identifying exact fault bus in the system.

Frequency	Changes obtained in δ			Remarks
	Healthy	f_{LL6}	f_{LL8}	
50 Hz	negative	positive	positive	Bus 8 is the exact fault bus
350 Hz	negative	negative	negative	Bus 6 is the exact fault bus
550 Hz	negative	negative	negative	Bus 8 is the exact fault bus

features have been extracted for depicting the fault bus in the system. The features extracted for depicting the exact fault bus in a system has been provided in Tables 2–3. The fault bus identification is dependent on the parametric change in the load angle and the operating zone of the feeders at normal and at fault in normal and fault P-δ plane.

4 SPECIFIC OUTCOME OF THE ANALYSIS

This analysis shows how a faulty load bus can be identified in a multi-bus power system from normal and fault P-δ planes. In this technique, advanced variable slack bus incorporated converged power flow analysis has been used for finding the bus voltages, load angles and real powers at normal and at fault. Table 3, clearly shows that, for 9–8 feeder the power flow direction reverses from healthy condition both for fault in bus 6 and bus 8, but if the feeder operating point can be assessed in P-δ plane it has been seen that the deviation of feeder operating point at fault in bus 8 is much more that for fault in bus 6. Thus it has been concluded that bus 6 is the exact fault bus in the system. Similarly for the system in presence of 7th and 11th harmonic frequencies, power flow direction remains same but the fault bus has been identified considering the position of the feeder operating points again in the normal and fault P-δ planes.

5 CONCLUSION

This paper presents a novel technique for identifying the faulty bus in a multi-bus power system, monitoring the deviation in the feeder operating points and corresponding load angles in normal and fault P-δ planes. It has also been seen that for line to line fault, occurring in a load bus, the harmonic and fault power magnitudes sometimes becomes larger than fundamental power magnitudes in P-δ planes and there lies a wide range of variation of the feeder operating points in P-δ planes from −90° to +90° for LL fault in different load buses of the system. Thus this analysis is well suitable for fault bus identification, monitoring the feeder operating points and the direction of power flow in a multi-bus power system.

ACKNOWLEDGEMENT

We are thankful to the EE Department, MCKV Institute of Engineering, Kolkata, India, for providing the ETAP and MATLAB software for analysis purpose.

REFERENCES

Abohagar, Abdulhamid. A., Mustafa, M.W., (2013). Identification of Asymmetrical Faults in Electrical Power Systems Based on Signal Processing and Neural Network, ARPN J. Engng and Applied Sciences, 8 (9), ISSN: 1819-6608, p. 699–702.

Anderson, P.M., Fouad, A. A., (2003). Power System Control and Stability, IEEE Press (eds), ISBN 978-0-471-23862-1, p. 38, New York.

Bakanagari, Sathish, Kumar, A. Mahesh, Cheenya, M, (2013). Three Phase Fault Analysis with Auto Reset for Temporary Fault and Trip for Permanent Fault, IJERA, 3 (6), ISSN: 2248-9622, p. 1082–1086.

Davidson, I.E, (2004). Modeling and analysis of a multibus reticulation network with multiple DG. Part II. Electrical fault analysis, Proc. AFRICON, 2 (1), ISBN (print): 0-7803-8605-1. DOI:10.1109/AFRICON.2004.1406797, p. 811–814.

Halpin, S. M., Grigsby, L. L, (1994). Fault analysis of multi-phase unbalanced non-radial power distribution systems, Proc. ICPS, ISBN (print): 0-7803-1877-3. DOI: 10.1109/ICPS.1994.303574, p. 203–209.

IEEE Standard Definitions for Measurement of Electric Power Quantities under Sinusoidal, Non-Sinusoidal, Balanced or Unbalanced Conditions, IEEE Power and Energy Society, Std. 1459–2010.

Jose, Prince., Bindu, V. R., (2014). Wavelet-Based Transmission Line Fault Analysis, IJEIT, 3 (8), ISSN: 2277-3754, pp. 55–60.

Kar Ray, D., Chattopadhyay, S., Das Sharma, K., Sengupta, S., (2015). Assessment of Harmonic Voltage angles in a multi-bus power system during symmetrical fault at certain bus, in IET Digital Library. DOI: 10.1049/cp.2015. 1638, pp. 42–47.

Kar Ray, D., Chattopadhyay, S., Das Sharma, K., Sengupta, S., (2016). Identification of Faulty Load bus in a Multi-Bus Power System, in IEEE Digital Library, pp. 289–293. DOI: 10.1109/CIEC.2016.7513670.

Morey, M. S., Ghodmare, Amit., Khomane, Vaibhav., Singh, A. K., Dewande, J., Shaikh, S. A. I., (2015). Microcontroller Based Three Phase Fault Analysis for Temporary and Permanent Fault, IRJET, 2 (1), e-ISSN: 2395-0056, p-ISSN: 2395-0072, pp. 59–63.

Stemmle, M., Neumann, C., Merschel, F., Schwing, U., (2007). Analysis of Unsymmetrical Faults in /High Voltage Power Systems with Superconducting Fault Current Limiters, In IEEE Transactions on Applied Superconductivity, 17 (2), ISSN: 1051-8223. DOI: 10.1109/TASC.2007.899136, pp. 2347–2350.

Ugale, U., Yadav, A., Ugle, S., Sarkar, J., (2016). Distribution Line Fault Detection & GSM Module Based Fault Signaling System, IJRASET, 4 (2), IC value: 13.98, ISSN: 2321-9653, pp. 452–455.

Computer, Communication and Electrical Technology – Guha, Chakraborty & Dutta (Eds)
© 2017 Taylor & Francis Group, ISBN 978-1-138-03157-9

Loss minimization of power network by reconfiguration using GA and BPSO

S. Pal, J.N. Bera & S. Sengupta

Department of Applied Physics, University of Calcutta, Kolkata, West Bengal, India

ABSTRACT: A multi-objective network reconfiguration based on Genetic Algorithm (GA) and Binary Particle Swarm Optimization (BPSO) approach is presented in this paper. The main objective of the network reconfiguration is to minimize the real power loss with limited deviation of the node voltages. The algorithm starts with categorization of the switching sequences for radial network configuration while observation of the P_{losses} and the profile of voltage were done. The final topology signified the minimized P_{losses} condition with all nodes receiving power with acceptable voltage profile. The result of simulations on IEEE 57 bus system is given and compared with the initial case which has all switches closed. This method proved that improvement of P_{losses} has been made by change of the switching topology. The simulation is done both with GA and BPSO technique in MATLAB environment. Comparison shows that BPSO gives a better optimized result in much less time than GA.

1 INTRODUCTION

Load varies time to time in a power distribution system which defines it as a dynamic system. With increase in load density, the operation becomes more and more complex. Power losses increase with decrease in voltage level and enhancement of current. Loss minimization is thus an important problem in this dynamic system, in which different approaches have been proposed which varies each other in problem formulation and selection of techniques. Change in the topology of the network or reconfiguration is one of the methods in reduction of losses and was first proposed by Merlin & Back (1975). The topology is changed with opening and closing of tie switches (normally open) and sectionalizing switches (normally closed). This helps to redirect the power flow when needed so that it can flatten the peak demands, improve voltage profile, increase reliability, balance loads, help in planned outages for maintenance and minimize loss.

The basic variable in reconfiguration problem is switching operations. Opening of certain switches and closing of certain other should ensure that the radial structure is maintained and that all the buses receive power. Since, the switch states are discrete in nature; the problem may be defined as discrete, constrained combinatorial optimization problem as by Zimmerman (1992). Again, there can be many candidate configurations of a network which gives optimized results. This is why many metaheuristic algorithms are used to solve this problem. Many such techniques have been improved over the years

and applied to the reconfiguration problem. For example, application of Genetic Algorithm was seen in the reconfiguration problem (Subburaj, Ramar, Ganesan & Venkatesh 2006, Nagy, Ibrahim, Ahmed, Adail & Soliman 2013, Pal, Sen & Sengupta 2015). Simulated Annealing was applied by Jeon, Kim J.C., Kim J.O., Shin & Lee (2002), while Ant Colony Optimization by Rao, Narasimham & Ramalingaraju (2008), Particle Swarm Optimization by Chang & Lu (2002) and many more. These algorithms give global solution but are difficult to be applied in the real time distribution automation because of the slow nature.

Study reveals that there are many similarities between PSO and GA. Both starts with randomly creating an initial set of solution. The evolutionary processes take place through iterations and finally an optimal value can be obtained. The difference between these two techniques is that PSO has no explicit selection, crossover or mutation processes as found in a work by Eberhart & Shi (1998). PSO is based on search space where each particle uses its best solution attained so far and the flock's current best solution to update its information like velocity and position. The concept of PSO is also simpler to implement than GA as given by Truong, Nguyen T.T., Nguyen L.T. & Pham (2015). However, typical PSO is for a continuous problem whereas reconfiguration by switching operation is a discrete problem. Therefore Binary Particle Swarm Optimization (BPSO) is used to solve this discrete problem as given in many works (Jin, Zhao, Sun, Li & Zhang 2004, Lee, Soak, Oh, Pedrycz & Jeon 2008).

In this paper, reconfiguration problem has been simulated with two different approaches and the results compared. The Genetic Algorithm is used with bit string type of population which takes care of the discrete nature of the switching conditions. For the same reason Binary PSO is used. The simulation is done on the IEEE-57 bus system, minimum loss conditions are achieved and the results along with time complexity are compared.

2 PROBLEM FORMULATION

The main aim of network operators is to minimize the total MW losses if the system is in normal state or after fault clearance. The problem can be formulated as:

$$Minimize\, f\left(x,c\right)\quad x \in S \tag{1}$$

where the variables are defined as follows:

"c" represents the voltage magnitude, angle and other network conditions at the time of operation; "x" denotes the variables, that is, operating state of the switches; "S" is the set of all possible configurations; and, "$f(x, c)$" is the sum of the real power losses.

Here, $f(x, c)$ is the objective function to be minimized which gives a measure of the real power loss in the current state c. In order for a configuration to be a valid solution to the problem it must satisfy certain constraints. The corresponding state c must be consistent with Kirchhoff's current and voltage laws to satisfy the electrical constraints and must satisfy the operational constraints of the system by not exceeding the physical limitations of any of the system components. It must also maintain bus voltages within appropriate bounds.

Any solution x satisfying the constraints is one of the best possible configurations. If the total number of tie and sectionalizing switches in the system is n_s, then the present operating state of the switches is represented as a vector $x = [x_1, x_2, ..., x_{ns}]$ where individual switch states $x_i \in \{0,1\}$, $1 \le i \le n_s$, where $x_i = 1$ indicates that switch i is closed, and $x_i = 0$ indicates that it is open.

In the present problem, $f(x, c)$ is the summation of the real power losses as in Eqn. (2) and S_{ij} and S_{ji} are calculated in Eqn. (3)–(4).

$$f(x,c)=\sum_{\substack{i=1\\j=1\\i\neq j}}^{n}Real(S_{ij}+S_{ji}) \tag{2}$$

$$S_{ij} = P_{ij} + jQ_{ij} = \bar{V}_i\bar{I}_{ij}^{*} \tag{3}$$

$$S_{ji} = P_{ji} + jQ_{ji} = \bar{V}_j\bar{I}_{ji}^{*} \tag{4}$$

where S_{ij} is the apparent power from node 'i' to node 'j'. The inequality constraints are:

$$V_i^{\min} \le V_i \le V_i^{\max} \tag{5}$$

$$I_i^{\min} \le I_i \le I_i^{\max} \tag{6}$$

3 OPTIMIZATION TECHNIQUES

3.1 Genetic algorithm

The algorithm starts with a current population, the first of which is called initial population. Selection process is executed on this current population to create intermediate population. Then crossover and mutation is applied on it to create next set of population.

In the present problem, the chromosome represents the variables which are the switching states. Population size is taken as twenty, where the initial population is randomly created. Due to discrete nature of switching state, bitstring type of population is considered. Each chromosome is evaluated through load flow by Newton-Raphson method and the value of the objective function calculated. Unless stopping criteria is met, uniform mutation and two-point crossover is performed to again get a new set of population. This goes on till the best chromosome is found to give the best optimized result.

3.2 Binary Particle Swarm Optimization

In PSO, each individual of a population set is treated as a particle in a search space. Each particle evaluates the function at each point it visits in space. Each particle or agent remembers the best value of the function so far by it (pbest) and its co-ordinates. Again, each agent knows the globally best position that one member of the swarm had found, and its value (gbest). Using the co-ordinates of the pbest and gbest each agent updates its new position and velocity and again evaluates the function unless best optimized result is obtained.

In case of BPSO, the trajectories are changes in the probability that a coordinate will take on binary value (0 or 1). The moving velocity is defined in terms of changes of probabilities that abit (position) will be in one state or the other. Thus a particle moves in a state space restricted to 0 and 1 on each dimension.

In the present problem, number of particle or population size is taken as twenty. The inertia weight is taken as 0.9 and c_1 and c_2 as 2 each.

4 SIMULATION AND RESULTS

The program was executed both by GA and BPSO creating random initial population with same population size and considering switches on every line of the IEEE-57 bus system.

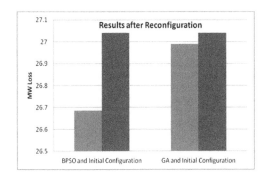

Figure 1. Comparison of results.

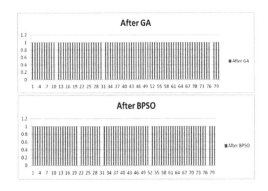

Figure 2. Switching states after simulation.

Table 1. Compared data after simulation.

Comparison	Initial configuration	GA	BPSO
MW loss	27.041	26.989	26.685
MVar loss	150.358	150.611	150.227
Time (sec)	–	133.25	51.59

Table 2. Switching states after simulation.

Test system	No. of switches	Open switches after GA	Open switches after BPSO
IEEE-57 Bus	80	Switches between buses	Switches between buses
		9–12, 21–20, 57–56	5–6, 21–20, 36–40, 57–56

As the foremost step, the active power loss was calculated with all the closed switches of the network. This was considered as the initial

Table 3. Voltage Profile after simulation.

Bus No.	At Initial Configuration (p.u.)	After GA (p.u.)	After BPSO (p.u.)
1	1.04	1.04	1.04
2	1.02	1.02	1.02
3	1.015	1.015	1.015
4	1.0085	1.0081	1.0099
5	1.0059	1.0058	0.998
6	1.01	1.01	1.03
7	1.0232	1.0232	1.031
8	1.055	1.055	1.055
9	1.01	1.01	1.01
10	1.0052	1.0053	1.0057
11	0.9982	0.9984	0.9966
12	1.025	1.025	1.025
13	0.9977	0.9979	0.9982
14	0.9891	0.9894	0.9903
15	1.007	1.0072	1.0078
16	1.0207	1.0206	1.0207
17	1.0213	1.0213	1.0213
18	1.0065	1.0032	1.0051
19	0.9814	0.9642	0.9662
20	0.9783	0.9528	0.9548
21	1.0277	1.031	1.0364
22	1.0297	1.031	1.0364
23	1.0282	1.0296	1.035
24	1.018	1.0189	1.0258
25	0.9727	0.9736	0.982
26	0.9787	0.9794	0.9863
27	1.0115	1.0119	1.0192
28	1.0301	1.0304	1.0379
29	1.0457	1.0458	1.0533
30	0.9553	0.9562	0.9652
31	0.9349	0.9356	0.9459
32	0.9585	0.959	0.9707
33	0.9563	0.9568	0.9685
34	0.9775	0.978	0.9902
35	0.9854	0.9859	0.9982
36	0.9956	0.9961	1.0084
37	1.0049	1.0049	1.0152
38	1.0331	1.0339	1.039
39	1.003	1.0023	1.0126
40	0.9928	0.9937	0.924
41	1.0219	1.0215	1.0072
42	0.9932	0.9944	0.971
43	1.035	1.0351	1.0298
44	1.037	1.0377	1.0419
45	1.0559	1.0563	1.0584
46	1.0806	1.0811	1.0832
47	1.0539	1.0546	1.0579
48	1.0479	1.0486	1.0524
49	1.057	1.0576	1.0603
50	1.0444	1.0448	1.0468
51	1.0728	1.073	1.0736
52	1.0088	1.0089	1.015
53	0.9954	0.9955	1.0008
54	1.0239	1.0239	1.0269
55	1.061	1.0611	1.0619
56	0.9938	0.9964	0.9645
57	0.9897	0.9913	1.0023

configuration with which the optimized results were compared. The time of execution for both the optimized techniques was compared.

Figure 1 shows the comparison of the MW losses which shows better result is obtained with BPSO. Figure 2 depicts the switching states.

Compared data after the simulation is shown in Table 1. It signifies that much less time is taken to execute BPSO and optimized result is also better. The switching states found when optimum loss is obtained are given in Table 2 whereas Table 3 shows the voltage profile after simulation.

5 CONCLUSION

Reconfiguration is an optimization problem for finding the minimum loss function subject to voltage profile maintenance. The results verify that the algorithms are able to optimize the objective function. GA has a very high time complexity and slow convergence rate due to its reproduction procedures like crossover and mutation. The implementation of the algorithm is also very complex compared to BPSO. Though both are metaheuristic optimization techniques and give one of the best results possible, time complexity is improved from GA to BPSO. Further research is possible in future for its improvement so that it can be applied to a real system.

ACKNOWLEDGEMENT

The authors acknowledge the usage of the laboratories in the Department of Applied Physics, University of Calcutta and financial assistance in the form of INSPIRE Fellowship to the first author from DST.

REFERENCES

Chang, R.F. & Lu, C.N., 27–31 Jan. 2002, "Feeder Reconfiguration for Load Factor Improvement", IEEE Power Engineering Society Winter Meeting, Vol. 2, pp. 980–984.

Eberhart, R.C. & Shi Y., 1998, "Comparison between Genetic Algorithms and Particle Swarm Optimization", The 7th Annual Conference on Evolutionary Programming, San Diego, USA.

Jeon, Young-Jae, Kim, Jae-Chul, Kim, Jin-O., Shin, Joong-Rin & Lee, Kwang Y., 2002, "An Efficient Simulated Annealing Algorithm for Network Reconfiguration in Large-Scale Distribution Systems", *IEEE Transactions on Power Delivery*, Vol. 17, No. 4, pp. 1070–1078. Retrieved from http://web.ecs.baylor.edu/faculty/lee/papers/journal/2002/200210.pdf.

Jin, X., Zhao, J., Sun, Y., Li, K., Zhang, B., 21–24 November 2004, "Distribution Network Reconfiguration for Load Balancing Using Binary Particle Swarm Optimization", 2004 International Conference on Power System Technology—POWERCON 2004 Singapore.

Lee, S., Soak, S., Oh, S., Pedrycz, W., Jeon, M., 2008, "Modified binary particle swarm optimization", *Progress in Natural Science 18 (2008)*, pp. 1161–1166. Retrieved from http://dx.doi.org/10.1016/j.pnsc.2008.03.018.

Merlin, A. & Back, H., 1975, "Search for a minimum loss operating spanning tree configuration for urban power distribution System," in Proc. 5th Power Syst. Computation Conf. (PSCC), Cambridge, U.K., Paper 1.2/6.

Nagy, S.A., Ibrahim, I.S., Ahmed, M.K., Adail, A.S. & Soliman, S., 2013, "Network Reconfiguration for Loss Reduction in Electrical Distribution System Using Genetic Algorithm", *Arab Journal of Nuclear Science and Applications*, 46(1), pp. 78–87. Retrieved from http://www.esnsa-eg.com/download/research-Files/_7_6.pdf.

Pal, S., Sen, S. & Sengupta, S., 12–13 September, 2015, "Power Network Reconfiguration for Congestion Management and Loss Minimization using Genetic Algorithm." Michael Faraday IET International Summit 2015, pp. 50(6).

Rao, R. Srinivasa, Narasimham, S.V.L., Ramalingaraju, M., 2008, "Optimization of Distribution Network Configuration for Loss Reduction Using Artificial Bee Colony Algorithm", *International Journal of Electrical Power and Energy Systems Engineering*. Retrieved from http://waset.org/publications/2630/optimization-of-distribution-network-configuration-for-loss-reduction-using-artificial-bee-colony-algorithm.

Subburaj, P., Ramar, K., Ganesan, L. & Venkatesh, P., 2006, "Distribution System Reconfiguration for Loss Reduction using Genetic Algorithm", *Journal of Electrical Systems*, (2–4), pp. 198–207. Retrieved from http://journal.esrgroups.org/jes/papers/2_4_2.pdf?acr_id=327.

Truong, A.V., Nguyen, T.T., Nguyen, L.T., Pham, C.T., 2015, "Comparison between Continuous Genetic Algorithms and Particle Swarm Optimization for distribution network reconfiguration", Article in *Mitteilungen Klosterneuburg,* 2015, pp. 65(10).

Zimmerman, Ray Daniel, May 1992, "Network reconfiguration for loss reduction in three-phase power distribution systems", Cornell University, New York.

Computer, Communication and Electrical Technology – Guha, Chakraborty & Dutta (Eds)
© 2017 Taylor & Francis Group, ISBN 978-1-138-03157-9

A comparative study of the polarization–depolarization current measurements on different polymeric materials

A. Kumar, N. Haque, R. Ghosh, B. Chatterjee & S. Dalai
Department of Electrical Engineering, Jadavpur University, Jadavpur, Kolkata, India

ABSTRACT: Polymeric materials are becoming increasingly popular in high-voltage systems because of their superior dielectric properties over conventional oil–paper insulation systems. However, condition monitoring and insulation diagnosis of polymeric insulation is still a critical issue. In this paper, Polarization–Depolarization Current (PDC) measurements, a popular method for condition monitoring of oil–paper insulation, is applied on different polymeric materials. An experimental setup was developed in the laboratory for this purpose. It was observed that the PDC measurements are highly dependent on material properties. The effect of temperature on polarization and depolarization processes was also studied and it was found that these processes are very sensitive to temperature. However, the degree of temperature dependence was different for different materials.

1 INTRODUCTION

The modern power network is a complex system that consists of a variety of electrical equipment such as power transformer, circuit breakers, potential transformer, current transformer, and high-voltage transmission lines and cables. Failure of any of these equipment may lead to interruptions in the power supply, which will further result in substantial loss of money and time. Hence, to ensure the reliability and quality of power system, it must be ensured that outages due to equipment failure are minimized. A major part of outages in a power network is caused by insulation failure. Environmental conditions such as temperature, humidity, and continuous electrical stress can degrade the insulation condition, which is vulnerable to electrical breakdown; therefore, condition monitoring-based maintenance of solid insulation is necessary.

There are various experimental techniques used for condition monitoring of solid insulating materials, which have been reported by various researchers. They include Thermally Stimulated Depolarization current (TSDS) (Bamji *et al.,* 1993, Martinez *et al.,* 1997), Laser Intensity Modulation Method (LIMM) (Wubbenhorst *et al.,* 1998), differential scanning calorimetry, Fourier transform infrared spectroscopy (FIIR), tensile strength measurement, and partial discharge measurement.

In this paper, polarization and depolarization current measurement techniques are used for condition monitoring of solid insulating materials used in high-voltage equipment. The polarization and depolarization current measurement test was successfully

used for the assessment of power transformers by researchers and the investigation of the moisture content and conductivity of oil insulation (Baral *et al,* 2013). This technique has also been applied for the assessment of the insulation of power cables (Bhumiwat *et al,* 2010) and XLPE insulation subjected to wet aging (*Abou et al. 2001*).

In this work, polarization–depolarization current measurements were made on samples of High-Density Polyethylene (HDPE), Polypropylene (PP), PTFE, and Polymethylate (PMMA) using an experimental setup developed in the laboratory. The effect of temperature on the polarization–depolarization processes was also investigated. Finally, conclusions were made on the feasibility of the presented method in the investigation of material properties and insulation diagnosis involving polymeric materials.

2 POLARIZATION AND DEPOLARIZATION CURRENT MEASUREMENT

In the Polarization and Depolarization Current (PDC) measurement, a steady excitation voltage is applied to the insulation under test (Houhanessial *et al.* 1998). The dipoles in the insulation try to align in the direction of the applied field and the polarization process starts. During polarization, monotonically decreasing current i_{pol} flows through the insulating media (as shown in Figure 1). These polarization processes are completed when all the dipoles are oriented in the direction of the applied field. Once the polarization process is completed, the polarization current becomes zero and

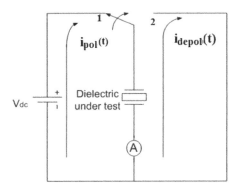

Figure 1. Basic experimental arrangement for the PDC measurement.

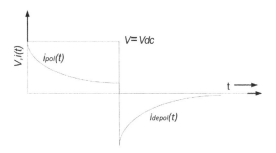

Figure 2. Typical nature of polarization and depolarization current.

conduction current flows through the medium. The magnitude of the conduction current depends on the insulation resistance of the dielectric media.

Now, if the dielectric specimen is short-circuited, the dipoles start to return to their original state, and the stored energy during polarization starts to release. This process is called de-polarization. Because of this process, a monotonically decreasing depolarization current, i_{depol}, flows in the direction opposite to i_{pol}. The typical nature of polarization and depolarization current is shown in Figure 2 (Gafvert et al., 2000).

If a steady voltage V_{dc} is applied to the insulation under test, the polarization current flowing through the insulation under test (as shown in Figure 2) can be obtained from equation (1) by replacing $V(t)$ by V_{dc}:

$$i_{pol}(t) = C_0 V_{ck} \left[\frac{\sigma_0}{\varepsilon_0} + \varepsilon_\infty \delta(t) + f(t) \right] \qquad (1)$$

where t_c represents the time span during which the field is applied to the test sample. After t_c, the excitation voltage source is removed and the test object is short-circuited. It may be observed that the displacement current ($\varepsilon_\infty \delta(t)$) has zero

contribution to the polarization current i_{pol} except at $t = 0$. Therefore, the resultant current is mainly composed of two components: conduction current, which is contributed by conductivity σ_0, and the dielectric response function $f(t)$, which can be modeled from the polarization current provided that the conduction current is known.

At the end of t_c, the voltage source is removed and the test object is short-circuited. The dipoles in the insulation start to relax and release the stored energy to return to the original state (Williams et al. 1985). The depolarization current i_{depol} starts to flow through the insulating material due to the re-orientation of the dipoles. This depolarization current can be expressed as:

$$i_{depol}(t) = -C_0 V_{dc} \left[f(t) - f(t - t_C) \right] \qquad (2)$$

It has already been mentioned that the dielectric response function $f(t)$ is a monotonically decreasing function. Therefore, for a sufficiently long charging time t_c, the magnitude of $f(t - t_c)$ is very low with the respect to $f(t)$ and can be neglected (Saha et al, 2005). Hence, the depolarization current in equation (2) can be re-written as equation (3):

$$i_{depol}(t) = -C_0 V_{dc} f(t) \qquad (3)$$

3 EXPERIMENTAL SETUP AND EXPERIMENTAL PROCEDURE

As discussed in the previous section, to measure polarization and depolarization currents in solid insulating materials, an experimental setup has been developed in the high-voltage engineering laboratory. It is capable of generating short time voltage pulses with varying pulse width and frequency, so that it can be adjusted according to the application requirements. The overall block diagram of this measurement of polarization and depolarization current of solid insulating material is shown in Figure 3. A critical part of the experimental setup is the excitation source which feeds the sample. The source needs to polarize and depolarize the sample and at the same time, stress it up to high voltage, so that the phenomena that are active particularly at high voltage levels are also reflected in the response current. In this circuit, the DC blocking capacitor helps to feed the sample with square wave excitation source, which is already at an elevated voltage. The sample is kept in a controlled oven in which the temperature and humidity can be adjusted. The response current is captured through a 16 bit data acquisition system Model X5133 from national instrument. This device is capable of capturing the analog signal

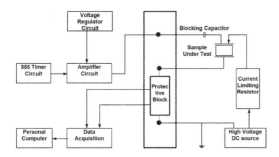

Figure 3. Block diagram of the experimental setup for measurement of polarization and depolarization current in solid insulating materials.

Figure 4. Photograph of the overall experimental setup for polarization and depolarization current measurement (1) High-voltage generator, (2) Electrodes and polymer sample, (3) PC, (4) A/D Converter system, (5) Data acquisition system.

and converts it into 16 bit equivalent digital data that can be stored in the computer.

After the setup is ready, the insulation samples are connected to the setup. The insulating samples are made of different insulating materials (LDPE, HDPE, PP, PMMA, TEFLON, and RUBBER) and are brought commercially in sheet sizes. After that, the sheets were cut down to the size of 4 × 4 cm. Electrodes made of brass with diameter of 5 cm are used in experiments. The samples were placed in between the electrode. Now the excitation source was kept on and 30 V pulses were applied on the sample. The high-voltage DC source as already mentioned was separated from the excitation source through a DC blocking capacitor. The polarization current data are recorded for 4 ms. After that, the setup is shifted to depolarization current mode and data are recorded for another 4 ms. After testing, the polarization and depolarization measurements are put into folders in the PC automatically. The photograph of the overall experimental setup for polarization and depolarization current measurement is shown in Figure 4.

4 RESULT AND DISCUSSIONS

4.1 Polarization current measurement

In this section, the results of polarization and depolarization current measurement of different insulating materials are shown in Figure 5(a) and 5(b). It can be observed that both peak polarization–depolarization current and settling time are highly dependent on the chosen materials. For HDPE, PP, PMMA, and PTFE, the peak polarization current varies in the range of 4.2–4.7 μA. In case of depolarization, for HDPE, PP, PMMA, and PTFE, the peak depolarization current is almost constant, around (4.5 μA).

4.2 Effect of temperature on PDC measurement of solid insulating materials

In order to investigate the effect of temperature on the insulating properties of the aforementioned insulating materials, the PDC measurement was made on the sample, keeping it in an environmental chamber. The environmental chamber used in these studies is procured from Thermotron, which offers controlled chamber environment from −40

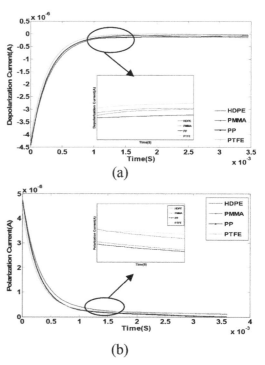

Figure 5. (a) Depolarization current of different insulating materials: (i) HDPE, (ii) PP, (iii) PMMA, (iv) PTFE; (b) Polarization current of different insulating materials: (i) HDPE, (ii) PP, (iii) PMMA, (iv) PTFE.

to 198°C. In addition to temperature, humidity can be controlled from 10% to 97% inside the chamber. The sample under study is kept in the environment chamber at a certain temperature and humidity for 1 h before starting the measurement and PDC measurements are taken subsequently.

In this study, the humidity was kept constant at 20% in all measurements and the chamber temperature was varied from 20 to 90°C. It was observed from Figure 6 that except for PTFE, the peak depo-

larization current increases with time, that is, for PMMA, HDPE, and PP. PMMA was associated with the highest increment. For PTFE, the change in depolarization current was minimal. Figure 7 shows that the time constant of depolarization current also varies with temperature. For HDPE and PTFE, there is little variation in the time constant with temperature. However, for PMMA and PP, the time constant increases with temperature. The increment was more in PMMA than PP.

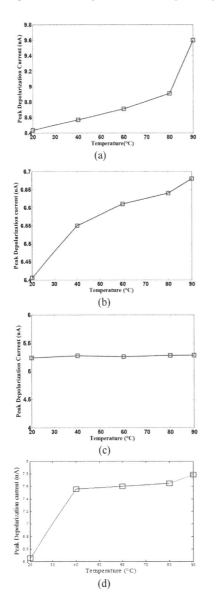

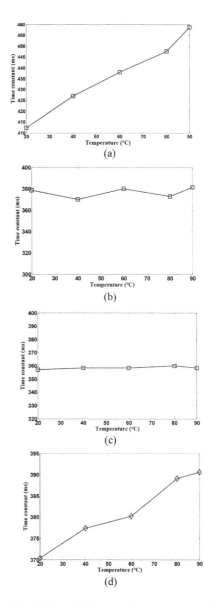

Figure 6. Relationship between peak depolarization current and temperature of different insulating materials: (a) PMMA, (b) HDPE, (c) PTFE, (d) PP.

Figure 7. Relationship between time constant and temperature of different insulating materials: (a) PMMA, (b) HDPE, (c) PTFE, (d) PP.

5 CONCLUSION

Solid polymeric materials are being increasingly used in various high-voltage equipment. Condition monitoring of these equipment is becoming highly important for power utilities as the approach of maintenance is changing from time-based to condition-based, for cost cutting and other economic reasons. In this work, studies were conducted on a few polymeric insulating materials to assess the feasibility of dielectric polarization–depolarization measurements on solid polymeric insulation used in high-voltage systems.

It was evident from obtained results that Polarization–Depolarization Current (PDC) measurements are highly dependent on material properties and temperature. The change in peak depolarization current and its time constant vary from material to material. This variation was found to be highest in Polypropylene. For Teflon, the effect of temperature was found to be minimum. In future works, the developed method will be applied for condition monitoring of aged polymeric insulations.

REFERENCES

Abou Dakka, M., S. S. Bamji, and A. T. Bulinski, (2001). *Conference on Electrical Insulation and Dielectric Phenomena Annual Report.* 2001, (pp. 123–126).

Bamji, S.S., Bulinski, A.T. and Chen. Y. (1993). Thermally stimulated current technique to evaluate polymer degradation due to water treeing. *IEEE Trans. on Electr. Insul.* 28, 299–302.

Baral, A. and Chakravorti, S. (2013). A Modified Maxwell Model for Characterization of Relaxation Processes within Insulation System having Non-uniform Aging due to Temperature Gradient. *IEEE Transactions on Dielectrics and Electrical Insulation.* 20(2). 524−534.

Baral, A. and Chakravorti, S. (2013). Assessment of Non-Uniform Aging of Solid Dielectric using System Poles of a Modified Debye Model for Oil-paper Insulation of Transformers. *IEEE Transactions on Dielectrics and Electrical Insulation, Vol. 20(5). 1922−1933.*

Baral, A. and Chakravorti, S. (2014). Condition Assessment of Cellulosic Part in Power Transformer Insulation using Transfer Function Zero of Modified Debye Model. *IEEE Transactions on Dielectrics and Electrical Insulation.* 21(5). 2028–2036.

Bhumiwat, S. A., Electrical Insulation (ISEI), (2010) *Conference Record of the 2010 IEEE International Symposium.* (pp. 1–5).

Gafvert, U., Adeen, L., tapper, M., Ghasemi, P. and Jonsson, B. (2000). Dielectric spectroscopy in time and frequency domain applied to diagnostic of power transformer. *Proceeding of the 6th international conference on properties and application of dielectric materials, Xi'an, china.*

Houhanessian, V. D. and Zaengl, W. S. (1998). On site diagnosis of power transformer by means of relaxation current measurement. *Proceeding conference on IEEE international symposium on electrical insulation New York, NY, USA.*

Martinez, J. J., Demont, P. and Lacabanne, C. (1997). Thermally stimulated creep spectroscopy as a method to characterize ageing in insulating materials. In *Proc. of the International Conference on Dielectric Materials, Measurements and Applications 363.* pp. 426–428.

Saha, T. K., Purkait, P. and Muller, F. (2005, Jan). Deriving an Equivalent Circuit of Transformer Insulation of Understanding the Dielectric Response Measurement. *IEEE Transactions on Power Delivery,* 20(1) 149–157.

Saha, T. K. and Purkait, P. (2004, Feb). Investigation of Polarization and Depolarization Current Measurement for the Assessment of oil–Paper Insulation of Aged Transformer. *IEEE Transaction on dielectric and electric insulation,* 11(1). 144–154.

Williams, G. (1985). Dielectric relaxation behavior of Amorphous Polymers and related materials. *IEEE transaction on electrical insulation,* 20(5). 843–857.

Wubbenhorst, M., Homsby, J., Stachen, M., Das Gupta, D.K., Bulinski, A. and Bamji, S. (1998). Dielectric properties and spatial distribution of polarization in polyethylene aged under ac voltage in a humid atmosphere. *IEEE Trans. Diel. Elec. Ins.* 5, 9–15.

Computer, Communication and Electrical Technology – Guha, Chakraborty & Dutta (Eds)
© 2017 Taylor & Francis Group, ISBN 978-1-138-03157-9

DTCWT based approach for power quality disturbance recognition

S. Chakraborty
Meghnad Saha Institute of Technology, Kolkata, India

A. Chatterjee & S.K. Goswami
Jadavpur University, Kolkata, India

ABSTRACT: This paper presents Dual Tree Complex Wavelet Transform (DTCWT) based approach for Power Quality (PQ) disturbance recognition. PQ disturbance recognition has been carried out using wavelet transform, S-transform for feature extraction combined with ANN or fuzzy logic for disturbance classification. But the proposed method for PQ event classification which combines DTCWT with ANN based binary classifiers efficiently detects the PQ disturbance events. The proposed method is feasible and promising for real applications.

Keywords: Power Quality (PQ) disturbance, Dual Tree Complex Wavelet Transform (DTCWT), Artificial Neural Network (ANN), classifier

1 INTRODUCTION

Electrical fault in a distribution network results in power quality disturbances (Chilukuri et al. 2004 & Zhao et al. 2007). These disturbances may be in the form of voltage sag, swell, voltage imbalances, transients, interruptions and harmonics. Real-time monitoring is required for identifying the signals of different events and making prompt decision for maintenance. In this context, a feature extraction tool can be used for retrieving the critical data representing any signal and reducing the dimensionality of the overall data. Wavelet transform based algorithms are widely used for feature extraction (Liao et al. 2009). Artificial Neural Networks (ANN), Fuzzy Logic (FL) based systems (Morsi et al. 2009) and Support Vector Machines (SVM) (Janik et al. 2006) are used for efficient event classifications. In recent years, S-Transform (ST) has been extensively used to extract features (Dash et. al. 2003 & Zhao et al. 2007) and has been combined with other pattern classifiers such as ANN (Lee et al. 2003, Mishra et.al. 2008), FL (Chilukuri et al. 2004), or SVM to classify power quality events. The present paper describes the development of a new power quality disturbance recognition system employing The Dual Tree Complex Wavelet Transform (DTCWT) (Chakraborty et al. 2014, Chakraborty et al. 2015, Selesnick et al. 2005), an improved version of the conventional Discrete Wavelet Transform (DWT). The DTCWT has a distinct advantage over the conventional DWT i.e. it's nearly invariant with a small shift

in the signal. The disturbance recognition module integrates the feature extraction methodology based on DTCWT with the conventional supervised neural network employing backpropagation based learning for several benchmark power quality signals. The proposed system has been implemented for several case studies and their performance evaluations in comparison with other competing similar systems demonstrate the usefulness of our proposed system.

The rest of the paper is organized as follows. Section 2 describes the overall system for power quality disturbance recognition using DTCWT. Section 3 presents a detailed discussion on the DTCWT and Section 4 a discussion on the supervised neural networks. Performance evaluation is presented in detail in Section 5 and conclusions in section 6.

2 POWER QUALITY DISTURBANCE RECOGNITION USING DTCWT

In the proposed method the power quality disturbance signals are classified using DTCWT combined with different BPNN variants. Firstly the DTCWT coefficients are generated from each disturbance signal under consideration. From these coefficients, certain values are selected, following any chosen philosophy, to reduce the dimension of each feature vector. These extracted features are used to train an ANN as a binary classifier i.e. the output set to +1 for the category of signal for which the network is being trained and −1 for all the other categories of signals. This process is carried out for all the types of PQ disturbance signals i.e. if there are Q such PQ disturbance categories, then Q such ANNs are separately trained, each for a particular chosen disturbance. The network is then tested for a particular disturbance signal using the same binary logic, in implementation phase. The output of the ANN is assigned to +1 for continuous output greater than 0 and −1 for output less than 0. Output value +1 means correct identification of the PQ signal under test whereas −1 means incorrect identification. Fig. 1 shows the flowchart of each such binary PQ disturbance recognizer developed in implementation phase.

3 DUAL TREE COMPLEX WAVELET TRANSFORM

The Dual Tree Complex Wavelet Transform (DTCWT) has been developed from the knowledge of Discrete Wavelet Transform (DWT) with the superior qualities of being shift invariant, directionally selective in any dimension and having lower redundancy factor compared to DWT

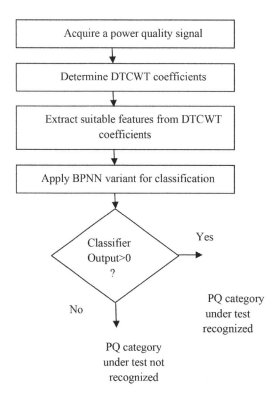

Figure 1. Flowchart of each binary DTCWT-BPNN classifier, in implementation phase.

(Chakraborty et al. 2014, Chakraborty et al. 2015, Selesnick et al. 2005). Wavelet transform has been successfully used for signal processing applications for the last 20 years (Kingsbury et al. 1997). DWT utilizes set of locally oscillating basis functions called wavelets which are stretched and shifted version of the fundamental, real valued bandpass wavelet $\psi(t)$. Wavelets combined with shifts of a real-valued low-pass scaling function ψ (t), gives an orthonormal basis expansion for one-dimensional (1-D) real-valued continuous-time signals (Daubechies 1992). Hence, any finite energy analog signal x(t) can be expressed in terms of wavelets and scaling functions as:

$$x(t) = \sum_{n=0}^{\infty} c(n)\phi(t-n) + \sum_{j=0}^{\infty}\sum_{n=-\infty}^{\infty} d(j,n)2^{j/2}\psi(2^j t - n) \quad (1)$$

The scaling coefficients $c(n)$ and wavelet coefficients $d(j,n)$ are obtained as

$$c(n) = \int_{-\infty}^{\infty} x(t)\phi(t-n)dt \quad (2)$$

$$d(j,n) = 2^{j/2}\int_{-\infty}^{\infty} x(t)\psi(2^j t - n)dt \quad (3)$$

The frequency content of the signal is obtained from the factor j and the time shift from the factor n. Two octave-band discrete time Filter Banks (FBs) utilizing a discrete-time low-pass filter $h_0(n)$, a high-pass filter $h_1(n)$, and upsampling and downsampling operations are used to obtain the coefficients $c(n)$ and $d(j, n)$. But the major drawbacks of the wavelet transform are the oscillation of the coefficients around singularities, significant disturbance in the oscillation of coefficients due to shift in signal and aliasing (Kingsbury et al. 1997, Kingsbury 1999 & Selesnick et al. 2005). But these disadvantages are not present in Fourier transform which primarily uses complex valued oscillating sinusoids given by

$$e^{j\Omega t} = \cos(\Omega t) + j\sin(\Omega t) \quad (4)$$

as compared to real valued oscillating wavelets used in DWT

Based on this a CWT (Kingsbury et al. 1997, Kingsbury et al. 1998, Kingsbury 1999) as (1)–(3) can be imagined with a complex-valued scaling function and complex valued wavelet

$$\Psi_c(t) = \Psi_r(t) + j\Psi_i(t) \quad (5)$$

The complex valued wavelet $\psi_r(t)$ is chosen as real and even and $j\psi_i(t)$ as imaginary and odd.

$\Psi_c(t)$ an analytic signal is formed from the Hilbert transform pair $\Psi_r(t)$ and $\Psi_i(t)$ which is supported on only one half of the frequency axis (Michak et al. 1999, Selesnick 2001). The projection of the signal onto $2^{j/2}\Psi_c(2^j t - n)$ gives rise to the complex valued wavelet coefficient as

$$d_c(j,n) - d_r(j,n) + j d_i(j,n) \qquad (6)$$

The magnitude and phase of $d_c(j,n)$ can be given as:

$$|d_c(j,n)| = \sqrt{|d_r(j,n)|^2 + |d_i(j,n)|^2} \qquad (7)$$

$$\angle d_c(j,n) = \arctan\left(\frac{d_i(j,n)}{d_r(j,n)}\right) \qquad (8)$$

The Dual Tree Complex Wavelet Transform (DTCWT) uses two Filter Bank (FB) trees and thus two bases (Kingsbury 2001 & Lang 1998). The two real DWTs give the real and imaginary part of the transform. The two sets of filters are designed simultaneously to obtain an analytic transform. Let $h_0(n)/h_1(n)$ denote the low-pass/high-pass filter pair for the upper FB, and let $g_0(n)/g_1(n)$ denote the low-pass/high-pass filter pair for the lower FB. Filters are designed to satisfy perfect reconstruction conditions. The two real wavelets associated with each of the two real wavelet transforms is designated as $\psi_h(t)$ and $\psi_g(t)$. The filters are so designed that the complex wavelet $\psi(t) = \psi_h(t) + j\,\psi_g(t)$ is approximately analytic. The dual tree CWT can be represented by the matrix

$$F = \begin{bmatrix} F_h \\ F_g \end{bmatrix} \qquad (9)$$

where F_h and F_g represent two real DWTs.

If the vector x represent a real signal, then $w_h = F_h x$ represent the real part and $w_g = F_g x$ represent the imaginary part of the dual-tree CWT. Hence the complex coefficients are given as $w_h + i w_g$. The inverse of F is given as

$$F^{-1} = \frac{1}{2}\left[F_h^{-1} F_g^{-1}\right] \qquad (10)$$

where

$$F^{-1}.F = I \qquad (11)$$

The two low pass filters are designed in such a way that one is a half sample shift of the other

$$g_0(n) \approx h_0(n - 0.5) \qquad (12)$$

This ensures that wavelets $\psi_g(t)$ and $\psi_h(t)$ approximately form a Hilbert transform pair, i.e. $\Psi_g(t) \approx H\{\Psi_h(t)\}$.

4 NEURAL NETWORK

Neural Networks (NN) have performed well as classifiers for image processing estimators for both linear and non linear applications (Kundu et al. 2010 & Lee et al. 2003). The backpropagation algorithm of ANN looks for the minimum of the error function in weight space using the method of gradient descent. In this work we consider three popular variants of BPNN, namely Levenberg-Marquardt (LM) learning based BPNN (BPNN-LM), resilient backpropagation (BPNN-RP) and conjugate gradient-Fletcher Powell based BPNN (BPNN-CGF) (Haykin 2004). Each BPNN classifier is developed as a binary classifier where the class level is chosen as either +1 (for the correct signal) or –1 (for the incorrect signal) (Kundu et al. 2010). Then, the binary classification problem for each classifier module can be formulated on the basis of a given data set (Ω), with x_i input features and d_i classification output, of the form

$$' = \{(\mathbf{x}_1, d_1), (\mathbf{x}_2, d_2), \cdots, (\mathbf{x}_N, d_N)\} \qquad (13)$$

where $x_i \in \Re^m$, $d_i \in \{+1, -1\}$, m the dimension of each feature vector and N the number of training samples/exemplars.

5 PERFORMANCE EVALUATION

Different types of power quality disturbance signals can be suitably recognized using our proposed DTCWT based neural network recognition system when the system remains energized with sufficiently high accuracy. Several experiments have also been conducted with different additive noise levels on the signal which primarily contaminates the signals in their digital acquisition. The evaluation has been carried out for two different case studies as described below.

CASE STUDY I:
In the first case study, eight different power quality disturbance signals stated in Table 4 are generated using the standard mathematical forms implemented in MATLAB programming. The DTCWT coefficients extracted from each of these signals are fed as input to a multilayer feedforward ANN with one input, one output and one hidden layer. The output of each such neural network developed recognizes the PQ events in binary form (that is output is +1 for the correct PQ event and −1 for

the other PQ events). After successful training of neural networks, they are implemented for testing the datasets. In each case, signals without any noise and noisy signals with Signal-to-Noise Ratio (SNR) level of 40 dB, 30 dB and 20 dB are considered. In the process of generating DTCWT coefficients from the PQ signals along-with the variation in the number of coefficients extracted, the suitability of utilizing a particular characteristic (e. g. max., min., mean, median, variance) of the generated coefficients to extract suitable feature has been evaluated. The 2*N characteristic features of the extracted DTCWT coefficients for N levels are fed as input to the binary ANN classifier. Each neural network has been trained in a batch mode. The results that have been tabulated for noisy data is the mean of the results obtained for the above mentioned noise contamination levels. The training has been carried with 800 PQ disturbance signals i.e. 100 signals from each category. Several trial and error runs optimizes each ANN hidden layer with 15 neurons. A total of 200 signals have been used in the testing phase for classification.

Table 1 shows the classification accuracy obtained with five levels of extraction employed with different characteristics mean (or max./min./median/variance) for each level of coefficients. The highest classification accuracy is obtained with the "max." value of the extracted coefficients in comparison to other similar characteristic features.

Table 2 shows the performance for N = 3, 4, 5, 6, 7 with the chosen characteristic feature type being "max" which indicates that an optimum choice of number of levels N = 5 would produce the highest possible average accuracy of 100% for signals without noise and 98.33% for noisy signals.

Table 3 depicts a performance comparison where we fix N = 5, and choose features using "max." characteristic and train all neural networks classifiers using three BPNN training variants i.e. BPNN-LM BPNN-RP and BPNN-CGF. In this detailed study BPNN-LM emerged as the best classifier using feature dimension of 10 with "max." characteristic chosen for each level.

The detailed analysis of eight types of PQ disturbance signals for optimum choice of feature level

Table 1. Performance comparison of the system for different characteristic features extracted from DTCWT coefficients with N = 5 levels of extraction.

Characteristic feature of the DTCWT coefficient	Average classification accuracy (%)	
	Without noise	With noise
Min	90.50	79.17
Mean	12.50	28.83
Median	55.50	36.33
Variance	96.00	88.42
Max	100.00	98.33

Table 2. Performance comparison of the system with change in the number of levels of DTCWT coefficient extraction (using "max." characteristic feature of the coefficients).

No. of characteristic features of the DTCWT coefficients	Average classification accuracy (%)	
	Without noise	With noise
6	99.00	64.67
8	100.00	78.17
10	100.00	98.33
12	84.00	70.17
14	98.50	93.83

Table 3. Performance comparison of the system with choice of different backpropagation training algorithms (with N = 5 and using "max." characteristic feature of the coefficients).

Backpropagation training algorithm	Average classification accuracy (%)	
	Without noise	With noise
BPNN-RP	92.50	80.00
BPNN-CGF	91.00	87.50
BPNN-LM	100.00	98.33

Table 4. Performance analysis of the system using 5 levels of DTCWT coefficient extraction with "max" characteristic feature of the coefficients and BPNN-LM algorithm.

Sl No.	PQ disturbance recognition method	Average classification accuracy			
		Without noise	With 40 dB noise	With 30 dB noise	With 20 dB noise
1.	Normal Sinusoid	100.00	100.00	100.00	100.00
2.	Sag	100.00	100.00	100.00	100.00
3.	Swell	100.00	100.00	100.00	100.00
4.	Harmonic Disturbance	100.00	100.00	100.00	100.00
5.	Sag with Harmonic	100.00	96.00	96.00	100.00
6.	Swell with harmonic	100.00	100.00	100.00	96.00
7.	Flicker	100.00	100.00	96.00	76.00
8.	Notch	100.00	100.00	100.00	100.00
Average		100.00	99.50	99.00	96.50

Table 5. Comparative study of performances vis-à-vis other proposed systems.

Sl No.	PQ disturbance recognition method	Average classification accuracy	
		Without noise	With noise
1.	S-Transform + Feedforward Network [9]	94.44	89.99
2.	S-Transform + Probabilistic Network [9]	94.00	90.89
3.	S-Transform + LVQ [10]	51.00	NA
4.	S-Transform + FFML [10]	88.00	NA
5.	S-Transform + PNN [10]	97.40	93.20
6.	Space phasor + SVM [7]	84.58	83.75
7.	Space phasor + RBF Network [7]	98.75	98.75
8.	S-Transform + Fuzzy Logic [10]	NA	90.26
9.	WT + ANN [5]	87.00	NA
10.	WT + NFCS [5]	99.40	92.80
11.	S-Transform [1]	NA	97.33
12.	Proposed method	100.00	98.33

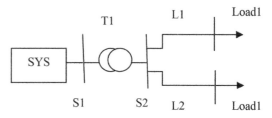

Figure 2. Single line dig. of simulated power distribution system.

of the cases of voltage sag, swell and momentary interruption were recognized using the proposed method with 100% accuracy.

6 CONCLUSION

In the present paper PQ disturbance recognition has been developed which successfully detects both simulated and real time signals. Unlike DWT the proposed method overcomes the effect of noise on actual signals. The DTCWT based approach combined with ANN based binary classifier for recognition of PQ events has an overall performance that has been shown by comparative studies to be superior to the existing approaches.

extraction and feature characteristic is tabulated in Table 4. Table 5 gives a comparative study between the proposed method and other established methods for PQ disturbance recognition establishing the fact that the proposed method could produce higher classification accuracy results both in no-noise and noisy signal acquisition situations.

CASE STUDY II:
In the second case study a power system model is built using the Power System Blockset in MATLAB environment shown in Fig. 2. The system (SYS) (Chilukuri et al. 2004) is described by initial short circuit apparent power $S_k'' = 3GVA$ and a voltage level of 110 kV. T1 is a two winding transformer 110/15 kV distribution transformer with s. L1 and L2 are typical overhead lines with lengths 1.5 an 5 km, respectively. Both lines supply the RL loads. Disturbances are simulated at the load end busbar of line L1 by creating (i) single-line-to-ground faults separately on the individual phases and (ii) line-to-line faults between a-b, b-c and c-a phases separately. Voltage swell signals have been generated by switching a shunt capacitor at the load end bus of line L1. Voltage signals are collected at the point of common coupling at bus S2 and are tested using the proposed method. A number of above mentioned simulations are performed by varying the length of the line. In each

REFERENCES

Chakraborty, S., A. Chatterjee & S. K. Goswami (2014). A sparse representation based approach for recognition of power system transients. *Engineering applications of Artificial Intelligence. 30*, 137–144.

Chakraborty, S., A. Chatterjee & S. K. Goswami (2015). A dual-tree complex wavelet transform based approach for recognition of power system transients, *Expert Systems. 32*, 132–140.

Chilukuri, M.V. & P. K. Dash (2004). Multiresolution S-transform based fuzzy recognition system for power quality events. *IEEE Trans. Power Delivery. 19*, 323–330.

Dash, P. K., B. K. Panigrahi & G. Panda (2003). Power quality analysis using S-transform. *IEEE Trans. Power Delivery. 18*, 406–411.

Daubechies, I. (*1992*). Ten Lectures on Wavelets. *PA: SIAM.*

Haykin, S. (2004). Neural Networks: A Comprehensive Foundation. *Pearson Education, India.*

Janik, P. & T. Lobos (2006). Automated classification of power quality disturbances using SVM and RBF networks. *IEEE Trans. Power Delivery. 21*, 1663–1669.

Kingsbury, N. G. & J. F. A. Magarey (1997). Wavelet transforms in image processing. *1st European Conf. Signal Anal. Prediction, Prague*, 23–24.

Kingsbury, N. G. (1998). The dual-tree complex wavelet transform: A new technique for shift invariance and directional filters. *8th IEEE DSP Workshop, Utah. 8.*

Kingsbury, N. G. (1999). Image processing and complex wavelets. *Philos Trans R. Soc. London A, Math. Phys. Sci. 357*, 2543–2560.

Kingsbury, N. G. (2001). Complex wavelets for shift invariant analysis and filtering of signals. *Appl. Comput. Harmon. Anal. 10,* 234–253.

Kundu, P. K., A. Chatterjee & P. C. Pnanchariya (2010). Electronic Tongue System for Water sample Authentication: A Slantlet-Transform–Based Approach. *IEEE Trans. Instrumenttion and Measurement. 60,* 1959–1966.

Lang, M. (1998). All pass filter design and applications. *IEEE Trans. Signal Processing. 46,* 2505–2514.

Lee, I.W. C. & P. K. Dash (2003). S-transform based intelligent system for classification of power quality disturbance signals. *IEEE Trans. Industrial Electronics. 50,* 800–805.

Liao, C. & H. T. Yang (2009). Recognizing noise influenced power quality events with integrated feature extraction and neuro-fuzzy network. *IEEE Trans. Power Delivery. 24,* 2132–2141.

Michak, M. K., I. Kozinstev, K. Ramachandran & P. Moulin (1999). Low complexity image denoising based on statistical modeling of wavelet coefficients, *IEEE Signal Processing Lett. 6,* 300–303.

Mishra, S. C., N. Bhende & B. K. Panigrahi (2008). Detection and classification of power quality disturbances using S- transform and probabilistic neural network. *IEEE Trans. Power Delivery. 23,* 280–287.

Morsi, W. G. & M. E. EI-Hawary (2009). Fuzzy-wavelet based electric power quality assessment of distribution systems under stationary and non-stationary disturbances. *IEEE Trans. Power Delivery. 24,* 2099–2106.

Selesnick, I. W. (2001). Hilbert transform pairs and wavelet bases, *IEEE Signal Processing Lett. 8,* 170–173.

Selesnick, I. W., R. G. Baraniuk & N. G. Kingsbury (2005). The dual-tree complex wavelet transform, *IEEE Signal Processing Magazine. 22,* 123–151.

Zhao, F. & R. Yang (2007). Power quality disturbance recognition using S-transform. *IEEE Trans. Power Delivery. 22,* 944–950.

Computer, Communication and Electrical Technology – Guha, Chakraborty & Dutta (Eds)
© 2017 Taylor & Francis Group, ISBN 978-1-138-03157-9

Optimised fractional order PID controller in automatic generation control

Asit Mohanty & Dillip Mishra
CET Bhubaneswar, India

Kesab Mohan
Sr Engineering, L&T ECC, India

Prakash K. Ray
IIIT Bhubaneswar, India

Sthita Pragyan Mohanty
CET Bhubaneswar, India

ABSTRACT: In the present work, the Differential Evolution (DE) algorithm optimized Fractional Order PID (FOPID) controller is proposed for Automatic Generation Control (AGC) of multi area multi source power system with parallel AC-DC link. The non-linearity such as time delay is included in the system model. The gains of the PID and FOPID controllers are optimized using an ITAE objective function. The superiority of FOPID controller over PID and Optimal controller is demonstrated by comparing dynamic performance of the same interconnected power system. Then sensitivity analysis is performed by changing the system parameters and loading condition from their nominal values. Finally proposed system is investigated under randomly varying load disturbances.

1 INTRODUCTION

In recent days a power system is found to be consisting of number of control areas interconnected with each other. For power system stability frequency control is an important aspect. Individual area maintains their generation speed to maintain frequency constant and the power angle to the pre specified values. During steady state operation, the sum total of the power generated by the generating sources is same as the power system load and losses. But in practical cases loads change randomly and quickly. The mismatch problem is taken care by the kinetic energy is available in the system and thereby the system frequency comes down. The main objective of AGC is to control power generated from different generating sources in order to keep the frequency at a specified limit.[1]

Generally bulk power transmission has been made possible due to HVDC lines. The advantages of HVDC line fast controllability and enhancement of transient stability.[2]

In present days a combination of different generating sources with their control areas having respective participation factors are combined with time delay and others communication channels.[3]

To achieve these objectives, several studies are carried put to regulate the AGC design. Thus stabilization of frequency oscillations becomes challenging and greatly expected in the prospect competitive market. As a result sophisticated control design is necessary in AGC in order to stabilizing frequency oscillation.

Fractional order controller as several important applications in engineering fields and other scientific areas. With introduction of fractional calculus Podlubny has given a more flexible structure $PI^\lambda D^\mu$ by extending to more in traditional areas of PID controllers[4–8]. DE optimization technique has been added to tune the FOPID[9], PID and output feedback controller for the AGC of multi area multi source power systems with HVDC link.

2 MATHEMATICAL MODELLING

This particular power system is having a two area interconnected AC-DC tie-lines based power system (shown in Fig. 1). Each area of the power system consists of hydro, gas and reheat thermal generating units. The controllers are having respective parameters and participation factor. For better analysis a transfer function system model has

Figure 1. Inter connected two area power systems through AC–DC link tie lines.

been considered. Each area of the power system is having a capacity of 2000 M and a loading 1000 MW. The thermal having 600 MW, hydro 250 MW and gas turbine 150 MW have been considered. The power ratings of the two areas are equal to P_{r1} and P_{r2} in MW unit respectively. For area-1 the constants K_{t1}, K_{h1} and K_{g1} are the shares of power generation from thermal, hydro and gas sources, respectively. P_{Gt1}, P_{Gh1} and P_{Gg1} are power generation in MW by thermal, hydro and gas sources in area-1 respectively.

$$P_{Gt1} = K_{t1}P_{Gt1}; P_{Gh1} = K_{h1}P_{G1}; P_{Gg1} = K_{g1}P_{Gg1} \tag{1}$$

During nominal operations, the total power, P_{G1} for area 1 has been

$$P_{G1} = P_{Gt1} + P_{Gh1} + P_{Gg1} \tag{2}$$

$$K_{t1} + K_{h1} + K_{g1} = 1 \tag{3}$$

The power flow from area-1 to area-2 with AC tie-line is mentioned as

$$P_{TieAC} = P_{12\max} \sin\left(\delta_1 - \delta_2\right) \tag{4}$$

During minute load change Eq. (4) is written as:

$$P_{TieAC} = T_{12\max}\left(\Delta\delta_1 - \Delta\delta_2\right) \tag{5}$$

Where the synchronizing coefficient T_{12} is given by:

$$T_{12} = P_{12\max} \cos\left(\delta_1 - \delta_2\right) \tag{6}$$

The small change in power flow through DC link is modeled during a minute change in frequency at rectifier side. For a small load change the AC tie-line ΔP_{TieAC} flow is given as (7)

$$P_{TieAC} = \frac{2\pi T_{12}}{s}\left(\Delta F_1 - \Delta F_2\right) \tag{7}$$

Small deviation in power flow, ΔP_{tie12} between area-1 to area-2 is given as $P_{tie12} = \Delta P_{tieAC}$ for small perturbation the DC tie-line flow, ΔP_{TieDC} can be given as:

$$P_{TieDC} = \frac{K_{DC}}{1 + sT_{DC}}\left(\Delta F_1 - \Delta F_2\right) \tag{8}$$

The sum power flow, P_{tie12} is given as:

$$P_{Tie12} = P_{TieAC} + P_{TieDC} \tag{9}$$

During minute load change:

$$\Delta P_{Tie12} = \Delta P_{TieAC} + \Delta P_{TieDC} \tag{10}$$

The area control errors ACE_1 and ACE_2 by taking AC/DC tie-line are given as:

$$ACE_1 = \beta_1\Delta F_1 + \Delta P_{TieAC} + \Delta P_{TieDC} \tag{11}$$

$$ACE_2 = \beta_2\Delta F_2 + \alpha_{12}\left(\Delta P_{TieAC} + \Delta P_{TieDC}\right) \tag{12}$$

Where β_1 and $\beta2$ stand for frequency biased parameters and area size ratio; α_{12} can be written as

$$\alpha_{12} = P_{r1} / P_{r2} = -1 \tag{13}$$

A. Optimal controller
Equation is represented as

$$\dot{x} = Ax + Bu \tag{14}$$

And

$$Y = Cx \tag{15}$$

$x = [x_1 x_2 x_3 \ldots\ldots x_{26}]^T$; is a state vector. $u = [u_1 u_2]^T$; is a control vector. $Y = [Y_1 Y_2 Y_3]^T$; is an output vector.
A, B and C are constant matrices with the dimensions of 26×26, 27×2 and 26×22, respectively. The output feedback controller is $U = -K_y$. Where K stands for output feedback gain matrix of dimension 26×2[2].

B. Differential Evolution (DE) algorithm
Differential Evolution (DE) algorithm is a newly introduced population-based stochastic optimization algorithm. The Advantages of DE include simplicity, efficiency and real coding, easy use, local searching property and speediness. The working of DE is based on two populations; old generation and new generation of the same population. The size of the population has been adjusted with parameter NP. The population is comprised of real valued vectors having dimension D which equals to the number of design parameters or control variables.

C. Fractional order PID controller

The fractional PID controller ($PI^\lambda D^\mu$) gives an extra two degree of freedom than the PID controller. The controller generalises the integer order PID controller and gives an expansion from the point to a plane. This expansion provides more flexibility in PID control design. The overall transfer function of the controller is given as

$$G(s) = K_P + \frac{K_I}{s^\lambda} + K_D s^\mu \qquad (16)$$

It is clear from equation (16) that, by selecting $\lambda = 1$ and $\mu = 1$, a classical PID controller can be considered. All these classical types of PID controllers are special cases of the $PI^\lambda D^\mu$ controller.

In an optimization problem objective function is defined depending on the desired specification and constraints.

The objective function J is given as

$$J = ITAE = \int_0^{t_{sim}} \left(|\Delta F_1| + |\Delta F_2| + |\Delta P_{Tie}| \right) \times t \times dt \qquad (17)$$

ΔF_1 and ΔF_2 stand for system frequency deviations at area-1 and area-2 respectively; P_{Tie} is the incremental change in the tie line power, t_{sim} is the time range of simulation.

3 SIMULATION RESULTS AND DISCUSSION

The model of the system under study shown in Fig. 2 is developed in MATLAB/SIMULINK environment and DE program is written (in .m file). Optimal/PID/FOPID controller is considered for each source. The developed model is simulated in a separate program (by .m file using initial pop-ulation/controller parameters) considering a 1% step load change in area 1. The objective function is calculated in the .m file and used in the optimization algorithm. In the present study, a population size of $N_P = 40$, generation number $G = 30$, step size $F = 0.2$ and crossover probability of $CR = 0.6$ have been used. Simulations were conducted on an Intel, core i-3 core CPU, of 2.4 GHz and 4 GB RAM computer in the MATLAB 7.10.0.499 (R2010a) environment. The optimization was repeated 20 times and the best final solution among the 20 runs is chosen as proposed controller parameters.

The best final solutions obtained in the 20 runs are shown in Tables 1 and 2 for the optimal gains and PID/FOPID controller respectively. The corresponding performance index in terms of ITAE value and settling times (2% band) in frequency and tie line power deviations are shown in Table 3. For proper comparison, it is clear from Table 3 that with same power system, minimum ITAE value is obtained with proposed FOPID controller (ITAE = 1.55) compared to PID (ITAE = 2.13) and optimal (ITAE = 9.11) controller.

Table 1. Gains of optimal controller.

Optimal feedback gain matrix (k)					
0.9897	1.4417	0.6582	0.5528	1.9245	1.7488
1.5555	1.6382	1.0931	1.5894	0.1065	1.2497
1.3039	1.5330	1.5438	1.0298	0.6788	0.5992
1.8149	1.2697	1.1891	1.5679	1.8270	0.7604

Table 2. Tuned controller parameters.

Controller gains	PID	FOPID
K_{P1}	−0.9448	−1.3644
K_{P2}	−1.6369	−0.0311
K_{I1}	−0.6430	−0.9035
K_{I2}	−0.1170	−0.5015
K_{D1}	−1.3872	−0.3163
K_{D2}	0.6692	−1.6662
λ_1	–	0.9031
λ_2	–	0.1051
μ_1	–	0.7451
μ_2	–	0.7294

Table 3. Performance index values for different cases.

Controller	Settling time in (Sec)			ITAE
	ΔF_1	ΔF_2	ΔP_{Tie}	
Optimal	58.66	55.54	48.56	9.11
PID	44.56	31.42	39.12	2.13
FOPID	13.07	12.42	2.044	1.55

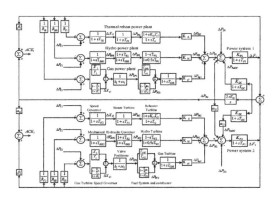

Figure 2. Transfer function model of multisource power system with HVDC link.

Hence, it can be concluded that proposed FOPID controller outperform the optimal and PID controller as minimum objective function value is obtained. To study the dynamic performance a step load increase in demand of 1% is applied at $t = 0$ s in area-1 and the system dynamic responses are shown in Figs. 3(a-c). For comparison, the simulation results with FOPID, PID and Optimal controllers for the same power system are also shown in Figs. 3(a)-(c).

Critical analysis of the dynamic responses clearly reveals that with the same power system significant improvement is observed with proposed FOPID controller compared to others controllers. Sensitivity analysis is carried out to study the robustness the system to wide changes in the operating conditions and system parameters. The various performance indexes (ITAE values and settling times) under normal and parameter variation cases are given in Table 4. Critical examination of Table 4 clearly reveals that ITAE and settling time values vary within acceptable ranges and are nearby equal to the respective values obtained with nominal system parameter. Hence the proposed controllers are robust and perform satisfactorily when the loading condition changes in the range ±25%. The dynamic performance of the system with the varied conditions of loading is shown in Fig. 4. To investigate the superiority of the proposed method, a random step load changes as shown in Fig. 5. The step loads are random both in magnitude and duration. The frequency deviation of area 1 for random load disturbances is shown in Fig. 6. It can be seen from Fig. 6 that proposed controller (FOPID) shows better transient responses than others. Also it is clearly observed from Fig. 6 that the proposed FOPID controller provides superior damping even in presence of a random load variations compare to others.

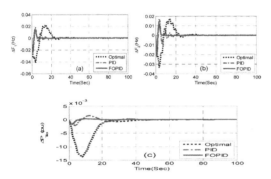

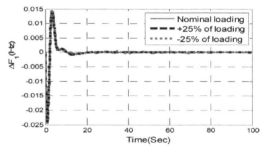

Figure 4. Frequency response of area 1 with variation of loading.

Figure 3. (a-c) Dynamic responses for 1% step load change in area 1. (a) Frequency deviations of area 1 (b) Frequency deviations of area 2 (c) Tie line power deviations.

Table 4. Sensitivity analysis.

| Parameters variations | % change | Settling time in (Sec) | | | |
		ΔF_1	ΔF_2	ΔP_{Tie}	ITAE
Nominal	0	13.078	12.420	2.044	1.5503
Loading	+25	13.094	13.094	1.971	1.5290
conditions	−25	12.494	12.494	1.903	1.5848
T_G	+25	12.576	12.576	1.869	1.5684
	−25	13.278	13.278	2.130	1.5782
T_T	+25	12.830	12.806	1.917	1.5508
	−25	12.608	12.608	2.014	1.5482
T_{RH}	+25	11.501	11.457	2.062	1.5041
	−25	11.484	11.454	1.877	1.5716
T_{CD}	+25	12.810	12.777	2.071	1.5794
	−25	13.239	13.239	2.052	1.5495
T_{12}	+25	12.303	10.951	1.792	1.5184
	−25	13.088	13.058	2.220	1.5524

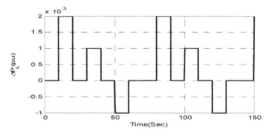

Figure 5. Variable step load.

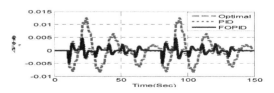

Figure 6. Area 1 frequency response of controllers to variable step load.

4 CONCLUSION

The study was conducted on a two area interconnected power system with identical plants having 25% generation from hydro, 60% from thermal with reheat turbines and 15% from gas turbine systems. Dynamic performance of FOPID, PID and optimal controller are shown for parallel AC-DC link. In order to make the power system more realistic, the non-linearity parameter such as time delay has been included in the system model. The gains of optimal/PID/FOPID controller are optimized employing a Differential Evolution (DE) technique. From simulation results, it is observed that the proposed controllers are robust and ensures satisfactory system performance in presence variation in system operating load conditions. Finally, the effectiveness and robustness of the proposed controller against random load variations were investigated.

APPENDIX

$$X_1 = X_2 = 0.6 \ s \ b_1 = b_2 = 0.05 \ s$$

$$Y_1 = Y_2 = 1.0 \ s \ T_{CD1} = T_{CD2} = 0.2 \ s$$

$$c_1 = c_2 = 1 \ K_{DC} = 1.0, T_{DC} = 0.2 \ s$$

$$T_{CR1} = T_{CR2} = 0.3 \ s \ T_F = 0.23 \ s$$

$$\beta_1 = \beta_2 = 0.425 \ p.u. \ MW/Hz \ T_{p1} = T_{p2} = 20 \ s$$

$$R_{T1} = R_{T2} = R_{H1} = R_{H2} = R_{G1} = R_{G2} = R_1$$
$$= R_2 = 24 \ Hz/p.u. \cdots T_{12} = 0.0433$$

$$T_{G1} = T_{G2} = 0.08 \ sec$$

$$T_{T1} = T_{T2} = 0.3 \ s \ \alpha_{12} = -1$$

$$K_{r1} = K_{r2} = 0.3 \ T_{R1} = T_{R2} = 5.0 \ s$$

$$T_{r1} = T_{r2} = 5 \ s \ T_{RH1} = T_{RH2} = 28.75$$

$$K_{P1} = K_{P2} = 120 \ Hz/p.u. \ MW \ T_{GH1} = T_{GH2} = 0.2 \ s$$

REFERENCES

[1] O.I. Elgerd, *Electric Energy Systems Theory an Introduction*, 2nd ed. New Delhi, India: Tata McGraw-Hill, 1983.

[2] Ibraheem, Nizamuddin and T.S. Bhatti. "AGC of two area power system interconnected by AC/DC links with diverse sources in each area," *International Journal of Electrical Power & Energy Systems,* vol. 55, pp. 297–304, 2014.

[3] S. Panda. "Differential evolution algorithm for SSSC-based damping controller design considering time delay," *J. Franklin Institute* vol. 348, no. 8 pp. 1903–1926, 2011.

[4] Bevrani, Robust Power System Frequency Control, Springer, 2009.

[5] O.I. Elgerd and C.E. Fosha, "Optimum megawatt-frequency control of multi-area electric energy systems," *IEEE Trans Power Appar Syst,* vol. 89, no. 4, pp. 556–63, 1970.

[6] P.K. Ibraheem and D.P. Kothari "Recent philosophies of automatic generation control strategies in power systems," *IEEE Trans Power Syst.* vol. 20, no. 1, pp. 346–357, 2005.

[7] Lasezlo Gyugyi, Kalyan K. Sen and C.D. Schavder "The Interline power flow controller concept: A new approach to power flow management in Transmission system," *IEEE transaction on power delivery;* pp. 1115–1123, 1989.

[8] R.K. Cavin, M.C. Budge and P. Rosmunsen "An Optimal Linear System Approach to Load Frequency Control," IEEE *Trans. on Power Apparatus and System,* PAS-90, pp. 2472–2482, 1971.

[9] I. Podlubny "Fractional-order systems and $PI^\lambda D^\mu$ controllers," *IEEE Transactions on Automatic Control,* vol. 44, no. 1, pp. 208–214, 1999.

Computer, Communication and Electrical Technology – Guha, Chakraborty & Dutta (Eds)
© 2017 Taylor & Francis Group, ISBN 978-1-138-03157-9

Fault detection in an IEEE 14-bus power system with DG penetration using wavelet transform

Prakash K. Ray, B.K. Panigrahi & P.K. Rout
Department of Electrical and Electronics Engineering, IIIT, Bhubaneswar, India
Department of Electrical Engineering, SOA University, Bhubaneswar, India

Asit Mohanty & Harishchandra Dubey
Department of Electrical Engineering, CET, Bhubaneswar, India
Department of ECE, The University of Texas at Dallas, USA

ABSTRACT: This paper proposes the application of wavelet transform for the detection of islanding and fault disturbances in a Distributed Generation (DG)-based power system. For this purpose, an IEEE 14-bus system with DG penetration is considered for the detection of disturbances under different operating conditions. The power system is a hybrid combination of a photovoltaic and wind energy system connected to different buses with different levels of penetration. The voltage signal is retrieved at the Point of Common Coupling (PCC) and processed through the wavelet transform to detect the disturbances. Furthermore, energy and Standard Deviation (STD) as performance indices are evaluated and compared with a suitable threshold to analyze a disturbance condition. Again, a comparative analysis between the existing and proposed detection is studied to prove the better performance of wavelet transform.

1 INTRODUCTION

Distributed Generations (DGs) are considered as small-scale power resources, which are basically installed near the loads or can be connected to the grid if required. Nowadays, these resources are gaining great popularity because of deregulation and restructuring of the modern power system (Onara et al. 2008, Kim et al. 2008). Indeed, the penetration levels of these DGs vary depending on the location and availability of natural resources. However, with the increase in the penetration level, some problems such as detection of islanding and fault disturbances become more vital and complex because of uncertain characteristics of renewable resources like solar and wind energy. Islanding is a phenomenon that usually occurs when the utility grid is being isolated from the DG because of some abnormal conditions; however, during the same condition, the DG continues to feed power to the loads connected to it. Similarly, a fault may occur at any bus or near any bus of the system because of symmetrical and unsymmetrical faults like L-G, L-L, L-L-G, L-L-L, L-L-L-G, etc. Therefore, once these faults occur, they have to be detected as quickly as possible and corrective measures have to be taken to protect the loads as well as the power system (Ray et al. 2010, Fernandez et al. 2002).

In the literature, many methods have been suggested to detect the disturbances (Yadav et al. 2014, Jang et al. 2004). The design and influence of multi-stage inverters in detecting the islanding events have been presented in Yadav et al. (2014). The effect of interfacing control and non-detection zones on the detection of islanding has been discussed in Zeineldin et al. (2006). Voltage Unbalance (VU) and Total Harmonic Distortion (THD) are also taken as indices to detect islanding (Jang et al. 2004). However, if a proper threshold is not selected, then it becomes difficult to detect under operating conditions.

Artificial intelligence techniques such as Artificial Neural Network (ANN), fuzzy network, Adaptive Neuro-Fuzzy Inference System (ANFIS), and Support Vector Machines (SVM) have been used by the researchers for the detection and classification of fault disturbances.

In this context, different signal processing techniques have been implemented, of which Wavelet Transform (WT) is an efficient tool for detection based on time–frequency localization (Dubey et al. 2011, Cheng et al. 2008). WT is very versatile in identifying disturbance features, and can easily detect any irregularities in the signal. A comparison of the detection ability between the existing methods such as Fourier Transform (FT), Short-Time

Fourier Transform (STFT), and wavelet transform is carried out using performance indices, standard deviation, and energy to prove the better performance of the proposed transform. This paper is organized as follows: modeling of an IEEE 14-bus system is described in Section 2. WT as a signal processing-based detection tool is presented in Section 3. Then, the MATLAB-based simulation results are discussed in Section 4. Finally, concluding remarks are presented in Section 5.

2 IEEE 14-BUS POWER SYSTEM

The penetration levels of different renewable power resources are increasing with time; therefore, the hybrid power system becomes more complex. Hence, the design, operation, and control of the power system are becoming more challenging. Suitable methods have to be adopted to identify the normal as well as the abnormal operating conditions. While operating in grid-connected mode, the grid may be disconnected due to some abnormal conditions, leading to the islanding condition where the DG still continues to supply power to the local loads. Similarly, fault disturbances may arise due to some type of faulty conditions between the phases and the ground.

Fig. 1 shows an IEEE 14-bus power system with DGs interconnected to some of its distribution buses. This model is developed in a MATLAB/Simulink environment.

2.1 Wind Energy Conversion System (WECS)

Wind turbine is a nonlinear, dynamic system that converts the wind kinetic energy into

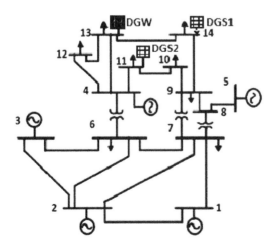

Figure 1. IEEE 14-bus power system with interconnected DGs.

mechanical energy, which is then processed to produce electrical energy by using wind generators. The characteristics of wind turbines are too uncertain, nonlinear, and complex to design and study. Different models of wind turbines are proposed with many Degrees of Freedom (DOFs) to ensure the energy conversion. The mechanical power output from the wind turbine in terms of the wind speed is expressed as (Ackermann 2005):

$$P_{mech} = \frac{\rho}{2} A_{wind} C_{wind} (\eta, \beta) V_{wind}^3 \qquad (1)$$

where P_{mech} is the mechanical power output from the wind (watts); ρ is the air density (kg/m³); C_{wind} is the power coefficient; η is the tip-speed ratio (V_t / V_{wind}); V_t is the blade tip speed (m/s) and V_{wind} is the wind speed (m/s); A_{wind} is the area covered by the rotor of the wind turbine (m²); and β is the pitch angle (degrees). The tip-speed ratio η is given as:

$$\eta = \frac{R\omega_r}{v_{wind}} \qquad (2)$$

where ω_r is the electrical speed (rad/s). The Doubly Fed Induction Generator (DFIG) model is used as part of the WECS (Tapia et al. 2003, Ray et al. 2010).

2.2 Photovoltaic (PV)

The solar PV module is formed by a combination of cells represented as the basic P–N junction diode. The PV cells convert the input solar radiation into electrical energy. These cells are connected in a series and parallel combination to vary the output power ratings. The output voltage V and the load current I of the PV system are expressed as:

$$I = I_L - I_0 \left[\exp\left(\frac{V + IR_s}{\alpha}\right) - 1 \right] \qquad (3)$$

where I_L is the PV current (amps); I_0 is the saturation current; R_s is the series resistance (ohms); and α is the thermal constant (Onara et al. 2008).

3 ISLANDING DETECTION METHODS

3.1 Wavelet Transform (WT) as a detection method

Wavelet transform is a tool applied to decompose a signal/function into different components and associate a frequency band with each of them. WT also changes the scales and frequencies of the analyzed signal and components. WT does not require a fixed basis function as in the case of Fourier Transform (FT), but different sets of possible basis functions can be formulated to improve the performance.

In this study, the voltage signals of the system are extracted and processed by Haar mother wavelet to decompose and detect the disturbances (Fernandez et al. 2002, Yadav et al. 2014). Indeed, the voltage is filtered out by a series of low- and high-pass filters to obtain approximation (A) and detail (D) coefficients. The decomposition of the approximation and detailed components are expressed as:

$$A_1(n) = \sum_k H(k-2n)C_0(k);$$
$$D_1(n) = \sum_k G(k-2n)C_0(k) \quad (4)$$

where $H(n)$ is the low-pass component and $G(n)$ is the high-pass component of the filter. For each decomposition scale, the approximate and detail coefficients $(A_1(n) \& D_1(n))$ are determined for the time–frequency analysis of the signal, $C_0(n)$.

3.2 STD and energy as performance indices

Based on Parseval's theorem, it is known that the energy of a signal $V(t)$ becomes the same both in time and frequency domains, which is given by:

$$E_{signal} = \frac{1}{T}\int_0^T |V(t)|^2 dt = \sum_{n=0}^K |V[n]|^2 \quad (5)$$

where T and K are the time period and the signal length, respectively, and $V[n]$ is the FT of the signal. Then, energy can be calculated by WT (Ackermann et al. 2005) as follows:

$$E_{signal} = \int |y(t)|^2 dt = \sum_{k=-\infty}^{\infty} |c(k)|^2 + \sum_{j=jo}^{\infty}\sum_{k=-\infty}^{\infty} |d_j(k)|^2 \quad (6)$$

The standard deviation and energy are calculated from the detail component (d1) of the voltage signal.

4 SIMULATED RESULTS

This section presents the simulation and analysis of the system and techniques for the detection of islanding and fault disturbances in the power system using wavelet transform under different operating conditions. The study is implemented in an IEEE 14-bus power system with DG penetration, which is a hybrid combination of a wind and solar photovoltaic energy system. The models of the power system under study are simulated in a MATLAB/SIMULINK environment. The parameters of various components used for simulation have been described in the literature (Zeineldin et al. 2006, Jang et al. 2004, Dubey at al. 2011, Cheng et al. 2008).

4.1 Islanding detection

The islanding detection study using wavelet transform is conducted in the 14-bus hybrid power system under different operating conditions. The voltage signal is captured from the 14-bus to which the solar DG1 (DGS1) system is connected along with the loads. The voltage signal is then processed by wavelet transform to detect the islanding disturbance. The simulation results of the coefficients of Haar wavelet transform is shown in Figure 2. Here, it can be seen that the signal is represented for 1000 sec, whereas the approximate and detail coefficients are represented for 500 sec, because the 1000 sec window is divided equally. Therefore, the islanding event that starts at about 610 sec is detected by the coefficients at about half of the 305 sec.

4.2 Fault disturbance detection

This section presents the study of the detection of fault disturbances. Similar to the previous section, the voltage signal at bus-13 to which the wind turbine is connected is measured and taken offline for processing. This signal is collected at the bus when there is an AG fault in the line connected between bus-13 and bus-14. Obviously, the voltage signals at both the buses are affected. However, the voltage signal at bus-13 is taken into account for study to determine the effect of the fault as well as the effect of the wind turbine on the voltage.

The voltage signal with its detection results by detail and approximate coefficients of Haar wavelet is shown in Figure 3. As seen from the figure,

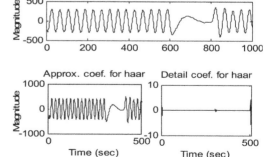

Figure 2. Islanding detection at bus-14 using wavelet transform.

223

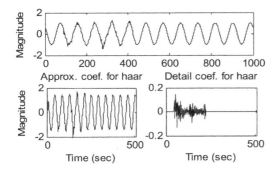

Figure 3. AG fault detection at bus-13 using wavelet transform.

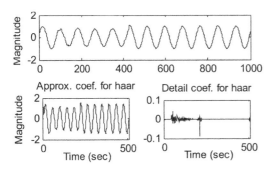

Figure 4. AB fault detection at bus-11 using wavelet transform.

the fault occurs at about 60 sec in the signal and the corresponding detection in the detail and approximate coefficients of Haar wavelet are reflected in figure. It has been observed from the detail coefficient that the instant fault occurrence is clearly detected with a sudden increase and oscillation of magnitude and when the fault is cleared, and the voltage and its coefficients become normal.

Similarly, the detection results for an AB fault in the connected line near bus-11 are shown in Figure 4. The voltage signal represents the variation, which is also detected by the wavelet coefficients. As soon as the fault is cleared, the signal as well as the coefficient return to the normal operating conditions.

4.3 *Energy and STD-based detection of islanding and fault disturbances*

The detection of islanding as well as fault disturbances based on the value of Performance Indices (PIs) like Standard Deviation (STD) and energy is presented in this section. The indices are evaluated from the output coefficients of wavelet. These values are then compared with the selected threshold values to determine whether they will be higher or lower.

Table 1. STD and energy for islanding faults.

(a) Haar wavelet transform

Scenario	Fault disturbance		Islanding	
	STD	Energy	STD	Energy
Bus-11 (AG)	1.39×10^{-5}	0.6676	1.49×10^{-5}	1.4657
Bus-12 (BG)	2.44×10^{-11}	0.8462	1.69×10^{-9}	2.3645
Bus-13 (CG)	2.74×10^{-5}	1.5670	2.68×10^{-4}	4.6767
Bus-11 (AB)	1.39×10^{-5}	1.6874	2.20×10^{-3}	5.6786
Bus-12 (BC)	1.39×10^{-5}	0.9761	1.67×10^{-5}	4.4656
Bus-13 (CA)	2.30×10^{-5}	1.5731	2.33×10^{-4}	4.6763

(b) Daubechies 4 (dB4) wavelet transform

Scenario	Fault disturbance		Islanding	
	STD	Energy	STD	Energy
Bus-11 (AG)	0.48×10^{-5}	0.0671	3.13×10^{-5}	1.1935
Bus-12 (BG)	1.83×10^{-11}	0.0892	2.37×10^{-4}	2.3843
Bus-13 (CG)	2.39×10^{-5}	0.2460	2.03×10^{-5}	3.0110
Bus-11 (AB)	1.23×10^{-5}	0.3834	2.35×10^{-5}	3.3921
Bus-12 (BC)	1.28×10^{-5}	0.2781	3.39×10^{-5}	2.9380
Bus-13 (CA)	2.06×10^{-5}	0.4830	2.29×10^{-4}	3.3921

(c) Comparison with different transforms

Technique	Fault		Islanding	
	STD	Energy	STD	Energy
FT	0.453	0.967	1.120	1.894
STFT	0.634	1.034	1.673	2.171
WT_dB1	0.842	1.495	1.784	2.742
WT_Haar	0.793	1.567	1.649	2.698
WT_Coif	0.830	1.675	1.801	2.822
WT_Demey	0.772	1.716	2.014	2.963

If the index is higher than the threshold, islanding is detected; otherwise, the fault disturbance is detected. The calculated values of energy and STD are presented in Table 1. However, the challenge lies in the selection of a proper threshold value, which may be complex depending on the type of network configuration as well as operating conditions. After finding the different values of PIs, as presented in Table, all the values are compared with the threshold value to identify the type of disturbance. A comparison between the different transforms is presented in (c), which suggests the increase in PI values for WTs, thereby increasing the detection accuracy.

5 CONCLUSIONS

A study on the detection of islanding and fault disturbances in an IEEE 14-bus hybrid power system with DGs is presented under different operating conditions. The disturbances were detected using

Haar and dB4 wavelet transforms. Case studies were shown for the detection of islanding and fault disturbances under various operating conditions. Furthermore, quantitative analysis using PIs in terms of STD and energy was also carried out for the detection of disturbances. Both qualitative and quantitative analyses showed the effectiveness and accuracy of the proposed transform in detecting the islanding and fault disturbances.

REFERENCES

Ackermann T. (2005) *Wind power in power systems*, Chichester, Wiley.

Cheng-Tao Hsieh, Jeu-Min Lin & Shyh-Jier Huang (2008). Enhancement of islanding-detection of distributed generation systems via wavelet transform-based approaches. *J. Electr. Power and Energy Syst., 30*, 575–580.

Dubey H. C., S. R. Mohanty, Nand Kishor & P. K. Ray (2011). Fault Detection in a Series Compensated Transmission Line using Discrete Wavelet Transform and Independent Component Analysis: a Comparative Study. *5th International Power Engineering and Optimization Conference (PEOCO), Shah Alam, Selangor, Malaysia.*

Fernandez ALO & NKI Ghonaim (2002). A novel approach using a FIRANN for fault detection and direction estimation for high voltage transmission lines. *IEEE Trans Power Deliv. 17*, 894–901.

Jang S. & K. Kim (2004). An islanding detection method for distributed generation algorithm using voltage unbalance and total harmonic distortion of current. *IEEE Trans Power Delivery 19(2)*, 745–752.

Kim, Seul-Ki, Jeon Jin-Hong, Cho Chang-Hee, Jong-Bo Ahn, & Kwon Sae-Hyuk (2008). Dynamic Modeling and Control of a Grid-Connected Hybrid Generation System With Versatile Power Transfer. *IEEE Trans. on Ind. Electr. 55(4)*, 1677–1688.

Onara, O. C., M. Uzunoglua & M. S. Alam (2008). Modeling, control and simulation of an autonomous wind turbine/photovoltaic/fuel cell/ultra-capacitor hybrid power system. *J. Power Sources. 185(2)*, 1273–1283.

Ray P. K., S. R. Mohanty & Nand Kishor (2010). Coherency determination in grid-connected distributed generation based hybrid system under islanding scenarios. *IEEE International Conference on Power and Energy (PECON), Kuala Lumpur, Malaysia, Nov 29– Dec. 1*, 10, 85–88.

Ray, P. K., H. C. Dubey, S. R. Mohanty, Nand Kishor & K. Ganesh (2010). Power quality disturbance detection in grid-connected wind energy system using wavelet and S-transform. *IEEE International Conference on Power, Control and Embedded Systems (ICPCES), November 29–Dec. 1*, 1–4.

Tapia, A., G. Tapia, J. X. Ostolaza & J. R. Saenz (2003). Modeling control of a wind turbine driven doubly fed induction generator. *IEEE Trans. Energy Convers. 18(2)*, 194–204.

Yadav A. & A. Swetapadma (2014). Improved first zone reach setting of artificial neural network-based directional relay for protection of double circuit transmission lines. *IET Gen Transm Distrib 8(3)*, 373–88.

Zeineldin H. H, Ehab F. El-Saadany & MMA. Salama (2006). Impact of DG interface control on islanding detection and non-detection zones. *IEEE Trans Power Delivery 21(3)*, 1515–1523.

Computer, Communication and Electrical Technology – Guha, Chakraborty & Dutta (Eds)
© 2017 Taylor & Francis Group, ISBN 978-1-138-03157-9

Detection of faults in a power system using wavelet transform and independent component analysis

Prakash K. Ray, B.K. Panigrahi & P.K. Rout
Department of Electrical and Electronics Engineering, IIIT, Bhubaneswar, India
Department of Electrical Engineering, SOA University, Bhubaneswar, India

Asit Mohanty & Harishchandra Dubey
Department of Electrical Engineering, CET, Bhubaneswar, India
Department of ECE, The University of Texas at Dallas, USA

ABSTRACT: Uninterruptible power supply is the main objective of power utility companies that identify and locate different types of fault as quickly as possible to protect the power system from complete blackouts using intelligent techniques. Therefore, this study presents a novel method for the detection of fault disturbances based on Wavelet Transform (WT) and Independent Component Analysis (ICA). The voltage signal is taken offline under fault conditions and is processed using wavelet and ICA for analysis. The time–frequency resolution of WT detects the fault initiation event in the signal. Again, a performance index is calculated from the ICA under fault conditions to detect fault disturbances in the voltage signal. The proposed approach is tested to be robust enough under various operating scenarios such as without noise, with 20-dB noise, and under frequency variation conditions. Furthermore, the detection study is carried out using a performance index, energy content, by applying the existing Fourier Transform (FT), Short-Time Fourier Transform (STFT), and the proposed wavelet transform. Fault disturbances are detected if the energy calculated in each scenario is higher than the corresponding threshold value. The study of fault detection is simulated in MATLAB/Simulink in a typical power system.

1 INTRODUCTION

Modern power utilities require an efficient protection scheme to protect the system itself as well as the connected equipment for improving better performance under normal and abnormal/faulty operating conditions. Nowadays, electromechanical relays are replaced by digital relays because of their characteristics such as faster operation, accuracy, and reliability. Fault detection through digital relays and the Fault Detector (FD) is very vital for implementing any real-time solutions (Phadke 1988, Dash 2000).

In the literature, many studies have reported on the monitoring and identification of faults in a power system. Fault situation can be identified by comparing the difference in the values between two consecutive cycles when they are higher than a threshold value based on a phasor (Sidhu et al. 2002, Sachdev et al. 1991). However, one disadvantage is the modeling of fault resistance. The Kalman filter is used for fault detection based on the estimation method (Chowdhury et al. 1991, Zadeh et al. 2010, & Girgis 1982). Wavelet Transform (WT) has been used for the detection of fault disturbances in a power system (Ukil &

R. Živanović 2007), i.e. it detects the changes in the signal. Then, a combination of adaptive filters and wavelet transform was used for fault identification in a power system (Ray et al. 2010). However, these techniques were affected by variation in frequency, noise, etc.

Therefore, this paper proposes an algorithm for fault detection using wavelet transform and independent component analysis. The sudden changes can be identified by using these detection techniques, so that suitable solutions can be adapted to protect the power system from any type of disturbances. The performances of the proposed method were analyzed under different operating scenarios, such as presence of noise, harmonics, and frequency variations. Wavelet Transform (WT) and Independent Component Analysis (ICA) are suitable techniques for the detection of fault disturbances because of their efficient time–frequency resolutions and reliability to extract the suitable features for identification purposes. In addition, the techniques are considered under different operating conditions to assess their robustness and accuracy (Pradhan et al. 2005).

The techniques used for analysis are presented in Section 2 followed by the implementation of the algorithm in Section 3. Then, the results of the analysis are discussed both qualitatively and quantitatively in Section 4, followed by the conclusions presented in Section 5.

2 TECHNIQUES FOR FAULT DETECTION

This section describes the techniques used for the identification of different types of faults in the wind system connected to the grid power system. The detection techniques WT and ICA are presented briefly with their mathematical modeling. Different operating scenarios are taken as case studies to test these techniques for fault identification. This helps to assess both normal and faulty operating conditions. The details of these methods are presented below.

2.1 Wavelet Transforms (WT)

The wavelet transform is a signal processing algorithm used for the detection of abnormal operating conditions based on the decomposition of power signals into different ranges of frequencies by means of a series of low-pass and high-pass filters. This usually paves way for a time–frequency multi-resolution analysis which can be used for identifying any sort of abrupt variations in electrical parameters such as voltage, phase, current, and frequency. Here, Daubechies 4 (dB4) is used as the mother wavelet basis function for the fault detection analysis (Ukil & R. Živanović 2007). The signal is normally divided into a set of approximate (a) and detail (d) coefficients that represent the low- and high-frequency bands, respectively. The decomposition is shown in Figure 1.

Denoting a voltage signal of the power system as $V(t)$, the Continuous Wavelet Transform (CWT) can be expressed as:

$$CWT(V,M,N) = \frac{1}{\sqrt{a}} \int_{-\infty}^{\infty} V(t) \Psi^* \left(\frac{t-N}{M} \right) dt \quad (1)$$

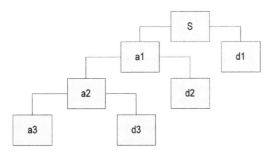

Figure 1. Wavelet decomposition tree.

where M and N are dilation and translation parameters and Ψ is the wavelet basis function. Now, WT in discrete form is given by:

$$DWT(V,M,N) = \frac{1}{\sqrt{M_0^m}} \sum_k V(k) \Psi^* \left(\frac{n-kM_0^m}{M_0^m} \right) \quad (2)$$

where M and N are replaced by M_0^m and kM_0^m, and k and m are integers. The scaling function in one stage is expressed as the sum of that in the next stage, which is given by:

$$\varphi(t) = \sum_{n=-\infty}^{\infty} h(n) \sqrt{2} \, \varphi(2t-n) \quad (3)$$

Using the above equation, the original voltage signal can be written as [8]:

$$V(t) = \sum_{k=-\infty}^{\infty} a_{jo}(k) 2^{jo/2} \phi(2^{jo}t-k) + \sum_{k=-\infty}^{\infty} \sum_{j=jo}^{\infty} d_j(k) 2^{j/2} \phi(2^j t-k) \quad (4)$$

where $J0$ is the coarse adjustment parameter in the scaling function. The detail and approximate coefficients can be written as:

$$a_j(k) = \sum_{m=-\infty}^{\infty} a_{j+1}(m) h(m-2k) \quad (5)$$

$$d_j(k) = \sum_{m=-\infty}^{\infty} c_{j+1}(m) h_1(m-2k) \quad (6)$$

2.2 Independent Component Analysis (ICA)

Independent Component Analysis (ICA) is an advanced or modified version of the Principal Component Analysis (PCA). It uses high-order statistical analysis-based de-correlation for the source input signal and provides important information regarding the presence of irregularities. In the ICA, the input data is modeled as linear coefficients that are mutually independent of each other. Based on the blind source separation algorithm, it can efficiently and accurately transform the input voltage signal into mutually and statistically independent components, thereby helping the detection process (Hyvärinen et al. 2001, Pöyhönen et al. 2003, Lopez et al. 2011 & Dubey et al. 2011).

The mutual information is nothing but the individual independency measurement of the input signal, and its entropy is Gaussian in nature which can be written as:

$$J(y) = h\left(v_{gaussian}\right) - h(v) \qquad (7)$$

Differential entropy H of an input signal y with density $p_v(\eta)$ is given by:

$$h(v) = -\int p_h(\eta) \log p_v(\eta) d\eta \qquad (8)$$

Here, the negentropy is estimated based on the estimation of probability functions of input signals. The negentropy can then be expressed as:

$$J(v_i) = J\left(e\left(w_i^T x\right)\right) = \left[e\left\{g\left(w_i^T x\right)\right\} - e\left\{g\left(v_{gaussian}\right)\right\}\right]^2 \qquad (9)$$

where e is the statistical expectation and G is a non-quadratic factor. For n linear mixtures taken as $x_1, x_2, \ldots x_n$ for n independent components, we can express:

$$x_j = a_1 s_1 + a_2 s_2 + \ldots + a_n s_n \quad \text{For all } j \qquad (10)$$

where x_j is a random mixture and s_k is a random independent variable. Assuming x as the random vector with elements $x_1, x_2, \ldots x_n$ and s as the random vector with elements $s_1, s_2, \ldots s_n$, if A is the matrix having elements a_{ij}, we can take column vectors x^T as the transpose of x. Then, we can write in matrix form as follows:

$$x = As \qquad (11)$$

where A is a column matrix of elements a j. Then, the model is expressed as:

$$x = \sum_{i=1}^{n} a_i s_i \qquad (12)$$

The ICA mathematical model is an iterative one, where the independent components are called latent variables. Here, the input mix matrix is considered to be unknown and the components s_i are statistically independent. Finally, once the matrix A is estimated, we can determine the inverse, W, and then evaluate the independent component by the following expression:

$$s = Wx \qquad (13)$$

3 FAULT DETECTION METHODOLOGY

This section explains the fault detection methodology for the grid-connected wind power system as follows:

i. A grid-connected wind power system is simulated in MATLAB where different types of

fault are created. The voltage signal is extracted at the PCC under faulty conditions. The signal is then passed through wavelet transform to carry out the time–frequency analysis for the identification of faulty situations.
ii. The voltage signal is processed to find the mean and de-correlation. This is the first level of processing.
iii. X data size is reduced and filtered for better data redundancy.
iv. The Principal Components (PC) of the input data are determined.
v. In this study, the independent components are calculated based on fixed-point iteration [12]–[15].
vi. The de-correlating matrix W_f and the independent component S_f of the voltage signal are determined to generate a matrix x_f.
vii. Finally, the pre-fault signal is considered for constructing the matrix x_n, which can be used to calculate the performance index, and the fault is detected when the index is higher than a specified threshold value. The performance index is given by:

$$PI(k) = \left(normal\left(absolute\left(W_f(k) * x_n(k) - s_f(k)\right)\right)^2\right) \qquad (14)$$

4 SIMULATED RESULTS

This section discusses the detection results for a wind system connected to the grid. The power system is simulated in a MATLAB/Simulink environment and the grid is of 230 kV, 50 Hz rating consisting of a thermal-based power plant. The grid-connected power system is shown in Figure 2 and, here, the sampling frequency is taken as 2 kHz.

4.1 Fault detection using wavelet transform

The voltage signal at the Point of Common Coupling (PCC) is taken offline under an AG fault. The signal is processed through WT that detects the fault event based on filtering through a set of low-pass and high-pass filters. The voltage signal with its detection result is shown in Figure 3.

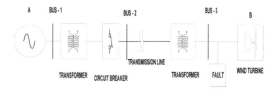

Figure 2. Grid-connected wind power system.

229

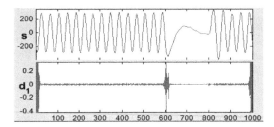

Figure 3. Detection of the phase-to-ground fault in the wind system connected to the grid (red, voltage signal; green, WT output).

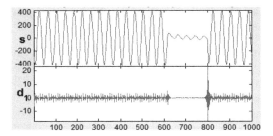

Figure 4. Detection of the phase-to-phase fault in the wind system connected to the grid (red, voltage signal; green, WT output).

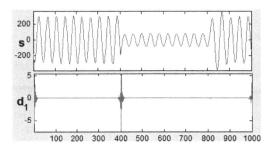

Figure 5. Detection of the three-phase-to-ground fault in the wind system connected to the grid (red, voltage signal; green, WT output).

Similarly, the detection result using the AB and ABCG faults is shown in Figure 4 and Figure 5, respectively. These results clearly show that WT detects the fault initiation event accurately. However, when the fault is cleared, as shown in the figures, the recovery to the normal waveform is not identified by wavelet transform.

4.2 Phase-to-ground fault detection using independent component analysis

A phase-to-ground fault is near to the point of common coupling and the corresponding voltage signal at the PCC is taken offline. Then, the index,

as explained in the ICA section, is calculated under different operating conditions such as without noise, with 20-dB noise, and under frequency variation conditions. The simulation of the performance index obtained from the ICA is shown in Figure 6 (a). It can be seen that the fault initiated at about 0.065 s is detected through a sudden increase in the index value. Before the occurrence of the fault, the index value is found to be almost zero. Once the fault occurs, the value increases suddenly, and if we set a threshold value, then the fault can be detected. However, in this case, we have not set the threshold value, but rather assessed the faulty condition based on a sudden increase.

4.3 Phase-to-phase fault detection using independent component analysis

Here we consider a phase-to-phase fault that is near to the point of common coupling and the corresponding voltage signal at the PCC. The index, as explained in the ICA section, is calculated under different operating conditions such as without noise, with 20-dB noise, and under frequency variation conditions. The simulation of the performance index obtained from the ICA is shown in Figure 6 (b). It can be seen that the fault initiated at about 0.065 s is detected through a sudden increase in the index value. Before the occurrence of the fault, the index value is found to be almost zero. Once the fault occurs, the value

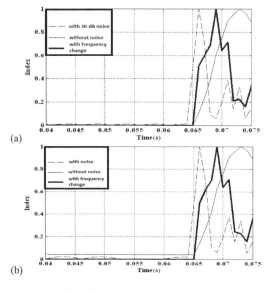

Figure 6. Detection of (a) phase-to-ground fault and (b) phase-to-phase fault in the wind system connected to the grid.

230

Table 1. Energy content using FT, STFT, and WT.

Scenario	Energy content		
	FT	STFT	WT
AG Fault	1.3672	1.7863	2.1663
BG Fault	1.4525	2.2414	2.7352
CG Fault	1.3324	1.8367	2.6538
AB Fault	2.2341	2.3532	3.2514
BC Fault	2.6342	3.1230	3.8724
ABC Fault	3.1302	3.8225	4.2431

increases suddenly, and if we set a threshold value, then the fault can be detected. However, in this case, we have not set the threshold value, but rather assessed the faulty condition based on a sudden increase.

4.4 Detection of fault disturbances using the Performance Index (PI)

This section describes the detection technique using a performance index called the energy content of the processed faulty signal. The voltage signal at the point of common coupling is extracted and processed through different signal processing transforms such as Fourier Transform (FT), Short-Time Fourier Transform (STFT), and wavelet transform. Then, the output waveform after processing is used to calculate the energy which is compared with a threshold value to determine the faulty or normal operating conditions in the power system. It can be observed that the proposed WT provides a higher energy value, so that a suitable threshold can be chosen to accurately detect the faulty conditions. The index in different cases is presented in Table 1.

5 CONCLUSIONS

This paper presents the detection of fault disturbances using wavelet transform and independent component analysis. Voltage signal extracted at the PCC was processed through the above techniques under different fault and operating scenarios. It was observed that WT detected the fault event very accurately. However, in some cases, it could not discriminate the disturbance condition because of noise or frequency variation. Therefore, the faults were accurately detected using ICA under all operating scenarios.

REFERENCES

Chowdhury, F. N., J. P. Christensen, & J. L. Aravena (1991). Power system fault detection and state estimation using Kalman filter with hypothesis testing. *IEEE Trans. on Power Delivery. 6*, 1025–1030.
Dash, P. K., A. K. Pradhan, & G. Panda (2000). A novel fuzzy neural network based distance relaying scheme. *IEEE Trans. on Power Delivery. 15*, 902–907.
Dubey, H.C., S. R. Mohanty, & N. Kishore (2011). Abrupt change detection of fault in power system using independent component analysis. *International Conference on Signal Processing, Communication, Computing and Networking Technologies (ICSCCN)*, 659–664.
Girgis A. (1982). A new Kalman filtering based digital distance relaying. *IEEE Trans. on Power Apparatus and Systems 101*, 3471–3480.
Hyvärinen, A., J. Karhunen, & E. Oja (2001). *Independent Component Analysis.* A Wiley Interscience Publication, John Wiley & Sons, Inc.
Lopez, M. G., H. M. Lozano, L. P. Sanchez, & L. N. O. Moreno (2011). Blind Source Separation of Audio Signals Using Independent Component Analysis and Wavelets. *21st International Conference on Electrical Communications and Computers (CONIELECOMP) Mexico*, 152–157.
Phadke, G. & J. S. Thorp (1988). *Computer Relaying for Power Systems*, New York, John Wiley.
Pradhan, A. K., A. Routray, & S. R. Mohanty (2005). A Moving Sum Approach for Fault Detection of Power Systems. *Electric Power Components and Systems 34 (4)*, 385–399.
Ray, P. K., H. C. Dubey, S. R. Mohanty, N. Kishor, & K. Ganesh (2010). Power quality disturbance detection in grid-connected wind energy system using wavelet and S-transform, *IEEE ICPCES*, 1–4.
Sachdev, M. S. & M. A. Nagpal (1991). Recursive least square algorithm for power system relaying and measurement applications. *IEEE Trans. on Power Delivery. 6*, 1008–1015.
Sanna, Pöyhönen., P. Jover, & H. Hyötyniemi (2003). Independent component analysis of vibrations for fault diagnosis of an induction motor. *Proceedings of IASTED International Conference Circuits, Signals, and Systems*, 19–21.
Sidhu, T. S., D. S. Ghotra, & M. S. Sachdev (2002). An adaptive distance relay and its performance comparison with a fixed data window distance relay. *IEEE Trans. on Power Delivery. 17*, 691–697.
Ukil, A. & R. Živanović (2007). Abrupt change detection in power system fault analysis using wavelet transform. *International Conference on Power Systems Transients, Montreal, Canada. June.* 19–23.
Ukil, A. and R. Živanović, (2007). Application of Abrupt Change Detection in Power Systems Disturbance Analysis and Relay Performance Monitoring. *IEEE Trans. on Power Delivery 22 (1)*, 59–66.
Zadeh, R.A., A. Ghosh, & G. Ledwich (2010). Combination of Kalman Filter and Least-Error Square Techniques. *IEEE Trans. on Power System 25 (4)*, 2868–2880.

Computer, Communication and Electrical Technology – Guha, Chakraborty & Dutta (Eds)
© 2017 Taylor & Francis Group, ISBN 978-1-138-03157-9

Effect of ultrasonic pretreatment on the osmotic drying of ash gourd during Murabba processing

D. Mandal, N. Nath & P.K. Sahoo
Department of Food Engineering, Faculty of Agricultural Engineering, Bidhan Chandra Krishi Viswavidyalaya, Mohanpur, Nadia, West Bengal, India

ABSTRACT: This paper presents the effect of ultrasonic pretreatment on osmotic drying of ash gourd and its associated micro structural changes. The ash gourd cubes of dimensions 15 mm × 15 mm × 15 mm were subjected to ultrasonic (US) waves at a frequency of $30 \pm 1\,kHz$ for 0, 10, 20, and 30 min. Then, osmotic drying was carried out by immersing the samples in 50, 60 and 70°C Brix sugar solution at various osmotic temperatures of 40°C, 50°C, and 60°C till equilibrium moisture contents are achieved in the samples. The microstructures of pretreated samples were examined using a scanning electron microscope. The results showed that an ultrasound pretreatment performed for more than 10 min had a positive effect on the mass exchange caused by osmotic drying. The creation of microchannels or the increase of tissue porosity in the samples was measured using *ImageJ* image processing software. Finally, the investigation showed that the ultrasound-assisted osmotic treatment prior to vacuum drying reduced the final drying time in the vacuum dryer by 47–70% and also reduced the total energy consumption by 56–63%.

1 INTRODUCTION

The ash gourd (*Benincasa hispida*), also called winter melon or white gourd, was originally grown in Southeast Asia and is now widely grown in East Asia and South Asia as well. However, it is not as commonly used as a vegetable in India as other vegetables such as potato, eggplant, cauliflower, cabbage, okra, etc. Since it possesses several nutritive and medicinal properties, it has a huge scope for value addition. In India, ash gourds are commercially utilized for manufacturing candy and wide varieties of sweet delicacies. One such product is called Murabba (Petha, in northern India). The Indian city of Agra (Uttar Pradesh) has become a renowned business center for the production of Murabba.

Traditionally, in the preparation of Murabba, ash gourds are processed in a series of steps: boiling, peeling, deseeding, followed by boiling in sugar syrup with the addition of rose water or other natural flavouring substances. Thus, ash gourds undergo osmotic dehydration in sugar syrup to form Murabba. Osmotic drying is a widely used method to remove water partially from fruits and vegetables by immersion in a hypertonic solution. However, its long processing time severely affects the quality of the product and also increases the total energy consumption (Shi *et al.* 1995). So, Murabba preparation needs new innovations to overcome the long processing time. To increase the rate of water removal, a novel technology like ultrasonication can be introduced prior to osmotic drying (Carcel *et al.* 2007).

Ultrasound is a mechanical wave with a frequency ranging from 20 kHz to 100 MHz that can propagate through a medium. The high-intensity and low-frequency ultrasound (called as power ultrasound) changes the physico-chemical properties of food products (Mulet *et al.* 2011, Zheng & Sun 2006). In solid-liquid systems, power ultrasound induces compression and expansion of the material, usually referred to as the sponge effect (Fernandes & Rodrigues 2007). It helps the flow of water out of the solid interface with simultaneous entry of the solute from the hypertonic solution. This effect creates microchannels that ease the moisture removal (Carcel *et al.* 2012).

The propagation of ultrasonic wave creates cavitations in the liquid medium, where small bubbles form, grow, and collapse due to pressure fluctuation (Mason 1998). Some cavitations that collapse asymmetrically when they are close to a solid surface usually generate a micro jet that hits the solid surface producing an injection of the fluid inside the solid and affecting the mass transfer (Carcel *et al.* 2012, Rastogi 2011). The occurrence of micro agitation at the solid-liquid interface due to ultrasonic wave enhances the mass transfer rate by a reduction of the solid diffusion boundary layer thickness (Mulet *et al.* 2011).

In recent years, ultrasonic pretreatments have been widely performed on fruits like melons, sapota, pineapples, papaya, strawberries, guavas (Fernandes *et al.* 2008a, b, c, Garcia-Noguera *et al.* 2010, Oliveira *et al.* 2011, Kek *et al.* 2013), etc. The results show that ultrasound has a significant effect on both water and solute transport. According to Shamaei *et al.* (2011), a low-frequency (i.e., 35 kHz) ultrasonic pretreatment during osmotic drying of cranberries was highly encouraging in removing water, decreasing hardness, and retaining color. Our present study attempts to evaluate the performance of Osmotic Drying (OD) of ash gourd coupling with low-frequency ultrasonic pretreatment. The micro structural changes associated with ultrasonic treatment was evaluated through image processing software and then linked with OD performance.

2 MATERIALS AND METHODS

2.1 *Materials*

Ash gourds (*Benincasa hispida*) of homogeneous sizes procured from Indian local markets were used as raw materials in the experiments. The samples were stored at a refrigerated temperature of 4 ± 1°C prior to the experiments. Prior to processing, the ash gourds were peeled using a hand peeler and then cut into 15 mm × 15 mm × 15 mm cubes using a knife.

2.2 *Experimental procedure*

2.2.1 *Ultrasonic (US) pretreatment*
The cubic samples of ash gourd were subjected to ultrasonic waves in a US water bath at a frequency of $30 \pm 1\,kHz$ for 10, 20, and 30 min. The instrument that was used to generate ultrasound (Rivotek Ultrasonicator) had the internal dimension of 150 mm × 100 mm × 100 mm. To avoid any flow out of the samples, the cubes were placed next to each other and covered with a net (Fig. 1). The pretreatment was carried out at a room temperature of about 27°C. After the treatment, the cubes were blotted with filter paper and taken for osmotic treatment.

2.2.2 *Osmotic Drying*
Osmotic Drying (OD) was carried out by immersing the ultrasonic pretreated ash gourd samples in 50°, 60°, and 70° Brix sugar solutions at osmotic temperature of 40°, 50°, and 60°C for a desired length of time. The desired concentrations of the sugar solutions were checked by using a refractometer. The osmotic solution to sample ratio was maintained at 6:1 (v/v). The osmotic temperature was controlled in a serological bath. The initial weights of samples were taken using a digital

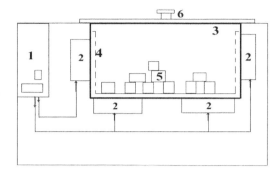

Figure 1. Schematic diagram of the experimental set up for indirect sonication: 1) ultrasound generator, 2) transducer, 3) stainless steel bath filled with water, 4) stainless steel container, 5) ash gourd sample cubes, and 6) lid.

weighing machine. The weighted samples were added to the sugar solution in a beaker. Then, the beaker was immersed in the serological bath at a particular temperature. The samples were removed from the solution at intervals of 15 min to monitor the moisture loss and solid gain of samples with the osmotic treatment time. The sample pieces that were removed from the solution at the end of the osmosis period were immediately placed on the blotting paper to remove the surface moisture. Then, those pieces were weighed using a digital weighing machine. To get oven-dried samples, the samples were placed in a hot oven dryer at 105°C for 24 hours.

Osmotic dehydration is a combination of simultaneous water and solute diffusion process, which occurs countercurrently. The Water Loss (WL) and Solute Gain (SG) associated with OD for each sample was determined according to the procedure described by Panagiotou *et al.* (1999) and Sahoo *et al.* (2007).

$$WL = \left[\frac{(M_i - M_s) - (M_{ot} - M_{os})}{M_i} \right] \qquad (1)$$

$$SG = \frac{M_{os} - M_s}{M_i} \qquad (2)$$

where M_i, M_s, M_{ot}, and M_{os} are initial mass of the sample, oven-dried mass of the fresh sample, mass of the osmotically treated sample after time t, and oven-dried mass of the osmotically treated sample after time t, respectively.

Calculation of the apparent mass Diffusivity of water (D_w) and that of the soluble solid (D_{ss}) follows Fick's second law, which has been expressed for cubical configuration (Matusek *et al.* 2008) as shown below:

$$M_r = \frac{(x_t - x_e)}{(x_0 - x_e)} = \sum_{n=1}^{\infty} C_n^3 exp\left[-D_w t q_n^2\left(\frac{3}{a^2}\right)\right] \quad (3)$$

$$S_r = \frac{(s_t - s_e)}{(s_0 - s_e)} = \sum_{n=1}^{\infty} C_n^3 exp\left[-D_{ss} t q_n^2\left(\frac{3}{a^2}\right)\right] \quad (4)$$

where x_0, x_t, and x_e are the initial moisture content (dry basis), the moisture content at any time t (dry basis), and the equilibrium moisture content (dry basis), respectively. s_0, s_t, and s_e are the initial soluble solid content (dry basis), the soluble solid content at any time t (dry basis), and the equilibrium soluble solid content (dry basis), respectively. Here, a is the geometric parameter of ash gourd cubes. C_n is expressed as

$$C_n = \frac{2\alpha(1+\alpha)}{(1+\alpha+\alpha^2 q_n^2)} \quad (5)$$

where q_n is a non-zero positive root of the equation $\tan q_n = -\alpha q_n$. Here, α is the ratio of volume of the solution to that of the sample cubes.

As per the procedure described by Matusek *et al.* (2008), D_w and D_{ss} can be calculated from the following equations:

$$D_w = \left[\{d(logM_r)/dt\}/\{d(logM_r)/d(F_{ow})\}\right] \quad (6)$$
$$(a^2/3)$$

$$D_{ss} = \left[\{d(logS_r)/dt\}/\{d(logS_r)/d(F_{os})\}\right] \quad (7)$$
$$(a^2/3)$$

where the Fourier numbers for moisture and solute diffusion are calculated as $F_{ow} = D_w t(3/a^2)$ and $F_{os} = D_{ss} t(3/a^2)$, respectively.

2.2.3 Scanning Electron Microscopy (SEM) analysis

For microstructural study, the ultrasound-treated and osmo-dehydrated ash gourd tissues of a fine section were first frozen at 0°C. Then, the samples were treated with a formaldehyde solution (2% formaldehyde, 70% ethanol, 5% acetic acid) for 15 min, dehydrated with ethanol (80%, 90%, and 99% each for 15 min), and finally vacuum dried for 2 h. The dried materials were mounted on SEM stubs and stored in desiccators with silica gel until fully dried before SEM imaging. Then, the samples were stored in desiccators with silica gel for at least 24 h to remove the remaining water. The glycerol-infiltrated samples were mounted on SEM stubs with conductive glue and examined without metal coating. Micrographs of the samples were taken using

the Field Emission SEM (FE-SEM Supra 55[Carl Zeiss, Germany]) instrument. The instrument was operated at an accelerating voltage of 30 kV. The process of imaging the micro porosity using SEM was as follows. First, the ash gourd samples were dehydrated in EM grade ethanol series consisting of 30, 50, 70, 80, 90, and 100% ethanol with an incubation time of 10 min. Then, the dehydrated samples were dried using the Critical Point Drying (CPD) unit. The sample was placed on an aluminum stub using a double-sided carbon tape and sputter coated with heavy metal, i.e., gold. Then, the samples were kept in the sample holding chamber of the SEM instrument and micrographs of each sample at various magnifications were obtained.

2.2.4 SEM image analysis

The images (.TIFF format) from the SEM were analyzed using the ImageJ 1.48v software (Fig. 2). Using the software, pore areas, i.e., the black areas of the image, were marked in a vector layer using *polygon selection* tool from the toolbar. Then, the area of the pores in pixel value was calculated using the *Analyze>Measure* command (ImageJ 1.48vUser Manual, 2011). The rectangular marked area was also calculated using the *Analyze>Measure* command. The total pore area in that rectangular section was calculated by summing the pixel area of the pores. Then, the porosity of the cross section was determined by the ratio of

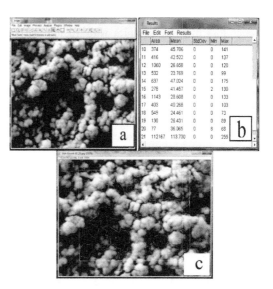

Figure 2. Measurement of tissue porous space across a sectional area in an SEM image using ImageJ software: a) raw image loaded in ImageJ environment, b) pore areas in pixel value, and c) marking the pore areas with a polygon selection tool with a rectangular marked section.

the pore area to total cross-sectional area of the rectangular marked section. In the same manner, more rectangular areas were marked on the same image and porosity of these rectangular areas were determined simultaneously. Then, the average porosity of that section was calculated.

2.2.5 *Vacuum drying*

After osmotic dehydration, the samples underwent final drying using a vacuum dryer (Simco Vacuum Tech.) at 54 kPa absolute pressure and 65°C temperature until 25% moisture content (wet basis) was achieved.

3 RESULTS AND DISCUSSION

3.1 *Solute Gain and Water Loss*

During the osmotic drying of food samples, the Solute Gain (SG) and Water Loss (WL) are usually affected by the concentration and temperature of the osmotic solution but highly affected by the ultrasound pretreatment prior to osmotic dehydration. At a particular solution concentration, SG and WL increase as we increase the pretreatment time.

It is quite evident that the increase in the ultrasonication time prior to OD results in a significant increase in SG and WL in ash gourd cubes due to the formation of micro channels and increase in tissue porosity (Figs. 3 & 4). After 2 h of OD, the samples with 30 min ultrasonication have about 79% more solute gain and about 166% more water loss over the samples with 10 min ultrasonication. A similar result of ultrasound-assisted osmotic dehydration treatment on papaya has been reported by Fernandes *et al.* (2008b).

3.2 *Microstructure of Murabba*

Osmotic dehydration along with ultra sound as pretreatment is known to cause changes in the

structure of plant tissue. Untreated tissue was characterized by a cell with uniform shape (Fig. 5a).

Conversely, the cells in treated tissue were irregular shaped, distorted, and showed numerous breakdown of cell wall along with the formation of microchannels (Fig. 5b to Fig. 5d). The microchannels had an average diameter of 85–95 μm. The alteration of cell structures caused by US were large when the longest ultrasound treatment time was applied. This finding is in accordance to those obtained by Fernandes & Rodrigues (2008c) and Fernandes *et al.* (2009) on melon and pineapple, respectively.

By analyzing the SEM images of ash gourd cubes with *ImageJ* software, we can see that with the increase of ultrasound treatment time, porosity of the sample tissues increase. Porosity of the studied tissues was increased from 12.69% to 35% due to the application of 30 min power ultrasound (Fig. 6). The increase of SG and WL was due to the increase of porosity (Fig. 7). The porosity pro-

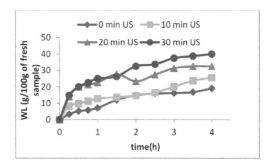

Figure 4. Water loss profiles for osmotic drying of ash gourd in 70° Brix sugar solutions at 60°C for different ultrasonication time.

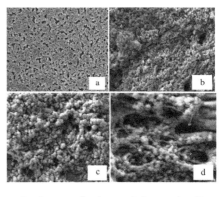

Figure 5. Images of ash gourd tissue using Scanning Electron Microscope a) untreated, b) after 10 min US treatment, c) after 20 min US treatment, and d) after 30 min US treatment with 15000X magnification showing pore areas marked with arrows.

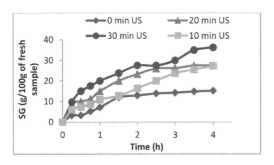

Figure 3. Solute gain profiles for osmotic drying of ash gourd in 70° Brix sugar solutions at 60°C for different ultrasonication time.

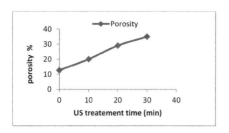

Figure 6. Effect of US treatment on porosity of the tissue.

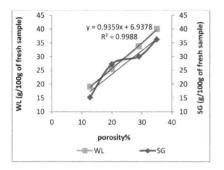

Figure 7. Effect of porosity on Solid Gain (SG) and Water Loss (WL).

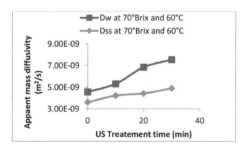

Figure 8. Effect of ultrasound pretreatment on apparent mass diffusivity of water and solute.

file with SG ($R^2 = 0.9988$) and WL ($R^2 = 0.9152$) shows a linear relationship. This is due to the fact that an increase in porosity reduces the resistance against mass diffusion.

3.3 Effect of ultrasound pretreatment on apparent mass diffusivity of water and solute

Ultrasound pretreatment increases the apparent mass diffusivity of water and solute. It affects the apparent mass diffusivity of water more than that of the solute at a particular process temperature (Fig. 8). It is possibly due to the molecular weight of sugar, which is higher than that of the water molecule. This is found to be similar to the results reported by Sahoo et al. (2007) and Rastogi et al. (2005).

3.4 Effect of ultrasonic pretreatment on final drying time and energy consumption

The vacuum drying time, total dehydration time, and energy consumption in final drying in terms of electrical energy for dried Murabba are presented in Table 1. The application of ultrasound pre-osmotic treatment prior to vacuum drying reduced the final drying time significantly by 46.8–70% when compared to the drying time needed for the vacuum drying with osmotic dehydration without ultrasonic pretreatment. The lowest drying time was observed when the ultrasound was applied for 30 min with energy consumption at the final drying time of 0.27 kW-h/kg dry initial mass. However, considering the total energy consumption, 20 min of US pretreatment has given an optimum condition for the whole process with 1.33 kW-h/kg dry initial mass and effective porosity of 29%. This is similar to the results of Kek et al. (2013) reported for the effect of ultrasound treatment on convective drying of guava slices.

Table 1. Effect of ultrasonic pretreatment on final drying time and energy consumption.

Process condition	MC of fresh sample	MC after OD	OD time	Avg.wt. of sample after OD	MC after vacuum drying	Vacuum drying time	Energy consumption in vacuum drying kW-h/kg dry initial mass	Energy consumption in US kW-h/kg dry initial mass	Total energy consumption kW-h/kg dry initial mass	Total drying time
OD + 0 mins US	91%	70%	4 h	14.4 g	25%	2.5 h	3.09	0	3.09	6.5 h
OD + 10 mins US	91%	62%	4 h	14.5 g	25%	1.33 h	1.07	0.29	1.36	5.33 h
OD + 20 mins US	91%	55%	4 h	14.1 g	25%	1 h	0.55	0.58	1.13	5 h
OD + 30 mins US	91%	48%	4 h	14.2 g	25%	0.75 h	0.27	0.87	1.14	4.75 h

4 CONCLUSIONS

Ultrasonic pretreatment for 10–30 min prior to osmotic drying could enhance the drying performance by reducing the total drying time by 47–70% compared to drying without ultrasonication. The total energy consumption in Murabba processing (including vacuum drying and ultrasonication) reduced by 56–63% due to the application of power ultrasound to the ash gourd samples. The SEM images of ultrasonicated ash gourd samples were analyzed by ImageJ software. The creation of microchannels and tissue porosity enhanced the mass diffusivity of both water and solutes and consequently enhanced the drying performance during Murabba processing.

REFERENCES

Carcel, J.A., Benedito, J., Rossello, C. & Mulet, A. 2007. Influence of ultrasound intensity on mass transfer in apple immersed in a sucrose solution. *Journal of food Engineering*, 78: 472–479.

Carcel, J.A., Garcia-Perez, J.V., Benedito, J. & Mulet, A. 2012. Food process innovation through new technologies: use of ultrasound. *Journal of Food Engineering*, 110: 200–207.

Fernandes, F.A.N. & Rodrigues, S. 2007. Ultrasound as pre-treatment for drying of fruits: Dehydration of banana. *Journal of Food Engineering*, 82: 261–267.

Fernandes, F.A.N., Linhares Jr., F.E. & Rodrigues, S. 2008a. Ultrasound as pretreatment for drying of pineapple. *Ultrasonics Sonochemistry*, 15: 1049–1054.

Fernandes, F.A.N, Oliveira, F.I.P. & Rodrigues, S. 2008b. Use of ultrasound for dehydration of papayas. *Food Bioprocess Technology*, 1: 339–345.

Fernandes, F.A.N., Gallao, M.I., & Rodrigues, S. 2008c. Effect of osmotic dehydration and ultrasound pretreatment on cell structure: Melon dehydration. *LWT-Food Science and Technology*, 41: 604–610.

Fernandes, F.A.N. & Rodrigues, S. 2008. Dehydration of Sapota (*Achrassapota L.*) using ultrasound as pretreatment. *Drying Technology*, 26: 1232–1237.

Fernandes, F.A.N., Gallao, M.I., & Rodrigues, S. 2009. Effect of osmosis and ultrasound on pineapple cell tissue structure during dehydration. *Journal of Food Engineering*, 90: 186–190.

Food Safety and Standards Regulations. 2009. Regulation 5.5.24: PP−439.

Garcia-Noguera, J., Oliveria, F.I.P., Gallao, M.I., Weller, C.L., Rodrigues, S., & Fernandes, F.A.N. 2010. Ultrasound assisted osmotic dehydration of strawberries: effect of pretreatment time and ultrasonic frequency. *Drying Technology*, 28: 294–303.

ImageJ 1.48v User Manual, 2011.

Kek, S.P., Chin, N.L. & Yusof, Y.A. 2013. Direct and indirect power ultrasound assisted pre-osmotic treatments in convective drying of guava slices. *Food and Bioproducts Processing. doi:10.1016/j.fbp.2013.05.003.*

Mason, T.J. 1998. Power ultrasound in food processing. The way forward. In: Povey, M.J.W., Mason, T.J. (Eds.), *Ultrasound in Food Processing. Chapman & Hall, London,* pp. 105–126.

Matusek, A., Czukor, B. & Meresz, P. 2008. Comparison of sucrose and fructo-oligosaccharides as osmotic agents in apple. *Innovative Food Science & Emerging Technologies*, 9: 365–373.

Mulet, A., Carcel, J.A., Garcia-Perez, J.V. & Riera, E. 2011. Ultrasound-assisted hot air drying of foods. In: Feng, H., Barbosa-Canovas, G.V., Weiss, J. (Eds.), *Ultrasound Technologies for Food and Bioprocessing. Spinger, New York,* pp. 511–534.

Oliveira, F.I.P., Gallao, M.I., Rodrigues, S. & Fernandes, F.A.N. 2011. Dehydration of malay apple (syzgium-malaccense L.) using ultrasound as pre-treatment. *Food Bioprocess Technology*, 4: 610–615.

Panagiotou, N. M., Karanthonos, V.T. & Maroulis, Z.B. 1999. Effect of osmotic rehydration of fruits. *Journal of Food Science and Technology*, 17: 175–189.

Rodrigues, S. & Fernandes, F.A.N. 2008. Ultrasound in fruit processing. In: Urwaye, A.P. (Ed), *New Food Engineering Research Trends. Nova Science Publishers Inc., New York,* pp. 103–135.

Rastogi, N.K., Raghavarao, K.S.M.S. & Niranjan, K. 2005. Development in osmotic dehydration. In: Sun, D.W. (Ed.), *Emerging Technologies for Food Processing. Elsevier Ltd, London, UK,* pp. 221–249.

Rastogi, N.K. 2011. Opportunities and challenges in application of ultrasound in food processing. *Crit. Rev. Food Science Nutrition,* 51: 705–722.

Sahoo, P.K., Mitra, J. & Chakraborty, S. 2007. Kinetics of apparent mass diffusivities on osmotic dehydration of carrots. *Journal of Food Science & Technology*, 44(1): 26–28.

Shamaei, S., Emam-Djomeh, Z. & Moini, S. 2011. Ultrasound assisted osmotic dehydration of cranberries: effect of finish drying methods and ultrasonic frequency on textural properties. *Journal of Texture Studies*, 43: 133–141.

Shi, X. Q., Fito, P. & Chiralt, A. 1995. Influence of vacuum treatment on mass transfer during osmotic dehydration of fruits. *Food Research International,* 28: 445–454.

Zheng, L. & Sun, D. W. 2006. Innovative applications of power ultrasound during food freezing processes-A review. *Trends in Food Science and Technology,* 17: 16−23.

Computer, Communication and Electrical Technology – Guha, Chakraborty & Dutta (Eds)
© 2017 Taylor & Francis Group, ISBN 978-1-138-03157-9

Analytical model of MEMS-based piezoresistive pressure sensor using Si_3N_4 diaphragm

Kakali Das
Department of Electronics and Communication Engineering, St. Thomas College of Engineering and Technology, Kolkata, India

Himadri Sekhar Dutta
Member, IEEE
Department of Electronics and Communication Engineering, Kalyani Government Engineering College, Kalyani, India

ABSTRACT: In this paper, Piezoresistive Pressure Sensor (PPS) with four Polysilicon piezoresistors on Si_3N_4 diaphragm with improved sensitivity is successfully designed by using MEMS technology. Sensing is accomplished via deposited polysilicon resistors like metal resistors. The analytical model of PPS is optimized for the geometry and different aspect ratios of the diaphragm. The system output sensitivity of the PPS is evaluated here by interpreting the proper selection of the geometry of a thin Si_3N_4 diaphragm. The maximum deflection, maximum induced stress and highest sensitivity for this sensor are obtained for the diaphragm when aspect ratio is minimum. It is found that sensitivity of the sensor is influenced more powerfully by diaphragm thickness. The applied pressure range is considered from 5 kPa to 40 kPa related to the equivalent minimum and maximum measurable blood pressure of human body.

1 INTRODUCTION

In the industrial applications, the pressure sensor is one of the most consumable and most widely used devices in the field of sensing. The pressure sensor has gained popularity in biomedical, automotive and avionics Industries (Eaton & Smith 1997). In the present work, modeling of piezoresistive MEMS based pressure sensors having both the Square and Rectangular diaphragms have been done using the analytical models through their best fitting to match with simulation results.

In this paper, the mechanical properties of Si_3N_4 are first highlighted to illustrate that Si_3N_4 is one of the important materials for realizing MEMS structures and then to develop a piezoresistive Pressure sensor followed by a parametric analysis of different effects on this sensor, optimizing mechanical and physical components with accuracies down to micron level. The piezoresistive behavior of Polysilicon is larger than that of typical metal film structures and smaller than that of single-crystal resistors.

1.1 Material property of pressure sensor

Silicon Nitride (Si_3N_4) is used as diaphragm material and Polysilicon is used as piezoresistor materials.

1.2 Silicon nitride as a diaphragm

In this design, silicon nitride is considered as diaphragm material instead of bulk silicon. The main reason behind for considering as diaphragm is its high strength (yield strength i.e. 14 GPa) which can withstand the maximum load without braking the diaphragm (Chaun & Can 2010). At the same time, higher mechanical sensitivity which is governed by the mechanical dimension of the diaphragm can be achieved by reducing the diaphragm size.

1.3 Key silicon nitride properties

High strength over a wide temperature range
High fracture toughness
High hardness
Outstanding wear resistance, both impingement and frictional modes
Good thermal shock resistance
Good chemical resistance

1.4 Polysilicon as a piezoresistive material

To convert mechanical stress generated in diaphragm due to external load, into an electrical signal, polysilicon is used as the sensing elements which offers significant advantages i.e. high piezoresistive coefficient, low hysteresis, long term stability (Liu et al. 2013).

1.5 Design of piezoresistive pressure sensor

In silicon pressure sensors, diaphragm and piezoresistors are made from same material i.e. silicon diaphragm is made by anisotropic etching at the back side of the bulk silicon whereas piezoresistors are embedded into diaphragm at front side using diffusion or ion implantation technique. In this paper, another family of diaphragm based pressure sensors is proposed which is based on insulating diaphragm like silicon nitride (Lou et al. 2012). The material properties of the diaphragm and piezoresistors are given in Tables 1 and 2.

It mainly consists of a diaphragm in which four piezoresistors are symmetrically placed on the four edges of the diaphragm as shown in Figure 1. The piezo resistors are arranged in the form of a Wheatstone bridge over the diaphragm to obtain the electrical output. The orientations of the piezoresistors are such that the two are placed to sense stress in the direction of their current axis and two are built to sense stress perpendicular to their current flow. When there is no applied pressure the bridge will be in balanced condition and there will not be any variations in the resistance values and hence output voltage will be zero. When an external pressure is applied there will be some deflection (deformation) in the diaphragm. As the diaphragm deforms there will

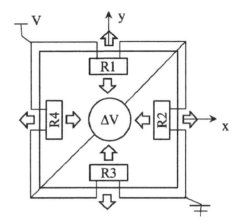

Figure 1. Top view of conventional diaphragm piezoresistive pressure sensor.

be strain produced on the diaphragm as well as on the piezo resistors. As a result the resistance of the piezo elements will change and the bridge is unbalanced. The unbalanced bridge results in an output voltage which is proportional to the applied pressure (Nisanth et al. 2014). Figure 1 shows the top view of a diaphragm piezoresistive pressure sensor in which R1 and R3 are longitudinal and R2 and R4 are transverse piezoresistors forming a Wheatstone bridge further, the longitudinal and transverse stresses both act on each resistor, *i.e.*, if a resistor experiences a stress σ_l length wise then it must also be subjected to a stress $v * \sigma_l$ along its width and vice versa, where v is the Poisson's ratio.

Table 1. Material properties of diaphragm.

Material property	Diaphragm
Material	Silicon Nitride (Si₃N₄)
Young Modulus (GPa)	250
Poisson's Ratio	0.28
Density (gm/cm³)	3.1
Electrical Resistivity (Ω-cm)	2.500
Specific Heat (J/g/°C)	0.710
Thermal Condition (W/cm/°C)	1.5

Table 2. Material properties of piezoresistor.

Material property	Piezoresistor
Material	Polysilicon
Young Modulus (GPa)	160
Poisson's Ratio	0.23
Density (gm/cm³)	2.3
Electrical Resistivity (Ω-cm)	5×10^{-5}
Specific Heat (J/g/°C)	0.710
Thermal Cond. (W/cm/°C)	1.5
Piezoresistive Co-efficient (Pa⁻¹)	$\pi_{11} = 6.6 \times 10^{-11}$
	$\pi_{12} = -1.1 \times 10^{-11}$
	$\pi_{44} = 138.1 \times 10^{-11}$

2 SENSOR DESIGN

The piezoresistive effect is given by

$$\frac{\Delta R}{R} = \frac{\Delta \rho}{\rho} + 1(1 + 2v)\varepsilon \qquad (1)$$

where ρ is the resistivity of the piezoresistor, ρ depends on the doping concentration of the piezoresistors, v is the Poisson's ratio and ε is the mechanical strain induced in the diaphragm.

The first part of the left hand side of equation 1 describes the piezoresistive effect and the second part defines the geometric effect. For piezoresistive materials like polysilicon, the piezoresistive effect dominates over geometric effect, so the second part of Equation 1 is neglected and the linear piezoresistive effect is stated by the superposition of the longitudinal and the transverse piezoresistive effect with the stress components σ_l in longitudinal and σ_t in transverse direction.

$$\frac{\Delta\rho}{\rho} = \frac{\Delta R}{R} = \pi_l \cdot \sigma_l + \pi_t \cdot \sigma_t \qquad (2)$$

where, π_l and π_t refers to longitudinal and transverse piezoresistive coefficient, σ_l and σ_t refers to longitudinal and transverse stress, ρ refers to resistivity of the material and usually given in $\Omega\ cm$, $\Delta\rho$ refers to change in resistivity of a piezoresistor.

The effective value for the longitudinal piezoresistive coefficient and the transverse piezoresistive coefficient can be calculated from the piezoresistive coefficient tensor π_{ij}. Here π is the piezoresistive coefficient which depends on temperature, doping concentration and band structure (Kanda 1982). π_l is the longitudinal piezoresistive coefficient and π_t is referred to as the transverse coefficient which can be expressed in terms of the basic piezoresistive coefficients π_{11}, π_{12}, and π_{44}.

$$\pi_{ij} = \begin{bmatrix} \pi_{11} & \pi_{21} & \pi_{31} & \pi_{41} & \pi_{51} & \pi_{61} \\ \pi_{12} & \pi_{22} & \pi_{32} & \pi_{42} & \pi_{52} & \pi_{62} \\ \pi_{13} & \pi_{23} & \pi_{33} & \pi_{43} & \pi_{53} & \pi_{63} \\ \pi_{14} & \pi_{24} & \pi_{34} & \pi_{44} & \pi_{54} & \pi_{64} \\ \pi_{15} & \pi_{25} & \pi_{35} & \pi_{45} & \pi_{55} & \pi_{65} \\ \pi_{16} & \pi_{26} & \pi_{36} & \pi_{46} & \pi_{56} & \pi_{66} \end{bmatrix} \qquad (3)$$

For anisotropic material, the number of independent components can be reduced due to symmetry effect. For example, in single-crystal silicon, there are only 12 non-zero coefficients instead of 36. Due to the cubic crystal symmetry, only three of them are independent those are π_{11}, π_{12}, and π_{44}.

For isotropic material, the number of independent components can be reduced to only two. The longitudinal and transverse piezoresistance coefficients of this sensor are:

$$\pi_l = \frac{1}{2}(\pi_{11} + \pi_{12} + \pi_{44}) \qquad (4)$$

$$\pi_t = \frac{1}{2}(\pi_{11} + \pi_{12} - \pi_{44}) \qquad (5)$$

The effective values for the longitudinal piezoresistive coefficient (π_l) and the transverse piezoresistive coefficient (π_t) can be calculated from the piezoresistive coefficient tensor π_{ij}.

The change of resistance reflects into a variation in the output voltage of the Wheatstone Bridge as given by the following Equation 6

$$\frac{\Delta V}{V} = \frac{r}{(1+r)^2}\left\{ \frac{\Delta R_1}{R_1} - \frac{\Delta R_2}{R_2} + \frac{\Delta R_3}{R_3} - \frac{\Delta R_4}{R_4} \right\} \qquad (6)$$

2.1 Stress-strain relations

For linear elastic, isotropy and plane stress conditions ($\sigma_z = \tau_{yz} = \tau_{zx} = 0$) stress-strain relations can be stated in the matrix form (Timoshenko & Krieger 1959)

$$\begin{Bmatrix} \sigma_x \\ \sigma_y \\ \tau_{xy} \end{Bmatrix} = \frac{E}{1-\nu^2} \begin{bmatrix} 1 & \nu & 0 \\ \nu & 1 & 0 \\ 0 & 0 & \frac{1-\nu}{2} \end{bmatrix} \begin{Bmatrix} \varepsilon_x \\ \varepsilon_y \\ \gamma_{xy} \end{Bmatrix} \qquad (7)$$

2.2 Kirchhoff Plate theory

Since the diaphragm thickness is very small like thin plate, Kirchhoff's Plate theory is applicable here. Transverse shear deformation is prohibited (Figure 2), and the resulting expression is given by

$$\begin{Bmatrix} \sigma_x \\ \sigma_y \\ \tau_{xy} \end{Bmatrix} = -\frac{Ez}{1-\nu^2} \begin{bmatrix} 1 & \nu & 0 \\ \nu & 1 & 0 \\ 0 & 0 & \frac{1-\nu}{2} \end{bmatrix} \begin{Bmatrix} \frac{\partial^2 w}{\partial x^2} \\ \frac{\partial^2 w}{\partial y^2} \\ 2\frac{\partial^2 w}{\partial x \partial y} \end{Bmatrix} \qquad (8)$$

z is any arbitrary value from neutral axis in z-direction ($z = \pm h/2$). Thus the moment-curvature relations for a homogeneous and isotropic Kirchhoff plate are,

$$\begin{Bmatrix} M_x \\ M_y \\ M_{xy} \end{Bmatrix} = -D \begin{bmatrix} 1 & \nu & 0 \\ \nu & 1 & 0 \\ 0 & 0 & \frac{1-\nu}{2} \end{bmatrix} \begin{Bmatrix} \frac{\partial^2 w}{\partial x^2} \\ \frac{\partial^2 w}{\partial y^2} \\ 2\frac{\partial^2 w}{\partial x \partial y} \end{Bmatrix} \qquad (9)$$

where,

$$D = \frac{Eh^3}{12(1-\nu^2)} \qquad (10)$$

D is called flexural rigidity and analogous to flexural stiffness EI and h is the thickness of the diaphragm (Co et al. 2007). The Figure 2 shows the

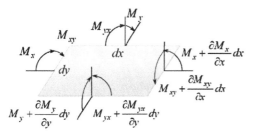

Figure 2. Moments under the lateral force on differential element.

different Moments under the lateral force on differential element.

The state of deformation and stress throughout a Kirchhoff plate are completely described by a single field, namely lateral deflection $w = w(x, y)$ of the mid-surface (Shaby 2015). The differential equation of the deflection surface following the equilibrium equation is given by,

$$\frac{\partial^4 w}{\partial x^4} + 2\frac{\partial^4 w}{\partial x^2 \partial y^2} + \frac{\partial^4 w}{\partial y^4} = \frac{P}{D} \tag{11}$$

For built-in edges and uniformly distributed load, the deflection is given by (Timoshenko & Krieger 1959)

$$w = \frac{4Pa^4}{\pi^5 D} \sum_{m=1,3,5,\ldots}^{\infty} \frac{(-1)^{\frac{m-1}{2}}}{m^5} \cos\frac{m\pi x}{a} \times$$
$$\left(1 - \frac{\alpha_m \tanh \alpha_m}{2\cosh \alpha_m}\cosh\frac{m\pi y}{a} + \frac{1}{2\cosh\alpha_m}\frac{m\pi y}{a}\sinh\frac{m\pi y}{a}\right) \tag{12}$$

where,

$$\alpha_m = \frac{m\pi b}{2a} \tag{13}$$

For rectangular plate with all edges built in under uniformly distributed load, the maximum deflection and moments are given by

$$w_{max} = \alpha\frac{Pa^4}{D} \tag{14}$$

$$(M_x)_{max} = \beta_1 Pa^2 \tag{15}$$

$$(M_y)_{max} = \beta_2 Pa^2 \tag{16}$$

The Table 3 describes the values of α and β for varying the ratios of lengths of larger side (b) to the smaller side (a) of the rectangular diaphragm.

Table 3. Coefficients of α, β for varying b/a ratios.

a/b	1.0	1.2	1.4	1.6	1.8	∞
α	0.00126	0.00172	0.00207	0.00230	0.00245	0.0026
β_1	0.0513	0.0639	0.0726	0.0780	0.0812	0.0833
β_2	0.0513	0.0554	0.0568	0.0571	0.0571	0.0571

The maximum induced stress on the square diaphragm is given by

$$\sigma_{max} = 6\frac{(M_{x\,or\,y})_{max}}{h^2} \tag{17}$$

Therefore, considering the Parallel–Normal Combination of Piezoresistors, according to the specific diaphragm (Meng and Zhao 2016), the equation for differential output voltage has been changed to

$$\frac{\Delta V}{V} = \frac{2r}{(1+r)^2}(1+\nu)(\pi_l - \pi_t)\sigma_{max} \tag{18}$$

Sensitivity 'S' of the sensor is given by,

$$S = \frac{\Delta V}{P} \tag{19}$$

3 PARAMETRIC STUDY

By MATLAB programming, the sensor simulates the results for diaphragm deflection and induced stress for the applied pressure range from 5 kPa (equivalent blood pressure of 40 mm of Hg of human body) to 40 kPa (equivalent blood pressure of 300 mm of Hg of human body). The Figure 3 shows that the four Si_3N_4 Square diaphragms having same cross sectional area but of different thicknesses deflect differently upon the application of same amount of pressure. From the analysis, it is seen that the diaphragm having comparatively smaller thickness shows maximum deflection. Out of four diaphragms, for the square diaphragm of length 400 μm and thickness of 4 μm shows maximum deflection about 1.2 μm for the maximum applied pressure of 40 kPa.

The Figure 4 shows the same analysis for the three Si_3N_4 Rectangular diaphragms having different cross sectional area [(400 μm × 400 μm), (400 μm × 480 μm), (400 μm × 560 μm)], it is concluded that with the increase of b/a ratio, the deflection on the diaphragm is increased for the application of the same applied pressure. The Figure 5 shows the stress profile for square diaphragms having same

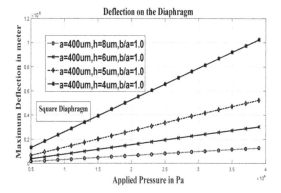

Figure 3. Variation of maximum deflection with different thicknesses of square diaphragm with the applied pressure.

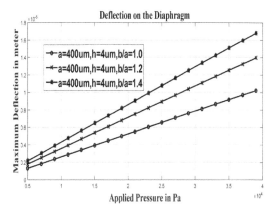

Figure 4. Variation of maximum deflection with different b/a ratio of the diaphragm with the applied pressure.

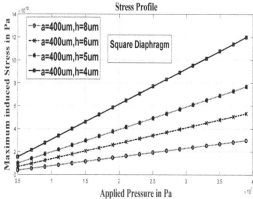

Figure 5. Plot of maximum induced stress on the diaphragm vs. applied pressure.

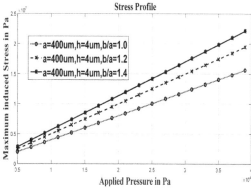

Figure 6. Plot of maximum induced stress on the diaphragm vs. applied pressure for different b/a ratio.

cross sectional area but of different thicknesses. It is observed that the maximum stress induced for the maximum applied pressure for the diaphragm for which the ratio of its length to thickness is maximum or aspect ratio (thickness to length ratio) is minimum. It is evident from the stress profile that the maximum stress induced for the diaphragm of length 400 μm and of thickness of 4 μm is about 120 GPa. The Figure 6 depicts that with the increase of b/a ratio of the diaphragms having same thickness, the induced stress is enhanced.

It is observed in Figure 7 that with the increase of b/a ratio of the diaphragms having same thickness, the differential output voltage is increased which indicates for the diaphragm having smaller aspect ratio explores better output voltage.

From the simulation as shown in Figure 8, it is visualized that by changing silicon nitride diaphragm geometries, the diaphragm having larger

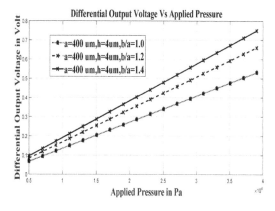

Figure 7. Variation of output voltage with different b/a ratio of diaphragm with the applied pressure.

243

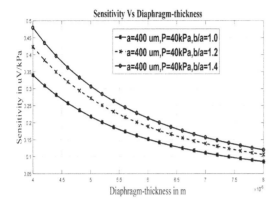

Figure 8. Variation of output sensitivity with different b/a ratio of diaphragm with the diaphragm thicknesses.

'b/a' ratio with the piezoresistors placed at the center of the edges of each side reflects comparatively better sensitivity, the sensitivity decreases with the increase of thickness of the respective diaphragm.

4 CONCLUSION

Finally, the investigations using MATLAB programming clearly indicates the outcome of using of different diaphragm dimensions and the position of piezoresistors. The maximum stress was induced at the centre of the edges of the diaphragm and accordingly the piezoresistors were placed to obtain maximum sensitivity. The diaphragm having same area and comparatively smaller thickness showed maximum deflection. The induced stress was found highest with the maximum applied pressure where the aspect ratio of the diaphragm is minimum. Out of the various diaphragm thicknesses used here, it was optimized that the diaphragm having smaller aspect ratio can be used as highly sensitive pressure sensor. It was depicted that using larger "b/a" ratio, Si_3N_4 diaphragm gives much better response with respect to others, so it can be treated as a highly sensitive piezoresistive micro pressure sensor.

REFERENCES

Chaun, Y. & Can, G. (2010). Investigation based on MEMS double Si_3N_4 resonant beams Pressure Sensor. *IEEE Proceedings on Nano/Micro Engineered and Molecular Systems*: pp. 5–8. Xiamen.

Co, H.S., Liu, W.C. & Gau, C. (2007). Novel fabrication of a pressure sensor with polymer material and evaluation its performance. *Journal of Micromechanics and Micro engineering*. vol. 17: pp. 1640–1648.

Eaton, W.J. & Smith, J.H. (1997). Micro Machined Pressure Sensor: review and recent developments, *Smart Materials and Structures*. vol. 6: pp. 530–539.

Kanda, Y. (1982). Graphical Representations of the piezoresistance coefficients in Silicon. *IEEE Transactions on Electron Devices*. vol. 29: pp. 64–70.

Liu, L., Tan, K. K., Teo, C. S. & Chen, S. L. (2013). Developement of an approach toward Comprehensive Identification of Hysteric Dynamics in Piezoelectric Actuators. *IEEE Transactions on Control Systems Technology*. vol. 21: pp. 1834–1845.

Lou, L., Zhang, S., Park, W.-T., Tsai, J.M., Kwong, D.-L. & Lee, C. (2012). Optimization of NEMS Pressure Sensors with a multilayered diaphragm using silicon nanowires as piezoresistive sensing elements. *Journal of Micromechanics and Micro Engineering*. vol. 22: pp. 2–15.

Meng, X., Zhao, Y. (2016). The design and optimization of a highly sensitive overload resistant piezoresistive pressure sensor. *Sensors*. vol. 16(3), 348.

Nisanth, A., Suja, K. J. & Komaragiri, R. (2014). Sensitivity enhancement of a silicon based MEMS pressure sensor by optimization of size and position of piezoresistor. *IEEE Proceedings on Electronics and Communication Systems*. pp. 1–5. Coimbatore.

Shaby, S. M. (2015). Design and analysis of silicon diaphragm of a MEMS Pressure Sensor. *ARPN Journal of Engineering and Applied Sciences*. vol. 10, no. 5.

Timoshenko, S. & Woinoushy-Krieger, S. (1959). *Theory of Plates and Shells*. New York: McGraw-Hill.

Computer, Communication and Electrical Technology – Guha, Chakraborty & Dutta (Eds)
© 2017 Taylor & Francis Group, ISBN 978-1-138-03157-9

Alcohol sensor-based cost-effective, simple car ignition controller

P. Deb & A. Seth
Department of Electrical Engineering, Guru Nanak Institute of Technology, Kolkata, India

S. Bhattacharya
Department of Electrical Engineering, RCCIIT, Kolkata, India

M. Chakraborty & A. Roy
Department of Electrical Engineering, Guru Nanak Institute of Technology, Kolkata, India

ABSTRACT: The proposed scheme is aimed at making vehicle driving safer by preventing accidents occurring because of drunken driving. The driver's condition in real time is monitored and a scheme is proposed for the detection of Blood Alcohol Content (BAC) using alcohol detector connected to ARDUINO-based controller. The system senses the level of BAC and if that level crosses a permissible limit, the vehicle ignition system will fail to start until the level of BAC decreases to its permissible limits. Until the driver is unfit for driving, ignition of car will not take place. Thus, such a smart alcohol sensor embedded in the car ignition system can prevent such accidents occurring because of drunken driving. The system implemented here is a simple yet cost-effective method for such an automobile ignition controller.

1 INTRODUCTION

India had earned the dubious distinction of having more number of fatalities due to road accidents in the world. Road safety is emerging as a major social concern around the world, especially in India. Drunken driving is already a serious public health problem, which is likely to emerge as one of the most significant problems in the near future [Fell, Beirness, Vons, Smith, Jonah, Maxwell, Price, & Hedlund (2016), Ferguson (2012)]. The system implemented here aims at reducing the number of road accidents due to drunken driving in the future [Ferguson (2012), Bareness & Marques (2010)].

The main aim of this paper is to design an electronic system for implementing an efficient alcohol detection system that will be useful to avoid accidents. The system determines the blood alcohol content of the drunken driver and immediately locks the engine of the vehicle. Hence, the implementation of such a system shall reduce the quantum of road accidents and fatalities due to drunken driving in future [Radun, Ohisalo, Rajalin, Radun, Wahde, & Lajunen, (2014), Azzazy, Chau, Wu, Tanbun-Ek (1995)]. Thus, there is an urgent need for an effective system to check drunken drivers [Azzazy, Chau, Wu, Tanbun-Ek (1995). Although there are laws to punish drunken drivers, they cannot be fully implemented because of the logistical limitations of manpower for checking each and every car driver whether he/she is drunk or not. This can be a major reason for accidents [Azzazy, Chau, Wu, Tanbun-Ek (1995), Anund, Anstonson, & Ihlstrom, (2015), Lahausse & Fildes (2009)]. Therefore, in order to avoid such accidents, a proposed scheme based on ARDUINO-based controller is presented and implemented on the prototype developed in our laboratory. In this system, a sensor circuit is used to detect the presence and level of alcohol consumed by the driver automatically whenever he/she occupies the driver seat and tries to start the ignition system. This system is designed in such a way that when alcohol concentration is detected, an alarm is raised and the ignition of the car shall fail. The sensor senses the BAC accurately, and if that level exceeds the permissible limits it sends a signal to the car ignition system to turn it off.

2 BAC AND ITS EFFECTS

Blood Alcohol Content (BAC) or blood alcohol concentration, which is actually blood ethanol concentration, is most commonly used as a measure of alcohol intoxication for legal or medical diagnosis. Percentage of ethanol in blood in units of mass of alcohol per volume of blood or mass of alcohol per mass of blood is actually a measure of BAC, which varies between countries as standardized according to their specific conditions. For example, it is 0.1 in North America, 0.15 in Taiwan, and 0.02 in Portugal. For instance, BAC of 0.1 (0.1% or one tenth of one percent) means that there are 0.10 g of alcohol in every dL of blood.

To calculate the Estimated peak Blood Alcohol Concentration (EBAC), a variation, including drinking period in hours, of the Widmark formula was used.

The formula is:

$$EBAC = \frac{0.86(SD \times 1.2)}{(BW \times Wt)} - (MR \times DP)$$

where 0.806 = a constant for body water in the blood (mean 80.6%); SD = number of standard drinks containing; 1.2 = factor to convert the amount in grams to Swedish standards set by the Swedish National Institute of Public Health; BW = body water constant; Wt = body weight (kg); MR = metabolism constant; DP = drinking period in hours. [link: http://www.wow.com/wiki/Excessive_blood level of alcohol]

There are several factors, and they vary between male and female. That is, female has higher Metabolism Rate (MR) than male and higher percentage of body fat than an average male. On the contrary, men are higher in average weight than women and others such as body water content. Therefore, such various parameters and variation in case to case has made the calculation of BAC complicated and may lead to erroneous result if factors are considered correctly. For this reason, most calculations of alcohol to body mass simply use the weight of the individual, not the body water content.

Blood Alcohol Concentration (BAC) is expressed in various units in different parts of the world, but each is defined as either a mass of alcohol per volume of blood or a mass of alcohol per mass of blood, but never in volume-to-volume ratio.

The amount of alcohol measured on the breath is generally accepted as proportional to the amount of alcohol present in the blood at a rate of 1:2100. Although a variety of units are used throughout the world, many countries use the g/L unit, which does not create confusion as percentages do. Blood Alcohol Content is used to define the level of intoxication and provides a rough measure of impairment as it requires for the purpose of law enforcement. Although the degree of impairment may vary among individuals with the same blood alcohol content, it can be measured objectively and is therefore legally useful and difficult to contest in court.

3 SYSTEM CONTROL MECHANISM

The main parts of the system are: ARDUINO UNO WITH ATMEGA328 controller, MQ3 Alcohol GAS sensor, and Relay SPDT 12 V, 5 A.

The MQ3 alcohol sensor circuit detects the alcohol level in the human breath and sends signal data to ARDUINO controller, which further controls the functioning of subsequent circuits. The schematic diagram of MQ3 alcohol sensor is shown in Fig. 1.

If the human breath contains alcohol, it will be detected by the alcohol sensor and its level is displayed on the LCD 16 × 2 displays.

The implementation of this work involves two portions: (a) hardware development and (b) software development. In Fig. 2, the sensor compares the breath of the driver with the standard reference safe limit of alcohol and generates the equivalent analog signal. The sensor output signal is fed to the

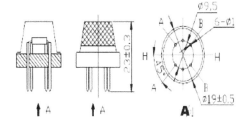

Figure 1. MQ3 sensor.

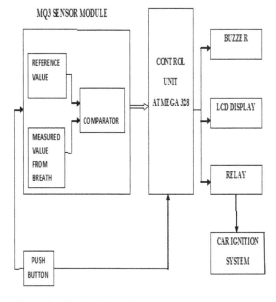

Figure 2. System block diagram.

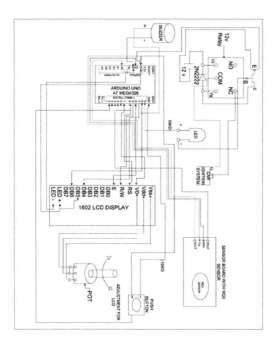

Figure 3. Circuit diagram.

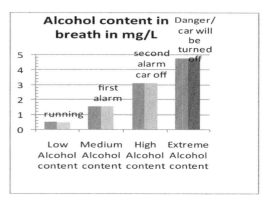

Figure 4. Barchart.

ARDUINO-based controller unit, where appropriate coding is done to control the output alarm devices connected to it. Car Ignition system is connected through a relay to the controller unit, which disables the ignition when high alcohol content is sensed from the drunken driver and ensures safe operation of the scheme.

ARDUINO boards are able to read inputs—light on a sensor, a finger on a button, or even a tweet—and turn it into an output—such as activating a motor, turning on an LED, or publishing something online. ARDUINO runs on Mac, Windows, and Linux. The advantages of using ARDUINO boards are: relatively less expensive than other microcontroller platforms, the ARDUINO software is published as open-source tools, and is available for extension by experienced programmers.

4 EXPERIMENTAL SETUP AND RESULTS

4.1 Circuit operation

Breath of driver is taken as input to the MQ3 sensor. This sensor produces analog electrical output signal AOUT from the sensor board based on calibration resistance pot of the MQ3 sensor. This analog signal is fed to the analog pin A0 of ARDUINO UNO ATMEGA 328. The value of the analog signal depends on the calibration

resistance of MQ3 sensor as well as the alcohol content (ethanol) in the breath of the tested person. DOUT (digital output pin) of the MQ3 sensor board is connected to D8 pin of ARDUINO. The DOUT will be set high when the AOUT value reaches the threshold value. This threshold value is set by the potentiometer of the sensor board of MQ3. The analog signal in the ARDUINO triggers the digital input/output pins 12 and 13. The threshold value is set by the user depending on the limit decided by the state police ARDUINO pins 12 and 13 are connected to buzzer and Single Pole Double Throw (SPDT) relay, respectively. The buzzer acts as audio alarm, whereas the 12 V SPDT switch performs the safety measures by disconnecting the vehicle ignition system in order to stop the vehicle. The vehicle ignition system is connected to the NC terminal of SPDT relay the moment the relay receives signal from ARDUINO pin 13 via one transistor 2N2222. One protective diode is used across SPDT relay in order to protect the relay from transient voltage due to switching. This transistor is used to amplify the current (100 mA) flowing through coil. The coil of relay would be energized to switch the contact to NO terminal and make the vehicle ignition system open-circuited. One LED is connected in series to the ARDUINO pin 13 to make sure the signal is passing from pin 13 to SPDT relay. LCD display is used to show the breath alcohol content amount by receiving the signal from ARDUINO pins 2, 3, 4, and 5. LCD's contrast is adjusted by using one potentiometer. One push button is used to make the ARDUINO reset by feeding the signal to ARDUINO pin 7. The whole control logic is embedded in the ATMEGA microcontroller chip, which is the core processor of ARDUINO UNO ATMEGA 328. The control logic is written in C language and inserted in the microcontroller chip in the form of hexodes.

4.2 *Results*

With the increase of alcohol consumption by manifold, Drink & Drive has reached an alarming state in India. Time has come up to say end to this dangerously irresponsible course of action. Awareness campaigning, proper policing, and administrative steps to incorporate alcohol sensor-based ignition controller will definitely minimize the untoward accidents due to drunken driver. This work is aimed toward a cost-effective and simple embedded technique to keep intoxicated driving at bay.

REFERENCES

Anund, A., H. Anstonson, & J. Ihlstrom. 2015. Stakeholders Opinions on a Future. In-Vehicle Alcohol Detection System for Prevention of Drunk Driving. *In Traffic Injury Prevention* 16(4), 336–344.

Andersson. Agneta, Wiréhn. Ann-Britt, Ölvander. Christina, Ekman. Diana, Bendtsen. Preben. 2009. Alcohol use among University students in Sweden measured by an electronic screening instrument. *BMC Public Health. 9.229.19594906.2724514.*

Azzazy, M. Chau, T. Wu, T. Tanbun-Ek. 1995. Remote Sensor to Detect Alcohol impaired Drivers. Lasers and Electro-Optics Society Annual Meeting. *IEEE Conference Publications.* 2, 320–321.

Beirness, D. J. & Marques P. R. 2010. Alcohol Ignition Interlock Programs. *In Traffic Injury Prevention* 5(3), 299–308.

Ferguson, S.A. 2012. Alcohol-Impaired Driving in the United States: Contributors to the Problem and Effective Countermeasures. *In Traffic Injury Prevention* 13(5), 427–441.

Flescher, Benjamin P. Nelson, Astoyao R. Adjetev, 2012. Design and Development of GSM/GPS based vehicle tracking and alert system for commercial intercity bus. *International Conference on Adaptive Science & Technology (ICAST).*

Lahausse, J.A. & B.N. Fildes, 2009. Cost benefit analysis of an alcohol interlock for installation in all newly registered vehicles. *In Traffic Injury Prevention* 10(6), 528–537.

link: http://www.wow.com/wiki/Excessive blood level of alcohol.

Computer, Communication and Electrical Technology – Guha, Chakraborty & Dutta (Eds)
© 2017 Taylor & Francis Group, ISBN 978-1-138-03157-9

Study of a contactless capacitive-type linear displacement sensor

Shikha Nayak
Department of Computer Science and Engineering, Meghnad Saha Institute of Technology, Kolkata, West Bengal, India

Subir Das & Tuhin Subhra Sarkar
Department of AEIE, Murshidabad College of Engineering and Technology, Berhampore, West Bengal, India

ABSTRACT: In order to measure and control the linear motion of different objects or machineries, a contactless capacitive sensor is proposed in this preliminary study. In this paper, a proof of concept of this sensor's design and performance is reported. In a traditional capacitive-type displacement sensor, the effect of two parallel plates' effective overlap area or the distance between them is considered as a variable sensing parameter, where the dielectric medium between them is considered as constant. Moreover, in order to obtain the signal based on this parameter, any single plate should be moved along with the object motion. Thus, electrical connectivity with this movable plate may lead to a mechanical error. To overcome this difficulty, the proposed design presents a concept where the dielectric medium between the two plates is moved along with the object and the plates remain fixed. Thus, its movable sensing part is completely separated from the electrical interface. Therefore, it is immune to the mechanical error. Here, the linear displacement is sensed by the effect of the push–pull mechanism. This state-of-the-art design is composed of two parallel conductive plates and a single dielectric plate, thus forming two capacitors. Moreover, the differential output of these two capacitors shows good linearity and sensitivity in terms of capacitance. Furthermore, a simple signal processing circuit is used for converting this capacitance into analog voltage. By using a laboratory prototype, a sensitivity of 18 mv/mm and a nonlinearity of ±2.6% have been achieved for the linear displacement in the range of ±30 mm.

1 INTRODUCTION

In order to measure and control the linear displacement of moving objects such as actuator or valve stem and automated color spraying machine arm, a displacement sensor is used. Traditionally, a capacitive sensor, a Linear Variable Differential Transformer (LVDT), or an optical-type Position Sensitive Device (PSD) is used for measuring this precise linear displacement. It can measure the displacement in the range of a few micrometers to centimeters. For a small range of measurement, a capacitive sensor is widely acceptable because of its simple structure, non-contact measurement, and absolute position measurement capability. However, its output linearity is quite limited because of the stray capacitance and fringe effects (Shieh et al. 2001, Lányi et al. 1998). Alternatively, a laser interferometer and an optical encoder have the longest measurement range and nanometer range of accuracy. However, their cost is high and they have a complex structure due to which they are poorly compatible with the measured object (Bobroff 1993, Eom et al. 2001, Yang et al. 2014).

In recent years, researchers have worked on the capacitive sensor (Ferri et al. 2015, Lee et al. 2009, Zeng et al. 2015, Yu et al. 2015, Smith et al. 2005, Ahn et al. 2008, Guo et al. 2016) to simplify its structure and improve its resolution and sensitivity. Normally, the basic principle of this sensor is to change the effective overlap area by moving any one of the capacitor plates with very small distance between them. The sensitivity of this sensor is drastically reduced because of a fringe effect around the edge of the plates. Thus, the resolution of this sensor counteracts with this problem. Thus, in order to improve the resolution of this sensor, the distance between the two plates should be reduced. However, it is difficult to reduce the distance in the range of micrometers because of the precision limit of the mechanical guide of the sensor. However, an interface cable between the plates and the processing unit may interrupt the movement of one plate, leading to a mechanical error. To overcome these difficulties, Bai et al. (2016) introduced a design of a capacitive sensor, where a ground shield aluminum window moved in a forward or backward direction between two parallel plates. This sensor provides a differential capacitive output based on the geometrical placement of parallel plates and the window structure. In addition, Geroge et al. (2008) & Gasulla et al. (2003) developed a contactless capacitive sensor for measuring the angular position of the rotating object by using the movement of the dielectric medium between two annular parallel plates. This phenomenon also provides a differential output with respect to the angular position.

These prior embodiments are able to provide high resolution, linearity, and sensitivity. However, its geometrical shape and interfacing circuit are very complex.

In this paper, a simple but reliable capacitive sensor for linear displacement measurement is designed. In this proposed design, the capacitance of the sensor is changed with the linear displacement of the dielectric material between two parallel plates. Here, a pair of parallel plates are arranged sequentially along the direction of the moving dielectric material. Thus, it is obvious that, during the displacement of the object, the dielectric material is moved accordingly due to the coupling between them. Consequently, the capacitance of these two parallel plates' capacitor will be changed in a push–pull manner. Thus, in this design, electrical connectivity with the moving parts of the sensor is not required. Thus, it is free from any mechanical error and more rigid compared with the prior design. In order to prove this concept, a prototype is designed, developed, and tested under laboratory conditions, where a satisfactory performance is achieved for linear displacement in the range of ±30 mm.

2 DESIGN OF THE PROPOSED CAPACITIVE SENSOR

A schematic of the capacitive sensor is shown in Figure 1. The sensing part of this sensor is composed of two elements: a pair of conductive parallel plates and a dielectric plate. Each conductive plate is fabricated over a copper sheet by using the PCB technology, where two plates are fabricated laterally with a space of 5 mm. The geometric shape of each plate is rectangular. In order to reduce the fringe effect and the stray capacitive effect, a 15 mm width guard ring is also fabricated between the two plates. The distance between the two parallel plates is only 7 mm. Thus, according to the conventional rule, the width of the guard ring is chosen as twice that of the distance. Here, a 2 mm-thick aluminum oxide plate (relative permittivity 9) acts as a dielectric medium between two parallel plates. The size (60 mm × 10 mm) of the dielectric plate is selected according to the size

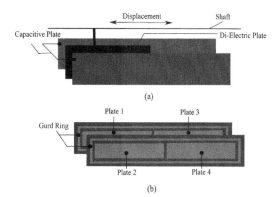

Figure 1. Schematic of the proposed capacitive sensor.

of a single fabricated plate. In order to sense the displacement by the push–pull mechanism, the dielectric plate should be kept initially at the middle position of each plate. Thus, it is apparent that, during the transition of the dielectric plate coupled with an object sensor, it will be easy to measure the linear displacement in the range of ±30 mm in terms of capacitance.

3 METHOD OF APPROACH

From Figure 1 it is obvious that plates 1 and 2 form the parallel plate capacitor of C_1 and similarly C_2 is formed between parallel plates 3 and 4. As the dielectric plate is moved forward and backward between the parallel plates, the capacitance of C_1 and C_2 will be changed accordingly. To illustrate the variation of capacitance, a graphical model is implemented, as shown in Figure 2.

Assume that the length of C_1, C_2, and dielectric plates is L. The width of these plates is 'w'. Thus, if the dielectric plate is present at the middle position of two capacitor plates, then it is obvious that the capacitors C_1 and C_2 will be separated individually into two equal parts. They are denoted as C_1' and C_1' for capacitor C_1, and C_2' and C_2'' for capacitor C_2.

Therefore, the dielectric medium of one part will be air (ε_0) and the other part will be aluminum (ε_r). Thus, the capacitance of each capacitor, C_1 and C_2, can be expressed as:

$$C_1 = C_1' + C_1'' \tag{1}$$

$$= \frac{\varepsilon_0 \varepsilon_r (x \times w)}{d} + \frac{\varepsilon_0 (L - x) \times w}{d} \tag{2}$$

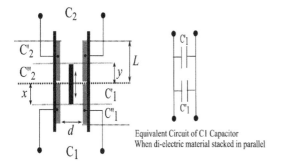

Figure 2. Graphical model of the proposed push–pull parallel plate capacitor.

$$= \frac{\varepsilon_0 (x \times w)}{d}[\varepsilon_r - 1] + \frac{\varepsilon_0 (L \times w)}{d} \qquad (3)$$

Likewise, $C_2 = C_2' + C_2''$ $\qquad (4)$

$$C_2 = \frac{\varepsilon_0 (y \times w)}{d}[\varepsilon_r - 1] + \frac{\varepsilon_0 (L \times w)}{d} \qquad (5)$$

where ε_0 is the permittivity of air and ε_r is the relative permittivity of the dielectric plate. The distance between the two plates is 'd'. Moreover, the available length of the dielectric medium in each capacitor, C_1 and C_2, is described as x and y, respectively. Thus, the combination of x and y will describe the actual length (L) of the dielectric plate.

Now, it may be concluded from (3) and (5) that, if the length of the available dielectric medium between C_1 and C_2 is equal, then the value of the capacitance of these two capacitors will also be equal accordingly. Furthermore, assume that the dielectric plate moves forward from C_1 to C_2, then during this transition, the relation between the available lengths of the dielectric medium between these capacitor plates would be $y > x$. Thus, the capacitance of the two capacitors will be changed by the relation $C_2 > C_1$. Therefore, the variation of the dielectric plate as well as the object may be determined by the difference between the two capacitance values. Therefore, it is confirmed that the capacitance of any capacitor will be maximum when the dielectric plates fully cover the entire area of the corresponding capacitor. Thus, it may be stated that the proposed sensor works according to the principle of the push–pull phenomenon between two capacitors.

4 EXPERIMENTAL SETUP AND RESULT ANALYSIS

In this preliminary study, a prototype is developed and tested in laboratory condition to observe its static characteristic. During the experiment, a stepper motor-driven mechanical slider is used for moving the dielectric plate between the parallel plates. In order to measure the precise displacement, a 1.5⁰ stepping angle-based motor is used so that the test rig is able to measure the smallest displacement of 2 mm against the measured range of ±30 mm. The entire experimental observations are carried out by the gradual displacement of dielectric plates in either a forward or backward direction between the capacitive plates with a space of 2 mm.

In the first step, the static characteristic of the capacitive sensing unit is found by following the linear displacement of the dielectric plate with the space of 2 mm. During this process, two digital LCR meters are used for measuring the capacitance value, and these are graphically plotted against the linear displacement, as shown in Figure 3(a). Here, the capacitance value of C_2 increases and that of C_1 decreases gradually due to the displacement of the dielectric plate towards C_2. This phenomenon occurs in reverse order while this plate moves towards C_1. Since the two capacitor values are changed like a push–pull manner, the variation of displacement of the dielectric plate can be determined by the difference between the two capacitances. The plot of the variation of the resultant capacitance against the linear displacement of the dielectric plate, illustrated in Figure 3(b), shows a linear relation between these two parameters. Therefore, it agrees with the mathematical approximation stated in (4) and (6), respectively. In addition, an error curve for the measured range is plotted in Figure 3(c) by comparing the experimental data with the theoretical one.

In the second step, the variation of C_1 and C_2 capacitances is converted into voltage by using the Signal Processing Circuit (SPC). In this circuit, an 8 kHZ triangular wave generator is used for stimulating one plate of this capacitor and the other plate receives this signal through the dielectric medium. Finally, it is converted into a pulse wave due to the placement of the capacitor into a differentiator circuit. Then, by using a precision rectifier, zero, and span adjustment circuit, a measureable voltage limit can be found. This output voltage is measured by a 4½ digit multimeter and the recorded data are plotted against the linear displacement of the dielectric plate; as shown in Figure 4(a).

Similar to a linear change in the capacitance of C_1 and C_2, the static characteristic of the

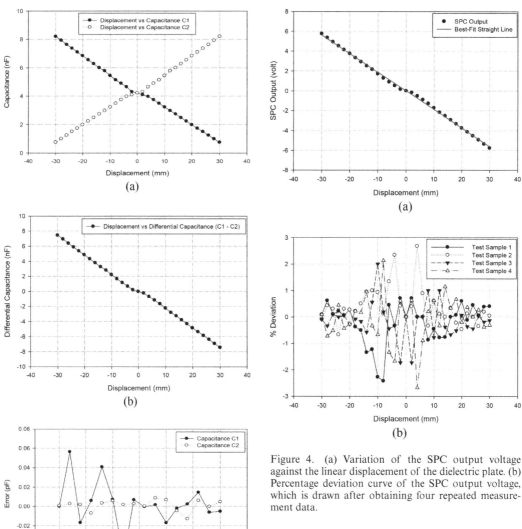

(a)

(b)

(c)

Figure 3. (a) Variation of capacitance against linear displacement. (b) Difference between C_1 and C_2 capacitances with respect to the displacement. (c) Measurement error in the capacitance value with respect to the theoretical value.

signal processing circuit is also found to be linear. Moreover, in order to verify its linearity, a percentage deviation curve from ideal linearity is plotted in Figure 4(b), where the ideal linearity data are obtained from the best fit regression model.

(a)

(b)

Figure 4. (a) Variation of the SPC output voltage against the linear displacement of the dielectric plate. (b) Percentage deviation curve of the SPC output voltage, which is drawn after obtaining four repeated measurement data.

In addition, the sensitivity of 18 mV/mm is achieved by this experiment.

5 CONCLUSION

In this paper, a preliminary study of a contactless capacitive sensor has been reported. Owing to the geometrical shape of this sensor, it is mostly suitable for the measurement of linear displacement of an object. Compared with the traditional capacitive-type linear displacement sensor, its sensing part is completely separated from the electrical interface. Thus, it will not be affected by any mechanical error. Moreover, its design is very simple and well protected against the fringe and stray capacitance effects. Unlike the recently developed push–pull mechanism-based capacitive sensor, its design is very reliable and requires a simple Signal Processing Circuit (SPC) to express its output in terms of analog voltage. A satisfactory performance of this proposed sensor has been achieved by using a laboratory prototype.

REFERENCES

Ahn, H.J., J.H. Park, C.Y. Um, & D.C. Han (2008), A disk-type capacitive sensor for five-dimensional motion measurements. *Meas. Sci. Technol.* 19, 334–340.

Bai. Y., Y. Lu, P. Hu, G. Wang, J. Xu, T. Zeng, Z. Li, Z. Zhang, & J. Tan (2016), Absolute position sensing based on a robust differential capacitive sensor with grounded shield window. *Sensors* 16, 680.

Bobroff, N (1993). Recent advances in displacement measuring interferometry. *Meas. Sci. Technol.* 4, 907–926.

Eom. T, & T. Choi (2001). Single Frequency Laser Interferometer with Subnanometer Accuracy. In *Proc. 2th Int. Conf. of European Society for Precision Engineering and Nanotechnology*, Turin, Italy, 1, 270–273.

Ferri, G., F. Parente, C. Rossi, V. Stornelli, G. Pennazza, M. Santonico, & A. D'Amico (2015), A Simplified Architecture for Differential Capacitance Sensors. In *Proc. XVIII AISEM Annual Conference*, Trento, Italy, 3–5.

Gasulla. M., X. Li, G.C.M. Meijer, L.V.D. Ham, & J.W. Spronck (2003), A contactless capacitive angular-position sensor. *IEEE sensor journal,* 3, 607–614.

George. B., M. Mohan, & V.J. Kumar (2008), A linear variable differential capacitive transducer for sensing planner angles. *IEEE Trans. Ins & Mes.* 57, 736–742.

Guo, J., P. Hu, & J. Tan (2016), Analysis of a segmented annular coplanar capacitive tilt sensor with increased sensitivity. *Sensors.* 16, 133.

Lányi, S (1998). Analysis of linearity errors of inverse capacitance position sensors. *Meas. Sci. Technol.* 9, 1757–1764.

Lee, J.-I., X. Huang, & P.B. Chu (2009), Nanoprecision MEMS capacitive sensor for linear and rotational positioning. *J. Microelectromech. Syst.* 18, 660–670.

Shieh, J., J. Huber, N. Fleck, & M. Ashby (2001). The selection of sensors. *Prog. Mater. Sci.* 46, 461–504.

Smith, P.T., R.R. Vallance, & E.R. Marsh (2005), Correcting capacitive displacement measurements in metrology applications with cylindrical artifacts. *Precis Eng.* 29, 324–335.

Yang, H., Y. Lu, P. Hu, Z. Li, T. Zeng, Q. He, Z. Zhang, & J. Tan (2014), Measurement and control of the movable coil position of a joule balance with a system based on a laser heterodyne interferometer. *Meas. Sci. Technol.* 25, 233–243.

Yu, H., L. Zhang, & M. Shen (2015), Novel capacitive displacement sensor based on interlocking stator electrodes with sequential commutating excitation. *Sens. Actuators A Phys.* 230, 94–101.

Zeng, T., Y. Lu, Y. Liu, & H. Yang (2015), A capacitive sensor for the measurement of departure from the vertical movement. *IEEE Trans. Instrum. Meas.* 65, 1–9.

Computer, Communication and Electrical Technology – Guha, Chakraborty & Dutta (Eds)
© *2017 Taylor & Francis Group, ISBN 978-1-138-03157-9*

Study of the liquid-level-sensing method based on a capacitive rotary sensor

Tuhin Subhra Sarkar, Arobindo Hore, Subir Das & Tamal Chowdhury
Murshidabad College of Engineering and Technology, Murshidabad, West Bengal, India

Badal Chakraborty
Bidhan Chandra Krishi Viswavidyalaya, Nadia, West Bengal, India

Himadri Sekhar Dutta
Kalyani Government Engineering College, Nadia, West Bengal, India

ABSTRACT: In this paper, a liquid-level measurement method is discussed by using a capacitive rotary sensor. The proposed arrangement consists of two parts: sensing module and signal processing circuit module. The sensing module composed of three segments: float arm, dielectric plate, and two copper plates. In this module, the float arm is mechanically connected to a dielectric plate, by which it can rotate in any direction due to variation in the liquid level in a tank. Then, two copper plates are arranged in parallel at either side of this plate, and due to the movement of the float arm, variation of capacitances was achieved by the tapping of the two plates. These capacitances are converted into voltage variation by using a signal processing circuit module, and through a universal serial bus, it is transmitted to a computer for analysis using the LabVIEW software. In this paper, a prototype of this method is designed to measure the liquid level in the range of 0 cm to 30 cm.

1 INTRODUCTION

Liquid-level measurement is one of the oldest process measurements techniques in any industry. There are various well-recognized techniques available for liquid-level measurement, which are broadly classified into two groups: direct method and indirect method. In the direct method, techniques such as gauge glass, float, and displacer are used. In the indirect method, the main techniques include capacitance, pressure sensing, ultrasonic, radiation absorption, and electrical conductivity. Indirect methods have some significant advantages in terms of characteristic properties over direct methods. In the direct method, the characteristic properties of sensing probes, such as float and displacer, have changed with the change in the liquid level. Here, this situation arises due to the physical or chemical reaction between the liquid level and the sensing probe, and thus their life cycle is limited. For this reason, here we focus on one of the popular indirect methods.

Capacitance-type liquid-level measurement techniques have been proposed by different researchers. A non-contact capacitance-type level transducer for a conductive liquid was developed by Bera et al. (2006). They have used the conducting liquid column as one electrode and a non-inductively wound short-circuited coil around a level-sensing cylinder made of an insulating material as the other electrode. By using this technique, they found good linearity, resolution, and accuracy. However, owing to the use of conducting liquid, a short circuit can occur between the circuit ground and the tank ground. To avoid this type of incident, a bridge circuit is introduced. This may increase the design complexity. In this context, Terzic, et al. (2010) introduced an artificial neural network approach to eliminate the effects of liquid slosh on the fluid-level measurement in an automotive fuel tank under a dynamic condition using a single tube capacitive sensor. Another research work (Canbolat 2009) introduced a sensor module, which consists of three parallel plate capacitive sensors. Of these sensors, two are used to develop a novel liquid-level measurement technique that is independent of the liquid type, air, and fluid dielectric constants in the tank. A microcontroller-based self-calibrating water-level measurement system based on an inter-digital capacitive sensor was proposed by Chetpattananondh, et al. (2014). It consists of a printed circuit board with configuration of two interpenetrating finger electrodes. Another research work conducted by Reverter, et al. (2007) was based on a remote grounded capacitive sensor. Here, a conductive liquid (tap water) level measurement system has been discussed. In addition, Bera, et al. (2014) designed a modified capacitance-type liquid-level transducer that eliminates the effect of self-inductance of the metallic sensing probe. They also found that this technique showed a good linearity. Within this field, another research work carried out by Biswas, et al. (2005) developed a finite element simulation program to predict the impedance behavior of a capacitive-type level sensor. This prediction analysis is required due to the double-layered polarizable liquid medium. The above-discussed level-sensing methods have shown good output characteristics such as linearity, resolution, and accuracy, but most of them have a very complex design and thus are cost-effective.

In this paper, we discuss a liquid-level measurement technique that is non-contact type, simple in design, of low cost, and of low power consumption by using a capacitive rotary sensor. Here, sensing embodiment consists of two parts: sensing unit and signal processing unit. The sensing unit is composed of three segments: 1) float arm, 2) dielectric material plate, and 3) two copper plates. In this sensing unit, the float arm is connected to a dielectric material plate through a shaft, which is moved due to the variation in the liquid level. Two copper plates are fabricated by the PCB technology for using fixed capacitive plates. Then, the dielectric material plate rotates between two capacitive plates due to the movement of the float arm that causes variations in the capacitance at the output. These capacitances are converted into voltage variation by using a simple signal conditioning circuit module, and through a universal serial bus, it is transmitted to a computer for analysis using the LabVIEW software. The proposed preliminary study demonstrates a proof of concept of this sensor by using a prototype. This is followed by a representation of the rotary sensor design, constructional details, the experimental investigation, and a series of discussion topics based on the experimental results.

2 ENCODER DESIGN AND METHOD DESCRIPTION

The main part of the proposed sensing unit is a capacitive rotary sensor, which consists of two half-circular copper plates (radius of each plate 1.75 cm) called the stator and a half-circular dielectric plate (material: aluminum oxide plate, radius 1.75 cm) called the rotor, as shown in Figure 1.

The rotor plate is placed between the two stator plates, so that it can move freely by using a shaft through a concentric hole of the two stator plates. Each annular shape stator plate is fabricated by using the standard PCB technology. To eliminate

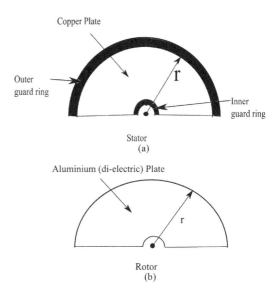

Figure 1. (a) Half-circular stator plate with inner and outer guard rings. (b) Half-circular rotor plate. Both plates have the same diameter.

the effects of stray capacitances, guard rings are also fabricated at the inner and outer surfaces of the stator plates by using the same technology, and the two stator plates are connected to the ground to keep them at the same potentials. Here, a float is mounted onto the wall of the tank, which is mechanically connected to a rotor plate. Then, the liquid level in a tank can be determined by measuring the movement of the rotor plate and calculating some mathematical expressions.

Here, the movement of the rotor plate causes the variation in capacitance at the sensor output. Therefore, the variation in the liquid level is directly proportional to the measured capacitance. To demonstrate the proof of this concept, a theoretical model is analyzed graphically, as shown in Figure 2.

At the initial condition, when the tank is empty, the half-circular rotor plate overlaps the half segment of two stator plates and makes a threshold capacitance at the output. Then, we assume that the liquid enters into the tank through the inlet port and the level of the liquid is increased in the tank by an H distance. Consequently, we assume that the float arm rotates the rotor (dielectric plates) in the anti-clockwise direction due to the pivot-like mechanism between the shaft and the float arm. Therefore, this rotational movement increases the overlapped segmented area of the two stator plates by the dielectric medium. This leads to an increase in the capacitance rather than the prior output.

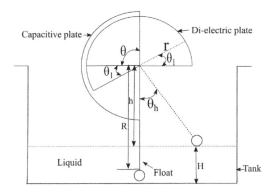

Figure 2. Cross-sectional view of the proposed method. Here, the dotted lines represent the changes in the fuel level, float movement, and shifting of the dielectric plate, respectively.

From Figure 2, it can be seen that the overlapped segment area varies with the angular displacement of the rotor plate. Thus, we assume that, at the initial position, the rotor plate covers the two stator plates by an angle of θ and then due to the increase in the liquid level by H, it is displaced by an angle θ_l from the previous position. Consequently, the value of capacitance may be expressed as:

$$C_{Total} = C_1 + C_{Air} \quad (1)$$

Here, it is apparent that the present position of the rotor plate inside the two stator plates makes two separate capacitors: one capacitor (C_l) having the dielectric medium of aluminum oxide (relative permittivity 9) and another one denoted as C_{air} having the dielectric medium of air. Therefore, it is obvious that the non-overlapping portion of the stator plate in which the air medium is present will be covered by the rotor plate as the liquid level in a tank increases accordingly. Thus, the total capacitance at the output may be defined as:

$$C_{Total} = C_1 + C_{Air} = \frac{1}{2}r^2(\theta + \theta_1) + \frac{1}{2}r^2(\pi - \theta - \theta_1) \quad (2)$$

Now, it is apparent that the liquid level (H) in a tank rotates the float arm in the anti-clockwise direction under the stator by an angle θ_h. This rotation of the float arm displaces the rotor plate in the same direction by an angle θ_l.

Thus, it can be written as:

$$\theta_1 = \theta_h \quad (3)$$

The liquid level can also be expressed as $H = R - h$, without considering the diameter of the float. The parameter R defines the length of

the float arm and h the length between the liquid surface and the center of the disc; thus, it can be rewritten as:

$$h = R\cos\theta_h$$

Therefore, $\cos\theta_h = \left(1 - \dfrac{H}{R}\right)$

Or, $\theta_h = \cos^{-1}\left(1 - \dfrac{H}{R}\right)$ \quad (4)

Finally, rearranging (3), (4), and (2), we obtain:

$$C_{Total} = \frac{1}{2}r^2\left[\theta + \cos^{-1}\left(1 - \frac{H}{R}\right)\right]$$
$$+ \frac{1}{2}r^2\left[\pi - \theta - \cos^{-1}\left(1 - \frac{H}{R}\right)\right] \quad (5)$$

Now, all the parameters on the right-hand side of (5) are constant except the value of H. Thus, it can be concluded that when the tank becomes empty (i.e., $H = 0$), the value of θ_h as well as θ_1 will be equal to zero. At this stage, we will obtain an initial (threshold) capacitance value at the output because of the presence of the rotor plate in the one half segment of the stator plates and the remaining area is covered by air. Then, while the tank is filled with liquid (i.e., $H = R$), θ_1 will have the maximum value, i.e., $\theta_1 = \theta$. This means that a maximum capacitance value can be obtained at the output because of the presence of the rotor plate in the entire area of the stator plates. Similarly, θ_h will reach its maximum value (since $\theta_1 = \theta$) when the tank is filled with liquid (i.e., $H = R$). Thus, it can be concluded that due to the variation in the liquid level, the rotor plate will cover the entire area of the intermediate stator plates gradually and, consequently, the capacitance of the rotary sensor will show a linear characteristic. Furthermore, it is obvious that, in this model, the maximum measureable height of the liquid level depends only on the length of the float arm (R). Therefore, a proper selection of the length of the float arm is very important for any desired range of liquid-level measurement at various heights of the tank.

3 EXPERIMENTAL SETUP AND RESULT ANALYSIS

An experimental prototype is used in the laboratory to evaluate the sensing performance of the proposed design. This prototype is composed of a PVC tank (52 cm × 54 cm) having an inlet and outlet ports through which the liquid can enter inside and be sucked out easily as required from time to time, a float (diameter 10 cm; material: PVC), a float arm (length 30 cm; material: stainless steel), capacitive rotary sensor, a Signal Processing Circuit (SPC), and a PC-based 10-bit Data Logger (DAL) unit. A graphical representation of the experimental setup is shown in Figure 3.

In this setup, the rotary sensor is mounted onto a tank wall and the dielectric plate is connected to a shaft through which it can rotate in any direction during the variation in the liquid level in a tank because of the buoyancy force exerted on the float. Here, a spherical-shaped float is used to ensure the effective contact between the float and the liquid surface. In addition, its diameter is set larger than the diameter of the capacitive disc. Thus, the liquid level cannot touch the capacitive rotary sensor at the maximum range.

Experimental observations are made only with water which was allowed to increase gradually in a tank until it reached the maximum limit (30 cm). Then, there is a vertically upward thrust on the float with the gradual increase in the fuel level and thus the rotor plate will rotate in the anti-clockwise direction because of the pivot-like mechanical arrangement between them. Consequently, there will be variations in capacitance at the output of the capacitive rotary sensor, and an equation can be used to calibrate the output of the capacitive rotary sensor due to an increase or decrease in the liquid level in the tank.

In the first step, static characteristics of the proposed rotary sensor are evaluated from three increasing and three decreasing modes of liquid-level variation with a space of 2 cm. In each mode, the output capacitance of the rotary sensor is measured using a digital LCR meter at every step of the liquid level. These values are graphically plotted against the liquid level, as shown in Figure 4(a). Here, it can be seen that when the

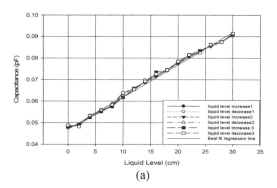

(a)

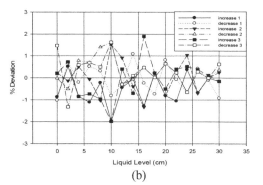

(b)

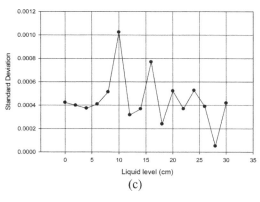

(c)

Figure 4. Static characteristic graph of the proposed rotary sensor: (a) rotary sensor response curve; (b) percentage deviation of rotary sensor linearity; and (c) standard deviation curve of the rotary sensor output for six repeated measurements.

liquid level is increased gradually in the tank, the output capacitance of the rotary sensor also increases or decreases accordingly while the liquid level decreases gradually. Therefore, it is consistent with the mathematical approximation as stated in (5). Although a linear change in capacitance is observed within the range of 0–30 cm, the sensitivity of the proposed sensor does not vary throughout the measuring range, which is determined as

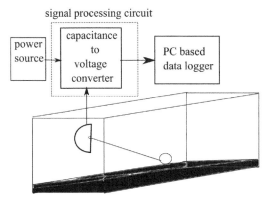

Figure 3. Block diagram representation of the proposed system.

0.01 pF/cm. In addition, the percentage deviation from ideal linearity is calculated for each data using (2) and the corresponding percentage deviation curve of the rotary sensor output is drawn against the liquid level, as shown in Figure 4(b). It can be seen from Figure 4(b) that the nonlinearity is in the range of ±2%. Similarly, a standard deviation curve is also drawn using the above data, as shown in Figure 4(c). It is plotted by the values of the standard deviation of each mode against the liquid level.

Next, the output of the rotary sensor, i.e., capacitance, is converted into voltage variation by a Signal Processing Circuit (SPC). This is followed by two steps of conversion: first, a capacitance to voltage converter circuit is used for obtaining a voltage variation in the range of 5.6 volt to 8.1 volt. Here, in order to measure small values of capacitance, the SPC is directly connected to the capacitance plate. Thus, the stray capacitance effect of the interface cable can be optimized. In addition, the SPC circuit having a high-frequency impedance amplifier is used for converting small variations in capacitance into voltage. The digitized data of these voltages are gradually stored in the hard drive of the PC using a 10-bit DAL, as shown in Figure 3. From this data, a real-time liquid-level monitoring software can be developed in the LabVIEW platform.

From Figure 5(a), it is obvious that the output voltage of the SPC unit is mostly linear. Therefore, by using this response curve, the sensitivity of the proposed system is determined, which is important for a transducer to indicate its maximum observable response per unit change in input stimulus. Since Figure 5(a) shows a straight line curve, it can be easily concluded that the sensitivity of that system does not vary in different zones of the entire measuring range. Here, we obtain a sensitivity of 120 mV/cm. To observe the linearity of the SPC unit, a percentage deviation curve from ideal linearity is drawn in Figure 5(b), which shows that the non-linearity is in the range of ±1%. A standard deviation curve is also drawn using this data, as shown in Figure 5(c). It is plotted by the values of the standard deviation of each mode against the liquid level. In the final step, the output of the SPC unit is connected to a 10-bit ADC to digitize this analog voltage, which is built on a DAL board. Thus, a maximum of 1024 numbers of input voltage levels can be sampled by the ADC. Therefore, by using this ADC, resolution of the proposed system is found as 0.04 cm/bit. From this, digitized data can be stored in the hard drive of the PC using an RS232 port; a LabVIEW software-based program is implemented to continuously monitor and record the fuel level in a tank. This software can display a minimum variation in the liquid level of 0.4 mm.

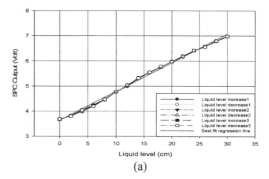

(a)

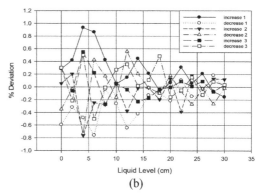

(b)

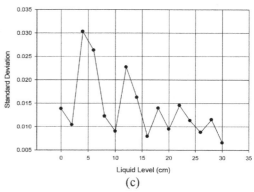

(c)

Figure 5. Static characteristic curve of a Signal Processing Circuit (SPC): (a) SPC output characteristic curve; (b) percentage deviation of SPC output linearity; and (c) standard deviation curve of the SPC output for six repeated measurements.

4 DISCUSSION

In this study, a capacitive rotary sensor has been proposed, designed, and realized for the measurement of the liquid level in the range of 0–30 cm. Here, the capacitive rotary sensor acts as a liquid level sensor, which is completely separated from the measuring liquid by the mechanical type of arrangement and fully enclosed by housing.

Hence, it will not be affected by liquid properties and the liquid itself. Moreover, sensitivity of the proposed sensor is independent of moisture and dust. Moreover, its static characteristic does not vary by increasing the measured level length, showing satisfactory performances in terms of linearity, sensitivity, and resolution. This measurement procedure is very simple due to its simple design, which is advantageous over the prior design (Bera 2006, Terzic 2010 & Canbolat 2009). In addition, the manufacturing cost of this embodiment (designed for the laboratory purpose only) is very low when compared with the prior design (Chetpattananondh 2014). The cost of the water level electrode with a signal conditioning circuit is $10/feet in the prior design. However, the proposed prototype manufacturing cost is just approximately $3. Both showed satisfactory performance of the output.

In this preliminary work, only one type of liquid has been considered during the experiments, but this methodology is applicable to any type of liquid, where the effect of the buoyancy force can be optimized by proper calibration of the SPC unit. Moreover, it will not be affected by small amplitudes of the liquid surface sloshing and tilt caused by vibration. During the experiments, sloshing occurred at the fuel surface, when the liquid entered into the tank through an inlet port. In order to optimize the error in the monitoring level due to sloshing, a software-based approach has been considered. Here, the uneven change in the voltage level within a least significant time instant is not considered in LabVIEW programming. However, if this scenario occurred for a long time, then it computed the moving average data of the level and displayed via an indicator showing the presence of very high sloshing at the liquid surface.

It is possible for fluid-level measurements because this process response is too slow. However, it is not suitable for a high-amplitude liquid sloshing environment like airborne application. The proposed method can measure the level in the range of a few millimeters to centimeters and the selected range is suitable for real-world applications such as a vehicle fuel storage tank, underground fuel tank, etc. However, the measured length can also be improved by modifying its float arm's length and encoded disc diameter.

REFERENCES

Bera S.C., J.K. Ray & S. Chattopadhyay (2006), A Low-Cost Noncontact Capacitance-type Level Transducer for a Conducting Liquid, *IEEE Transactions on Instrumentation and Measurement*, 55, 778–786.

Bera S.C., H. Mandal, S. Saha & A. Dutta (2014), Study of a Modified Capacitance-Type Level Transducer for Any Type of Liquid, *IEEE Transactions on Instrumentation and Measurement*, 63, 641–649.

Biswas K., S. Sen & P.K. Dutta (2005), Modeling of a capacitive probe in a polarizable medium, *Sensors and Actuators A*, 120, 115–122.

Canbolat H. (2009), A Novel Level Measurement Technique Using Three Capacitive Sensors for Liquids, *IEEE Transactions on Instrumentation and Measurement*, 58, 3762–3768.

Chetpattananondh K., T. Tapoanoi, P. Phukpattaranont & N. Jindapetch (2014), A self-calibration water level measurement using an interdigital capacitive sensor, *Sensors and Actuators A*, 209, 175–182.

Reverter F., X. Li & G.C.M. Meijer (2007), Liquid-level measurement system based on a remote grounded capacitive sensor, *Sensors and Actuators A*, 138, 1–8.

Terzic E., C.R. Nagarajah & M. Alamgir (2010), Capacitive sensor-based fluid level measurement in a dynamic environment using neural network, *Engineering Applications of Artificial Intelligence*, 23, 614–619.

Computer, Communication and Electrical Technology – Guha, Chakraborty & Dutta (Eds)
© 2017 Taylor & Francis Group, ISBN 978-1-138-03157-9

Comparison of two and four electrode methods for studying the impedance variation during cucumber storage using Electrical Impedance Spectroscopy (EIS)

Atanu Chowdhury & D. Ghoshal
Department of Electronics and Communication Engineering, National Institute of Technology, Agartala, Tripura, India

Tushar Kanti Bera
King Abdullah University of Science and Technology, Saudi Arabia

Badal Chakraborty
Faculty of Agricultural Engineering, Bidhan Chandra Krishi Viswavidyalaya, West Bengal, India

M.L. Naresh Kumar
Indian Institute of Technology, Powai, Mumbai, India

ABSTRACT: Electrical Impedance Spectroscopy (EIS) has been used to study the impedance variation in cucumber during its storage over a wide range of signal frequencies. The electrical impedance measurement of biological tissues is generally conducted either with a two electrode method or four electrode method. The objective of the research is to perform impedance measurement on cucumber during storage using 2-electrode and 4-electrode method and to compare variation in impedance between these two methods. The experimental results show that impedance measurement performed on cucumber by four electrode method gives better results in terms of modulus of impedance and impedance phase angle as compared to two electrode method. It is observed that the problem of electrode impedance associated with two electrode method is eliminated in four electrode method.

1 INTRODUCTION

Electrical Impedance Spectroscopy (EIS) (Sammer et al. 2014, Birgersson et al. 2012) as a non-invasive method has been widely used to study the electrical properties of biological materials especially fruits and vegetables (Keshtkar, 2007, Jose et al. 2014, Bera et al. 2014, Bera & Nagaraju, 2011, Bera, 2014, Bera et al. 2016, Rothlingshofer et al. 2011). Fruits and vegetables are made up of biological cells which has complex bioelectrical impedance depending on the stimulus used for impedance measurement. A low amplitude alternating current (Jesus et al. 2008) signal is applied to measure the potential difference developed between different parts of the biological tissue (Gomez et al. 2012) and it is then used for impedance calculation using Ohms law. The stimulus signal is applied through the surface electrodes attached to the biological material. Surface electrodes are necessary in EIS in order to make electrical contact with the biological material under study. Therefore, an appropriate electrode array is required to be chosen for

successful electrical impedance analysis (Li G & Peng M, 2013) Silver-silver chloride (Ag/AgCl) electrodes are widely used for non-invasive Electrical impedance measurements. These electrodes use a conductive and sticky gel which reduces electrode polarization impedance (Z_{ep}) (Ruiz, 2013) and helps to attach the electrode to the surface of biological tissue. The electrical impedance of biological material is generally measured either with a two-electrode method or four electrode method. The two-electrode method uses two electrodes which are used both for current injection and voltage measurement and hence, it is known as two-electrode technique (Netter FH, 1997). As the alternating current is passed through the electrodes, impedance generated due to electrode polarization gets added to the measurement of sample and electrode tissue interface. As a results, there is an over estimation of sample impedance. In four electrode method, one pair of electrode (outer electrodes) is used for current injection and another pair of electrode (inner electrodes) is used for voltage measurement. As a result, the electrode polarization impedance (Z_{ep}) is eliminated in four electrode technique. This is because voltage is measured with a very high input impedance so as to prevent the flow of current in sensing or voltage electrodes. Electrical impedance of vegetables depends on the tissue properties which again depend on tissue composition and tissue health. Measuring the bioimpedance of vegetables the freshness of the vegetables could be detected.

In this work, a performance comparison of two electrodes and four electrode method of Electrical Impedance Spectroscopy (EIS) has been made by measuring impedance magnitude (Z) and its phase angle (θ) of cucumber during its storage. Electrical impedance of cucumber (Z_{cucu}) and its phase angle (θ_{cucu}) are measured using small sinusoidal signal over a wide frequency range with two electrode and four electrode schemes. Ag/AgCl electrodes are used to make electrical contact with cucumber samples for both of these two kinds of electrode set up.

2 IMPEDANCE MEASUREMENT TECHNIQUE

Electrical Impedance Spectroscopy (EIS) is an electrical impedance measurement method which estimates the complex electrical impedance and its phase angle at different frequency points. For a biological material, the impedance is related to capacitive reactance and the resistive components of the biological tissues. The magnitude impedance and the phase angle of electrical impedance are dependent on the resistive effects and capacitive effects of the biological tissues. The impedance can be measured by applying either current or voltage and measuring the voltage or current respectively.

In EIS technique, a low amplitude ac current or voltage signal is applied to the object under test over a wide range of frequency. At every frequency point, impedance is measured as a ratio of voltage measured (or voltage applied) across the sample to the current applied (current measured).

$$Z = \frac{V}{I}$$

Magnitude and phase angle of impedance are expressed as

$$|Z| = \sqrt{(Z_{real})^2 + (Z_{img})^2}$$

where $Z_{real} = |Z| Cos\phi$ and $Z_{img} = |Z| Sin\phi$ and

$$\phi = \tan^{-1}\left(\frac{Z_{img}}{Z_{real}}\right)$$

The measurement of complex electrical impedance of biological material is conducted either with a two-electrode method (Fig. 1) or four electrode method (Fig. 2).

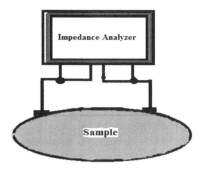

Figure 1. Two-electrode impedance measurement technique.

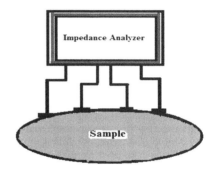

Figure 2. Four-electrode impedance measurement technique.

The two electrode method of electrical impedance measurement technique is highly dependent on the electrode contact impedance (Bragos et al. 2006) since same two electrodes are used both for current injection and voltage measurement. Therefore, electrode impedance is required to be considered in the two electrode method of EIS analysis. The measurement error produced of electrode contact impedance associated with two electrode method is eliminated in four electrode method of impedance measurement.

3 MATERIALS AND METHOD

3.1 Materials

The experiments were carried out on a matured fresh cucumber samples which were brought from local market. A normal size cucumber was chosen and stored at room temperature and the impedance of cucumber (Z_{cucu}) and phase angle (θ_{cucu}) are measured during its storage. All the experiments were carried out for 5 days. Electrical contact with the cucumber was made using Ag/AgCl electrodes for carrying out impedance measurement.

Figure 3. Two electrode setup with Ag/AgCl electrode.

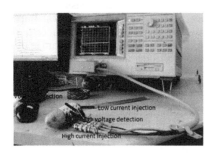

Figure 4. Four electrode setup with Ag/AgCl electrodes.

The position of electrode was arranged properly for accurate reading.

3.2 *Electrical impedance measurement*

For both two electrodes and four electrode methods, impedance measurement process was conducted using Agilent precision impedance analyzer (4294 A). The impedance and phase angles were measured at 100 frequency points between a frequency range of 50 Hz to 1 MHz by applying a 1 mA sinusoidal current signal injected through Ag/AgCl electrodes attached to cucumber samples.

4 RESULTS AND DISCUSSION

Fig. 5 and Fig. 6 show Impedance vs frequency curves for different days and phase vs frequency curves for different days obtained with two electrode method during cucumber storage.

It is seen from Fig. 5 that on 1st day, the impedance magnitude increases at lower frequencies and decreases with the increase in frequency. The similar variation in impedance magnitude with respect to frequency is also fund for remaining storage days. The variation of impedance magnitude from Day 1 to day 5 is larger at lower frequencies, but the variations are very small at higher frequencies.

It is observed from Fig. 6 that on 1st day, phase angle decreases with increase in frequency and

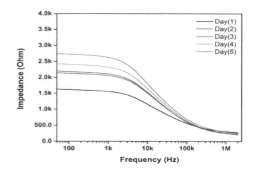

Figure 5. Impedance vs frequency curves for different days obtained with two electrode methods.

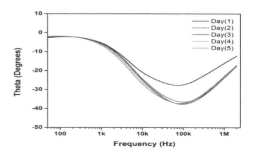

Figure 6. Phase angle vs frequency curves for different days obtained with two electrode method.

then reaches valley at the frequency of 100 KHz and there after increases at higher frequencies. The same variation in phase angle with respect to frequency are also observed for Day 2 to Day 5.

Fig. 7 and Fig. 8 show Impedance Vs frequency curves for different days and phase vs frequency curves for different days obtained with Four- electrode method during Cucumber storage.

It is seen from the Fig. 7 that impedance magnitude increases day by day at lower frequencies and decreases at higher frequencies, like two electrode methods. It is seen from Fig. 5 and Fig. 7 that in case of four electrode method, magnitude of impedance is less as compared to two electrode method. On the 1st day, impedance magnitude in four electrode methods as shown in Figure 7 at 50 Hz is 955 Ω but in two electrode method, it is 1638 Ω as shown in Fig. 5. Thus there is a difference of 685 Ω and this is due to electrode impedance. Similarly, there is a difference of 1158 Ω on 2nd day, 955 Ω on 3rd day, 1103 Ω on 4th day, 1273 Ω on 5th day. But these impedance difference decreases as frequency increases because electrode impedance is negligible at higher frequencies.

It is observed from Fig. 8 that the phase angle decreases with the increase in frequency. It is also

263

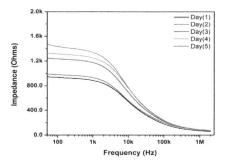

Figure 7.   Impedance vs frequency curves for different days obtained with four-electrode method.

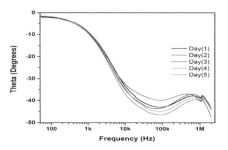

Figure 8.   Phase angle vs frequency curves for different days obtained with four-electrode method.

observed that phase angle attains a valley at the frequency of 100 KHz and there after increases and again decreases. The maximum phase angle obtained by four electrode method is higher than two electrode method. Thus, it is seen from Fig. 5 and Fig. 7 that the impedance magnitude in two electrode method is greater than the four electrode method as the two electrode method includes the electrode impedance. This is as because same two electrodes are used for current injection and voltage measurement.

It is obvious that, the measurement error occurred due to the contact impedance in two probe method is reduced in the four electrodes technique. As observed in the EIS studies of the cucumber samples, the patterns of the impedance variation with the signal frequency as well as with the number of days are found more or less same for both two-electrode and four-electrode configurations. Apparently it may indicate that the quality assessment of the cucumber could be successfully done with two electrode method as the relative degradation in the quality could be found similar in the impedance data obtained for two or four electrode method. But, it is important to note that through the four electrode method does not have contact impedance error but the two electrode method significantly suffer with the contact impedance of the electrodes.

The vegetable skin profile significantly changes with the storage of cucumber which also changes the contact impedance with time. As the contact impedance is also depend on frequency, the variation in contact impedance will change the impedance response significantly in two probe method. Therefore, four probe measurements are always preferred.

5 CONCLUSION

A performance comparison of two electrode and four electrode impedance measurement schemes has been performed during Electrical Impedance Spectroscopy (EIS) study of cucumber during its storage. The impedance and phase angle obtained with two-electrode and four-electrode measurement using Ag/AgCl electrodes show larger variations at lower frequency and mid frequency range, but slight variations at high frequencies. It is observed from the results that impedance decreases with increasing frequency but increases continuously with days. As the two electrode method includes the electrode impedance, the impedance magnitude in two electrode method is greater than the four electrode method.

REFERENCES

Bauchot, A.D., F.R Harker, W.M. Arnold (2000). The Use of Electrical Impedance Spectroscopy to assess the Physiological Condition of Kiwifruit. Postharvest Biology and Technology, 18, 9–18.

Bean R.C., J. P. Rasor, G. G. Porter (1960). "Changes in electrical Characteristics of avocados during storage," California Avocado Society, Yearbook, 44, 75–78.

Birgersson, U. H., E. Birgersson, and S. Ollmar (2012). Estimating electrical properties and the thickness of skin with electrical impedance spectroscopy: Mathematical analysis and measurements. J Electr Bioimp, 3(1), 51–60.

Bragós R, E. Sarró, A. Fontova, A. Soley, J. Cairó, A. Bayés-Genís, and J. Rosell (2006). Four Versus Two-Electrode Measurement Strategies for Cell Growing and Differentiation Monitoring Using Electrical Impedance Spectroscopy", Proceedings of the 28th IEEE EMBS Annual International Conference New York City, USA.

Chowdhury A., T.K. Bera, D. Ghoshal, B. Chakraborty (2015). Studying the Electrical Impedance Variations in Banana Storage Using Electrical Impedance Spectroscopy (EIS)" 3rd International Conference on Computer, Communication, Control and Information Technology (C3IT-2015), West Bengal, 1–4, February, 2015,

Chowdhury A., T.K. Bera, D. Ghoshal, B. Chakraborty (2016). Electrical Impedance Variations in Banana Storage: An Analytical study with Electrical Impedance Spectroscopy. Journal of Food Process Engineering. DOI:10.1111/jfpe.12387.

Gomez F, J. Bernal, J. Rosales, T. Cordova (2012), Modeling and simulation of equivalent circuits In description of biological systems—A fractional Calculus Approach, Journal of Electrical Bioimpedance, 3, 2–11.

Jackson, P.J., F.R. Harker (2000). Apple Bruise Detection by Electrical Impedance Measurement. Hort Science 35(1), 104–107.

Jesus I.S., J.A. Machado T and J.B. Cunha (2008). Fractional Electrical Impedances in Botanical Elements Journal of Vibration and Control, 14 (9–10), 1389–1402.

Li. G, Milao P. (2013). Electrochemical analysis of proteins and cells, 42–69.

Netter F.H., (1997). Atlas of Human Anatomy, Rittenhouse Book Distributors Inc.; 2nd edition.

Ruiz J.M. (2013). Sensor-Based Garments that Enable the Use of Bioimpedance Technology: Towards Personalized Healthcare Monitoring. Doctoral Thesis, Stockholm, Sweden.

Sammer, M., B. Laarhoven, E. Mejias, D. Yntema, E. C. Fuchs, G. Holler, G. Brassure and E. Lankmayr, (2014). Biomass measurement of living Lumbriculus variegatus with impedance spectroscopy. J Electr Bioimp, 5(1), 92–98.

Varlan, A.R. and W. Sansen, (1996). Nondestructive electrical Impedance analysis in fruit: normal storage and injuries characterization. Electro-Magneto biology 15, pp. 213–227.

Vozáry E, and P. Benkő (2010). Non-destructive determination of impedance spectrum of fruit flesh under the skin. Journal of Physics: Conference Series, 224,

Computer, Communication and Electrical Technology – Guha, Chakraborty & Dutta (Eds)
© 2017 Taylor & Francis Group, ISBN 978-1-138-03157-9

WHT-based tea quality prediction using electronic tongue signals

P. Saha & S. Ghorai
Department of AEIE, Heritage Institute of Technology, Kolkata, India

B. Tudu & R. Bandyopadhyay
Department of Instrumentation and Electronics Engineering, Jadavpur University, Kolkata, India

N. Bhattacharyya
Centre for Development of Advanced Computing (CDAC), Kolkata, India

ABSTRACT: Quality of finished Cut-Tear-Curl (CTC) tea mainly depends on biochemical components like Thearubigin (TR) and Theaflavin (TF). Traditional estimation of TF and TR requires analytical instruments. These are expensive, and require long time and laborious effort to determine TF and TR. This paper presents an effective method to estimate the content of TF and TR of tea samples using Electronic Tongue (ET) response. A regression model is developed using the features extracted from the ET signals to predict TF and TR content of tea. Energy values of the signals from different sensors of ET computed by the coefficients of Walsh Hadamard Transform (WHT) are used as feature to develop regression models. Two different models such as Support Vector Regression (SVR) and Vector-Valued Regularized Kernel Function Approximation (VVRKFA) are used to justify the performance of the proposed method. High prediction accuracy ensures the usefulness of the proposed method for the prediction of TF and TR using ET signals.

1 INTRODUCTION

Tea quality significantly depends on the tea processing techniques, as it maintains the desired liquor quality of finished tea produces from plucked green tea leaves. There are different stages for the processing of CTC tea. These are, according to the order, plucking, withering, CTC, fermentation, and drying. The taste and aroma of tea leaves are increased during the fermentation stage. In fact, in the fermentation stage, the biochemical components, like polyphenols, in the tea leaves are oxidized due to the presence of air and, as a result, oxidative enzymes are extracted from the tea leaves. This is responsible for the formation of two important biochemical compounds like Thearubigin (TR) and Theaflavin (TF) (Mahanta 1988, Robertson 1992), which are highly responsible for taste attributes suchas briskness and strength of the tea liquor. Table 1 represents the contribution of major biochemical constituents to the taste of tea liquor (Mahanta 1988). The amounts of TF and TR present in the tea leaves are 0.5–2.0% and 6–18% of dry weight, respectively, based on the tea processing parameters.

TF is responsible for brightness and briskness, whereas TR is responsible for color and mouth-feel satisfaction (Robertson 1983). Table 1 shows

Table 1. Biochemical constituents of tea.

Compounds	Taste
Theaflavin	Astringent
Thearubigin	Ashy and slight astringent
Amino acids	Brothy
Caffeine	Bitter
Polyphenol	Astringent

that TR has an ashy and slightly astringent taste, while TF has very high astringent. Thus, overall astringency of tea liquor is significantly influenced by TF. Obandaa et al. (2004) and Ngurea et al. (2009) established that the TF content correlates positively with the brightness of the tea liquor, whereas the TR content correlates negatively with both the tea liquor and taste, although it increases mouth-feel satisfaction of the tea liquor. Among the different biochemical components that contribute multi dimensionally to the taste of tea, the TR and TF contents have higher influence on the briskness, strength, color, mouth-feel satisfaction, brightness, and overall quality of finished CTC tea (Hazarika et al. 2002, Mo et al. 2008, Wright et al. 2002). In biochemical methods, tea quality can be estimated by evaluating TF and TR contents using instruments such as high-speed counter current chromatography (Degenhardt et al. 2000), high-performance liquid chromatography (Robertson & Bendall 1983), spectrophotometry (Kumar et al. 2011), and near-infrared reflectance spectroscopy (Hall, Robertso, & Scotter 1988). Limitations of these methods are time consuming, expensiveness, and requirement of trained staff. Tea quality can also be evaluated by some professionals, known as tea tasters. However, this method is subjective and thus the quality indicated by them is less accurate and non repeatable.

In order to overcome all these limitations, the researchers are motivated to explore some electronic instruments such as Electronic Nose (EN) (Bhattacharyya et al. 2008, Dutta et al. 2003) and Electronic Tongue (ET) (Palit et al. 2010, Saha et al. 2014) to evaluate tea quality. The output of these devices is calibrated with respect to the quality index of tea provided by the tea tasters that are purely biased in nature. In order to overcome this problem, it is required to establish a correlation model between the ET response and the concentration of biochemical components that are responsible for the taste of tea. In this paper, we proposed a method to develop such a correlation model. We have used a voltammetric electronic tongue with Large Amplitude Pulse Voltammetry (LAPV) (Ivarsson et al. 2001, Xiao and Wang 2009) for experimentation. A characteristic response of ET for LAPV method is shown in Fig. 1. Time series response from voltammetric measurements is represented by a large number of measured points, which are very difficult to characterize properly. It is very important to reduce the dimensionality of the data set by extracting relevant and informative features to increase the discrimination between different grades of tea.

Haddad et al. (2007) proposed an effective feature extraction algorithm from the transient response of electronic nose. Ding et al. (2005) reported a feature extraction method from the response of a semiconductor-type gas sensor using wavelet transform. A feature extraction method for qualitative analysis of water from voltammetry ET signal using Discrete Cosine Transform (DCT) was proposed by Scozzari et al. (2007). Valle et al. (2012) reported a feature extraction method based on Principal Component Analysis (PCA) for the qualitative analysis of water. Discrete Wavelet Transforms (DWT) with PCA techniques were proposed by Palit et al. (2010) and Ghosh et al. (2012), to extract relevant features from ET signal for the estimation of tea quality. TF and TR contents in black tea from ET response using wavelet features were estimated by Ghosh, Tamuly, Bhattacharyya, Tudu, Gogoi, & Bandyopadhyay (2012).

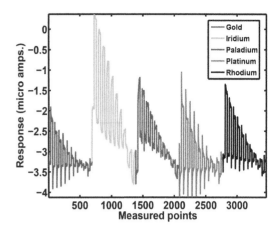

Figure 1. Characteristic response of voltammetric ET based on LAPV.

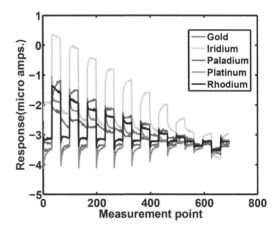

Figure 2. Multidimensional ET response.

In this paper, instead of one-dimensional ET signal, we considered a multidimensional ET response as shown in Fig. 2 to collect the instantaneous characteristic features from the response of individual sensors. The method proposed here is based on Walsh Hadamard Transform (WHT) (Fino and Algazi 1976, Pratt et al. 1969), which is used to represent the time series signal in frequency domain. A subset of WHT coefficients for the response of each individual sensoris considered to find the signal energy. These energy values are considered as the features of the ET response. Support Vector Regression (SVR) (Vapnik 1998, Smola and Schölkopf 1998) and Vector-Valued Regularized Kernel Function Approximation (VVRKFA) (Ghorai et al. 2010, Ghorai et al. 2012) algorithms are used to develop regression models to establish correlation between ET response and TF and TR contents.

2 EXPERIMENTAL SETUP

We considered 46 different types of tea samples for experimental verification. These samples are collected from Tea Research Association (TRA), Tocklai, India. TF and TR contents of each individual samples are determined by spectrophotometric analysis, and these values are used to calibrate the developed regression models.

2.1 Spectrophotometric analysis

A tea sample (6 g) is mixed with 250 ml of boiled water, and the mixture was kept for 10 min in a thermoflask. The mixture is filtered and cool down to room temperature, and 6 ml of this cooled tea liquor is then added to the mixture of 6 ml of 1% (w/v) di-sodium hydrogen phosphate and 10 ml of ethyl acetate. The upper layer of this mixture containing the TR and TF fractions is collected. This is added to 5 ml of ethyl acetate and this extract is used to prepare sample solution and reference solution. The spectrophotometric analysis results are described in our previous research work (Ghosh, Tudu, Tamuly, Bhattacharyya, & Bandyopadhyay 2012, Ghosh, Tamuly, Bhattacharyya, Tudu, Gogoi, & Bandyopadhyay 2012).

2.2 Electronic Tongue(ET) setup

The voltammetric ET, which is developed in our laboratory, consists of five working electrodes, namely gold, iridium, palladium, platinum, and rhodium; the reference electrode is made of Ag/AgCl (saturated KCl, Gamry Instruments Inc., USA) and the counter-electrode is made of platinum (PH Ionics, India). Interested readers are referred to the description of ET setup in Ghosh, Tudu, Tamuly, Bhattacharyya, & Bandyopadhyay (2012) and Ghosh, Tamuly, Bhattacharyya, Tudu, Gogoi, & Bandyopadhyay (2012).

2.3 Sample preparation for electronic tongue

Tea liquor samples are prepared by mixing 1 g of dry tea sample with 200 ml of deionized boiled water and keeping them for 5 min. Then, the tea leaves are separated from the tea liquor,which is then allowed to cool down to room temperature. Finally, this tea liquor is presented to the electronic tongue for the collection of responses.

3 PRELIMINARIES

3.1 Regression models

In this experiment, we considered two types of regression models to validate the performance of the proposed method. Brief description of the regression models is given below.

3.1.1 SVR

Support Vector Machine (SVM) (Vapnik 1998) is an extensively used binary supervised classifier that constructs a separating hyperplane by maximizing the separation distance between the two classes. Support Vector Regression (SVR) (Smola & Schölkopf 1998) works on SVM algorithm, but is applied for fitting a function or regression purpose. SVR is able to form a linear hyper surface in the high-dimensional feature space to fit the mapped training patterns using kernel trick. This linear hyper surface in the feature space corresponds to a nonlinear fitting surface in the input space. During the training process, it can tolerate a small error to fit the data set, and for this an ε-insensitive loss function is included to its constrained quadratic objective function. The trained model is used to predict the target values of the test samples.

3.1.2 VVRKFA

Vector Valued Regularized Kernel Function Approximation (VVRKFA) was introduced by Ghorai et al. (2010), where regression or function approximation method is used for multiclass data classification. In VVRKFA, the input training samples are mapped into a high-dimensional feature space by using kernel trick. These feature vectors are fitted to model the target values of the samples in low-dimensional label space. The same approach is employed in this work with a modification of its target values. For classification, the target values are of a vector with elements equal to the number of classes. Here, for general case, we considered it as a vector of continuous values consisting of multiple targets. If the vector consists of only one element, it reduces to a model of fitting a single target. In this paper, we considered the TF and TR contents as targets, separately. The details of the formulation and VVRKFA and its solution procedure can be obtained in Ghorai et al. (2010) and Ghorai et al. (2012).

3.2 Proposed feature extraction method

In this experiment, dimensionality reduction is done through a transformation of the time series ET signal into frequency domain using Walsh Hadamard Transform (WHT) (Fino and Algazi 1976, Pratt et al. 1969). WHT is used in many applications such as signal processing, filtering, and power spectrum analysis, as it has high energy compaction capability and is associated with very fast implementation. Thus, it can reduce computational complexity drastically. Walsh Hadamard transformation actually means either Walsh transformation or ordered Hadamard transformation, as their basis functions are identical. The WHT $W(k)$ of a time series signal $f(x)$ with length N (total measured points) can be expressed as:

$$W(k) = \frac{1}{N}\sum_{x=0}^{N-1} f(x)\prod_{i=0}^{n-1}(-1)^{b_i(x).b_{n-1-i}(k)},\qquad(1)$$

where k represents the frequency index, N represents the number of samples, n represents the number of bit, and $b_m(z)$ represents mth bit in binary representation of z. Fig. 3 shows the variation of energy with the number of coefficients. It is evident from Fig. 3 that the first 200 transformed coefficients contain above 95% of the total energy of the signal collected for each of the five electrodes. Therefore, we considered only the first 200 coefficients of each transformed signal to determine the energy of the signal. The energy value of each sensor response is considered as the characteristic feature. Thus, the final feature vector F can be represented by combining the energy values of the output of all the five sensors, that is, $F = [f_1\ f_2 ..... f_5]^T$, which is a five-dimensional feature vector. Here, f_i is the feature value of the i-th sensor output, which is calculated as:

$$f_i = \sum_{k=1}^{5} W(k)^2.\qquad(2)$$

Experimental setup is made such that it collects 694 samples for response of each electrode. The ET response of five electrodes for a single run can be represented by $1 \times (5 \times 694)$ samples, that is, a vector of 1×3470. In the feature domain, this response is represented by a 1×5 vector.

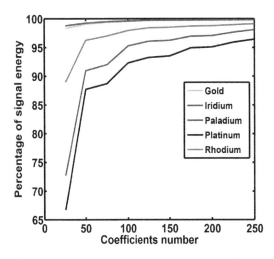

Figure 3. Variation of signal energy with transformed coefficients.

4 RESULTS AND ANALYSIS

Voltammetric Electronic Tongue (ET) based on LAPV method is used in this experiment, where 46 different types of tea samples are used. TF and TR contents of each samples are determined using spectrophotometry, which are used as target values. Table 2 presents the description of tea samples. Tea samples collected over the first season have higher TF and TR contents than that of the samples collected over the second season. This indicates that the quality of the first season tea samples with higher briskness and brightness are better than that of the second season tea samples. We recorded 38 responses for each of 46 tea samples, where each sample response contains $5 \times 694 = 3470$ sample values. This reduces to a training feature set of $(38 \times 46) \times 5 = 1748 \times 5$ with rows representing the number of observations.

4.1 Experimental procedure

To evaluate the performance of the proposed method, SVR and VVRKFA regression models are implemented in MATLAB with Gaussian kernel. A proportion of 70% of the total 1748 samples are used to train the models while the remaining 30% are used to test. The results presented here is the average of 50 such runs. The samples are randomly permuted before each run. The training feature set is normalized in the range of [01], while the testing set is normalized with respect to the minimum and maximum values of the training features in each run. The optimal values of the Gaussian kernel parameter γ and regularization parameter c for the regression models are selected from the set of values $\{2^i \,|\, i = -9,-8,...,8,9\}$ and $\{10^i \,|\, i = -8,-7,...,7,8\}$, respectively, by the performance of the respective regression model on a tuning set consisting 20% of training data, while the remaining 80% training data are used to train the model (Mitchell 1997). Performance of each model is evaluated on testing data set by using only selected optimal parameters.

4.2 Experimental results

Performance of the regression models to predict the value of TF and TR using the proposed method of voltammetric electronic tongue based on LAPV method is shown in Table 3. The table also presents the average Root Mean Square Error (RMSE) and standard deviation of 50 runs. Furthermore, the table shows the minimum RMSE obtained in a particular experimental run and p-values of the test results by performing t-test. It is evident from the table that the VVRKFA model performs better than the SVR model to predict TF values. VVRKFA provides an average RMSE of 4.29% with a standard deviation of ±0.39%,

Table 2. Details of tea samples used in the experiment.

| Number of collected tea samples | Duration of season | Range of | | |
		TR content (%)	TF content (%)	Quality index
19	April–May	13.11–19.67	0.97–2.57	6–9
27	September–October	10.1–15.6	0.95–1.39	5–8

Table 3. Performance of the proposed method.

Regression model type	Biochemical compounds	Average % RMSE (± std.dev.) on 50 runs	Number of run to minimum RMSE	Minimum % RMSE (p-value)
SVR	TF	6.81(0.683)	07	5.64(0.0159)
	TR	5.05(0.38)	19	4.44(0.0054)
VVRKFA	TF	4.29(0.39)	03	3.52(0.0988)
	TR	5.91(0.57)	45	4.69(0.0208)

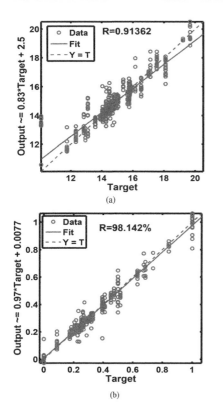

Figure 4. Fitted responses for (a) TR by SVR model and (b) TF by VVRKFA model.

whereas SVR provides an average RMSE of 6.81% with a standard deviation of ±0.683%. For prediction of TR values, SVR performs slightly better than VVRKFA with average RMSEs of 5.05% and 5.91%, respectively. The results of two-sided t-test show that some of the regression models are adequate at a significance level of 0.01 for the prediction of TF and TR values, respectively, by VVRKFA and SVR models.

Among the 50 runs of VVRKFA model to predict the TF value, the lowest RMSE of 3.52% was obtained in the third run, while SVR provided the lowest RMSE of 4.44% in the 19th run for TR value prediction. Thus, the proposed method can predict TF and TR values very efficiently. How well the predicted data are fitted to the actual data is described in Fig. 4, which shows the plot of the fitted response to predict TF and TR values on testing data set versus the observed response along with the R-squared values. For an ideal case, the model outputs should be equal to the target values, that is, the data should be placed along a line having slope of 1. R-squared value indicates the degree of linear dependence between the observed values and their predicted values. The R-squared values for the prediction of TF by VVRKFA and TR by SVR models are 0.98142 and 0.91362, respectively. Thus, high R-squared values ensure closeness of the predicted values to the target values.

5 CONCLUSION

In this paper, WHT-based feature extraction method from ET signal is described to evaluate the concentrations of TF and TR for the estimation of tea quality. The proposed method ensures that the regression models developed by VVRKFA and SVR methods are highly efficient to estimate the quality of tea through the prediction of TF and TR values. The higher accuracy for both the regression models justified the effectiveness of the proposed method for the evaluation of black CTC tea quality by determining the TF and TR values using ET signal. Thus, the proposed method gets rid of the disadvantages of biochemical instruments required to estimate the two most responsible biochemical compounds for the taste of tea.

REFERENCES

Bhattacharyya, N., R. Bandyopadhyay, M. Bhuyan, B. Tudu, D. Ghosh, & A. Jana (2008). Electronic nose for black tea classification and correlation of measurements with "tea taster" marks. *IEEE Transactions on Instrumentation and Measurement 57*, 1313–1321.

Degenhardt, A., U. H. Engelhardt, A. S. Wendt, & P. Winterhalter (2000). Isolation of black tea pigments using high-speed counter current chromatography and studies on properties of black tea polymers. *Journal of Agricultural and Food Chemistry 48*, 5200–5205.

Ding, H., H. Ge, & J. Liu (2005, June). High performance of gas identification by wavelet transform-based fast feature extraction from temperature modulated semiconductor gas sensors. *Sensors and Actuators B: Chemical 107*(2), 749–755.

Dutta, R., K. R. Kashwan, M. Bhuyan, E. L. Hines, & J. W. Gardner (2003). Electronic nose based tea quality standardization. *Neural Networks 16*, 847–853.

Fino, B. J. & V. R. Algazi (1976, November). Unified matrix treatment of the fast walsh-hadamard transform. *IEEE Transactionson Computers 25*(11), 1142–1146.

Ghorai, S., A. Mukherjee, & P. K. Dutta (2010, June). Discriminant analysis for fast multiclass data classification throughregularized kernel function approximation. *IEEE Trans. N. Nets. 21*(6), 1020–1029.

Ghorai, S., A. Mukherjee, & P. K. Dutta (2012). *Advancesin Proximal Kernel Classifiers*. Germany: LAP LAMBERT Academic Publishing.

Ghosh, A., P. Tamuly, N. Bhattacharyya, B. Tudu, N. Gogoi, & R. Bandyopadhyay (2012). Estimation of theaflavin content in black tea using electronic tongue. *Journal of Food Engineering 110*, 71–79.

Ghosh, A., B. Tudu, P. Tamuly, N. Bhattacharyya, & R. Bandyopadhyay (2012). Prediction of theaflavin and thearubigincontent in black tea using a voltammetric electronic tongue. *Chemometr. Intell. Lab. Sys. 116*, 57–66.

Haddad, R., L. Carmel, & D. Harel (2007, January). A feature extraction algorithm for multi-peak signals in electronic noses. *Sensors and Actuators B: Chemical 120*, 462–472.

Hall, M. N., A. Robertso, & C. N. G. Scotter (1988). Nearinfrared reflectance prediction of quality, theaflavin content and moisture content of black tea. *Food Chemistry 27*, 61–75.

Hazarika, M., M. R. Goswami, P. Tamuly, S. Sabhapondit, S. Baruah, & M. N. Gogoi (2002). Quality measurement in tea- biochemist's view. *Two and a bud 49*, 3–8.

Ivarsson, P., S. Holmin, N. Hojer, C. Krantz-Rilcker, & F. Winquist (2001). Discrimination of tea by means of a voltammetric electronic tongue and different applied waveforms. *Sensorsand Actuators B: Chemical 76*, 449–454.

Kumar, R. S. S., N. N. Muraleedharan, S. Murugesan, G. Kottur, M. P. Anand, & A. Nishadh (2011). Biochemical quality characteristics of ctc black teas of south India and their relation to organoleptic evaluation. *Food Chemistry 129*, 117–124.

Mahanta, P. K. (1988). *Modern method of plant analysis*, Chapter Colour and flavor characteristics of made tea, pp. 221–295. Berlin, Germany: Springer-Verlag.

Mitchell, T. M. (1997). *Machine Learning*. The McGRaw Hill Companies, Inc.

Mo, H., Y. Zhu, & Z. Chen (2008). Microbial fermented tea—a potential source of natural food preservatives. *Trends in Food Science and Technology 19*, 124–130.

Ngurea, F. M., J. K. Wanyokob, S. M. Mahungua, & A. A. Mahungua (2009). Catechins depletion patterns in relationto theaflavin and thearubigins formation. *Food chemistry 115*(1), 8–14.

Obandaa, M., P. O. Owuora, R. Mang'okab, & M. M. Kavoi (2004). Changes in thearubigin fractions and theaflavin levels due to variations in processing conditions and their influenceon black tea liquor brightness and total colour. *Food Chemistry 85*, 163–173.

Palit, M., B. Tudu, N. Bhattacharyya, A. Dutta, P. K. Dutta, A. Jana, R. Bandyopadhyay, & A. Chatterjee (2010). Comparison of multivariate preprocessing techniques as applied to electronic tongue based pattern classification for black tea. *Anal. Chim. Acta 675*(1), 8–15.

Palit, M., B. Tudu, P. K. Dutta, A. Dutta, A. Jana, J. K. Roy, R. Bhattacharyya, N. ande Bandyopadhyay, & A. Chatterjee (2010). Classification of black tea taste and correlation with tea taster's mark using voltammetric electronic tongue. *IEEE Transactions on Instrumentation and Measurement 59*(8), 2230–2239.

Pratt, W. K., J. Kane, & H. C. Andrews (1969). Hadamard transform image coding. In *Proc. IEEE*, Volume 57, pp. 58–68.

Robertson, A. (1983). Effects of physical and chemical conditionson the in vitro oxidation of tea leaf catechins. *Phytochemistry 22*, 896–903.

Robertson, A. (1992). *Tea: Cultivation to consumption*, Chapter The chemistry and biochemistry of black tea production, then on volatiles, pp. 555–601. London: UK: Chapman and Hall.

Robertson, A. & D. S. Bendall (1983). Production and hplc analysis of black tea theaflavins and thearubigins during in vitro oxidation. *Phytochemistry 22*, 883–887.

Saha, P., S. Ghorai, B. Tudu, R. Bandyopadhyay, & N. Bhattacharyya (2014, October). A novel method of black tea quality prediction using electronic tongue signals. *IEEE Transactionson Instrumentation and Measurement 63*(10), 2472–2479.

Scozzari, A., N. Acito, & G. Corsini (2007, December). A novel method based on voltammetry for the qualitative analysis of water. *IEEE Transactions on Instrumentation and Measurement 56*(6), 2688–2697.

Smola, A. J. & B. Schölkopf (1998). A tutorial on support vector regression. Technical Report 3, Royal Holloway College, Neuro COLT Tech. Rep. TR-1998–030, London, UK.

Valle, S., W. Li, & S. J. Qin (2012). Selection of the number of principal components: the variance of the reconstruction error criterion with a comparison to other methods. *Ind. Eng. Chem. Res. 38*(11), 4389–4401.

Vapnik, V. N. (1998). *The Nature of Statistical Learning Theory*. New York: John Willy & Sons, Inc.

Wright, L. P., N. K. Mphangwe, H. Nyirenda, & Z. Apostolides (2002). Analysis of the theaflavin composition in black tea (camellia sinesis) predicting the quality of black tea produced in central and southern Africa. *Journal of the Science of Food and Agriculture 82*, 517–525.

Xiao, H. & J. Wang (2009). Discrimination of xihulongjing teagrade using an electronic tongue. *African Journal of Biotechnology 8*, 6985–6992.

Multiple inhomogeneity phantom imaging with a LabVIEW-based Electrical Impedance Tomography (LV-EIT) System

Tushar Kanti Bera
King Abdullah University of Science and Technology (KAUST), Thuwal, Saudi Arabia

Sampa Bera & J. Nagaraju
Indian Institute of Science (IISc), Bangalore, India

Badal Chakraborty
Faculty of Agricultural Engineering, Bidhan Chandra Krishi Viswavidyalaya, Mohanpur, West Bengal, India

ABSTRACT: Practical phantom studies are essential to test, assess and calibrate an EIT system. Medical EIT systems are always prescribed to thoroughly study with practical phantoms to test and calibrate the systems for estimating the system accuracy and ensuring the medical safety. As the multiple inhomogeneity imaging is very important to characterize the multiple tumor in human subject or to image the multiple object geometries in several applications in non-medical fields. In this direction multiple inhomogeneity phantoms are developed and studied with an EIT system. The boundary potentials are collected with a LabVIEW based EIT (LV-EIT) system by injecting a 1 mA 50 kHz constant amplitude sinusoidal current to the multiple inhomogeneity phantoms and the resistivity images are reconstructed in EIDORS. Experimental results demonstrate that the LV-EIT system successfully reconstructs the double, triple and quadruple inhomogeneity configurations. The object position, object geometry and the background profile all are properly reconstructed for all the multiple inhomogeneity phantoms.

1 INTRODUCTION

Electrical Impedance Tomography (EIT) [Webster J.G., 1990, Bera *et al.* 2014, Bayford R.H., 2006] is an image reconstruction technique in which the electrical conductivity (σ) or resistivity (ρ) of a conducting domain (Ω) is reconstructed from the set of current-voltage data measured at the domain boundary ($\partial\Omega$) (Fig. 1). Being a non-invasive, non-radiating, non-ionizing and inexpensive methodology, electrical impedance tomography has been extensively researched in clinical diagnosis and medical fields (Holder D.S. 1993, Kwon H. *et al.* 2013, He W. *et al.* 2012) as well as in other areas like industrial process control (Dicfin F. *et al.* 1996), chemical engineering (Stephensona D.R. *et al.* 2009), geotechnical research (Kotre C.J. *et al.* 1996) and biotechnology (Linderholm P. *et al.* 2008) and other fields of applied science, engineering and technologies (Yao A. *et al.* 2013). EIT technology has been found with a number of advantages over other computed tomographic techniques (Bushberg J.T. *et al.* 2001). But still the poor Signal to Noise Ratio (SNR) (Farrell J.A. *et al.* 2007, Beckmann N. *et al.* 1989) of the boundary potential data and poor spatial resolution (Yoshida E. *et al.* 2013,

Amin I.J. *et al.* 2008, Wang Yun-Heng *et al.* 2014) are the challenges to the researchers in EIT technology. In medical EIT systems, the SNR of the boundary potential data increases with the amplitude of the current signal but the maximum current limit is restricted below a certain level for patient safety (Webster J.G., 1990). A number of research groups are studying EIT to improve the spatial resolution and image quality by improving the electronic instrumentation and the image

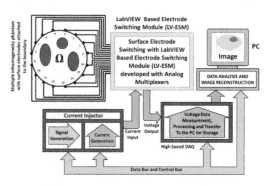

Figure 1. The schematic of the Electrical Impedance Tomography (EIT) system with a multiple inhomogeneity phantom.

reconstruction algorithms (Bera T.K. *et al.* 2014, Lionheart WRB 2004, Bera T.K. *et al.* 2011). A thorough image reconstruction study using practical phantoms (Bera T.K. *et al.* 2011, Griffiths H. 1995, Holder D.S. *et al.* 1994, Bera T.K. *et al.* 2011) is essential to assess and calibrate the EIT systems. It is, also, highly recommended to conduct a profound study on the impedance imaging of the newly developed medical EIT system using the practical phantoms to assess the system's efficiency, reliability and factor of safety prior to conduct the diagnostic imaging on the patients. For non-medical EIT systems are also suggested to be tested evaluated and calibrated by the laboratory studies conducted with practical phantoms before applying the systems in the practical fields of applications.

As the multiple inhomogeneities imaging is very important in EIT to characterize the multiple tumors in human subject or to image the multiple object geometries in several applications in non-medical fields, multiple inhomogeneity practical phantoms are essential to be developed and studied. To evaluate the multiple imaging efficiency of an EIT system and to characterize the multiple object geometry such as multiple tumors or other multiple objects, the multiple inhomogeneity imaging is very important. In this direction multiple inhomogeneity phantoms are developed and multiple inhomogeneity images studied with a LabVIEW [Travis J. *et al.* 2006, Wang Z. *et al.* 2013, Alegria F.C. 2009] base EIT (LV-EIT) system.

2 MATERIALS AND METHODS

2.1 *Multiple tumor*

Tumor or cancer cells may grow in a single region or multiple regions. Sometimes the tumor may originate in a particular region in the body and may spread over the other normal tissues. This process is called the metastasis (Poste *et al.* 1980, Steeg P.S. 2003) and the cancerous cells or the tumor is called metastatic tumor (Guba Markus *et al.* 2002, Gilbert *et al.* 1978) whereas the cancer is called metastatic cancer. Thus the metastatic cancer can be understood as a cancer which has started in a particular body part of the subject (the primary site) and then has spread from the primary site to other parts of the body. The metastatic cancer can be formed in breast, lung, kidney, skin, bladder, colon, or elsewhere in the body. In many types of cancers, the metastatic cancer is also called stage IV (four) cancer. A number of medical imaging methods have been applied to image the metastatic cancers (Kinkel *et al.* 2002, Ogunbiyi *et al.* 1997) such as ultrasound imaging

(US), X-Ray computed tomography (X-Ray CT), Magnetic Resonance Imaging (MRI), Positron Emission Tomography (PET) imaging. Being a radiation free, fast, portable technology EIT can also be applied for metastatic cancers imaging.

2.2 *EIT Instrumentation*

The LV-EIT system is developed with a LabVIEW based Constant Current Injector (LV-CCI), Programmable Electrode Switching Module (P-ESM) and a LabVIEW based Data Acquisition System (LV-DAS) (Bera T.K. *et al.* 2013). The LV-DAS is developed with a NIUSB-6251 DAQ card which is interfaced with PC and the EIT hardware through a NISCB-68 Connector Module (Bera T.K. *et al.* 2013). A LabVIEW based Graphical User Interface (LV-GUI) is develop to control the current injection by Constant Current Injector (LV-CCI) (Bera T.K. *et al.* 2010) and data acquisition by LV-DAS. Multiple inhomogeneity practical phantoms are developed and the boundary data are collected from the phantoms by injecting a constant amplitude sinusoidal current using LV-EIT system. The boundary data are collected from the multiple inhomogeneity phantoms with different inhomogeneity configurations and different current injection patterns (Bera T.K. *et al.* 2012, Bera T.K. *et al.* 2013) and the impedance images are reconstructed in EIDORS (Electrical Impedance and Diffuse Optical Reconstruction Software) (Polydorides N. *et al.* 2002, Bera T.K. *et al.* 2012). The LV-EIT system injects a 1 mA 50 kHz constant amplitude sinusoidal current with different multiple inhomogeneity configurations and with different current protocols and the boundary potentials are collected with a LabVIEW based EIT (LV-EIT). Boundary data collected by the DAS is processed through the signal processing blocks and then sent to the PC for image reconstruction in EIDORS.

2.3 *Multiple inhomogeneity phantoms*

Multiple inhomogeneity phantoms with different inhomogeneity configurations are developed (Fig. 2) with stainless steel rectangular electrodes (100 μm thick) fixed on the inner wall of a shallow plastic tank (inner diameter 228 mm) filled with 0.05% (w/v) NaCl solution (Fig. 2a). All the electrodes are cut form a 100 μm thick marine grade stainless steel sheet to avoid the localized pitting corrosion leading to the creation of small holes in the EIT electrodes. All the electrodes are electrically connected with the EIT instrumentation through the low resistive flexible multi-strand copper wires (Fig. 2a) of equal lengths for maintaining an identical impedance path for each electrode. The surface electrodes touching the KCl solution boundary act

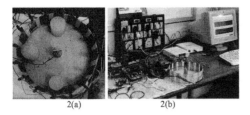

2(a)   2(b)

Figure 2. Multiple inhomogeneity phantom imaging with a LabVIEW-based electrical impedance tomography (LV-EIT) system (a) a multiple inhomogeneity phantom developed with a plastic tank (228 mm dia.) and three nylon cylinders near electrode 5, 9 and 13, (b) boundary data collection from the phantom with LV-EIT instrumentation.

as the EIT sensors allowing the LV-CCI instrumentation to send the current signal to the phantom and sending back the potential data to the LV-DAS. A cylindrical CME is placed at the phantom center (Fig. 2a) and connected to the ground point of the EIT hardwires to reduce the common mode error (Smith J.R. 1994) of the electronic circuits.

The phantom tank is filled with a 0.9% (w/v) NaCl solution and two or more nylon cylinders and other high resistive objects are put inside the NaCl solution at different electrode positions as the inhomogeneities and a number of multiple inhomogeneity phantoms configurations are obtained. Electrical impedance of NaCl solution is measured with a test signal of 1 mA, 50 kHz using QuadTech LCR meter to calculate the NaCl solution conductivity.

2.4 Current injection and boundary potential measurements

In the present study, 1 mA, 50 kHz constant sinusoidal current is injected through the current electrodes or driving electrodes and the boundary potentials are measured on the voltage electrodes or sensing electrodes (Fig. 2b) with opposite and neighbouring current injection pattern [Smith J.R. 1994]. In all the current projections in both the current patterns, the boundary potentials are measured on all the electrodes with respect to the analog ground point. Though the voltage measurement on current electrodes sometime are avoided [Webster J.G., 1990] for contact impedance problems, but to obtain the greatest sensitivity to the resistivity changes in the domain the voltages (RMS) on all the electrodes are measured. The current injection and boundary data collection strategies followed in the present study, have been explained in the Fig. 3. As shown in Fig. 3, for an EIT system with sixteen surface electrodes (E_1, E_2, E_3, E_4, E_5, E_6, E_7, E_8, E_9, E_{10}, E_{11}, E_{12}, E_{13}, E_{14}, E_{15} and E_{16}), both the opposite and neighbouring current patterns yields

sixteen current projections (P_1, P_2, P_3, P_4, P_5, P_6, P_7, P_8, P_9, P_{10}, P_{11}, P_{12}, P_{13}, P_{14}, P_{15} and P_{16}) and each of which produces sixteen voltage data (V_1, V_2, V_3, V_4, V_5, V_6, V_7, V_8, V_9, V_{10}, V_{11}, V_{12}, V_{13}, V_{14}, V_{15} and V_{16}) which are measured with respect to analog ground. For the present sixteen electrode EIT system, both the opposite and neighbouring method produces sixteen current projections yielding sixteen boundary potential data and hence the complete scan yields 256 boundary potential data (Fig. 3). Boundary potentials are collected for a number of multiple inhomogeneity phantoms with different object configuration and the resistivity imaging is studied in EIDORS.

2.5 Image reconstruction

Resistivity images are reconstructed from the boundary data using EIDORS using a symmetric

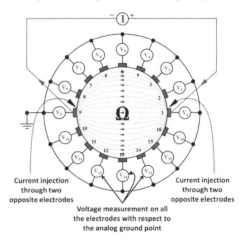

Figure 3. Surface electrode switching and boundary voltage data collection from the multiple inhomogeneity phantoms by injecting a constant amplitude sinusoidal current signal with opposite current pattern.

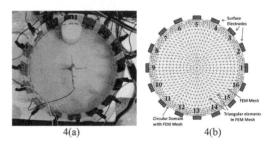

4(a)   4(b)

Figure 4. Domain discretization with FEM mesh (a) practical phantom with a circular domain with a diameter of 228 mm, (b) the imaging domain discretized with a FEM mesh containing 1968 triangular elements and 1049 nodes showing the electrode positions at the domain boundary.

275

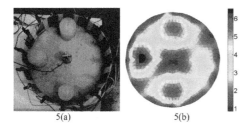

5(a)   5(b)

Figure 5. Resistivity imaging of a triple inhomogeneity phantom with LV-EIT system (a) triple inhomogeneity phantom containing three nylon cylinders (40 mm diameter each) placed near electrode number 5, 9 and 13, (b) resistivity image of the phantom shown in Fig. a reconstructed in EIDORS with a FEM mesh containing 1968 triangular elements and 1049 nodes.

FEM mesh discretizing the phantom domain (Fig. 4a). EIDORS, which is an open source software, aims to reconstruct the 2D-images from electrical or diffuse optical data. In the present study of the image reconstruction with EIDORS, the circular domains (diameter 228 mm) are discretized with a symmetric triangular element mesh containing 1968 elements and 1049 nodes (Fig. 4a–4b), is used in forward solution and inverse solution. The inverse problem is solved with applying the Levenberg-Marquardt Regularization (LMR) technique. The electrode positions are identified in FEM mesh in the algorithm and the boundary conditions are applied accordingly.

3 RESULTS

Resistivity imaging of multiple inhomogeneities RE-BDRP is studied with different current injection methods. Boundary data are collected from multiple inhomogeneity multiple inhomogeneity phantoms using LV-EIT instrumentation. Resistivity images are reconstructed in EIDORS. A triple inhomogeneity RE-BDRP phantom containing three nylon cylinders (40 mm diameter each) placed near electrode number 5, 9 and 13 is shows in Fig. 5a. Boundary data are collected by injecting a 1 mA, 50 kHz constant current signal using LV-EIT instrumentation. Fig. 5b shows the resistivity image of the phantom shown in Fig. 5a reconstructed in EIDORS with LMR regularization. Results show that the inhomogeneities are successfully reconstructed.

4 CONCLUSIONS

Practical phantom studies are essential to test, assess and calibrate a medical and non-medical EIT system. Multiple inhomogeneity imaging is very important to image and characterize the multiple tumor in human subject or to image the multiple object geometries in several applications in non-medical fields and hence multiple inhomogeneity practical phantoms are essential to be developed and studied. In this direction, a number of multiple inhomogeneity phantoms are developed and the resistivity imaging is studied using a using a LabVIEW based EIT (LV-EIT) system. Boundary potentials, developed for a 1 mA, 50 kHz constant current injected to the multiple object phantoms with different configurations, are collected for different inhomogeneity configurations and the resistivity imaging is studied in EIDORS with LMR techniques. Experimental results demonstrate that the LV-EIT system successfully reconstructs the double, triple and quadruple inhomogeneity configurations in saline tank phantoms. The object position, object geometry and the background profile of all the phantom configurations are properly reconstructed in EIDORS.

REFERENCES

Alegria F.C., E. Martinho, F. Almeidac. (2009). Measuring soil contamination with the time domain induced polarization method using LabVIEW, Measurement. 42, 1082–1091.

Amin I.J, A.J. Taylor, F. Junejo, A. Al-Habaibeh, R.M. Parkin. (2008). Automated people-counting by using low-resolution infrared and visual cameras, Measurement. 41, 589–599.

Bayford R.H. (2006). Bioimpedance Tomography (Electrical Impedance Tomography), Annu. Rev. Biomed. Eng. 8, 63–91.

Beckmann N, S. Müller, J. Seelig. (1989). Comparison of the signal-to-noise ratio at 1.5 and 2.0 T using a whole-body system. Magn Reson Med. 9, 391–394.

Bera T.K. and J. Nagaraju. (2010). A multifrequency constant current source suitable for Electrical Impedance Tomography (EIT), Proceedings of the IEEE International Conference on Systems in Medicine and Biology, India, 2010, 278–283.

Bera T.K. and J. Nagaraju. (2011). Resistivity Imaging of A Reconfigurable Phantom with Circular Inhomogeneities in 2D-Electrical Impedance Tomography, Measurement, 44, 518–526.

Bera T.K. and J. Nagaraju. (2011). A Chicken Tissue Phantom for Studying An Electrical Impedance Tomography (EIT) System Suitable for Clinical Imaging, Sensing and Imaging: An International Journal, Springer, 12, 95–116.

Bera T.K. and J. Nagaraju. (2012). Studies and Evaluation of EIT Image Reconstruction in EIDORS with Simulated Boundary Data, International conference on soft computing for problem solving (SocProS 2012), Jaipur, India.

Bera T.K. and J. Nagaraju. (2012). Studying the Resistivity Imaging of Chicken Tissue Phantoms with Different Current Patterns in Electrical Impedance Tomography (EIT), Measurement, 45, 663–682.

Bera T.K. and J. Nagaraju. (2013). A LabVIEW Based Data Acquisition System for Electrical Impedance Tomography (EIT). International conference on soft computing for problem solving, IIT Roorkee, India, 26–28.

Bera T.K. and J. Nagaraju. (2013). A MATLAB-Based Boundary Data Simulator for Studying the Resistivity Reconstruction Using Neighbouring Current Pattern, Journal of Medical Engineering, 2013, Article ID 193578, 1–15.

Bera T.K. and J. Nagaraju. (2014). Electrical Impedance Tomography (EIT): A Harmless Medical Imaging Modality, Research Developments in Computer Vision and Image Processing: Methodologies and Applications, Chapter 13, pp. 224–262, IGI Global USA.

Bera T.K., S.K. Biswas, K. Rajan and J. Nagaraju. (2014). Projection Error Propagation-based Regularization (PEPR) method for resistivity reconstruction in Electrical Impedance Tomography (EIT), Measurement, 49, 329–350.

Bushberg J.T., J.A. Seibert, Jr. E.M. Leidholdt, J.M. Boone. (2001). The Essential Physics of Medical Imaging, Lippincott Williams & Wilkins; 2nd edition.

Dickin F. and M. Wang (1996). Electrical resistance tomography for process applications, Measurement Sc. and Technology, 7, 247.

Fabrizi L, A. McEwan, T. Oh, E.J. Woo and D.S. Holder. (2009). An electrode addressing protocol for imaging brain function with electrical impedance tomography using a 16-channel semi-parallel system, Physiol. Meas. 30, 85–101.

Farrell J.A., B.A Landman, C.K. Jones, S.A Smith, J.L. Prince, P.C. van Zijl, S. Mori. (2007). Effects of signal-to-noise ratio on the accuracy and reproducibility of diffusion tensor imaging-derived fractional anisotropy, mean diffusivity, and principal eigenvector measurements at 1.5 T., J Magn Reson Imaging. 26, 756–767.

Gilbert, W. Robert, Kim Jae-Ho, and Posner B Jerome. (1978). "Epidural spinal cord compression from metastatic tumor: diagnosis and treatment." Annals of neurology, 3.1, 40–51.

Griffiths H. (1995). A Cole phantom for EIT, Physiol. Meas. 16, 29–38.

Guba, Markus, P. Von Breitenbuch, M. Steinbauer, G. Koehl, S. Flegel, M. Hornung, C.J. Bruns, C. Zuelke, S. Farkas, M. Anthuber M, K.W. E.K. Jauch, Geissler. (2002). Rapamycin inhibits primary and metastatic tumor growth by antiangiogenesis: involvement of vascular endothelial growth factor. Nature medicine, 8, 128–135.

He W., P. Ran, Z. Xu, Li Bing, and Li Song-nong. (2012). A 3D Visualization Method for Bladder Filling Examination Based on EIT, Computational and Mathematical Methods in Medicine, Article ID 528096, 1–9.

Holder D.S. (1993). Clinical and Physiological Applications of Electrical Impedance Tomography, Taylor & Francis; 1 edition.

Holder D.S. and A. Khan. (1994). Use of polyacrylamide gels in a saline-filled tank to determine the linearity of the Sheffield Mark 1 Electrical Impedance Tomography (EIT) system in measuring impedance disturbances, Physiol. Meas. 15, 45–50.

Kinkel, Karen, Y. Lu, M. Both, R.S. Warren, R.F. Thoeni. (2002). Detection of hepatic metastases from cancers of the gastrointestinal tract by using non-invasive imaging methods (US, CT, MR imaging, PET): a meta-analysis. Radiology 224(3), 748–756.

Kotre C.J. (1996). Subsurface electrical impedance imaging: measurement strategy, image reconstruction and in vivo results, Physiol. Meas. 17, 197–204.

Kwon H., A.L. McEwan, T. Oh, A. Farooq, E.J. Woo, and J.K. Seo. (2013). A Local Region of Interest Imaging Method for Electrical Impedance Tomography with Internal Electrodes, Computational and Mathematical Methods in Medicine, Article ID 964918, 1–9.

Linderholm P., L. Marescot, M.H. Loke, and P. Renaud. (2008). Cell Culture Imaging Using Microimpedance Tomography, IEEE Transactions on Biomedical Engineering, 55, 138–146.

Lionheart W.R.B. (2004). EIT reconstruction algorithms: pitfalls, Challenges and recent developments, Review Article, Physiol. Meas., 25: 125–142.

Ogunbiyi O.A., F.L Flanagan, F. Dehdashti, B.A. Siegel, D.D. Trask, E.H. Birnbaum, J.W. Fleshman, T.E. Read, G.W. Philpott, I.J. Kodner (1997). "Detection of recurrent and metastatic colorectal cancer: comparison of positron emission tomography and computed tomography." Annals of surgical oncology 4(8), 613–620.

Polydorides N., and W.R.B. Lionheart. (2002). A Matlab toolkit for three-dimensional electrical impedance tomography: a contribution to the Electrical Impedance and Diffuse Optical Reconstruction Software project. Measurement Sc. and Technology, 13, 1871–1883.

Poste, George, and F.J. Isaiah. (1980). The pathogenesis of cancer metastasis. Nature, 283, 139–146.

Smith J.R. (1994). The reduction of common-mode errors in impedance spectroscopy, Meas. Sei. Technal. 5, 333–336.

Steeg, Patricia S. (2003). Metastasis suppressors alter the signal transduction of cancer cells. Nature Reviews Cancer 3(1), 55–63.

Stephensona D.R., T.L. Rodgersa, R. Manna and Yorkb T.A. (2009). Application of Three-Dimensional Electrical Impedance Tomography to Investigate Fluid Mixing in a Stirred Vessel, 13th European Conference on Mixing, London.

Travis J, J. Kring, LabVIEW for Everyone: Graphical Programming Made Easy and Fun (3rd Edition), Prentice Hall; 3 edition.

Wang Yun-Heng, J. Qiao, Jun-Bao Li, P. Fu, Chu Shu-Chuan, J.F. Roddick. (2014). Sparse Representation-Based MRI Super-resolution Reconstruction, Measurement, 47, 946–953.

Wang Z, Y. Shang, J. Liu, X. Wu. (2013). A LabVIEW based automatic test system for sieving chips. Measurement, 46(1), 402–410.

Webster J.G. (1990). Electrical impedance tomography. Adam Hilger Series of Biomedical Engineering, Adam Hilger, New York, USA.

Yao A, C.L. Yang, J.K. Seo, and M. Soleimani. (2013). EIT-Based Fabric Pressure Sensing, Computational and Mathematical Methods in Medicine, 2013, Article ID 405325, 1–9.

Yoshida E, Y. H, H. Tashima, S. Kinouchi, M. H, M. Suga, and T. Yamaya. (2013). De-sign Study of the DOI-PET Scanners with the X'tal Cubes Toward Sub-Millimeter Spatial Resolution, J. Med. Imaging Health Inf. 3, 131–134.

Computer, Communication and Electrical Technology – Guha, Chakraborty & Dutta (Eds)
© 2017 Taylor & Francis Group, ISBN 978-1-138-03157-9

Electrical Impedance Spectroscopy (EIS) based fruit characterization: A technical review

Tushar Kanti Bera
Department of Mechanical Engineering, KAUST, Thuwal, Saudi Arabia

Sampa Bera
Department of Instrumentation and Applied Physics, IISc, Bangalore, India

Atanu Chowdhury & Dibyendu Ghoshal
Department of ECE, National Institute of Technology, Agartala, Tripura, India

Badal Chakraborty
Faculty of Agricultural Engineering, Bidhan Chandra Krishi Viswavidyalaya, West Bengal, India

ABSTRACT: Fruit physiology and composition changes during their maturity and ripening process. The variation in physiological and physiochemical composition of fruits produce a change in their electrical response which can be probed by measuring the bioelectrical impedance of the fruits for their noninvasive characterization. In this direction Electrical Impedance Spectroscopy (EIS) has been applied on a number of fruits to study the physiology and ripening noninvasively. As the composition and physiological changes occurred inside the fruits varies from species to species the impedance response for different fruits will be different. A thorough study on EIS analysis on different fruits and their ripening is essential to understand the correlation between the physiological changes and the corresponding impedance response. In this paper, a technical review on the EIS techniques applied for the fruit physiology and their ripening analysis have been conducted and presented. The physiology and ripening process of a number of fruits and the relation between the fruit impedance and the ripening states are summarized. After introducing the bioimpedance and equivalent circuit modelling of fruit tissues a detail discussion has been made on the EIS technique. A detail review has been presented on the EIS based studies conducted on the fruit ripening and fruit characterization. The advantages and disadvantages of the EIS technology for fruit analysis along with the recent trends have been summarized at the end.

1 INTRODUCTION

Plant tissues (Hopkins, 1999; White, 1943; Curits and Clark, 1951; Meyer et al., 1960; Eames and MacDaniels, 1947) are developed with the three-dimensional arrangements of cells embedded in an extracellular matrix (Alberts *et al.*, 2002). The cells in plan tissues are composed of cell organelles suspended in an intra cellular matrix (Alberts *et al.*, 2002) surrounded by cell envelop like cell membrane (Alberts *et al.*, 2002) and cell wall (Alberts *et al.*, 2002). Intra cellular matrix called intracellular fluid (ICF) (Wu *et al.*, 2008, Bera, 2014, and Bera, *et al.* 2016) and extracellular matrix called extracellular fluid (ECF) (Wu *et al.*, 2008, Bera, 2014, and Bera, *et al.* 2016) are developed with highly conductive and less conductive (resistive) materials respectively which produce electrical resistance to an alternating electrical signals. On the other hand, the cell membrane produces some capacitive reactance due to its protein lipid protein bi-layer structure (Wessels, 1996 and Mouritsen *et al.*, 1993). Thus under an alternating electrical excitation, plant tissue exhibits a complex electrical impedance which is called electrical bioimpedance (Bernstein, *et al.* 1986; Shoemaker, *et al.* 1984; Ohmine, *et al.* 1999) or bioelectrical impedance (Lukaski, *et al.* 1985; Kushner, *et al.* 1986) or simply bioimpedance (Z_b) (Bera, 2014, and Bera, *et al.* 2016; Martinsen and Grimnes, 2011; Lukaski, *et al.* 1985). The bioimpedance of the plant tissues (Liu, 2006; Zhang, *et al.* 2002; Väinölä, *et al.* 2000; Ando, *et al.* 2014; Cao, Yang, *et al.* 2011; Repo, *et al.* 2002; Repo, *et al.* 1997) depends on the tissue composition (Repo, *et al.* 2004; Laarabi, *et al.* 2014; Inaba, *et al.* 1995; Vicente, Ariel R., *et al.* 2005), tissue structure (Vozáry, Eszter, *et al.* 2011) and tissue health (Ozier-Lafontaine and Bajazet, 2016; Bera, *et al.* 2016; Azzarello, Elisa, *et al.* 2006) which can be the physical assessed or evaluated from the experimental measurement of its impedance responses. Electrical Impedance Spectroscopy (EIS) (Macdonald, 1992; Macdonald, 2005; [33]. Lasia, 2002; Chang & Park, 2010; Bera & Nagaraju, 2011) is an impedance analysing technique which estimates the electrical impedance of an object at different frequencies (ω) by measuring the boundary voltage-current data using an array of surface electrodes. Thus the EIS method applied for any biological tissue (Dean, *et al.* 2008; Gabriel, *et al.* 1996; Da Silva, *et al.* 2000; Gabriel, *et al.* 2009; Kyle, *et al.* 1999; Wu *et al.*, 2008, Bera, 2014, Bera, *et al.* 2015; and Bera, *et al.* 2016) can provides the frequency response of the electrical bioimpedance $Z_b(\omega)$ and its phase angle ($\theta_b(\omega)$) as the signature of the tissue composition and tissue health.

Fruit ripening (Lelièvre 1997; Brady, 1987; Kader, 1997) is a biochemical process (Seymour *et al.* 2012) in which a matured raw fruit undergoes through several physiological and physiochemical changes (Lester & Dunlap, 1985; Amorós, *et al.* 2003; Shackel, *et al.* 1991; Hoeberichts, 2002) over a certain time called ripening period (Motilva, *et al.* 2000). The fruit ripening which may be either natural ripening (ripening of fruits attached to the plant) (Burg & Burg, 1965) or are artificial ripening or post-harvesting ripening (ripening of fruits after harvesting and during their storage) (Herath, *et al.* 1998). The quality of the fruits (Kader 1997; Mitchell, J. P., *et al.* 1991) generally depends on the fruit ripening (Gierson and Kader 1986; Kader 1997; Chowdhury, *et al.* 2015; Chowdhury, *et al.* 2016). A proper and optimized ripening of fruit (Gierson and Kader 1986; Kader 1997; Chowdhury, *et al.* 2015; Chowdhury, *et al.* 2016) is essentially required for getting the physiological developments of all the necessary nutrients. Generally, the proper ripening of most of the fruits ensures the better taste along with the maximum availability of all the desired nutrients (Chowdhury, *et al.* 2016). Therefore, the process of fruit ripening is not only very important and essential to be studied to identify an optimum ripening state for obtaining ripened fruits with all the essential and important nutrients (Chowdhury, *et al.* 2016). Also, the monitoring and analysing ripening process helps us to obtain a number of information about the physiological and biochemical changes inside the fruits undergoing the ripening (Chowdhury, *et al.* 2016) procedure.

During the ripening process, as the physiological and physiochemical changes occur in fruits which changes the electrical properties of the fruit tissues, using the EIS studies, the bioelectrical impedance of fruits (Z_f) can be found as the signature on the physiological and physiochemical status during ripening. Many researchers (Chowdhury, *et al.*

2016; Vozary, *et al.* 1999; Fang, *et al.* 2007; Euring, *et al.* 2011; Bauchot, *et al.* 2000; Harker & Dunlop, 1994; Harker, *et al.* 2000; Jackson & Harker, 2000; Caravia, *et al.* 2015; Bean, *et al.* 1960; Juansah, *et al.* 2012; Juansah, *et al.* 2012; Harker & Maindonald, 1994; Harker, *et al.* 1997; Benjakul, *et al.* 2013) have studied EIS on the fruits anatomy, physiology and the ripening processes and investigated frequency response of the fruit bioimpedance as an effective, noninvasive method to study the ripening states and fruit quality by correlating the electrical impedance with their physiological and physiochemical status during the ripening process. In this paper a technical review on the EIS studies on the fruits and their ripening has been presented. The fruits physiology, ripening process, the relation between the fruit impedance and the ripening states are discussed followed by the detail discussion about the EIS methods. After introducing the bioimpedance a detail discussion has been made on the EIS instrumentation and the process. A detail review has been presented on the studies conducted on the fruit ripening and fruit characterization. The advantages and disadvantages of the EIS technology along with the recent trends have been summarized at the end.

2 METHODS

2.1 *Biological tissues and bioimpedance (Z_b)*

Plant tissues are composed of plant cells (Figure 1a) which are arranged in a three dimensional (3D) space to develop the 3D structure of the plant tissues (Figure 1b). The plant cells are composed of ICF which contains cell organnelles and vacuoles enclosed within cell envelopes. The plant cells (Figure 1b) are enclosed with an another envelop called cell wall to provide the mechanical strength and the protection to the cells and the tissues. The plan cells are embedded in an extracellular matrix or ECF which is developed with the extracellular molecules to provide the structural and biochemical support to the cells in the tissue.

Like other biological cells, the Intracellular Fluid (ICF) or protoplasm (Wilson, 1899) of the plant cells is enclosed by the Cell Membrane (CM) (Figure 2a) and a Cell Wall (CW) (Figure 2b). The protoplasm is composed of the cytoplasm (Thaine, 1962) and cell organelles (Pridham, 2012) such as nucleus with nucleolus, chloroplast, golgi bodies, vacuole, cytoskeleton, chromatin, mitochondrion, plasmodesma etc. The cytoplasm matrix or cytosol is developed with composed of water, salts, and proteins whereas the ECF is developed with a noncellular 3D macromolecular network composed of collagens, proteoglycans, elastin, fibronectin, laminins, and several other glycoproteins. The cell membrane structure could be understood by the Fluid Mosaic Model (Singer, *et al.* 1972) which described it as a plasma membrane developed with a mosaic of phospholipids, cholesterol molecules, proteins and carbohydrates. The protein is hydrophilic and the lipid is hydrophobic (Figure 2a). The composition of the cell wall (Carpita, *et al.* 1993; Vorwerk *et al.* 2004; Heredia, *et al.* 1995) of land plants is composed of the polysaccharides cellulose, hemicellulose and pectin whereas the cell wall of the in bacteria is composed of peptidoglycan.

Neglecting the individual electrical properties of the cell organelles for their small volume, the electrical properties of the protoplasm is found as resistive in nature towards an alternating electrical excitation. As the ICF and ECF are made of conducting materials (such as water, protein and salt), under the excitation with an alternating electrical signal, they both behave like the normal resistors (Wu *et al.*, 2008; Bera, 2014, and Bera, *et al.* 2016) (Figure 3a). As the cell membrane is made up of insulating lipid bi-layer sandwiched between two conductive protein layers, the cell membrane behaves as a capacitor (Wu *et al.*, 2008; Bayford & Tizzard, 2012; Bera, *et al.* 2016) which can be modeled as a capacitor in series with a resistor (Wu *et al.*, 2008; Bayford & Tizzard, 2012; Bera, *et al.* 2016) as shown in the Figure 3a. Thus, the bioimpedance of a single plant cell suspended in ECF can be modeled by the extracellular resistance (R_{ECF}) (Wu *et al.*, 2008; Bera, *et al.* 2016), intracellular fluid resistance (R_{ICF}) (Wu *et al.*, 2008; Bera, *et al.* 2016) and cell membrane capacitance (C_{CM}).

As the electrical current pass through partly through the ICF and partly through ECF depending on the C_{CM} and signal frequency, the electric current faces two paths on through R_{ECF} and another through the intracellular impedance Z_{ICF} (which is formed by the series combination of R_{ICF} and C_{CM}). Therefore, the equivalent model of CM enclosed ICF within ECF is a parallel combination of R_{ECF} and the Z_{ICF}. As the plant cells are enclosed by the cell walls, which also provide some resistance (R_{CW}) to the applied electrical signal, the complete model of a plant cell will have the R_{ECF} in parallel to the R_{ECF} (Wu *et al.*, 2008; Bera, *et al.* 2016). Thus under an alternating electrical excitation, the plant cell impedance (Z_{PC}) depends not only depends on R_{ECF}, R_{CW} and R_{ICF} and C_{CM} but also it varies with applied signal frequency. Thus, the degree of penetration of the current or the conduction paths of the electrical current through the cells (Figure 3a) within a plant tissue, depend on R_{ECF}, R_{CW} and R_{ICF} and C_{CM} (Wu *et al.*, 2008; Bera, *et al.* 2016). Also, as the capacitive reactance ($|X_{CM}| = 1/(\omega C_{CM})$) depends on the applied signal frequency, the degree of penetration of the electrical signal to a plant tissue depends on frequency (Figure 3b). As shown in Figure 3b, at low frequencies (ω), the membrane capacitance produces a large capacitive reactance (as $|X_{CM}| \propto 1/\omega$) and hence the Z_{PC} becomes very high which does not allow the current signal to pass through (Figure 3a). But, at sufficiently high frequencies, the capacitive reactance of the cell membrane becomes very small which reduces the ZPC significantly and hence allows the current signal to pass (Figure 3a) through the cells by penetrating the cell membranes (Bera, *et al.* 2016).

The equivalent electrical impedance of a single isolated plant cell suspended in the ECF can be

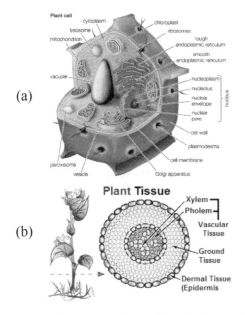

(a)

(b)

Figure 1. Structure of plant cells and tissues: (a) anatomy of a plant cell, (b) anatomy of a plant tissue [Photo courtesy: Figure a is taken from https://www.asps.org.au/research/cell-biology (applied for permission for using the art work); Figure b is taken from http://www.edu-resource.com/biology/plant-and-animal-tissues.php (applied for permission for using the art work).

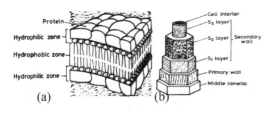

(a)   (b)

Figure 2. Structure of the plant cell membrane and cell wall (a) cell membrane [Fig. 2a has been taken from Bera *et al.* 2011], (b) anatomy of a plant cell wall.

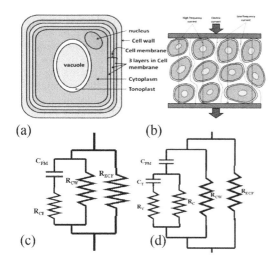

$$Z_p(\omega) = R_p(\omega) - jX_p(\omega) \qquad (1)$$

$$\theta_p(\omega) = tan^{-1}\left(\frac{X_p(\omega)}{R_p(\omega)}\right) \qquad (2)$$

where $R_p(\omega)$ and $X_p(\omega)$ are the real and imaginary parts of Z_p and j is the complex number $\sqrt{(-1)}$. Both $R_p(\omega)$ and $X_p(\omega)$ depend on the signal frequency and tissue composition such that

$$R_p(\omega) = |Z_p(\omega)|cos(\theta_p(\omega)) \qquad (3)$$

$$X_p(\omega) = |Z_p(\omega)|sin(\theta_p(\omega)) \qquad (4)$$

In EIS procedure, Z_p and its phase angle (θ_p) of a test subject are measured at different frequency points, ω_i (ω_i: ω_1, ω_2, ω_3, ... ω_n), by injecting a constant amplitude, Alternating Current (AC) signal ($I(\omega)$) and by measuring the boundary potentials ($V(\omega_i)$) through using an array of surface electrodes attached to the subject (Figure 1). Thus EIS provides the spectra of bioimpedance of the plant tissue which could be utilized for analyzing tissue properties, developing the equivalent tissue circuit model for noninvasive tissue characterization. The equivalent electrical circuit models or Equivalent Circuit Models (ECM) [6, 13] of biological tissue under test are very useful to understand the anatomical, physiological and compositional aspects of biological tissues. Impedance spectra, such as Nyquist plots (Imaginary part of Z versus Real part of Z), impedance (Z) versus frequency (ω) plots (Bode plots), conductivity (σ) versus frequency (ω) plots and permittivity (ε) versus frequency (ω) plots, obtained from EIS can be used to study the overall anatomy and physiology of plant tissues. In particular, Nyquist plots present the graphical expression of imaginary part of the Z_p versus the real parts of Z_p and serve as a guide to define equivalent electrical circuit parameters.

The EIS measurements are conducted, generally, by either two electrode method (Bera, 2014; Bera, *et al.* 2016; Alonso & Cliquet, 2016) or four electrode method (Bera, 2014; Bera, *et al.* 2016; Alonso & Cliquet, 2016) using an impedance analyzer or an impedance measurement instrumentation (He, *et al.* 2011). In two electrode method (Figure 1a) of impedance measurement current injection and voltage measurement are conducted with same electrodes (two surface electrodes) and hence it is known as two electrode method or two probe method. Four electrode methods (Figure 4b) uses two separate electrode pairs for current injection and voltage measurement and therefore this impedance measurement technique is known as four electrode method or four probe method.

Figure 3. The frequency dependent impedance properties of plant cells and their equivalent circuit modeling (a) The anatomy of an isolated plant cell showing Intracellular Fluids (ICF), Extracellular Fluids (ECF), the Cell Membrane (CM) and Cell Wall (CW), (b) electrical current conduction through the biological cells in an plant tissue, (c) lumped model proposed by Hayden *et al.* (1969), (d) double shell model proposed by Zhang *et al.* (1990) showing vacuole sap resistance (R_{VS}), tonoplast capacitance (C_{TP}).

represented either as a single cell model (Bera, *et al.* 2016; Damez *et al.* 2007; Zhang *et al.* 1995; Azzarello, *et al.* 2012; Hayden *et al.* 1969) proposed by Hayden *et al.* (1969) as shown in the Figure 3b or as a double cell model (Zhang *et al.*, 1990; Zhang *et al.*, 1991; Zhang *et al.*, 1992) proposed by Zhang *et al.* (1990) showing the vacuole impedance (Z_V) developed by the vacuole sap resistance (R_{VS}) and tonoplast capacitance (C_{TP}) (Figure 3c) or else.

2.2 *Electrical Impedance Spectroscopy (EIS)*

Electrical impedance spectroscopy conducted on a plant tissue sample provides the electrical impedance (Z_p) and its phase angle (θ_p) of the tissue sample as a lumped parameters at different frequency points. The impedance value found at each frequency is the lumped impedance value of the tissue sample under test which depends on the dimension of the sample. The lumped impedance measured by EIS is contributed by the each cell within the tissue interacting with the applied current signal passing through it. Thus the plant tissue impedance obtained from the EIS could be expressed by a complex, frequency-dependent bioelectrical impedance, $Z_p(\omega)$ (Bera, *et al.* 2016) and its frequency-dependent phase angle, $\theta_p(\omega)$ are given by:

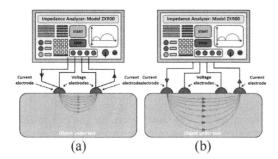

(a)   (b)

Figure 4. EIS measurement process with impedance analyzer (a) EIS measurements using two electrode method (b) EIS measurements with four electrode method.

In two probe methods the measurement data contains the errors produced by the electrode skin contact impedance (Z_c) whereas the, four-probe method (Figure 4b) is found free from the contact impedance error and hence the four-probe based impedance measurement process is usually preferred. As shown in Figure 4a and 4b, in four-probe method based EIS, a constant amplitude sinusoidal current signal ($I_f(\omega)$) is injected through the outer electrodes of a linear array of four electrodes (Figure 4) which are known as the current electrodes or the driving electrodes (red electrodes in Figure 4b). The surface potential developed across the inner two electrodes, called voltage electrodes or sensing electrodes (blue electrodes in Figure 4b), is measured as the frequency-dependent AC voltage ($V_f(\omega)$). The fruit tissue impedance ($Z_f(\omega)$) is calculated as the transfer function of the tissue sample under test from the voltage-current data by dividing $V_f(\omega)$ by $I_f(\omega)$ using the Ohm's law:

$$Z_f(\omega) = \frac{V_f(\omega)}{I_f(\omega)} \qquad (5)$$

3 EIS BASED FRUIT CHARACTERIZATION

The physiology and the ripening of a number of fruits have been studied by EIS. The following section tried to review the research works conducted on the EIS technology applied for studying the fruit physiology and their ripening process.

3.1 Studying avocado maturity and ripening

Maturity and ripening of avocado fruits have been studied (Bean, *et al.* 1960) with impedance spectroscopy using with alternating currents from 600 Hz to 60 kHz. The EIS measurements indicated

that almost two-thirds of the impedance of the avocado may be due to capacitive reactance and an inverse correlation was observed between the changes in impedance and respiration during the ripening of avocados. Results demonstrated that the electrical bioimpedance decreased during the maturation and the ripening of the avocado, but the individual variation between fruits tended to be too large in comparison with the change during maturation.

3.2 EIS studies of nectarines during coolstorage and ripening

The electrical impedance variations in nectarine fruit were studied (Harker & Dunlop, 1994) with EIS between 50 Hz and 0.1 MHz using a laboratory based impedance measurement set up developed with a function generator (Model F32, Interstate Electronics Corporation), an Oscilloscope (Model 50103 N, Tektronix) and array of 4 parallel stainless steel (0.45 mm diameter) or silver (0.60 mm diameter) electrodes. The EIS studies were conducted on the ripe and unripe nectarine fruits at harvest and after 3 and 8 weeks storage at 0°C. Results demonstrated that the electrode impedances with stainless steel electrodes were higher compared to the silver electrodes. It is found that the resistance of whole fruit was lower than that the tissue blocks excised from the same fruit. Results also show that the tissue resistance variation were greatest at low frequencies (50–100 Hz) which increased between 0 and 3 weeks storage and decreased between 3 and 8 weeks storage.

3.3 Electrical impedance studies of nectarines

Using EIS, electrical impedance measurements were conducted on nectarines (*Prunus persica*) to study the changes in intracellular and extracellular resistance as well as changes in the condition of membranes during ripening (Harker & Maindonald, 1994). The impedance measurements obtained from EIS were related to changes in fruit texture assessed by flesh firmness and apparent juice content. The electrical model obtained from the EIS data indicated that, during ripening (d 1–5) of freshly harvested fruit, the resistance of the cell wall and vacuole declined by 60 and 26%, respectively, and the capacitance of the membranes decreased by 9%. An additional resistance component, which declined by 63% during ripening, was thought to be associated with either the cytoplasmic or membrane resistance. The variation in nectarines tissue resistance measured at low frequencies was closely related to flesh firmness. After storage at 0°C for 8 weeks, the nectarines developed a woolly (dry) texture during ripening at 20°C and the resistance of the cell wall was higher in woolly tissue (4435 Ω after 5 d at 20°C) than in nonwoolly tissue (2911 Ω after 5 d at 20°C).

3.4 Ripening and chilling injury in persimmon fruit

Electrical impedance spectroscopy was conducted on persimmon fruit (*Dyospyros kaki*) between 50 Hz and 1 MHz (Harker & Forbes, 1997) to follow ripening and chilling injury development. Results demonstrated that, the impedance responses of the fruit to both ripening at 20°C and storage in modified atmosphere at 7°C were distinct and easily detected using EIS and found as a series semicircular arcs. During ripening, the arcs dilated between Days 1 and 21, and then contracted, until at Day 35 they were smaller than at Day 1. It was observed from the electrical modelling that the dilation occurred as a result of a 43, 115, and 17% increase in cell wall resistance, cytoplasm resistance, and vacuole resistance, respectively. The cytoplasm resistance of the chill-injured fruit was significantly lower than other fruits upon removal from storage, although it rapidly increased when fruit were transferred to 20°C for ripening. The physiological changes occurred during ripening and development of chilling injury in persimmon.

3.5 Assessing the physiology of Kiwifruit

The EIS studies have been conducted kiwifruit (*Actinidia deliciosa*) to study its electrical impedance variation during fruit ripening (Bauchot, *et al.* 2000) using either a Hewlett-Packard Model HP4194A impedance analyser (Hewlett-Packard, Hyogo, Japan), or a Hewlett-Packard Model HP4284A Precision LCR meter (Hewlett-Packard, Hyogo, Japan). Alternating current at frequencies between 50 Hz and 1 MHz was passed through fruit and tissue samples, and the kiwifruit impedance was measured for whole fruit, and tissues excised from the outer pericarp, inner pericarp and core. The complex impedance spectra were separated into the resistances of the apoplast, cytoplasm and vacuole, and capacitances of the plasma membrane and tonoplast. The differences in apoplast resistance (R50 Hz) and total tissue resistance (R1 MHz) between tissues were explained in terms of the anatomy and composition of the respective tissues. Though research studies conducted on nectarine, persimmon and tomato fruit showed a considerable reduction in impedance during ripening, there was little change in the kiwifruit impedance characteristics during ripening, despite a 10-fold decrease in firmness. The results suggested that the mobility of electrolytes within the cell wall did not change

during kiwifruit ripening and hence physicochemical interactions that take place within the cell wall may have a major impact on the impedance of kiwifruit tissue.

3.6 Studying the physical and mechanical changes in strawberry

The effect of CO_2 treatments on the cell-to-cell adhesion in 'Pajaro' strawberry (*Frageria* × *ananassa* Duch) fruit was studied (Harker *et al.* 2000) and correlated with the corresponding variation of electrical response of the apoplast and symplast in the fruits using electrical impedance spectroscopy. The fruits were exposed to 5–40% CO_2 for 0–3 days stored at 0°C for up to 3 weeks. The cell-to-cell adhesion was measured by the application of tensile tests to plugs of tissue, followed by the examination of fracture surfaces using low temperature scanning electron microscopy. Results demonstrated that cell-to-cell adhesion increased by 60% as a result of CO_2 treatments, however, there were no differences in the density, electrolyte leakage, propensity for cells to rupture in hypertonic solutions, water potential, osmotic potential or turgor of CO_2-treated and control fruit. EIS studies showed that CO_2 treatments reduced the apoplast resistance (resistance at 50 Hz), but did not affect the resistance of the symplast (resistance at 1 MHz). Author speculated that CO_2-induced firmness enhancement in strawberry is due to high concentrations of H^+ and $HCO3^-$ in the apoplast which changes the apoplast pH which may promote the precipitation of soluble pectins and enhance the cell-to-cell bonding in strawberry fruit.

3.7 Apple Bruise detection

Electrical impedance spectroscopic studies were conducted on apple fruit (*Malus domestica*) before and after bruising (Phillipa *et al.* 2000) to determine the extent of tissue damage that occurred as a result of bruising. Impedance measurements were made at 36 frequencies points between 50 Hz and 1 MHz and Nyquist characteristics (plots of reactance against resistance) were then related to bruise weight. Results showed that the change in resistance that occurred as a result of fruit impact ($\Delta R50$ Hz) was the best predictor of bruise weight. Before bruising, resistance of fruit was higher in 'Splendour' than in 'Granny Smith' ($P < 0.001$), and at 0°C than at 20°C ($P < 0.001$), but was not influenced by fruit weight. As the influence of apple cultivar and temperature on electrical impedance may cause difficulties when implementing these measurements in a commercial situation, further development of proposed method may require to be improved for assessing bruise weight without having to wait for browning of the flesh.

3.8 Non-destructive evaluation of citrus fruits

Garut citrus fruits acidity was investigated using electrical impedance spectroscopy (Juansah, *et al.* 2012) and the measurements of electrical properties of Garut citrus fruits were conducted at various levels of pH and frequencies. The electrical parameters of citrus fruits such as electrical impedance, resistance, reactance, inductance and capacitance were measured using an LCR meter (3532–50 LCR HiTESTER, Hioki, Tokyo, Japan) at 0.1 kHz, 1.0 kHz, 0.01 MHz, 0.1 MHz and 1 MHz. The pH estimations by using multiple regressions of electrical parameters were highly linear especially at frequency of 1 MHz. Resistance per weight, reactance per weight, inductance per weight, and impedance per weight of Garut citrus fuits declined when the pH is increased whereas the capacitance per weight of the Garut citrus fruits increases as pH increases.

3.9 Cell vitality studies on Shiraz berries

The electrical impedance of the *Shiraz* berry has been studied (Caravia, *et al.* 2015) as a function of the state of cells within, to detect the loss of cell vitality. Electrical impedance of Shiraz berries (774 berries) from two locations (one of them over two seasons) from veraison to harvest was studied by EIS method performed between 100 Hz and 1 or 2 MHz. The proportion of the living tissues for each berry was estimated with fluorescein diacetate and the cell death in the mesocarp of fruits has been correlated with berry mass loss that occurs late in ripening. The results demonstrated that, from veraison to the onset of cell death, electrical impedance of the berry fruits follows the accumulation of Total Soluble Solid (TSS). And the impedance decreases proportionally to the extent of berry cell death hereafter. Using impedance spectroscopy the changes in cell vitality of the Shiraz grape berry can be objectively determined. EIS studies on grape berry demonstrated the technique as a promising application for measuring berry composition.

3.10 Ripening of banana fruit

EIS studies have been conducted to study the electrical impedance variations during banana ripening using Agilent 4294A impedance analyzer (Chowdhury *et al.* 2016). The electrical impedance of banana samples is measured by at different states of ripening and the ripening states are correlated with their corresponding impedances. The banana impedance, phase angle, real part and imaginary part of the impedance are studied to analyze the ripening phenomena in terms of banana impedance. The results demonstrated that the banana impedance and tits real part and imaginary part all increase with the progresses in the ripening process. Statistical analysis conducted on the banana samples from a same bunch showed that the electrical EIS technique can be applied for noninvasive characterization of banana ripening.

4 CONCLUSIONS

Fruit physiology and ripening process can be studied with EIS method noninvasively. Electrical impedance of the fruits changes with the physiological and physiochemical changes occurred during the ripening of the fruits which can be probed by EIS noninvasively to characterize either the fruit physiology or the ripening process. Though the EIS technology provides a number of advantages but still the technology is found very sensitive to the several measurement errors produced by electrode contact impedance, cable impedance, stray capacitance etc. Though few commercial impedance analyzer or LCR meters are available with high measurement accuracy, but for the impedance measurement with a wide frequency band development of prototype standalone laboratory work bench is, generally found essential. As the composition and physiological changes occurred inside the fruits varies from species to species there may not be similar impedance response for all the fruits. Therefore a common analytical model is not possible to be developed. Also as the electrical impedance depends on the geometry of the sample, the EIS studies requires very accurate observation for assessing the entire fruit species with the response obtained from a limited number of samples. Statistical analysis may be required to develop a solution to this practical problem. As the effect of contact impedance during impedance measurement play a significant role four electrode method is preferable to study the fruits physiology and ripening. Though the housing of four electrodes on the fruit surface is found, sometimes, difficult for the fruits of small volume the microelectronics technology can be adopted for small fruit analysis as the future research direction.

REFERENCES

Alonso, K. C., & A. Jr. Cliquet (2016). Body composition assessment by bioelectrical impedance analysis and body mass index in individuals with chronic spinal cord injury. J Electr Bioimp, 7(1), 2–5.

Amorós, A., P. Zapata, M. T. Pretel, M. A. Botella, M. Serrano (2003). Physico-chemical and physiological changes during fruit development and ripening of five loquat (Eriobotrya japonica Lindl.) cultivars. Food Science and Technology International 9.1, 43–51.

Ando, Yasumasa, K. Mizutani, and N. Wakatsuki (2014). Electrical impedance analysis of potato tissues during drying. Journal of Food Engineering 121, 24–31.

Azzarello, Elisa, Mugai, C. Pandoifi, E. Masi, S. Mancuso (2006). Stress assessment in plants by impedance spectroscopy. Floriculture, Ornamental and Plant Biotechnology 3, 140–148.

Azzarello, E., E. Masi, S. Mancuso (2012). Electrochemical Impedance Spectroscopy. In Plant Electrophysiology. 205–223 Springer Berlin Heidelberg.

Bauchot, Anne D., F. Roger Harker, and W. Michael Arnold (2000). The use of electrical impedance spectroscopy to assess the physiological condition of kiwifruit. Postharvest Biology and technology 18.1, 9–18.

Bayford, R., & A. Tizzard (2012). Bioimpedance imaging: an overview of potential clinical applications. Analyst, 137(20), 4635–4643.

Bean, Ross C., P. John Rasor, and G. Gerald Porter (1960). Changes in electrical characteristics of avocados during ripening. Yearbook Calif. Avocado Soc 44, 75–78.

Benjakul, Suppawut, T. Eadkhong, W. Limmun S. Danworaphong (2013). Probability of finding translucent flesh in mangosteen based on its electrical resistance and capacitance. Food Science and Biotechnology 22.2 (2013): 413–416.

Bera T. K. and J. Nagaraju (2011). Electrical Impedance Spectroscopic Study of Broiler Chicken Tissues Suitable for The Development of Practical Phantoms in Multifrequency EIT, Journal of Electrical Bioimpedance. 2, 48–63.

Bera, Tushar Kanti, S. Bera, K. kar, S. Mondal (2016). Studying the Variations of Complex Electrical Bio-Impedance of Plant Tissues During Boiling. Procedia Technology 23, 248–255.

Bera, Tushar Kanti, J. Nagaraju, and G. Lubineau (2015). Electrical impedance spectroscopy (EIS)-based evaluation of biological tissue phantoms to study multi frequency electrical impedance tomography (Mf-EIT) systems. Journal of Visualization, 1–23.

Bera, Tushar Kanti (2014). Bioelectrical impedance methods for noninvasive health monitoring: A review. Journal of Medical Engineering, 1–28, Article ID 381251.

Bernstein, Donald P. (1986). Continuous noninvasive real-time monitoring of stroke volume and cardiac output by thoracic electrical bioimpedance. Critical care medicine 14.10, 898–901.

Brady, C. J. (1987). Fruit ripening. Annual review of plant physiology 38.1, 155–178.

Bruce, Alberts A. Johnson, J. Lewis, M. Raff, K. Roberts, and P. Walter (2002). Molecular Biology of the Cell, 4th edition, New York: Garland Science.

Burg, S. P., & E. A. Burg (1965). Ethylene action and the ripening of fruits ethylene influences the growth and development of plants and is the hormone which initiates fruit ripening. Science, 148(3674), 1190–1196.

Cao, Yang, T. Repo, R. Silvennoinen, T. Leheto and P. Pelkonene (2011). Analysis of the willow root system by electrical impedance spectroscopy. Journal of experimental botany 62.1, 351–358.

Caravia, L., C. Collins, and S. D. Tyerman. (2015). Electrical impedance of Shiraz berries correlates with decreasing cell vitality during ripening. Australian Journal of Grape and Wine Research 21.3, 430–438.

Carpita, Nicholas C., and David M. Gibeaut. (1993). Structural models of primary cell walls in flowering plants: consistency of molecular structure with the physical properties of the walls during growth. The Plant Journal 3.1, 1–30.

Chang, B. Y., & S.M. Park, (2010). Electrochemical impedance spectroscopy. Annual Review of Analytical Chemistry, 3, 207–229.

Chowdhury, Atanu, T.K. Bera, D. Ghosal and B. Chakraborty (2016). Electrical Impedance Variations in Banana Ripening: An Analytical Study with Electrical Impedance Spectroscopy. Journal of Food Process Engineering, DOI:10.1111/jfpe.12387

Chowdhury, A., T.K. Bera., D. Ghosaland B. Chakraborty (2015). Studying the electrical impedance variations in banana ripening using electrical impedance spectroscopy (EIS). 3rd International Conference Computer, Communication, Control and Information Technology (C3IT), IEEE, Hoogly, West Bengal.

Curits, Otis F., and C.G. Daniel (1951). An introduction to plant physiology. Soil Science 71.1, 78.

Da Silva, J. Estrela, J.P. Marques De Sá, and J. Jossinet (2000). Classification of breast tissue by electrical impedance spectroscopy. Medical and Biological Engineering and Computing 38.1, 26–30.

Damez, J. L., S. Clerjon, S. Abouelkaram, & J. Lepetit (2007). Dielectric behavior of beef meat in the 1–1500 kHz range: Simulation with the Fricke/Cole–Cole model. Meat Science, 77(4), 512–519.

Dean, D. A., T. Ramanathan, D. Machado, and R. Sudararajan (2008). Electrical impedance spectroscopy study of biological tissues. Journal of electrostatics 66.3, 165–177.

Eames, Arthur Johnson, and L.H. MacDaniels (1947). An introduction to plant anatomy. An introduction to plant anatomy. 2nd edition.

Euring, Frank, W. Russ, and W. Wilke (2011). Development of an impedance measurement system for the detection of decay of apples. Procedia Food Science 1, 1188–1194.

Fang, Q., X. Liu, and I. Cosic (2007). Bioimpedance study on four apple varieties. 13th International Conference on Electrical Bioimpedance and the 8th Conference on Electrical Impedance Tomography. Springer Berlin Heidelberg.

Gabriel, S., R. W. Lau, and C. Gabriel (1996). The dielectric properties of biological tissues: II. Measurements in the frequency range 10 Hz to 20 GHz. Physics in medicine and biology 41.11, 2251.

Gabriel, C., A. Peyman, and E. H. Grant (2009). Electrical conductivity of tissue at frequencies below 1 MHz. Physics in medicine and biology 54.16, 4863.

Gierson, D., & A.A. Kader (1986). Fruit ripening and quality. In The tomato crop., 241–280, Springer Netherlands.

Harker, F. Roger, and J.H. Maindonald. (1994). Ripening of nectarine fruit (changes in the cell wall, vacuole, and membranes detected using electrical impedance measurements). Plant physiology 106, 165–171.

Harker, F. R., and J. Dunlop (1994). Electrical impedance studies of nectarines during cool storage and fruit ripening. Postharvest biology and technology 4.1–2, 125–134.

Harker, F. Roger, H.J. Elger, C.B. Watkins, P.J. Jackson (2000). Physical and mechanical changes in strawberry fruit after high carbon dioxide treatments. Postharvest Biology and Technology 19.2, 139–146.

Harker, F. Roger, and Shelley K. Forbes (1997). Ripening and development of chilling injury in persimmon fruit: an electrical impedance study. New Zealand Journal of Crop and Horticultural Science 25.2, 149–157.

Hayden, R. I., C.A. Moyse, F.W. Calder, D.P. Crawford, & D.S. Fensom, (1969). Electrical impedance studies on potato and alfalfa tissue. Journal of Experimental Botany, 20(2), 177–200.

He, Jian-xin, Z. Wang, Y. Shi, L. Huang (2011). A prototype portable system for bioelectrical impedance spectroscopy. Sensor Letters 9.3, 1151–1156.

Heredia, Antonia, A. Jiménez, and R. Guillén (1995). Composition of plant cell walls. Zeitschrift für Lebensmittel-Untersuchung und Forschung 200.1, 24–31.

Hoeberichts, Frank A., H.W. Linus, Van Der Plas, and Ernst J. Woltering (2002). Ethylene perception is required for the expression of tomato ripening-related genes and associated physiological changes even at advanced stages of ripening. Postharvest Biology and Technology 26.2, 125–133.

Hopkins, William G. (1999). Introduction to plant physiology. No. Ed. 2. John Wiley and Sons.

Inaba, Akitsugu, T. Manabe, H. Tsuji, T. Iwamoto (1995). Electrical impedance analysis of tissue properties associated with ethylene induction by electric currents in cucumber (Cucumis sativus L.) fruit. Plant physiology 107.1, 199–205.

Jackson, Phillipa J., and F. Roger Harker (2000). Apple bruise detection by electrical impedance measurement. Hort Science 35.1, 104–107.

Juansah, Jaja, I.W. Budiastra, K. Dahlan, K.B. Seminar (2012). The prospect of electrical impedance spectroscopy as non-destructive evaluation of citrus fruits acidity. International Journal of Emerging Technology and Advanced Engineering 2.11, 58–64.

Kader, Adel A. (1997). Fruit maturity, ripening, and quality relationships. International Symposium Effect of Pre- & Postharvest factors in Fruit Storage 203–208.

Kyle, Alastair H., Carmel T.O. Chan, and Andrew I. Minchinton (1999). Characterization of three-dimensional tissue cultures using electrical impedance spectroscopy. Biophysical journal 76.5, 2640–2648.

Kushner, Robert F., and Dale A. Schoeller (1986). Estimation of total body water by bioelectrical impedance analysis. The American journal of clinical nutrition 44.3, 417–424.

Lasia, A. (2002). Electrochemical impedance spectroscopy and its applications. In Modern aspects of electrochemistry 143–248, Springer US.

Lelièvre, J. M., A. Latchè, B. Jones, M. Bouzayen, and J.C. Pech (1997). Ethylene and fruit ripening. Physiologia plantarum, 101(4), 727–739.

Lester, G. E., and J. R. Dunlap (1985). Physiological changes during development and ripening of 'Perlita' muskmelon fruits. Scientia Horticulturae 26.4, 323–331.

Liu, Xing. (2006). Electrical impedance spectroscopy applied in plant physiology studies.

Lukaski, Henry C., P.E. Johnson, W.W. Bolonchuk, G.I. Lykken (1985). Assessment of fat-free mass using bioelectrical impedance measurements of the human body. The American journal of clinical nutrition 41.4, 810–817.

Laarabi, Saïd. (2014). Characterization of short-term stress applied to the root system by electrical impedance measurement in the first leaf of corn (Zea mays L.) and Pumpkin (Cucurbita maxima L.). American Journal of Plant Sciences 5.9, 1285–1295.

Macdonald, J. Ross. (1992). Impedance spectroscopy. Annals of biomedical engineering 20.3, 289–305.

Macdonald, J. R., & W.B. Johnson (2005). Fundamentals of impedance spectroscopy. Impedance Spectroscopy: Theory, Experiment, and Applications, Second Edition, 1–26.

Margo, C., Katrib, J., Nadi, M., & A. Rouane (2013). A four-electrode low frequency impedance spectroscopy measurement system using the AD5933 measurement chip. Physiological measurement, 34(4), 391.

Martinsen, Orjan G., and Sverre Grimnes. Bioimpedance and bioelectricity basics. Academic press, 2011.

Meyer, Bernard Sandler, Donald B. Anderson, and Richard H. Böhning (1960). Introduction to plant physiology. Introduction to plant physiology.

Mitchell, J. P., C. Shennan, S.R. Grattan, D.M. May (1991). Tomato fruit yields and quality under water deficit and salinity. Journal of the American Society for Horticultural Science 116.2, 215–221.

Motilva, M. José, M.J. Tovar, M.P. Romero, S. Alegre (2000). Influence of regulated deficit irrigation strategies applied to olive trees (Arbequina cultivar) on oil yield and oil composition during the fruit ripening period. Journal of the Science of Food and Agriculture 80.14, 2037–2043.

Mouritsen O.G., and M. Bloom (1993). Models of Lipid-Protein Interactions in Membranes, Annual Review of Biophysics and Biomolecular Structure. 22, 145–171.

Ohmine, Yuken, T. Morimoto, Y. Kinouchi, T. Iritani, M. Takeuchi, Y. Monden (1999). Noninvasive measurement of the electrical bioimpedance of breast tumors. Anticancer research 20.3B (1999): 1941–1946.

Ozier-Lafontaine, Harry, and Thierry Bajazet (2005). Analysis of root growth by impedance spectroscopy (EIS). Plant and Soil 277. 1–2, 299–313.

Pridham, J., ed. (2012). Plant cell organelles. Elsevier.

Repo, Tapani, P. Hiekkala, T. Hietala, L. Tahvanainen (1997). Intracellular resistance correlates with initial stage of frost hardening in willow (Salix viminalis). Physiologia plantarum 101.3, 627–634.

Repo, T., D. H. Paine, and A. G. Taylor (2002). Electrical impedance spectroscopy in relation to seed viability and moisture content in snap bean (Phaseolus vulgaris L.). Seed Science Research 12.01, 17–29.

Repo, Tapani, E. Oksanen, and E. Vapaavuori (2004). Effects of elevated concentrations of ozone and carbon dioxide on the electrical impedance of leaves of silver birch (Betula pendula) clones. Tree physiology 24.7, 833–843.

Seymour, G. B., Taylor, J. E., & Tucker, G. A. (Eds.). (2012). Biochemistry of fruit ripening. Springer Science & Business Media.

Shackel, Kenneth A., C. Greve, J.M. Lavavitch, and H. Ahmadi (1991). Cell turgor changes associated with ripening in tomato pericarp tissue. Plant Physiology 97.2, 814–816.

Shoemaker, William C., C.C. Wo, M.H. Bishop, J.M. Ven De water, G.R. Harrington, X. Wang, R.S. Patil (1994). Multicenter trial of a new thoracic electrical bioimpedance device for cardiac output estimation. Critical care medicine 22.12, 1907–1912.

Singer, S. J., and Garth L. Nicolson (1972). The fluid mosaic model of the structure of cell membranes. Membranes and Viruses in Immunopathology; Day, SB, Good, RA, Eds, 7–47.

Thaine, R. (1962). A translocation hypothesis based on the structure of plant cytoplasm. Journal of Experimental Botany 13.1, 152–160.

Väinölä, Anu, and Tapani Repo (2000). Impedance spectroscopy in frost hardiness evaluation of Rhododendron leaves. Annals of Botany 86.4, 799–805.

Vicente, Ariel R., M.L. Costa, G.A. Martinez, A.R. Chaves, P.M. Civello (2005). Effect of heat treatments on cell wall degradation and softening in strawberry fruit. Postharvest Biology and Technology 38.3, 213–222.

Vorwerk, Sonja, Shauna Somerville, and Chris Somerville (2004). The role of plant cell wall polysaccharide composition in disease resistance. Trends in plant science 9.4, 203–209.

Vozary, Eszter, Peter Laslo, and Gabor Zsivanovits (1999). Impedance parameter characterizing apple bruise. Annals of the New York Academy of Sciences 873.1, 421–429.

Vozáry, Eszter, I. Jocsak, M. Droppa, K. Boka (2011). Connection between structural changes and electrical parameters of pea root tissue under anoxia. Edited by Pamela Padilla 131.

Wessels J.G.H. (1996). Fungal hydrophobins: proteins that function at an interface. Trends in Plant Science 1, 9–15.

White, Philip R. (1943). A handbook of plant tissue culture. Soil Science 56.2, 151.

Wilson E.B., (1899). The structure of protoplasm. Science 10, 33–45.

Wu, L., Y. Ogawa, & A. Tagawa (2008). Electrical impedance spectroscopy analysis of eggplant pulp and effects of drying and freezing–thawing treatments on its impedance characteristics. Journal of Food Engineering, 87(2), 274–280.

Zhang, M. I. N., & J. H. M. Willison (1990). Electrical conductance of red onion scale tissue during freeze–thaw injury. Acta botanica neerlandica, 39(4), 359–367.

Zhang, M. I. N., & J. H. M. Willison (1991). Electrical impedance analysis in plant tissues: a double shell model. Journal of Experimental Botany, 1465–1475.

Zhang, M. I. N., & J. H. M. Willison (1992). Electrical impedance analysis in plant tissues: the effect of freeze-thaw injury on the electrical properties of potato tuber and carrot root tissues. Canadian Journal of Plant Science, 72(2), 545–553.

Zhang, M. I. N., T. Repo, J. H. M. Willison, & S. Sutinen (1995). Electrical impedance analysis in plant tissues: on the biological meaning of Cole-Cole (in Scots pine needles). European Biophysics Journal, 24(2), 99–106.

Zhang, Gang, Aija Ryyppö, and Tapani Repo (2002). The electrical impedance spectroscopy of Scots pine needles during cold acclimation. Physiologia Plantarum 115.3, 385–392.

Computer, Communication and Electrical Technology – Guha, Chakraborty & Dutta (Eds)
© 2017 Taylor & Francis Group, ISBN 978-1-138-03157-9

Congestion constraint corrective rescheduling in the competitive power market with the integration of a wind farm

Sadhan Gope
Department of Electrical Engineering, Mizoram University, Tanhril, Aizawl, India

Arup Kumar Goswami & Prashant Kumar Tiwari
National Institute of Technology, Silchar, Assam, India

ABSTRACT: A deregulated power market creates a huge pressure to the System Operator (SO) to ensure a congestion-free transmission network for the continuity of power supply with minimum cost. This work describes Congestion Management (CM) with the integration of a Wind Farm (WF) in a double auction electricity market to minimize fuel cost, system losses, and Locational Marginal Price (LMP). The optimal location of a WF is identified based on the value of the Bus Sensitivity Factor (BSF); also, one load bus is selected for double auction bidding with the help of BSF. A modified 39-bus New England test system is used to demonstrate the effectiveness of the approach presented here.

1 INTRODUCTION

All over the world, privatization and deregulation of the power sector has greatly affected the power system. The competition in power market trading has exponentially increased the transaction in electricity. GENCOs and DISCOs, in a deregularized market, can independently or in pair transmit power. This may lead to unanticipated direction, volume, and length of power flow through the transmission networks. Such trade of power may not be accommodated by the present transmission network. As a result, congestion occurs in the system, creating a violation of the system security as well. Congestion usually occurs in modern power system transmission lines due to an increase in the electric power demand or line outage of the system (Sudipta 2008).

Literature has discussed the optimal bidding strategy of a supplier considering double-sided bidding (Arvind 2015). For uninterrupted power supply, it is essential to maintain the security of a power system under a deregulated power market. An optimal bidding strategy for electricity suppliers under the congestion environment has also been discussed (Tengshum 2003). Bidding strategy is utilized to maximize profit by using the Refined Immune Algorithm. The electrical market is settled based on the Locational Marginal Price (LMP) of the system. Optimal GENCOs bidding strategies of the electricity market has been established using an agent-based approach and Numerical Sensitivity Analysis (NSA) technique (Mahvi 2011).

Research has described bidding strategy in a joint spinning reserve market and day-ahead energy market with the integration of a micro grid (L. Shi 2014). The optimal cost of energy and spinning reserve bids are obtained by solving the bi-level bidding model using the interior point algorithm. Optimal bidding for the power market depends on the bidding revenue, expected imbalance, and operation cost. Ricardo *et al.* (2014) have presented an auction-based market for consumer payment minimization under a pool-based day-ahead electricity market. A bi-level programming method has been used to characterize the LMP. Demand-side bidding strategies are applied in the electricity market to minimize the cost of purchasing power (Rico 2012 and Faria 2015). Flexible AC Transmission System (FACTS) devices like TCSC and SSSC are used to maximize social welfare in a double-sided auction market (Nabavi 2012).

Wind energy sources are one of the rapidly increasing energy sources in the world. Wind power heavily relies on environmental conditions and hence produces unpredictable power output. Earlier works have showed that wind and hydro power generation plays an important role in the deregulated electricity market (Agustin 2013, Mahmoud 2013, and Edgardo 2004). Three bidding strategies are used to formulate the day-ahead bidding model. Research works such as those by Subhasish (2015) consider wind energy for congestion management. Until now, to the best of our knowledge, no research work has carefully studied congestion constraint corrective rescheduling in a competitive power market with integration of a Wind Farm (WF). Therefore, in this paper, WFs are integrated considering double auction bidding in power markets to minimize the congestion as well as to minimize the LMP of the system. By considering the cost function, GEN-COs and DISCOs participate in the pool market and maximize their generation as well as minimize the overall generation cost of the system. WF position in the bus is identified based on the BSF value. In this paper, a modified 39-bus New England test system has been used as a test system for solving the proposed method.

2 PROBLEM FORMULATION

2.1 Objective function

Consider that N number of WFs have been installed in an existing power system network. As a result, the proposed objective function consists of three terms. Mathematically, the objective function is to minimize the total fuel cost of the thermal generating unit, congestion cost, and investment cost of the WFs. Mathematically, the minimum objective function of the given approach is given by

$$OF = \sum_{i=1}^{NG} C_i(P_{Gi}) + \sum_{j=1}^{NG} C_j\,(\Delta P_G) + \sum_{k=1}^{N} C_{WF}(k) \quad (1)$$

where NG is equal to the number of generators, NL is equal to the number of loads, $C_i\,(P_{Gi})$ is equal to the generating cost of the thermal unit, $C_j\,(\Delta P_G)$ is equal to the congestion cost, and C_{WF} is equal to the installation cost of the WF.

Constraints
The following two types of constraints are used in the optimal power flow problem.

1. *Equality constraints*
a. *Real power balanced*

$$\sum_{i=1}^{NG} P_{Gi} + P_{WF} - P_{loss} - P_D = 0 \quad (2)$$

$$\sum_{i=1}^{NG} \Delta P_{Gi} = 0 \quad (3)$$

$$P_{loss} = \sum_{J=1}^{TL} G_J \left[|V_i|^2 + |V_j|^2 - 2|V_i||V_j|Cos\left(\delta_i - \delta_j\right) \right] \quad (4)$$

where P_G is the active power generation of the thermal generating unit, P_{WF} is the total wind power generation of the WF, P_{loss} is the transmission loss, P_D is the total load, TL is the number of transmission lines, G_J is the conductance of the

line between buses i and j, |Vi| is the voltage magnitude, and δ_i is the voltage angle.

b. *Power flow equations*

$$P_i - \sum_{k=1}^{BN} |V_i V_k Y_{ik}| Cos\left(\theta_{ik} - \delta_i + \delta_k\right) = 0 \quad (5)$$

$$Q_i + \sum_{k=1}^{BN} |V_i V_k Y_{ik}| Sin\left(\theta_{ik} - \delta_i + \delta_k\right) = 0 \quad (6)$$

where P_i and Q_i are the active and reactive powers injected into the system, BN is the number of buses, Y_{ik} is the bus admittance matrix connected between the ith row and k^{th} column, and θ_{ik} is the bus admittance matrix connected between the ith row and k^{th} column.

2. *Inequality constraints*
a. *Bus voltage limits*

$$V_i^{min} \le V_i \le V_i^{max} \quad i = 1,2,3,\ldots\ldots N \quad (7)$$

where V_i^{min} is the lower limit of the bus voltage, V_i^{max} is the upper limit of the bus voltage, N is number of buses.

b. *Power limits of the generator unit*

$$Q_{Gi}^{min} \le Q_{Gi} \le Q_{Gi}^{max} \quad i = 1,2,3,\ldots\ldots N \quad (8)$$

$$P_{Gi}^{min} \le P_{Gi} + \Delta P_{Gi} \le P_{Gi}^{max} \quad i = 1,2,3,\ldots\ldots N \quad (9)$$

where P_{Gi}^{min} and P_{Gi}^{max} are the thermal unit's minimum and maximum active power limits and Q_{Gi}^{min} and Q_{Gi}^{max} are the thermal unit's minimum and maximum reactive power limits.

c. *Security limits of the transmission line*

$$\left| MVA\ flow_{i,j}^0 \right| \le MVA\ flow_{i,j}^0 \max_0 \\ MVA\ flow_{i,j},\, i \ne j \quad (10)$$

$$\left| MVA\ flow_{i,j}^k \right| \le MVA\ flow_{i,j}^k \max_k \\ MVA\ flow_{i,j},\, i \ne j \quad (11)$$

where MVA $flow_{ij}$ max_0 and MVA $flow_{ij}$ max_k are the maximum powers that can flow through the line connecting the buses i and j, respectively, during the pre-congestion and post-congestion states.

d. *Discrete tap settings of the transformer*

$$T_i^{min} \le T_i \le T_i^{max} \quad i = 1,2,3,\ldots\ldots N \quad (12)$$

e. *Ramp limit of the generator unit*

$$P_{Gi} - P_{Gi}^{min} = \Delta P_{Gi}^{min} \le \Delta P_{Gi} \le \Delta P_{Gi}^{max} = P_{Gi}^{max} - P_{Gi} \\ i = 1, 2, 3 \ldots\ldots N \quad (13)$$

where P_{Gi}^{min} and P_{Gi}^{max} are the thermal unit's minimum and maximum active power limits.

2.2 Bus Sensitivity Factor (BSF)

The bus sensitivity factor is calculated for a congested line l by using following formula (Subhasish 2015):

$$BSF_k^l = \frac{\Delta P_{ij}}{\Delta P_k} \qquad (14)$$

where ΔP_{ij} is the change in the real power flow of line k and BSF_k^l is the change in active power flow in the congested line l due to active power injection at bus k. Active power flow a line k, which is connected between bus i and bus j, is equal to P_{ij} and can be written as

$$P_{ij} = |V_i||V_j||Y_{ij}|\cos(\theta_{ij} - \delta_i + \delta_j) - V_i^2 Y_{ij}\cos\theta_{ij} \quad (15)$$

where V_i is the voltage magnitude, δ_i is the voltage angle, and Y_{ij} and θ_{ij} are the magnitude and angle of ijth element of Y_{Bus} matrix. Ignoring higher-order terms and using the Taylor series approximation method, equation (15) can be written as

$$\Delta P_{ij} = \frac{\partial P_{ij}}{\partial \delta_i}\Delta\delta_i + \frac{\partial P_{ij}}{\partial \delta_j}\Delta\delta_j + \frac{\partial P_{ij}}{\partial V_i}\Delta V_i + \frac{\partial P_{ij}}{\partial V_j}\Delta V_j \quad (16)$$

The simplified form of equation (16) can be written as

$$\Delta P_{ij} = \alpha_{ij}\Delta\delta_i + \beta_{ij}\Delta\delta_j + \gamma_{ij}\Delta V_i + \eta_{ij}\Delta V_j \quad (17)$$

where

$$\alpha_{ij} = V_i V_j Y_{ij}\sin(\theta_{ij} + \delta_j - \delta_i) \qquad (18)$$

$$\beta_{ij} = -V_i V_j Y_{ij}\sin(\theta_{ij} + \delta_j - \delta_i) \qquad (19)$$

$$\gamma_{ij} = V_j Y_{ij}\cos(\theta_{ij} + \delta_j - \delta_i) - 2V_i Y_{ij}\cos\theta_{ij} \qquad (20)$$

$$\eta_{ij} = V_i Y_{ij}\cos(\theta_{ij} + \delta_j - \delta_i) \qquad (21)$$

The relationship of Newton-Raphson Jacobian is

$$\begin{bmatrix} \Delta P \\ \Delta Q \end{bmatrix} = [J]\begin{bmatrix} \Delta\delta \\ \Delta V \end{bmatrix} = \begin{bmatrix} J_{11} & J_{12} \\ J_{21} & J_{22} \end{bmatrix}\begin{bmatrix} \Delta\delta \\ \Delta V \end{bmatrix} \quad (22)$$

Equation (22) can be written as follows by neglecting the coupling between ΔP & ΔV and ΔQ & $\Delta\delta$:

$$\Delta P = [J_{11}][\Delta\delta] \qquad (23)$$

$$\Delta Q = [J_{22}][\Delta V] \qquad (24)$$

Equation (23) can be written as

$$\Delta\delta = [J_{11}]^{-1}[\Delta P] = [M][\Delta P] \qquad (25)$$

From equation (25), we can write

$$\Delta\delta_i = \sum_{t=1}^{k} m_{it}\Delta P_t \qquad i = 1, 2, \ldots\ldots, k, \quad i \neq s \qquad (26)$$

where k is the number of buses and s is the slack bus. Neglecting the coupling between ΔP and ΔV, equation (17) can be written as

$$\Delta P_{ij} = \alpha_{ij}\Delta\delta_i + \beta_{ij}\Delta\delta_j \qquad (27)$$

From equation (26) and equation (27), we get

$$\Delta P_{ij} = \alpha_{ij}\sum_{t=1}^{k} m_{it}\Delta P_t + \beta_{ij}\sum_{t=1}^{k} m_{jt}\Delta P_t \qquad (28)$$

$$\Delta P_{ij} = (\alpha_{ij}m_{i1} + \beta_{ij}m_{j1})\Delta P_1 + (\alpha_{ij}m_{i2} + \beta_{ij}m_{j2})\Delta P_2 + \ldots\ldots\ldots\ldots + (\alpha_{ij}m_{ik} + \beta_{ij}m_{jk})\Delta P_k \qquad (29)$$

Equation (27) can be written as follows:

$$\Delta P_{ij} = BSF_1^l\Delta P_1 + BSF_2^l\Delta P_2 + \ldots\ldots + BSF_k^l\Delta P_k \qquad (30)$$

Therefore, the congested line BSF can be calculated by using following mathematical expression:

$$BSF_k^l = \alpha_{ij}m_{ik} + \beta_{ij}m_{jk} \qquad (31)$$

3 RESULT AND DISCUSSION

The proposed concept has been illustrated on a modified 39 bus New England Test System, which has 10 generators and 29 load buses. A single line diagram of the modified 39 bus New England test system is shown in Figure 1.

The test data for the modified 39 bus New England system is given in the work by K.R. Padi-

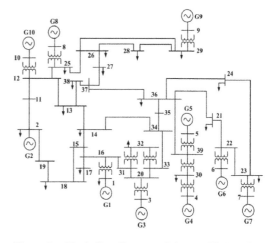

Figure 1. Single line diagram of the modified 39 bus New England test system.

yar (1996). In the proposed method, a violation on lines 15 and 16 has been created due to 14–34 line outages in the system. In the given approach, 50 MW of wind power is connected based on the BSF value. It has been found from a reference paper (Subhojit 2016) that the investment cost of wind power generator (for 1 MW capacity) is

Table 1. Bus Sensitivity Factor (BSF) for congested line (15–16).

Bus no	BSF	Bus no	BSF	Bus no	BSF
1	0	14	–0.2536	27	0.0531
2	–0.0334	15	–0.0440	28	0.0325
3	0.1987	16	–0.0047	29	0.0330
4	0.1526	17	–0.0215	30	0.1539
5	0.1533	18	–0.0293	31	0.1300
6	0.1547	19	–0.0317	32	0.1936
7	0.1543	20	0.1943	33	0.2575
8	–0.0181	21	0.1542	34	0.4196
9	0.0332	22	0.1543	35	0.2326
10	–0.0376	23	0.1543	36	0.1536
11	–0.0354	24	0.1537	37	0.0796
12	–0.0379	25	–0.0193	38	0.0242
13	–0.0655	26	0.0307	39	0.1538

Table 2. Congested line power flow before and after bidding when 14–34 line outage occurs.

Power low	Before SAB (MVA)	After SAB (MVA)	After SAB with WF (MVA)	After DAB (MVA)	After DAB with WF (MVA)
15–16	576.92	496.43	496.20	496.43	496.93

3.75 \$/hr (approx.). So, for 50 MW wind power, a cost of 187.50 \$/hr has been added to the bidding optimal cost of the wind power condition. Two most sensitive buses, i.e, one positive and one negative bus have been selected for analyzing the proposed method. Out of this two, one sensitive bus is selected for the double auction bidding and another one is connected with a Doubly Fed Induction Generator (DFIG) based WF. The congestion problem is solved by using a Sequential Quadratic Programming (SQP) approach, which is also used as an optimization tool for minimizing the fuel cost of the thermal generating unit. Based on the bus sensitivity factor, a DFIG-based WF is connected at bus number 14 and double auction bidding at bus number 34.

In the present study, violation occurs in L15–16 due to a line outage in L14–34. So, the line L15–16 is called a congested line. It is not always correct that only one line will be congested after a single line outage. More than one line may be congested after a single line outage. In that case, the most severe line determined by the line Performance Index (PI) can be chosen for analysis. But, in the present study, only one line (i.e., L 15–16) is congested after L 14–34 outage; other lines are not congested since the power flows of the remaining lines are within their thermal limit. BSF is calculated for the congested line L 14–34. For the congested line 15–16, BSF is shown in Table 1 for the modified 39 bus New England test system.

With the presence of a WF, Double Auction Bidding (DAB) has been done at bus number 34 with a load of 60 MW. Table 2 shows the power flow through the congested line before and after Single Auction Bidding (SAB) and DAB with and without the presence of a WF. From the result, it can be stated that after bidding, the active power flow

Table 3. Optimal cost for SAB and DAB with and without presence of a WF for 14–34 line outage.

Gen output	Result [Lei 2012]	SAB without outage	SAB with outage	SAB with WF and outage	DAB with outage	DAB with WF and outage
PG1	604.47	677.00	593.76	609.80	572.80	590.55
PG2	646.00	689.42	737.20	715.24	733.09	711.74
PG3	715.41	673.21	610.16	619.84	591.88	603.05
PG4	652.00	646.73	652.00	648.04	651.41	641.23
PG5	508.00	508.00	508.00	508.00	508.00	508.00
PG6	687.00	657.45	670.48	658.63	662.30	651.55
PG7	580.00	580.00	580.00	580.00	580.00	580.00
PG8	564.00	564.00	564.00	564.00	564.00	564.00
PG9	667.79	637.50	671.89	655.13	667.08	651.17
PG10	674.44	669.10	721.46	698.14	718.26	695.48
Gen Cost (\$/h)	41941.34	41932.73	42169.21	41571.81	41399.62	40798.26
Loss (MW)	44.88	48.180	54.715	52.586	54.583	52.524

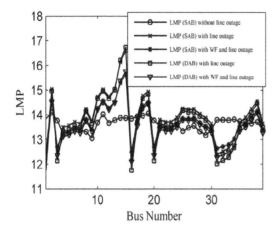

Figure 2. Locational Marginal Price (LMP).

through the congested line (L 15–16) is reduced from the maximum level to the reliable margin.

Table 3 shows the optimal fuel cost of the thermal generating unit with and without considering a WF for SAB and DAB.

Generators submit the price offers to the power market in SAB and the buyers neutralize the seller's market power by the resale value of their demand in DAB. It can be seen from this table that power losses are also less in DAB. In the case of DAB, the total generation cost is very less compared to SAB. The losses of the system after optimization are 52.586 MW for SAB with a WF and 52.524 MW for DAB with a WF. Figure 2 shows the LMP for SAB and DAB with and without the presence of a WF. The LMP is calculated by the bids/offers submitted by buyers and sellers of the market or market participant. It has been seen that LMP decreases the connection of the WF for SAB as well as DAB.

4 CONCLUSION

This paper investigates the effectiveness of the Double Auction Bidding (DAB) strategy with the integration of a wind farm to alleviate transmission congestion. LMP is found to be minimum in the presence of a wind farm in the case of DAB when compared to single auction bidding. Sequential Quadratic Programming (SQP) is used for implementing the proposed method and it is applied on modified 39 bus New England test systems. It is observed that fuel cost and losses are minimum in DAB in the presence of a wind farm compared to that of single auction bidding in the presence of a wind farm for mitigating transmission congestion.

REFERENCES

Agustín A. Sánchez de la Nieta, Javier Contreras, and José Ignacio Muñoz, (2013). Optimal Coordinated Wind-Hydro Bidding Strategies in Day-Ahead Markets. *IEEE Transactions on Power Systems*, vol. 28, pp. 798–809.

Arvind Kumar Jain, Suresh Chandra Srivastava, Niwas Singh, Laxmi Srivastava, (2015). Bacteria Foraging Optmization Based Bidding Strategy under Transmission Congestion. *IEEE System Journal*, vol. 9, no. 1, pp. 141–151.

Edgardo D. Castronuovo, and J. A. Peças Lopes, (2004). On the Optimization of the Daily Operation of a Wind-Hydro Power Plant. *IEEE Transactions on Power Systems*, vol. 19, pp. 1599–1606.

Faria Nassiri Mofakham, Mohammad Ali Namatbakhsh, Ahmad Baraani Dastjerdi, Nasser Ghasem Aghaee, Ryszard Kowalczyk, (2015). Bidding Strategy for Agent in Multi-Attribute Combinational Double Auction. *Expert Systems with Application*, vol. 42, pp. 3268–3295.

Lei Tang and James D. McCalley, (2012). An Efficient Transient Stability Constrained Optimal Power Flow using Trajectory Sensitivity. *North American Power Symposium (NAPS)*, pp. 1–6.

Mahmoud Ghofrani, Amirsaman Arabali, Mehdi Etezadi-Amoli, and Mohammed Sami Fadali, (2013). A Framework for Optimal Placement of Energy Storage Units within a Power System with High Wind Penetration. *IEEE Transactions on Sustainable Energy*, vol. 4, pp. 434–442.

Mahvi M., M.M. Ardehali, (2011). Optimal Bidding Strategy in a Competitive Electricity Market Based on Agent Based Approach and Numerical Sensitivity Analysis. *Energy*, vol. 36, pp. 6367–6374.

Nabavi S.M.H., A. Kazemi and M.A.S. Masoum, (2012). Social Welfare Maximization with Fuzzy Based Genetic Algorithm by TCSC and SSSC in Double-Sided Auction Market. *Scientia Iranica, Computer Science & Engineering and Electrical Engineering*, vol. 19, no. 3, pp. 745–758.

Padiyar K. R., Power System Dynamics, (1996). Stability and Control. *NewYork: Wiley*, p. 601.

Ricardo Fernández Blanco, José M. Arroyo, Natalia Alguacil, (2014). Network-Constrained Day-Ahead Auction for Consumer Payment Minimization. *IEEE Transactions on Power Systems*, vol. 29, no. 2, pp. 526–536.

Rico Herranz, Antonio Munoz San Roque, Jose Villar, Fco Alberto Campos, (2012). Optimal Demand-Side Bidding Strategies in Electricity Spot Market. *IEEE Transaction on Power System*, vol. 27, no. 3, pp. 1204–1213.

Shi L., Y. Luo, G.Y. Tu, (2014). Bidding Strategy of Micro Grid with Consideration of Uncertainty for Participating in Power Market. *Electrical Power and Energy Systems*, vol. 59, pp. 1–13.

Subhasish Deb, Sadhan Gope, Arup Kumar Goswami, (2015). Congestion Management Considering Wind Energy Sources using Evolutionary Algorithm. *Electric Power Components and Systems*, vol. 43, pp. 723–732.

Subhojit Dawn, Prashant Kumar Tiwari, (2016). Improvement of Economic Profit by Optimal

Allocation of TCSC & UPFC with Wind Power Generators in Double Auction Competitive Power Market. *Electrical Power and Energy Systems*, vol. 80, pp. 191–201.

Sudipta Dutta and S.P Singh, (2008). Optimal Rescheduling of Generators for Congestion Management Based on Particle Swarm Optimization. *IEEE transaction on power system,* vol. 23, pp. 1560–1569.

Tengshum Peng, Kevin Tomsovic, (2003). Congestion Influence on Bidding Strategies in an Electricity Market. *IEEE Transaction on Power System*, vol. 18, no. 3, pp. 1054–1061.

Computer, Communication and Electrical Technology – Guha, Chakraborty & Dutta (Eds)
© 2017 Taylor & Francis Group, ISBN 978-1-138-03157-9

Survey on solar photovoltaic system performance using various MPPT techniques to improve efficiency

B. Pakkiraiah & G. Durga Sukumar

Department of Electrical and Electronics Engineering, Vignan's Foundation for Science Technology and Research University, Guntur, Andhra Pradesh, India

ABSTRACT: Nowadays, to meet the increase in power demand and to reduce global warming, renewable energy sources-based systems are used. Out of the various renewable energy sources, solar energy is the main choice. But, compared to other sources, the solar panel system converts only 30–40% of solar irradiation into electrical energy. To get the maximum output from a Photovoltaic (PV) panel system, an extensive research has been underway for a long time to assess the performance of the PV system and to investigate the various issues related to the effective use of solar PV systems. This paper, therefore, presents different types of PV panel systems, maximum power point tracking control algorithms, usage of power electronic converters with control aspects, various controllers, usage of filters to reduce harmonic content, and usage of battery systems for PV systems. Attempts have been made to highlight the current and future issues involved in the development of a PV system with improved performance. This paper has an appended list of 105 research publications on this.

1 INTRODUCTION

As the earth's natural resources are decreasing day by day, to meet the increase in power demand, the power sector is looking at alternate energy resources. With the usage of renewable energy sources, the carbon content in the atmosphere can be reduced, thus taking a step to overcome the problem of global warming. Out of the various renewable sources, the solar photovoltaic (PV) system is most used. The various structures of PV panel systems and their suitability for specific locations has been discussed in past research (Saadi A. et al. 2003, Ait Cheikh S. et al. 2007, Fangrui liu et al. 2008, Roberto Faranda et al. 2008, Adel Mellit et al. 2008, Anna Mathew et al. 2011). The efficiency of the PV system can be increased using a maximum power point controller. The extraction of the maximum available power from a PV module is done using a Maximum Power Point Tracking (MPPT) controller. This might also increase the efficiency of the system. Several algorithms have been developed to track the maximum power point efficiently. Most of the existing MPPT algorithms suffer from the drawback of being slow at tracking. Due to this, the utilization efficiency is reduced. Various methods of maximum power point technique algorithms such as Incremental Conductance (INC), Perturb & Observe (P&O), etc. have been discussed. The output obtained from the MPPT controller contains harmonics due to the closed-loop tracking of sunlight. This

can be eliminated by using filter circuits. From this, the obtained output is given to a DC-DC converter and inverter. Literature thus far has discussed various power electronic converter circuits and their control techniques (Chin S. et al. 2003, Patel S. et al. 2011, Huusari J. et al. 2012, Trejos A. et al. 2012).

In this paper, we have surveyed various aspects of the solar PV fed system drives. Section II presents the various available types of PV panel systems, their performances, and suitability. Section III discusses MPPT and its performance. Section IV presents the review of power electronic circuits and their controlling techniques with MPPT. Section V presents the harmonic reduction of a PV panel system using a filter circuit. Section VI discusses the battery and inverters. Section VII presents the concluding remarks.

2 PHOTOVOLTAIC SYSTEM

A PV system is a solid-state semiconductor device that generates electricity when exposed to the light. The building block of a solar panel is the solar cell. A PV module is formed by connecting multiple solar cells in series and in parallel. PV modules are connected in series to get maximum output voltage and connected in parallel to obtain maximum output current. Solar PV power systems have been commercialized in many countries due to their merits such as long-term benefits and very little

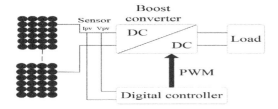

Figure 1. Block diagram of a PV generation system.

maintenance. The major challenge in using PV power generation systems is tackling the nonlinear characteristics of a PV array. The PV characteristics depend on the level of irradiance and temperature. A PV array experiences different irradiance levels due to passing clouds and neighboring buildings or trees. Figure 1 shows the block diagram of a PV generation system.

PV systems are mainly classified into grid-connected and standalone systems, which are designed to provide DC and AC power service to operate independent of the utility grid that is connected with other energy sources and storage systems.

2.1 Grid-connected PV system

Grid-connected PV systems are designed to operate in parallel and connected with the electric utility grid system. They mainly consist of an inverter that converts dc into ac power with voltage and power quality requirements of the utility grid. A bi-directional interface is made between the PV system AC output circuits and electric utility network at the on-site distribution panel or service entrance to allow the AC power to either supply or back feed the grid when the PV system output is greater than the load demand (Kamaruzzaman S. 2009). Figure 2 shows the block diagram of a grid-connected PV system.

A PV system is capable of operating in grid-connected and standalone modes using a multi-level inverter and boost converter for extracting maximum power and feeding it to the utility grid and standalone system (Liu C. et al. 2004).

2.2 Standalone PV systems

These are mainly designed to operate independent of the electric utility grid and designed size to supply certain DC or AC electric loads. The simplest type of standalone PV system is a direct-coupled system. This operates only during sunlight hours and energy will not be stored in the battery system, like in ventilation fans, water pumps, and small pumps for solar thermal water heating systems (Kamaruzzaman S. et al. 2009). Figure 3 shows the block diagram of a direct coupled PV system.

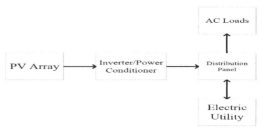

Figure 2. Block diagram of a grid-connected PV system.

Figure 3. Block diagram of direct coupled PV system.

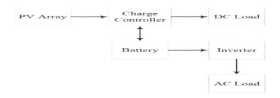

Figure 4. Standalone PV system with a battery storage powering DC and AC loads.

For the direct-coupled PV system given above, several studies have investigated topics such as continuous variations of solar radiation, quality of load matching, geographical location, climatic, degree of utilization, and accurate sizing to various loads (Appelbaum J. 1987, Saied M. M. et al. 1989, Khouzam K. Y. 1990, Khouzam K. Y. et al. 1991, Kou Q. et al. 1998). Lorendana Cristaldi et al. (2012) have observed the behavior of the PV panel from a direct-coupled system with a switched-mode converter and MPP algorithm. They compared the V-I characteristics of the panel with and without the MPP algorithm. The design methodology in this is mainly based on the variation of solar radiation; due to this, the results could not be safely applied over a period of time for a specific input radiation time series. Figure 4 shows another type of standalone PV system powering DC and AC loads with a battery storage system (Hund T. D. & Thomson B. 1997).

3 MAXIMUM POWER POINT TRACKING

The MPPT control technique is used mainly to extract the maximum capable power of the PV modules with the respective solar irradiances and temperatures at a particular instant of time by the MPPT controller. Most of the existing MPPT algorithms suffer from the drawback of being slow at tracking. Due to this, the utilization efficiency is reduced. There are several MPPT control techniques that can be used to improve the efficiency, such as Incremental Conductance (INC), Hill Climbing or Perturbation & Observation (P&O), Artificial Neural Network (ANN) with back propagation technique, Fuzzy Logic Controller Intelligent Control (FLCIC) with DC-DC converter, Particle Swarm Optimization, Open Circuit Voltage Control (OCVC), Short Circuit Current Control (SCCC), Feedback of power variation with voltage technique, Feedback of power variation with current technique, Single input Fuzzy controller for tracking MPP, Ant Colony Optimization (ACO), and Genetic Algorithm (GA) methods. Azli N. A. et al. (2008) proposed that the MPPT performance improves with the temperature of the PV system; 4% of the efficiency is affected by the variations of the fill factor with the climatic conditions and the geographic region. Muhammad Tauseef et al. (2012) have presented the modeling of the MPPT with a buck converter with oscillations of less than 0.5% in the output power. Also, the PV cell with two-diode model for the MPPT controller relies on the fact that the ratio of V_{mpp}/V_{oc} does not strongly depend on environmental conditions.

Xiaojin Wu et al. (2009) and Pakkiraiah B. et al. (2016) proposed the MPPT method performances to its structure and cost in the common insolation condition to reach the efficiency of 98% of the real maximum output power. Michael Ropp E. et al. (2009) have discussed the aspects of PV inverter accuracy consisting of detection behavior and the action of MPP tracker response to grid voltage and frequency fluctuations. Brando G. et al. (2007) have explained the sensor-less control of a H bridge multi-level converter for MPPT in grid-connected PV systems to deliver maximum power for variations of incident irradiations on PV arrays with the H-bridge 5-level converter to reduce the current oscillations to up to 1% though the voltage slope changes. Shakil Ahmad khan et al. (2010) have explained the implementation of the MPP algorithm in an 8-bit microcontroller to generate optimized code in C.

3.1 Incremental Conductance (INC)-based MPPT techniques

It has been proposed to obtain the MPP operating point for adaptive voltage step changes based on the slope of the PV curve. To get the changes in the voltage step value from the PV curve, acceleration and deceleration factors are applied in the next iteration steps. The adaptive voltage step change enables the PV system to quickly track

the environment condition variations. In this way, more solar energy can be harvested from the PV energy systems. It is easy to implement since it does not require the knowledge of I-V characteristics of PV panels and the parameters (Harada K. et al. 1993, Hussein K. H. et al. 1995, Irisawa K. et al. 2000, Kim T. Y. et al. 2001, Kuo Y. C. et al. 2001, Wu W. et al. 2003, Kobayashi K. et al. 2003, Koizumi H. et al. 2005, Meng Yue et al. 2014, Pakkiraiah B. et al. 2015). The MPP is tracked by searching the peak of the P-V curve. This algorithm uses the instantaneous conductance I/V and dI/dV for MPPT. Using these two values, the algorithm determines the location of the operating point of the PV module in the P-V curve and also shows the operation at the MPP along with left and right sides of the MPP in the P-V curve (Fangrui L. et al. 2008, Safari A. et al. 2011, Mei Qiang et al. 2011, Kok Soon Tey et al. 2014). Figure 5 shows the flow chart of the algorithm.

Fangrui L. et al. (2008) presented the variable step size INC algorithm to increase the tracking speed of the MPPT controller. The INC method determines the radiation direction to change the voltage under a rapidly changing condition. In addition, it also calculates the MPP. Thus, the oscillation problem of the P&O algorithm around MPP is eliminated. For a uniform radiation condition, there is no significant difference between the efficiencies of these two methods (Hohm D. P. et al. 2003, Esram T. et al. 2007). The INC method was determined to operate with more efficiency under randomly generated conditions (Roman E. et al. 2006).

However, the cost of the INC method is high due to the requirements of high-sampling compliance and speed control as a result of the complex structure. Classically, the INC method is the most-used technique as a part in the hill-climbing algorithm, but it has the drawback in decision making as the

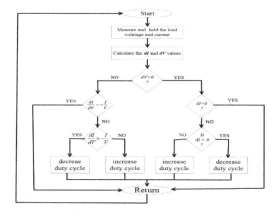

Figure 5. Flowchart of IC algorithms for MPPT.

speed increases in proportion to the step size of the error. However, a higher-error step size reduces the efficiency of the MPPT and direction errors under rapid atmospheric changes (Kazmi S. et al. 2009). Adly M. et al. (2012) have tested the INC method for the PV models under fast-changing environmental conditions by using a DC-DC converter as the isolation stage through the load-connected PV at the required maximum power point voltage. Through this, 15% energy extraction losses can be saved.

Yu G. J. et al. (2002) discussed the dependency of MPP on temperature and solar isolation. In this different MPPT control methods to draw maximum power from solar array such as the constant voltage control, Incremental Conductance conditions two mode MPPT control algorithm. Ahmad Al Nabulsi et al. (2011) have presented hill climbing using INC for tracking the maximum power point to get the maximum output power of a PV panel. Here, the optimum duty cycle results in the maximum power delivered from the PV panel through DC/DC converter to the load. But, the operating point of the PV array oscillates around the MPP causing power loss.

3.2 *Perturb and observation based MPPT algorithm*

The P&O MPPT algorithm is easy to implement. Here, the PV array is perturbed of a radiation of direction. If the power drained from the array increases, the operating point varies toward the MPP, which in turn suites the working voltage in the similar direction. If the power drained from the PV array decreases, the operating point varies away from the MPP; thus, the direction of the working voltage perturbation has to be overturned (Nicola Femia et al. 2005, ajay Patel et al. 2013, Quamuzzam M. et al. 2014, Tuffaha T. H. et al. 2014, Pakkiraiah B. et al. 2016). A disadvantage of the P&O MPPT method is the steady state. The operating point oscillates in the region of the MPP, giving rise to the waste of energy. A number of improvements of the P&O algorithm have been deliberated to decrease the oscillations in steady state, but this slows down the speed of response of the algorithm during atmospheric changes. Song I.K (2011) discussed power oscillations in the P&O algorithm is that the array terminal voltage is perturbed every MPPT cycle. Liu C. et al. (2004) have proposed to solve decoupling the PV power fluctuations caused by the hill climbing P&O process from the variations of the irradiance to solve the oscillation problem. In this, the incremental change in power ΔP is measured. If the ΔP value is positive, the operating voltage is increased to get MPP; if the value of ΔP is negative, the direction

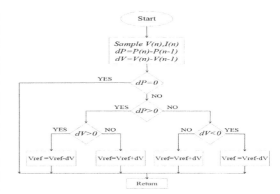

Figure 6. Flow chart of the perturb and observation method.

of the voltage adjustment is reversed and the operating point in trying to make it is closed to the MPP (Yafaoui A. 2007). Figure 6 shows the flow chart algorithm.

However, the applications of analog circuits using the P&O algorithm is presented in Enrique J. M. et al. (2010). Al-Amoudi A. et al. (1998) have proposed a variable step-size that gradually moves toward MPP. But this method is not adaptive because the steps are varied in a predetermined way. A model-based approach is given by Zhang L. et al. (2000) to measure the PV array temperature and irradiance.

Liang-Rui Chen et al. (2008) have proposed the biological swam chasing the PV MPPT algorithm to improve the MPPT performance. On comparing with a typical P&O algorithm, the efficiency of the MPPT is improved to 12.19% in the transient state. Wang Nian Chun et al. (2011) have proposed the PV model-based mathematical models of the PV array on the basis of the P&O method. In this, the characteristics of PV arrays are analyzed. Norhasniza Md Razali et al. (2011) presented the experimental works on a standalone solar system with the P&O MPPT algorithm.

3.3 *Artificial neural network based MPPT techniques*

In this technique, a multi layered feedback neural network with a back propagation trained network is used. A two-stage off-line trained ANN-based MPPT with two cascaded ANNs is used to estimate the temperature and irradiance levels from the PV array voltage and current signals. This technique gives the better performance even under rapidly changing environmental conditions for both steady and transient instants with reducing the training set usually a three layer RBFN NN is adopted for implementing the MPPT. Lucio Ciabattoni et al. (2013) have presented the home energy management system using neural network that monitors the home loads, forecasts PV production, home consumptions, and influences the users on their energy choice. Lian Jiang et al. (2013) have implemented the two-stage MPPT to improve non-uniform irradiance on the PV modules. In this solar array model, a blocking diode is connected in series to the PV string to prevent reverse current flow from the load; a bypass diode is also used to improve the power capture and prevent hotspots.

Hong Hee Lee et al. (2010) have presented the ANN-based MPPT 2-stage method for maximum power point; this algorithm which is independent of time dependency and trade property due to this MPP can be tracked without time increment through PV characteristic changes. Muhammad Sheraz et al. (2012) showed how to overcome the problem of nonlinear characteristics of PV array with rapidly changing irradiation and temperature using Differential Evolution (DE) and ANN along with conventional MPPT to track the maximum power point. Phan Quoc Dzung et al. (2010) have implemented the ANN-based MPPT and INC method for searching maximum power point based using feedback voltage and current without PV array characteristics. This method solves the time dependence and trade off as tracking time is very fast. Experimental valuation is carried out using dSpace 1104.

Zhao Yang et al. (2012) have presented the ANN-based MPPT control to track the MPP at different weather and irradiance. This network is used to overcome slow tracking speed, more output oscillations, and power oscillations. Ramaprabha R. et al. (2009) discussed about the three-layered ANN with back propagation algorithm based MPPT for boost converter for a standalone PV system to minimize the long-term system losses and to increase the conversion efficiency. Even under variable temperature, it gave optimum output voltage. Figure 7 shows a Multi-layered feed forward neural network based MPPT algorithm.

3.4 Fuzzy logic controller based MPPT techniques

A fuzzy logic control is a convenient way to map an input space to the output space. Fuzzy logic uses the fuzzy set theory, in which a variable is a member of one or more sets with a specified degree of membership. A fuzzy logic controller basically includes three blocks named as fuzzification, inference, and de-fuzzification. Hamza Afghoul et al. (2013) implemented the hybrid adaptive neuro-fuzzy inference system to find the operating point near the MPP. Shilpa Sreekumar et al. (2013) implemented fuzzy logic to obtain the MPP operating voltage fluctuations using Mamdani's method. This method tracks the maximum power rapidly compared to the conventional algorithm.

Mellit A. et al. (2007) predicted the daily total solar radiation data from sunshine duration and ambient temperature based on the ANFIS model. Here, the Mean Relative Error (MRE) produces a very accurate estimation between the actual and predicted data not exceeding 1%. Hossain M.I. et al. (2011) proposed the intelligent method for the MPPT of a PV system under variable temperature and insolation conditions. To overcome the nonlinearity characteristics of PV cell, fuzzy logic control is implemented; the duty cycle for the converter is calculated based on the fuzzy logic control algorithm.

Ahmad Al Nabulsi et al. (2011) made a comparison between two controllers as hill Climbing, two-input fuzzy controller, and single-input fuzzy controller for tracking the MPP. The basic idea behind these MPPT techniques is to find the optimum duty cycle that results in maximum power delivered from the PV panel through a DC/DC converter to the load. In this, hill climbing controls the operating point of the PV array oscillates around the MPP causing power loss; fuzzy with 2 inputs requires more calculation with N*N rules in addition to two inputs and output gains, but the single-input fuzzy controller is implemented with the least settling time, acceptable oscillations around MPP.

3.5 Particle swarm optimization based MPPT algorithm for PV system

This algorithm is used to reduce the steady-state oscillation to practically zero once the MPP is located. Furthermore, it has the ability to track the MPP for extreme environmental conditions like large fluctuations of insolation and the partial shading condition. The MPP tracker based on particle swarm optimization for PV module arrays is capable of tracking global MPPs of multi-peak characteristic curves where the fixed values were adopted for weighing within the algorithm, the

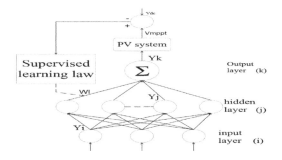

Figure 7. ANN-based MPPT.

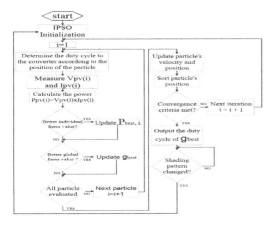

Figure 8. Flowchart of the IPSO-based MPPT algorithm.

tracking performance lacked robustness, causing low success rates when tracking the global MPPs.

Though the MPPs were tracked successfully, the dynamic response speed is low. The IPSO-based MPPT controller algorithm for various environmental conditions like fully shaded conditions and partially shaded conditions to find new global MPP with re-initialization of particles can be observed (Abdulkadir A. et al. 2014). Figure 8 shows the IPSO-based flow chart algorithm.

The PSO is a simple structure that is easily implemented and has fast computation capability. It is able to locate the MPP for any type of P-V curve regardless of the environmental variations; it can also track the PV system as the search space of the PSO is reduced and the time required for the convergence can be greatly reduced. Interestingly, the searching speed through adaptive learning factors and inertia weight were not satisfied by researchers (Zhang L. et al. 2000, Ishaque K. et al. 2012, Liu Y. H. et al. 2012, Yau H. T. et al. 2013, Chao K. H. et al. 2013) because of the linear in line with increasing iteration numbers were adopted for weighing the PSO formulas.

4 MPPT CONTROLLER WITH DC-DC CONVERTERS

Chin S. et al. (2003) presented a DC-DC converter to transfer maximum power from the solar PV module to the load. This was achieved by the module by changing the duty cycle of a converter so that the load impedance was varied and matched at the point of peak power with the source. Spiazzi G. et al. (2009) have discussed about the Ripple Correlation Control (RCC) technique on correlation existing between the PV panel power, voltage

and current ripples and can exploit the AC signals caused by the switching action of the converter. It is mainly used to the equilibrium point in case of non-sinusoidal and to analyze the simple effects of reactive parasitic components like panel shunts capacitance and series inductance. Finally, it is observed that the combined effects of the reactive parasitic components can lead to unacceptable errors in the MPP tracking capability where a high-frequency DC-DC converter is used.

Huusari J. et al. (2012) proposed the distributed MPPT to overcome the effect of shading, which was used to reduce the output power. Each PV module was connected by a DC-DC converter to extract maximum power where its output was connected to the grid-connected inverter. It also discussed the terminal constraints, topological constraints, and dynamic constraints to get the maximum output voltage. Weiping Luo et al. (2009) studied the MPPT with an MCU control system in a grid-connected PV generation system by introducing DC-DC conversion techniques. Patel S. et al. (2011) gave an analysis of the PV module characteristics and determination of short-circuit current to get converging MPPT control. It is observed by connecting the DC-DC converter between the PV source and the load with pulse width modulation control of the converter. In this, the short circuit current, open circuit voltage, and load voltage at MPP was determined. Here, the boost converter was turned to ensure the operation of the PV array at its MPP regardless of the atmospheric conditions and load variations. Adriana Trejos et al. (2012) presented the DC link voltage oscillation compensations in grid-connected PV systems based on three different DC/DC converters such a Cuk converters, SEPIC converters, and Zeta converters. These high-order converters are suitable for reducing the current ripple injected in the PV array and load.

5 REDUCTION OF HARMONICS

Jain S. et al. (2007) proposed that the voltage obtained from a PV system is low and a DC-DC boost converter is necessary to generate a regulated higher DC voltage. A voltage source inverter interfaces the PV module with the standalone application to provide an AC output voltage without transformer. For lower generating systems, the conventional pulse width modulated inverter was used because of its simple and easy control mechanism (Calais M. et al. 2002). PWM inverters usually involve a fast switching of semiconductor switches due to which a high-frequency noise gets generated. Also, all PWM methods inherently generate harmonics that originate from the high dv/dt and di/dt semiconductor switching transients (Bhattacharya I. et al. 2010). These drawbacks resulted in distorted output; traditionally passive filters have been used to remove the harmonics from the line current. These filters improve the power factor of the system (Rivas D. et al. 2003, Yousif S. N. A. L. et al. 2004), but the passive filters have many drawbacks such as tuning problems, series and parallel resonance, etc. (Chandani M. et al. 2012). Typical shunt and series active filter topologies along with passive filters were proposed to provide high-quality power for the PV system (Blofan A. et al. 2010). The passive filter design is carried out with LC, LCL, and LLCC filter topologies. The work evaluates the total harmonic distortion using each passive and a comparison is done with respect to the power quality improvement.

According to Khoucha F. et al. (2013), the main objective was to minimize the inductor size, capacitor current/voltage ripples, and harmonic content through the special structured interleaved boost converter, to increase the converter efficiency. A study of prototype hardware also took place with DSP-based controlled PV emulator, to make the operating point always at the MPP. Nousiaanen L. et al. (2011) presented the dynamic analysis of the converter with the one-phase grid-connected inverter applications by considering the dynamic properties of the PV generator. In this, the converter gives the automatic high-power de-coupling that eliminates the ripple at twice the grid frequency at the output.

Methods that work with high switching frequencies have many commutations for the power semiconductors in one period of the fundamental output voltage. A very popular method in industrial applications is the classic carrier-based Sinusoidal PWM (SPWM) that uses the phase-shifting technique to reduce the harmonics in the load voltage. Another interesting alternative is the SVM strategy, which has been used in three-level inverters. Methods that work with low switching frequencies generally perform one or two commutations of the power semiconductors during one cycle of the output voltages, generating a staircase waveform. Representatives of this family are the multilevel selective harmonic elimination and the Space-Vector Control (SVC) (Tolbert L. et al. 1999, Nehrir M. H. et al. 2000, Kroposki B. et al. 2008).

6 USAGE OF BATTERY AND INVERTER SYSTEMS

An inverter is a power electronic device that converts DC to AC. The inverter is essential for a PV system designed to be as a grid-tied or off grid.

6.1 Grid-tied inverters

In this, the solar panels are usually wired in series to produce very high voltages, usually between 200–600 volts using Cu wire; this is expensive nowadays due to the rise in the price of copper. Grid-tied inverters are designed to take high DC voltages as input that are common in the systems. They are also designed such that if the grid goes down for any reason, the inverter will turn off the supply from the solar panels also this is worker at the utility because side should not get shock.

6.2 Standalone inverters

These are designed for off-grid systems. They are designed to receive power both from the solar panels and the battery backup system. Off-grid inverters usually have a number of features designed to optimize the performance of your battery system. In addition, they are designed to work with lower voltages given that the batteries are usually operating at between 12 and 48 volts.

6.3 Grid-tied inverters with battery backup

These are used in hybrid systems that are tied but also have battery banks as backup if ever the grids go down. They tend to have features that are found in both grid-tied and off-grid inverters (http://www. Energybible.com/solarenergy/inverters.html). The PV system architecture can be categorized into three basic classes: central inverter, string or multistring inverter, and module integrated converter. Although the central inverter can operate at high efficiency with only one DC/AC power conversion stage, this structure has some disadvantages; each module may not operate at its MPP, which results in less energy harvested. Additional losses are introduced by string diodes and junction box and the single point of failure and mismatch of each string of PV panel affects the PV array efficiency greatly (http://www.epia.org, Xue Y. et al. 2004, Kjaer S. B. et al. 2005, Quan L. et al. 2008).

The PV stations make use of smart PV inverters as proposed by Seal (Seal B. 2010). Low-voltage batteries require a voltage boost for the medium voltage grid connection, which is achieved by special power electronic topologies (Thomas S. et al. 2009, Blank T. et al. 2010, Logeswaran T. et al. 2013, Durga Sukumar et al. 2014). The application of grid support with EV batteries requires the combination of both. High-load functions and continuous operation over many years are challenging requirements from a reliability and lifetime perspective. Therefore, the number of allowed cycles over time has a remarkable influence on the cost and lifetime of such a system (Mangor D. et al. 2009).

7 CONCLUSION

The use of solar energy is essential for providing solutions to the environmental problems and also for energy demand. The vast development to improve the efficiency by the MPPT algorithms encouraged the domestic generation of power from solar panels. The available MPPT techniques based on the number of control variables involved, types of control strategies, circuitry, and applications are possibly useful for selecting an MPPT technique for a particular application for a grid-tied or standalone mode of operations. This review has included many recent hybrid MPPT techniques along with their benefits for mismatched conditions such as partial shading, non-uniformity of PV panel temperatures, and dust effects.

We have also observed that perturb & observe and incremental conductance methods are simple and used by many researches, but they have slow tracking and low-utilization efficiency. To overcome the drawbacks, fuzzy and neural network techniques are used currently, which increase the efficiency. To boost the voltage, various DC-DC converters are used along with battery storage systems in order to store the excessive energy from the solar PV panel. The harmonic content is reduced from the output of DC-DC converters using filter circuits. Passive filters such as LC, LCL, and LLCC are used for harmonic distortion as well as to improve the power quality. Filter capacitors are used to reduce a high-frequency current ripple.

This DC is again fed to the inverter for converting the DC to AC with various PWM techniques. These PWM inverter techniques yield better AC outputs, which are used to connect the grid inter connections and standalone AC loads. Multilevel inverters with sinusoidal PWM and SVM are used to reduce the harmonics in the load voltage even in a low switching frequency. Grid-tied inverters with battery backup are preferred in hybrid systems for backup even if the grid goes down for both grid-tied and off-grid systems.

ACKNOWLEDGMENT

The funding support was given by SERB, Department of Science and Technology (DST), Government of India with vides SERB order No: SERB/ET-069/2013 for the solar-based project.

REFERENCES

Abdulkadir, M., A.H.H. Yatin, & S.T. Yusuf (2014). An Improved PSO-based MPPT Control Strategy for photovoltaic Systems. *International Journal of Photoenergy. 2014(2014)*, 1–11.

Adly, M., M. Ibrahim, & H. El Sherif (2012, March). *Comparative Study of Improved Energy Generation Maximization Techniques for Photovoltaic Systems.* in Proceedings of the IEEE Asia-Pacific Power and Energy Engineering Conference (APPEEC '12), Shanghai, China, 1–5.

Afghoul, H., F. Krim, & D. Chikouche (2013, March). *Increase the photovoltaic conversion efficiency using Neuro-Fuzzy control applied to MPPT.* IEEE International Renewable and Sustainable Energy Conference (IRSEC'13), Ouarzazate, Morocco, 348–353.

Ait Cheikh, S., C. Labres, G.F. Tchoketch Kebir, & A. Zerguerras (2007). Maximum Power Point Tracking Using a Fuzzy Logic Control Scheme. *Revue. Des. Energies. Renouvelables. 10*, 387–395.

Al Nabulsi, A., R. Dhaouadi and H. Rehman (2011, April). *Single Input Fuzzy Controller (SFLC) based maximum power point tracking.* 4th International Conference on IEEE Modeling, Simulation and Applied Optimization (ICMSAO'11), Kuala Lumpur, Malaysia, 1–5.

Al-Amoudi, A., & L. Zhang (1998, September). *Optimal control of a grid-connected PV system for maximum power point tracking and unity power factor.* in Proceedings of the IEEE 7th International Conference on power Electronics Variable Speed Drives, *456*, London, UK, 80–85.

Appelbaum, J. (1987). The Quality of Load Matching in a Direct Coupling Photovoltaic System. *IEEE Transactions on Energy Conversion. 2(4)*, 534–541.

Azli, N. A., Z. Salam, A. Jusoh, M. Facta, B.C. Lim, & S. Hossain (2008, December). *Effect of fill factor on the MPPT performance of a grid-connected inverter under Malaysian conditions.* in proceedings of the 2nd IEEE International Power and Energy conference (PECon'08), Johor Bahru, Malaysia, 460–462.

Bhattacharya, I., Yuhang Deng, & S.Y. Foo. (2010, August). *Active filters for harmonics elimination in solar photovoltaic grid connected and stand alone systems.* IEEE In proceeding of the 2nd Asia symposium on quality electronic design (ASQED '10), Penang, Malaysia, 280–284.

Blank, T., S. Thomas, C. Roggendorf, T. Pollok, I. Trintis, & D. U. Sauer (2010, June). *Design and Construction of a Test Bench to Characterize Efficiency and Reliability of High Voltage Battery Energy Storage Systems.* IEEE International Telecommunications Energy Conference (INTELEC'10), Orlando, Fla, USA, 1–7.

Blorfan, A., D. Flieller, P. wira, G. Sturtzer, & J. Merckle (2010). A new approach for modeling the photovoltaic cell using orcad comparing with the model done in Matlab. *International Review on modeling and simulations. 3(5)*, 948–954.

Brando, G., A. Dannier, & R. Rizzo (2007, May). *A Sensor Less Control of H-bridge Multilevel Converter for Maximum Power Point Tracking in Grid Connected Photovoltaic Systems.* IEEE Proceedings of international conference on Clean Electrical Power (ICCEP'07), Capri, Italy, 789–794.

Calais, J.M.A. Myrzik, T. Spooner, & V.G. Agelidis (2002, June). *Inverters for single-phase grid connected photovoltaic systems.* IEEE 33rd Annual Power Electronics Specialists Conference (PESC'02), *4*, Cairns, Australia, 1995–2000.

Chao, K. H., L.Y. Chang, & H.C. Liu (2013). Maximum power point tracking method based on modified particle swarm optimization for photovoltaic systems. *International journal of Photoenergy. 2013(2013)*, 1–6.

Chen, B.-C., & C.-L. Lin (2011, June). *Implementation of maximum-power-point-tracker for photovoltaic arrays.* in Proceedings of the 6th IEEE Conference on Industrial Electronics and Applications (ICIEA'11), Beijing, China, 1621–1626.

Chen, L. R., C.-H. Tsai, Y.-L. Lin, Y.-S. Lai (2010). A Biological Swarm Chasing Algorithm for Tracking the PV Maximum Power Point Energy conversion. *IEEE Transactions on Energy Conversion. 25(2)*, 484–493.

Chin, S., J. Gadson, & K. Nordstrom (2003). Maximum power point tracker. *Tufts University Department of Electrical Engineering and Computer Science.* 1–6.

Chovitia, C. M., Narayan P, Gupta, Preeti, & N Gupta (2012). Harmonic Mitigation using shunt active filter at utility end in grid connected to renewable source of energy. *International Journal of Emerging technology and Advanced Engineering. 2(8)*, 230–235.

Ciabattoni, M. Grisostomi, G. Ippoliti, & S. Longhi (2013, June). *Neural Networks based Home Energy Management System in Residential PV Scenario.* 39th IEEE Photovoltaic Specialists Conference (PVSC'13), Tampa, FL, 1721–1726.

Cristaldi, L., M. Faifer, M. Rossi, & S. Toscani (2012, May). *MPPT Definition and validation a new model based approach.* IEEE Instrumentation and Measurement Technology Conference, Graz, Austria, 594–599.

Enrique, J. M., J.M. Andujar, & M.A. Bohorquez (2010). A reliable, fast and low maximum power point tracker for photovoltaic applications. *Solar Energy. 84(1)*, 79–89.

Esram, T., & P.L. Chapman (2007). Comparison of photovoltaic array maximum power point tracking techniques. *IEEE transactions on Energy Conversion. 22(2)*, 439–449.

European Photovoltaic Industry Association Global market outlook for photovoltaics until 2016 (2016). *http://www.epia.org.*

Faranda, R., & Sonia Leva (2008). Energy Comparison of MPPT techniques for PV systems. *WSEAS Transactions on Power Systems. 3(6)*, 446–455.

Femia, N., G. Petrone, G. Spagnuolo, & M. Vitelli (2005). Optimization of Perturb and Observe Maximum Power Point Tracking Method. *IEEE Transactions on Power Electronics. 20(4)*, 963–973.

Go, S., S. Ahn, J. Choi, W. Jung, S. Yun, and I. Song (2011). Simulation and analysis of existing MPPT control methods in a PV generation system. *Journal of International Council on Electrical Engineering, 1(4)*, 446–451.

Harada, K., & G. Zhao (1993). Controlled power interface between solar cells and AC source. *IEEE Transactions on Power Electronics. 8(4)*, 654–662.

Hohm, D. P., & M.E. Ropp (2003). Comparative study of maximum power point tracking algorithms. *Progress in Photovoltaics: Research and Applications. 11(1)*, 47–62.

Hossain, M. I., S.A. Khan, M. Shafiullah, & M.J. Hossain. (2011, March). *Design and Implementation of MPPT Controlled Grid Connected Photovoltaic System.* IEEE Symposium on Computers and informatics (ISCI'11), Kuala Lumpur, Malaysia, 284–289.

http://www.Energybible.com/solarenergy/Inverters.html.

Hund, T. D., & B. Thomson (1997, September). *Amp-Hour Counting Charge control for Photovoltaic Power Systems.* 26th IEEE Photovoltaic Specialists Conference, Calif, USA, 1281–1284.

Hussein, K. H., I. Mutta, T. hoshino, & M. Osakada (1995). Maximum photovoltaic power tracking an algorithm for rapidly changing atmospheric conditions. *IEE Proceedings: Generation, Transmission and Distribution. 142(1)*, 59–64.

Huusari, J., & T. Suntio (2012, September). *Interfacing Constraints of Distributed Maximum Power Point Tracking Converters in Photovoltaic Applications.* IEEE Power Electronics and Motion Control Conference (EPE/PEMC'12), Novi Sad, Serbia, DS3d.1-1-DS3d.1-7.

Irisawa, K., T. Saito, I. Takano, & Y. Sawada (2000, September). *Maximum power point tracking control of photovoltaic generation system under non-uniform isolation by means of monitoring cells.* 28th IEEE Conference on Photovoltaic Specialists, Anchorage, AK, 1707–1710.

Ishaque, K., Z. Salam, M. Amjad, & S. Mekhilef (2012). An Improved particle swarm optimization-based MPPT for PV with reduced steady-state oscillation. *IEEE Transactions on Power Electronics. 27(8)*, 3627–3638.

Jain, S., & V. Agarwal (2007). A single-stage grid connected inverter topology for solar PV systems with maximum power point tracking. *IEEE Transactions on power electronics. 22(5)*, 1928–1940.

Jiang, L. L., D.R. Nayanasri, D.L. Maskell, & D.M. Vilathgamuwa (2013, November). *A Simple and Efficient Hybrid Maximum Power Point tracking Method for PV Systems Under Partially Shaded Conditions.* 39th annual conference on IEEE Industrial Electronics Society (IECON'13), Vienna, Austria, 1513–1518.

Kamaruzzaman, S. (2009). *Optimization of a Stand—Alone wind/PV Hybrid System to Provide Electricity for a household in Malaysia.* In Proceedings of 4th IASME/WSEAS International Conference on Energy and Environment (EE'09), Bangi, Selangor, Malaysia, 435–438.

Kazmi, S. M. R., H. Goto O, Ichiokura, & H.J. Guo (2009, September). *An Improved and very efficient MPPT controller for PV systems subjected to rapidly varying atmospheric conditions and partial shading.* In Proceedings of the Australasian Universities Power Engineering Conference (AUPEC'09), Adelaide, Australia, 1–6.

Khan, S. A., & M.I. Hossain (2010, December). *Design and Implementation of Microcontroller Based Fuzzy Logic Control for Maximum Power Point Tracking.* IEEE International Conference on Electrical and Computer Engineering (ICECE'10), Dhaka, Bangladesh, 322–325.

Khoucha, F., A. Benrabah, O. Herizi, A. Kheloui, & M.E.H. Benbouuzid (2013, May). *An improved MPPT interleaved boost converter for fast solar electric vehicle applications.* IEEE power engineering, energy and electrical drives (POWERENG'13), Istanbul, Turkey, 1076–1081.

Khouzam, K. Y. (1990). Optimum Load Matching in Direct Coupled Photovoltaic Power Systems Applications to Resistive Loads. *IEEE Transactions on Energy Conversion. 5(2)*, 265–271.

Khouzam, K. Y., & P. Groumpos (1991). Optimum Matching of Ohmic Loads to the Photovoltaic Array. *Solar Energy. 46(2)*, 101–108.

Kim, T. Y., A. Ho-Gyun, P. Seung Kyu, & L. Youn-Kyun (2001, June). *A Novel maximum power point tracking control for photovoltaic power system under rapidly changing solar radiation.* In proceedings of the IEEE International Symposium on Industrial electronics (ISIE'01), Pusan, South Korea, 1011–1014.

Kjaer, S. B., J. K. Pedersen, & F. Blaabjerg (2005). A review of single phase grid connected inverters for photovoltaic modules. *IEEE Transactions on Industrial Applications. 41(5)*, 1292–1306.

Kobayashi, K., I. Takano, & Y. Sawada (2003, July). *A study on a two stage maximum power point tracking control of a photovoltaic system under partially shaded insolation conditions.* IEEE power Engineering Society General Meeting, *4*, Toronto, Canada, 2612–2617.

Koizumi, H., & K. Kurokawa (2005, June). *A Novel Maximum Power Point Tracking Method for PV Module Integrated Converter.* IEEE 36th Power Electronics Specialists Conference (PESE'05), Recife, Brazil, 2081–2086.

Kou, Q., S.A. Klein, & W.A. Beckman (1998). A Method for Estimating the Long Term Performance of Direct Coupled PV Pumping Systems. *Solar Energy. 64(1)*, 33–40.

Kroposki, B., R. Lasseter, T. Ise, S. Morozumi, S. Papatlianassiou, & N.B. Hatziargyriou (2008). Making micro grids work. *IEEE Power Energy Magazine. 6(3)*, 40–53.

Kuo, Y. -c., T.-J. Liang, and J.-F. Chen (2001). Novel maximum-power-point-tracking controller for photovoltaic energy conversion system. *IEEE Transactions on Industrial Electronics, 48(3)*, 594–601.

Lee, H. H., Le Minh Phuong, Phan Quoc Dzung, Nguyen Truong Dan Vu, & Le Dinh Khoa (2010, November). *The New Maximum Power point Tracking Algorithm using ANN-based Solar PV Systems.* IEEE region 10 conference (TENCON'10), Fukuoka, Japan, 2179–2184.

Liu, C., B. Wu, & R. Cheung (2004, August). *Advanced Algorithm for MPPT Control of Photovoltaic Systems.* In Proceeings of the Canadian Solar Buildings Conference, Montreal, Canada, 1–7.

Liu, F., S. Duan, F. Liu, B. Liu, & Y. Kang (2008). A variable step size INC MPPT method for PV systems. *IEEE Transactions on Industrial Electronics, 55(7)*, 2622–2628.

Liu, F., Yong Kang, Yu Zhang, & Shaxu Duan (2008, June). *Comparison of P&O and Hill Climbing MPPT Methods for Grid Connected PV Converter.* In Proceedings of the 3rd IEEE Conference on Industrial Electronics and Applications (ICIEA'08), Singapore, 804–807.

Liu, Y.H., S.C. Haung, J.W. Huang, & W.C. Liang (2012). A particle swarm optimization based maximum power point tracking algorithm for PV systems operating under partially shaded condition. *IEEE Transactions on Energy Conversion. 27(4)*, 1027–1035.

Logeswaran, T., & A. Senthilkumar (2014, December). A Review of Maximum Power Point Tracking Algorithms for Photovoltaic Systems under Uniform and Non-Uniform irradiances. *4th International Conference on Advances in Energy Research, 54*, 228–235.

Luo, W., & G. Han (2009, November). *Tracking and Controlling of Maximum Power Point Application in Grid Connected Photovoltaic Generation System.* IEEE 2nd International Symposium on Knowledge Acquisition and Modeling (KAM'09), Wuhan, China, *3*, 237–240.

Magnor, D., J.B. Gerschler, E.M. Merk, & D.U. Sauer (2009, September). *Concept of a Battery Aging Model for Lithium-ion Batteries Considering the Lifetime Dependency on the Operation Strategy.* In Proceedings of 24th European Photovoltaic Solar Energy Conference and Exhibition, Hamburg, Germany, 3128–3134.

Mathew, A., & A. Immanuel Selvakumar (2011, March). *MPPT based Standalone Heater Pumping System.* IEEE International Conference on Computer, Communication and Electrical Technology (ICCCET'11), Tamilnadu, India, 455–460.

Mei, Q., M. Shan, L. Liu, & J. M. Guerrero (2011). A novel improved variable step-size incremental-resistance MPPT method for PV systems. *IEEE Transactions on Industrial Electronics, 58(6)*, 2427–2434.

Mellit, A., & Soteris A. Kalogirou (2008). Artificial Intelligence Techniques for Photovoltaic Applications. A Review *Progress in Energy and Combustion Science. 34(5)*, 574–632.

Mellit, A., A. Hadjarab, N. jhorissi, & H. Salhi (2007, June). *An ANFIS based forecasting for solar radiation data from sunshine duration and ambient temperature.* *IEEE* Proceedings of the Power Engineering Society General Meeting, Tampa, Fla, USA, 1–6.

Nabulsi, A. A., R. Dhaouadi, & H. Rehman (2011, April). *Single Input Fuzzy Controller (SFLC) based maximum power point tracking.* IEEE 4th International Conference on Modeling, Simulation and Applied Optimization (ICMSAO'11), Kuala Lumpur, Malaysia, 1–5.

Nehrir, M.H., B.J. Lameres, G. Venkataramanan, V. Gerez, & L.A. Alvarado (2000). An approach to evaluate the general performance of standalone wind photovoltaic generating systems. *IEEE Transactions on Energy Conversion. 15(4)*, 433–439.

Nousiaainen, L., & T. Suntio (2011, September). *Dynamic Characteristics of Current Fed Semi Quadratic Buck-Boost Converter in Photovoltaic Applications.* IEEE 3rd Annual Energy Conversion Congress and Exposition (ECCE'11), Phoenix, Ariz, USA, 1031–1038.

Onat, N. (2010). Recent Development in Maximum Power Point Tracking Technologies for Photovoltaic System. *International Journal of Photoenergy. 2010(2010)*, 1–11.

Pakkiraiah, B., & G. Durga Sukumar (2015, November). *A New Modified MPPT Controller for Solar Photovoltaic System.* 2015 IEEE International Conference on Research in Computational Intelligence and Communication Networks (ICRCICN'15), Kolkata, India, 294–299.

Pakkiraiah, B., & G. Durga Sukumar (2016). A New Modified MPPT Controller for Improved Performance of an Asynchronous Motor Drive under Variable Irradiance and Variable Temperature. *International Journal of Computers and Applications-Taylor & Francis.* 1–14.

Pakkiraiah, B., & G. Durga Sukumar (2016). Research Survey on Various MPPT Performance Issues to

Improve the Solar PV System Efficiency. Journal of Solar Energy. *2016(2016)*, 1–20.

Patel, A., V. Kumar, & Y. Kumar (2013). Perturb and observe maximum power point tracking for Photovoltaic cell. *Innovative Systems Design and Engineering, 4(6)*, 9–15.

Patel, S., & W. Shireen (2011, July). *Fast converging digital MPPT control for Photovoltaic applications.* IEEE power and Energy Society General Meeting, San Diego, Calif, USA, 1–6.

Phan, Q. D., Le Dinh Khoa, Hong Hee Lee, Le Minh Phuong, & Nguyen Truong Dan Vu. (2010, October). *The New MPPT Algorithm Using ANN-Based PV.* IEEE International forum on Strategic Technology (IFOST'10), Ulsan, Republic of Korea, 402–407.

Quamuzzam, M., & K.M. Rahaman (2014). A Modified Perturb and Observe Maximum Power Point Tracking Technique for Single Stage Grid Connected Photovoltaic Inverter. *WSEAS transaction on power systems. 9*, 111–118.

Quan, L., & P. Wolfs (2008). A review of the single phase photovoltaic module integrated converter topologies with three different dc link configurations. *IEEE Transactions on Power Electronics. 23(3)*, 1320–1333.

Ramaprabha, R., B.L. Mathur, & M. Sharanya (2009, June). *Solar Array Modeling and Simulation of MPPT using Neural Network.* IEEE International Conference on Control Automation, Communication and Energy Conservation (INCACEC'09), Perundurai, Tamilnadu, India, 1–5.

Razali, N. M., & N.A. Rahim (2011, June). *DSP-Based Maximum Peak Power Tracker using P&O Algorithm.* IEEE First Conference on Clean Energy and Technology (CET'11), Kuala Lumpur, Malaysia, 34–39.

Rivas, D., L. Moran, L.W. Dixon, & J.R. Espinoza (2003). Improving passive filter compensation performance with active techniques. *IEEE Transactions on Industrial Electronics. 50(1)*, 161–170.

Roman, E., R. Alonson, P. Ibanez, S. Elorduizapatarietxe, & D. Goitia (2006). Intelligent PV module for grid connected PV systems. *IEEE Transactions Industrial Electronics. 53(4)*, 1066–1073.

Ropp, M. E., & S. Gonzalez (2009). Development of a MATLAB/Simulink Model of a Single-Phase Grid-Connected Photovoltaic System. *IEEE Transactions on Energy Conversion. 24(1)*, 195–202.

Saadi, A., & A. Moussi (2003). Neural Network use in the MPPT of photovoltaic pumping system. *Rev. Energy. Ren.* 39–45.

Safari, A., & S. Mekhilef (2011). Simulation and hardware implementation of incremental conductance MPPT with direct control method using Cuk converter. *IEEE Transactions on Industrial Electronics. 58(4)*, 1154–1161.

Saied, M. M., & M.G. Jaboori (1989). Optimal Solar Array Configuration and DC Motor Field Parameters for Maximum Annual Output Mechanical Energy. *IEEE Transactions on Energy Conversion. 4(3)*, 459–465.

Seal, B., (2010, October). *Specification for Smart inverter Interactions with the Electric Grid using International Electrochemical Commission 61850.* Electric Power Research Institute, Knoxville, Tenn, USA, 1–86.

Sheraz, M., & M.A. Abido (2012, December). *An Efficient MPPT Controller using Differential Evolution and Neural Network.* IEEE International conference on Power and Energy (PECon'12), Kota Kinabalu, Malaysia, 378–383.

Spiazzi, G., S. Buso, & P. Mattavelli (2009, September–October). *Analysis of MPPT Algorithms for Photovoltaic Panel Based on Ripple Correlation Techniques in Presence of Parasitic Components.* IEEE 2009 Brazilian Power Electronics Conference, Bonito-Mato Grosso do Sul, 88–95.

Sreekumar, S., & A. Benny (2013, July). *Maximum Power Point Tracking of Photovoltaic System using Fuzzy Logic Controller based Boost Converter.* IEEE International conference on Current Trends In Engineering and Technology (ICCTET), Coimbatore, Tamilnadu, India, 275–280.

Sukumar, D., J. Jitendranath, & S. Saranu (2014). Three-level Inverter-fed Induction Motor Drive Performance Improvement with Neuro-fuzzy Space Vector Modulation. *Electrical Power Components and Systems-Taylor & Francis. 42(15)*, 1633–1646.

Tauseef, M., & E. Nowiki (2012, May). *A Simple and Cost Effective Maximum Power Point Tracker for PV Array Employing as Novel Constant Voltage Technique.* In Proceedings of the 25th IEEE Canadian conference on Electrical and Computer Engineering, Montreal, Canada, 1–4.

Tey, K. S., & S. Mekhilef (2014). Modified Incremental Conductance MPPT Algorithms to Mitigate Inaccurate Responses under Fast-changing Solar Irradiation Level. *Power Electronics and Renewable Energy Laboratory PEARL, Science Direct Solar Energy (101).* 333–342.

Thomas, S., T. Blank, D. U. Sauer, & R. W. D. Doncker (2009, October). *DC/DC Converter with High Transformation Ratio for Characterization of High-Voltage Batteries Up to 6 KV for Modular Energy Storage Systems in Medium-Voltage Grids.* International ETG-Kongress, *27(28)*, Germany, 1–6.

Tolbert, L. M., F.Z. Peng, & T. Habetler (1999). Multilevel Converters for Large Electric Drives. *IEEE Transactions on Industrial Applications. 35(1)*, 36–44.

Trejos, A., C.A. Ramos-Paja, & S. Serna (2012, October). *Compensation of DC link voltage oscillations in grid connected PV systems based on high order DC/DC converters.* IEEE International Symposium on Alternative Energies and Energy Quality (SIFAE'12), Barranquilla, Colombia, 1–6.

Tuffaha, T. H., M. Babar, Y. Khan, & N.H. Malik (2014). Comparative study of different hill climbing MPPT through simulation and experimental test bed. *Research Journal of Applied sciences, engineering and technology. 7(20)*, 4258–4263.

Wang N. C., WU MeiYue, & Shi GuoSheng (2011, March). *Study on characteristics of photovoltaic cells based on Matlab Simulation.* IEEE Asia-Pacific Power and Energy Engineering Conference, Wuhan, 1–4.

Wu, W., N. Pongratananukul, Q. Weihong, K. Rustom, T. Kasparis, & I. Batarseh (2003, February). *DSP-based multiple peak power tracking for expandable power system.* 18th annual IEEE Applied Power Electronics Conference and Exposition (APEC'03), Miami Beach, Fla, USA, 525–530.

Xiaojin Wu, Zhiqiang Cheng, & Xueye Wei (2009, November). *A study of maximum power point tracking in novel small-scale photovoltaic LED lighting systems.*

IEEE Artificial Intelligence and Computational Intelligence (AICI'09), Shanghai, China, 40–43.

Xue, Y., L. Chang, S. B Kjaer, J. Bordonau, & T. Shimizu (2004). Topologies of single phase inverters for small distributed power generators: an overview. *IEEE Transactions on Power Electronics. 19(5),* 1305–1314.

Yafaoui, A., B. Wu, & R. Cheung (2007, June). *Implementation of maximum power point tracking algorithm for residential photovoltaic systems.* In proceedings of the 2nd Canadian Solar Building Conference, Calgary, Canada, 1–6.

Yau, H. T., C.J. Lin, & Q.C. Liang (2013). PSO based Pi controller design for a solar charger system. *The Scientific world Journal. 2013(2013), 1–13.*

Yong, Z., L. Hong, L. Liqun, & G. XiaoFeng (2012, August). *The MPPT Control Method BY using BP Neural Networks on PV Generating Systems.* IEEE International conference on Industrial Control and Electronics Engineering (ICICEE'12), Xi'an, China, 1639–1642.

Yousif, S. N. A. L., M.Z.C Wanik, & A. Mohamed. (2004, November). *Implementation of different passive filter designs for harmonic mitigation.* In proceedings of the IEEE National Power and Energy Conference (PEcon'04), Kuala Lumpur, Malaysia, 229–234.

Yu, G. J., Y.S. Jung, J.Y. Choi, J.H. Song, & G.S. Kim (2002, May). *A Novel Two-Mode MPPT Control Algorithm based on Comparative Study of Existing Algorithms.* 29th IEEE Photovoltaic Specialists Conference, 1531–1534.

Yue, M., & X. wang (2014, May). A Revised Incremental Conductance MPPT Algorithm for Solar PV Generation Systems. *Computational Engineering, Finance and Science,* http://arxiv.org/abs/1405.4890.

Zhang, A. Al-amoudi, & Y. Bai (2000, September). *Real time maximum power point tracking for grid connected photovoltaics systems.* In proceedings of IEEE 8th International Conference Power Electronics Variable Drives, London, UK, 124–129.

Zhou, Y., F. Liu, J. Yin, & S. Duan (2008, June). *Study on realizing MPPT by Improved Incremental Conductance Method with variable step size.* 3rd IEEE Conference on Industrial Electronics and Applications, Singapore, 547–550.

Computer, Communication and Electrical Technology – Guha, Chakraborty & Dutta (Eds)
© 2017 Taylor & Francis Group, ISBN 978-1-138-03157-9

An improved strategy of energy conversion and management using PSO algorithm

MD.T. Hoque, A.K. Sinha & T. Halder
Department of Electrical Engineering, Kalyani Government Engineering College, Kaylani, West Bengal, India

ABSTRACT: This paper shows that the improved energy conversion strategies and management using Particle Swarm Optimization (PSO) are highly adoptable as a cost-effective solution for the power transportation and energy storage systems. The wider application of maximum demand control of the distribution and the Electrical Vehicle (EV) charging stations are important to establish the green technology and achieve satisfactory performance improvements based on the present status of the feeders, huge invitations of renewable energy, smart grid technology, and eco-friendly impacts of the power conversion as optimal solutions of the power market.

1 INTRODUCTION

The demand side management plays a significant role in the energy management, optimal load allocation, and loss optimization of power systems using PSO technique. This method is not only productive to reduce power line congestion, load forecasting, and abnormal loading of the power distributions, but it also reduces the peak demand of the system in terms of the high-quality uninterrupted power, which highly attracts revenue collection to make profitable enterprise (Andries P. Engelbrecht, et al., 2007).

The innovative approach of this paper is to charge the Electric Vehicles (PEV) for supporting the operational norms of the power grid in terms of good voltage profile, power loss allocation of the distribution feeders, expansion of the transmission lines and maximum bi-directional power flow, and suitable utilization to PEVs (Halder, et al., 2012).

A methodology and objective function is proposed by the unique, real-time pricing of electricity in the PEVs to preserve operational characteristics at optimal cost of power generation and load forecasting in association with load. Furthermore, the improved strategy is highlighted to use the smart configurations, special approach power management, and automotive infrastructures of the PEV system as well as power grids. The storage batteries of the PEV system are mostly used to change its battery during the off peak hour of the load, when the grid power is more than the demand of electricity. On the contrary, the storage battery will hurl the excess storage power to nearby power grids to balance generation and load demand

during the peak hour when the electricity demand is at the open market to crack the monopole business (Halder, et al., 2012).

The restructure of the feeders with the storage batteries of PEVs facilitates load balance of the grid at which the battery of the car is charged by charging station and discharged in a specific time of a day to improve the demand response. These parameters are determined by the PSO optimization technique for optimal power loss indoctrinate to reduce the cost of electricity as proper energy management and green power cultivation to meet the energy crisis. The techniques of power supply to consumers are encouraged by a cost-effective solution of the optimization for the power utilities. The overloading of transformers, feeders, and devices is tied up to capitalize the features of the smart grid with some innovative PEV technologies (Zachmann, et al., 2008).

The competence of intelligent condition monitoring and communication gives rise to a potential resource of a smart grid, which provides utilities robust control over all aspects of operations of the power system (i.e. generation to allocation) for smart metering and billing as better revenue collection techniques.

As a result, the power utilities administer carefully how EV charging takes place, by observing its smart metering data and specific prices by the Demand Response (DR) programs based on online information of the consumers, power utilities of the EV charging status and billing impacts, and data compilation of the greenhouse gases or carbon credits as government financial support Plug in Hybrid Electric Vehicles (PHEV) (Acharya, Ghosh & Halder, et al., 2012).

2 PLUG IN ELECTRIC VEHICLE

Gross domestic product increases with the development of a country. The per capita electricity consumption increases due to quick enlargement and living standards of people. The expansion of the power grid and distribution sites needs to restructure for the loss reduction and risk management of the power system. However, heating and transportation initiate a significant height for the flexible and usual consumption or load pattern throughout a day. The power transmission and distribution sector are being renovated under the Accelerated Power Developments Reform Program (APDRP). At present, electricity is an essential commodity for the society. The revolution of the society started with uninterrupted power. The PEV has solved that prospect and practices in the open power market.

The environmental issues are now a bargaining question for the irregular effects of Greenhouse Gases (GHG) and temperature rise of the earth randomly. India is the largest crude oil consumer in the world. This matter has been re-initiated by energy conservation and huge introduction of PEV in energy sector as attractive motion in the past few years.

The study of the present status of the substantial use of the PEVs is dominating to save electricity and industry in the near and far future to establish fresh environment inviting the green power technology. The alternative approach is to use biodiesel instead of the conventional petroleum product-based fuel for transportation in every division of the society or complex.

A large number of residential service connections are changing to populate residential feeders especially in the United States and Australia to increase awareness of the common people. A major impact in the society is due to the high penetration levels of the PEVs allied with power distribution zone. It is most likely that the PEVs will charge and discharge during the overall peak load period for the stringent of twig congestions, crooked peak power demands, haphazard voltage fluctuations, abnormal loss in the system, and poor power quality due to improper coordination and maintenance works.

Some singular studies are conducted to observe the existing distribution system and infrastructure to become a robust support of a low PEV penetration level for the island mode of grid operations using some special power converter-based systems and huge investments for uninterrupted, high-quality power.

Several technologies of PEV introduction are used effectively to overcome problems, and special approaches have been incorporated to charging and discharging processes of PEVs, which are controlled by the bi-directional energy flow (i.e. Grid to Vehicle (G2V) or Vehicle to Grid (V2G)).

Substantial PEV management techniques are optimistic based on the deterministic, stochastic, and dynamic programming methods, which are encouraged to predict the improved performance. The adaptation and prediction of the rapid PEV charging profiles and vehicle range reliability are recorded by memory vehicle and RAM. It designing guidelines and the PSO-based low cost and load scheduling algorithm forecasted electricity price and PEV power demands.

Many countries have ventured into smart metering and smart appliances to improve the load profile and reduce peak demand to overcome these problems of the system, so that Demand Side Management (DSM) is implemented by load control and proper power management in the power grid in the automotive fashions adopting the PEV technology to provide auxiliary services, counting energy storage, and frequency regulation policy. The additional effects of the PEVs enable the power grids to hurriedly heal and self-regulate under any abnormal conditions for emergency perusal, thereby improving system security and reliability and efficiently managing energy, power delivery, and consumption.

The major part of open strategies on load control and power management may be treated as load forecasting and individual entities. Even, the load sharing in the same load characteristics are considered in a group of loads rather than individual loads by categorizing loads into a relatively small number of load types. The size of the proposed optimization problem does not change as the load population increases, which is a valuable feature for large-scale load management in this scheme.

3 A PSO TECHNIQUE FOR THE ENERGY CONSERVATION AND MANAGEMENT

The PSO technique is used to harvest energy conservation and management using two objective functions to accomplish the goal.

The first objective function is written by two variables here to simplify the model of the system as:

$$f_1(x_1, x_2) = x_1^2 + x_2^2 \tag{1}$$

Furthermore, the search variable function representing the set of free variables of the given function is:

$$f_2(x_1, x_2) = x_2 sin(4\pi x_1) + x_1 sin(4\pi x_2 + \pi) + 1 \tag{2}$$

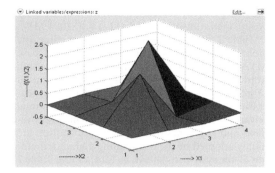

Figure 1. Plotting of the function of f(X1, X2).

The two objective functions of the system may be plotted in terms of its variables as shown in Fig. 1.

It is lucid from the global minimum of the objective function $f_1(x_1, x_2) \& f_2(x_1, x_2)$ at $(0,0)$ yields as:

$$f_1(0, 0) \& f_2(0, 0) = (0, 0) \tag{3}$$

The origin of the objective function in the search space is shown in Fig. 1 as a unit-model function, which has only one minimum. However, in order to find the global optimum, it is not so easy for multi-model functions, which have also multiple local minima. In the Fig. 1 shows of the second objective function, $f_2(x_1, x_2)$, which has a rough search space with multiple peaks with so many agents have to start from different initial locations and continue exploring the search space until at least one negotiator arrives at the global optimal position.

During this procedure, the entire agents may converse and allocate their information among themselves. This feature highlights how to resolve the multi-model function problems throughout the software simulations. Further, the PSO algorithm represents a multi-agent parallel search practice, which continues a swarm of particles as energy conservation and each particle represents a potential solution in the swarm as energy management. All particles fly through a multi-dimensional search space where the position of each particle can be adjusted as per its own experience and that of neighbors'. As a consequence, if x_i^t denotes the position vector of the particle (i) in the multi-dimensional search space (i.e. R_n) at the time step, then the position of each particle is updated in the search space as:

$$x_i^{t+1} = x_i^t + v_i^{t+1} \tag{4}$$

In order to explain the significance of equation (4), a condition is to be imposed for the optimal solutions of the objective functions as:

$$x_i^0 \sim U(x_{min}, x_{max}) \tag{5}$$

Therefore, in a PSO technique, all the particles are started with random manner and weighed up to work out the fitness for finding the individual best (best value of each particle) and global best (best value of particle in the entire swarm).

A loop is initiated to find an optimal solution with the help of quickness up-gradation of the computational loop taking first into account the particles and velocity to determine the location by the personal and global bests. Each particle occupies an updated location to achieve the current velocity of the iterations under the massive operations in the new loop and to form another loop for the up-gradation.

This proposed criterion is imposed on two PSO algorithms in terms of global best (G_{Best}) and local best (L_{best}) A PSO which is different in the dimension of their neighborhoods has been developed. These algorithms are discussed to following the special method of programming algorithm.

4 GLOBAL BEST PSO FOR THE ENERGY CONSERVATION AND MANAGEMENT

The global best (G_{Best}) PSO is a technique in which the position of each particle of the total swarm is exclusively prejudiced by the best fit for the particle in the entire swarm. A star is used to denote the people of a society, where the public information is appeared from all the particles in the swarm group signifying the following in the equality of the function as:

$$i \in [1 \cdots n] \times n \geq 1 \tag{6}$$

Equation (6) represents on a current position in the search space in terms of the position (x_i) and current velocity of the same space (v_i) in accordance with individual best position, which communicates in the search space where the particle had minimal value as indentified by the objective functions as mentioned in equation. (10) and (2). If it is treated as optimization function in association with the position of the particle giving the lowest value among the individual best position, it will be also called as the global best position, which is considered by the following set of equations as:

$$v_{ij}^{t+1} = v_{ij}^t + C_1 r_{1j}^t \left[P_{best,i}^t - x_{1j} \right] + C_2 r_{2j}^t \left[G_{best,i}^t - x_{ij}^{t\_} \right] \tag{7}$$

$$p_{best,t}^{t+1} = \left\{ p_i^t \ if \ f\left(x_i^{t+1}\right) > p_{best,i}^t \right\} \tag{8}$$

$$G_{best} = Min\left\{ p_{best,i}^t \right\} \tag{9}$$

And imposing a global condition for the optimal solution of the objective function is yielded as:

$$i \in [1,2,3\ldots\ldots\ldots n] \qquad (10)$$

5 LOCAL BEST PSO TECHNIQUE

The soft computing approach of the local best PSO (p_{best}) manner only permits each particle to be subjective.

Further, it replicates a circle of the social network inviting the basic information swapped over the neighborhood of the particle describing the energy conservation and management terms of the local knowledge based on the atmosphere. In this way, the velocity of particle i is calculated by the following equation:

$$v_{ij}^{t+1} = v_{ij}^t + C_1 r_{1j}^t \left[P_{best,i}^t - x_{1j} \right] + C_2 r_{2j}^t \left[L_{best,i}^t - x_{ij}^t \right] \quad (11)$$

The best fitness for particle selected from its locality of equation (11) represents the optimal solutions.

6 RESULTS

A soft computing method using PSO solution and the substantial simulation results are shown to envisage the G_{best} solution with respect to adequate iteration as software-based soft computing solution HIG reduces but LIG increases in Table 1. When the power of the power system increases in

Table 1. Optimal power loss computations for different iterations.

PSO: 1/100 iterations, G_{Best} = 809.38457000170501.
PSO: 20/100 iterations, G_{Best} = 804.5451968064732.
PSO: 40/100 iterations, G_{Best} = 804.49144215759554.
PSO: 60/100 iterations, G_{Best} = 804.48864845232106.
PSO: 80/100 iterations, G_{Best} = 804.48845781267528.
PSO: 100/100 iterations, G_{Best} = 804.48845781267528.
P1 = 48.8283 21.5189 22.2218 12.1916 12.0000

F1 = 804.4885
P = 176.7479 48.8283 21.5189 22.2218 12.1916 12.0000

VV = Columns 1 through 16
[1.0600 1.0430 1.0243 1.0158 1.0100 1.0140 1.0045
1.0100 1.0526 1.0464 1.0820 1.0580 1.0710 1.0385
1.0352 1.0460]

Columns 17 through 30
[1.0410 1.0270 1.0253 1.0298 1.0335 1.0338 1.0188
1.0186 1.0177 1.0001 1.0257 1.0119 1.0059 0.9945]

T_L = 10.1084

the LIG group by decreasing the power of the HIG group, then transmission loss (T_L) = 10.1084 MW.

Table 2 is also incorporated to strengthen the global and local best solutions for the 100 base MVA, and a large number of iterations using Matlab software and simulating platform are used to achieve the optimal solution and reliable generation, transmission, and distribution, as optimal power loss allocation in the system. This improvement and guidelines of the power distribution are highly adopted by the overloading and underloading of the system, as shown in Tables 1 and Table 2

Table 2. Optimal power Loss computations for different iterations.

PSO: 1/100 iterations, G_{Best} = 810.29244733188489.
PSO: 20/100 iterations, G_{Best} = 803.49567015852961.
PSO: 40/100 iterations, G_{Best} = 803.49163779834112.
PSO: 60/100 iterations, G_{Best} = 803.49163779834112.
PSO: 80/100 iterations, G_{Best} = 803.4909546808484.
PSO: 100/100 iterations, G_{Best} = 803.49089179654709.
P1 = 48.6422, 21.4113, 22.1936, 12.3032, 12.0000

F1 = 803.4909
P = 176.6699 48.6422 21.4113 22.1936 12.3032 12.0000

VV = Columns 1 through 16
[1.0600 1.0430 1.0245 1.0161 1.0100 1.0139 1.0045
1.0100 1.0518 1.0451 1.0820 1.0592 1.0710 1.0407
1.0388 1.0470]

Columns 17 through 30
[1.0403 1.0289 1.0261 1.0301 1.0329 1.0335 1.0282
1.0225 1.0199 1.0022 1.0268 1.0121 1.0070 0.9956]

TL = 9.8202

Table 3. Optimal power Loss computations for different iterations and power flows.

PSO: 1/100 iterations G_{Best} = 807.48729999374723.
PSO: 20/100 iterations G_{Best} = 801.9286684991755.
PSO: 40/100 iterations G_{Best} = 801.84493001932765.
PSO: 60/100 iterations G_{Best} = 801.84458274916324.
PSO: 80/100 iterations G_{Best} = 801.84407311952407.
PSO:100/100 iterations, G_{Best} = 801.84365471582908.
P1 = 48.8344 21.4689 21.6845 12.0513 12.0000

F1 = 801.8437
P = 176.7376 48.8344 21.4689 21.6845 12.0513 12.0000

VV = Columns 1 through 16
[1.0600 1.0430 1.0254 1.0171 1.0100 1.0148 1.0050 1.0100
1.0530 1.0467 1.0820 1.0599 1.0710 1.0450 1.0402
1.0471]

Columns 17 through 30
[1.0415 1.0304 1.0277 1.0317 1.0345 1.0350 1.0296
1.0237 1.0203 1.0027 1.0269 1.0128 1.0071 0.9957]

Table 4. Optimal power flow for the different groups of HIG, MIG, and LIG at optimal loss allocation.

No	Code	Mag.	Degree	MW	Mvar	MW	Mvar	Q_{min}	Q_{max}	Mvar
1	1	1.06	0.0	0.0	0.0	0.0	0.0	0	0	0
2	2	1.043	0.0	11.70	12.7	40.0	0.0	−40	50	0%HIG
3	0	1.0	0.0	2.4	1.2	0.0	0.0	0	0	0%HIG
4	0	1.06	0.0	7.6	1.6	0.0	0.0	0	0	0%HIG
5	2	1.01	0.0	94.2	19.0	0.0	0.0	−40	40	0%HIG
6	0	1.0	0.0	0.0	0.0	0.0	0.0	0	0	0%HIG
7	0	1.0	0.0	22.8	10.9	0.0	0.0	0	0	0%MIG
8	2	1.01	0.0	30.0	30.0	0.0	0.0	−10	60	0%MIG
9	0	1.0	0.0	0.0	0.0	0.0	0.0	0	0	0%HIG
10	0	1.0	0.0	13.8	2.0	0.0	0.0	−6	24	19%HIG
11	2	1.082	0.0	0.0	0.0	0.0	0.0	0	0	0%LIG
12	0	1.0	0	11.2	7.5	0	0	0	0	0%MIG
13	2	1.071	0	0.0	0.0	0	0	−6	24	0%LIG
14	0	1	0	10.2	1.6	0	0	0	0	0%LIG
15	0	1	0	8.2	2.5	0	0	0	0	0%MIG
16	0	1	0	1.5	1.8	0	0	0	0	0%LIG
17	0	1	0	9.0	5.8	0	0	0	0	0%MIG
18	0	1	0	3.2	0.9	0	0	0	0	0%MIG
19	0	1	0	9.5	3.4	0	0	0	0	0%MIG
20	0	1	0	2.2	0.7	0	0	0	0	0%MIG
21	0	1	0	17.5	11.2	0	0	0	0	0%HIG
22	0	1	0	0	0.0	0	0	0	0	0%HIG
23	0	1	0	3.2	1.6	0	0	0	0	0%LIG
24	0	1	0	8.7	6.7	0	0	0	0	4.3%LIG3
25	0	1	0	0	0.0	0	0	0	0	0%LIG
26	0	1	0	3.5	2.3	0	0	0	0	0%LIG
27	0	1	0	0	0.0	0	0	0	0	0%LIG
28	0	1	0	0	0.0	0	0	0	0	0%MIG
29	0	1	0	2.4	09	0	0	0	0	0%LIG
30	0	1	0	10.6	1.9	0	0	0	0	0%LIG

as simulation results. These managements and systems are solved by the Matlab software to improve line congestion and provide high-quality, uninterrupted power supply.

The automation and computerized architectures to obtain better flexibility, safety, reliability, and efficiency are the integral part of the electrical system using the PSO-based optimal solution.

Table 4 shows the conveying for the analysis of the transmission and distribution capacity of loss allocation in the power lines in terms of the load allocation and EV charging station to crack down the monopole business of the power utilities.

The overloading and underloading capacities of the system are shown in Table 2. When the power in the HIG group is increased by decreasing the power in the LIG group, the transmission loss becomes T_L = 9.8202 MW. Therefore, customers charge their vehicles for the battery storage in the LIG zone during off peak hour, and in the peak hour, they disconnect their home supply by connecting the battery storage system for domestic supply or may put up for sale surplus

power to nearby grid for the tariff optimization of electricity.

Operations and uninterrupted power management costs of the power utilities are effective to reduce costs for the consumers, as shown in Table 4.

The reduced peak demand, which will also harvest lower electrical energy tariffs and higher integration of large-scale renewable energy systems, better integrates customer to owner systems, thereby improving security and reliability. Table 4 of the electric vehicle are not popularized for the substantial reduction in peak demand of the HIG group as cost-effective solution of electricity pricing. The lower cost of electricity is to be supplied to the consumers at optimal power flow and loss of the system. The overloading of the transformers during peak hours is also eliminated as improvement of the demand response curves in the power system.

The speedy charging and discharging electrical system is now considered a green revolution of the green power management in the power sector.

7 CONCLUSIONS

The smart strategies with the PSO and Plugs in Electric Vehicles (PEVs) are the clean and green energy source for the energy conservation, associated with minimal power loss of transportation and no greenhouse gas emission. The peak demand of the battery storage system decreases, line overcrowding reduces, and the power quality is improved. On the contrary, the consumer can get financial benefits by selling the excess electricity to grid by discharging their PEVs during peak hours using Vehicle to Grid (V2G) mode as open electricity pricing in the open market.

REFERENCES

Acharya, S., R. Ghosh & T. Halder, (2012) "Adverse Effect of the Harmonics for the Power Quality Issues" IEEE ICCTICT, 16, pp. 306–311.

Andries P. Engelbrecht, (2007) Computational Intelligence: An Introduction: John Wiley and Sons, ch. 16, pp. 289–358.

Halder, T. (2011) "Some New Loss reduction Techniques of Distribution and Transmission Networks in Power System", National Conference on Energy System Planning Implementation and Operation, ESPIO-2011, organized by IEEE Calcutta Section, pp. 38–43.

Halder, T. (2012) "Recent trends of High Voltage Direct Current (HVDC)" National conference on Recent Trends in Communication Measurement & Control, CMC-2012, Organized by the Institution of Engineers (I), pp. 52–59.

Zachmann, G. (2008) "Electricity Wholesale Market Prices in Europe: Convergence?" Energy Economics 30: 1659–1671.

Computer, Communication and Electrical Technology – Guha, Chakraborty & Dutta (Eds)
© 2017 Taylor & Francis Group, ISBN 978-1-138-03157-9

A study on variability of anthropometric data of various pinna patterns of human ear for individualization of head related transfer function

S. Chaudhuri & D. Dey
Department of Electrical Engineering, Jadavpur University, Kolkata, India

S. Bandyopadhyay
Department of Biomedical Engineering, University of Florida, Gainesville, USA

A. Das
Department of Electrical Engineering, Indian Institute of Science, Bangalore, India

ABSTRACT: The human ear can be of a myriad shape. All these shapes contribute to reflection, scattering and diffraction of the incoming sound waves and hence alter the Head Related Transfer Function (HRTF) between the receiver and the source. This paper aims at classifying the human pinna into different shapes and providing the dimensions of the anthropometric parameters in these different classes to help in individualizing the HRTFs for making user-specific hearing aids or Virtual Auditory Displays (VADs). The data used for the purpose is taken from the public domain CIPIC and the Listen Project HRIR datasets. A study has been performed to see the morphological variations of the pinna by taking all combinations from the anthropometric dataset. The mutually uncorrelated parameters are clustered to demonstrate the different possible pinna shapes. This study is a prelude to the future research on individualization of HRTFs based on the different classes of each anthropometric variable.

1 INTRODUCTION

The way in which a person hears a certain sound is dictated by the reflection, scattering and diffraction off the head, body (torso) and especially the pinna. This causes selective amplification and attenuation of various frequencies of the incoming sound which excite the human auditory nerves to produce the sensation of sound. These signatures are present in the Head-Related Transfer Function (HRTF) which is unique in each ear of a person (Grijalva et al. 2014). The Head-Related Transfer Function can be defined as the frequency and space-dependent acoustic transfer functions between the sound source and the eardrum (Geronazzo & Spagnol 2010). It is obvious that the different anatomical footprints of the pinna like its size, shape of helix, width of antihelix, depth of concha, and the ear pinna rotation and flare angles have a significant role to play in reconstructing this HRTF for each person. A lot of work has been carried out on discerning the major anatomical landmarks in the human pinna (Zotkin et al. 2003, Ju et al. 2010); but literature on classification of each of these major parameters into groups has been found lacking. Such classification is important as knowing the range and nature of variations of each anthropometric variable will help in individualizing

the HRTF for different patients and help provide individualized hearing aids (Katz 2001). The use of individualized HRTFs are not only limited to hearing aids, but also to many other modern and emerging applications like, development of individualized Virtual Auditory Displays (VAD), auditory human-computer interfaces etc (Iwaya et al. 2012, Pralong & Carlile 1999, Paquier & Koehl 2015, Geronazzo et al. 2013).

The low-cost and straightforward mean of providing sound reproduction by speakers of hearing aids presently used for commercial production is by recording non-individualized HRTF sets by constructing "dummy-heads" or mannequins using average anthropometric measurements of the ear (Durant & Wakefield 2002, Kahana & Nelson 2007, Brown & Duda 1998, Avanzini & Crosato 2006, Algazi et al. 2001, Qu et al. 2009). However, non-individualized HRTFs are known to produce evident sound localization errors, including incorrect perception of elevation and front-back reversals (Zahorik et al. 2006). Hence, individualization of HRTFs plays a key role in assisted binaural hearing (Kimberly et al. 2014, Moller 1992, Gierlich 1992). Literature delineating morphological variations of the human pinna is mostly found in the field of biometrics where the same is used for forensic studies and identity verification (Verma et al. 2014, Kumar & Chenye 2011, Mokhtari et al. 2007). But a detailed quantitative classification of each anthropometric pinna landmark into groups by specifying their dimensions for the purpose of constructing a unique HRTF for each such group is unheard of (Brown 1996). In this paper such classification has been done based on the CIPIC dataset and the Listen Project HRIR dataset, which is a shared-cost RTD project in the Information Society Technologies (IST) Program of the European Commission's Fifth Framework Program (CIPIC HRTF database files 2001, LISTEN HRTF DATABASE, 2003). An automatic clustering technique is used in this work, where the optimal number of clusters can be found by the clustering algorithm itself, no prior knowledge is required (Yongguo et al. 2011). Such algorithms are already in use for many problems as mentioned in (Yongguo et al. 2011, Bandyopadhyay et al. 2002). In total the measurement of the different anthropometric parameters of 86 human subjects has been taken, and combinations of this data are clustered to study the variability of the pinna shapes present in the dataset.

2 EAR ANATOMY

The human ear consists of three parts, the outer ear or pinna, the middle ear and the inner ear (Batteau 2014). Of these the HRTF depends only on the structure of the pinna. "Fig. 1" shows some different types of pinna shapes present in humans.

The pinna consists of undulated cartilage covered by skin. The interlinked structures in the pinna help to guide the incident sound waves towards the eardrum (Nishimura et al. 2011, Spagnol et al. 2013, Mokhtari et al. 2011). "Fig. 2" shows the various important anatomical landmarks of the pinna.

The parameters shown in "Fig. 2" are as follows – 1) d_1 is the length of cavum concha, 2) d_2 is the cymba concha height, 3) d_3 is cavum concha width, 4) d_4 is the width of the antihelix, 5) d_5 is the pinna height, 6) d_6 is the pinna breadth, 7) d_7 is the width of intertragal incisura, 8) d_8 is the depth of concha, 9) θ_1 is the pinna rotation angle and 10) θ_2 is the pinna

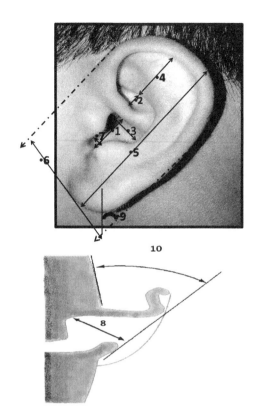

Figure 2. Anthropometric parameters of human pinna.

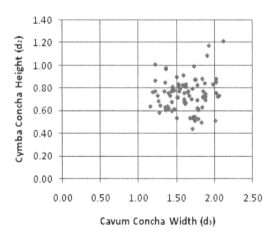

Cavum Concha Width (d₃)

Figure 3. Scatter plot between d_2 and d_3 showing no correlation (Correlation Coefficient = 0.078).

flare angle. In order to reduce the computational burden of the study, the attributable parameters are dispensed away with and only the anthropometric parameters which are grossly independent of each other are taken into account (Algazi

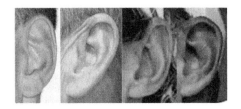

Figure 1. Different shapes of pinna.

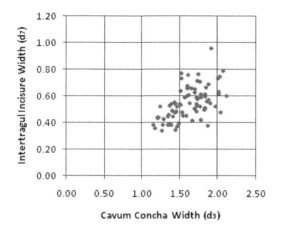

Figure 4. Scatter plot between d_7 and d_3 showing distinct positive correlation (Correlation Coefficient = 0.502).

et al. 2001, Xie, second edition, chapter-7, pp. 233, Morimoto 2001). These are listed as d_1, d_2, d_3, d_4, θ_1. The independence between the parameters is tested by judging their scatter plots and by computing their correlation coefficient. "Fig. 3" and "Fig. 4" depicts the scatter plots of such correlated and uncorrelated anthropometric parameters.

3 METHOD OF COMPUTATION

The computation is carried out on an anthropometric dataset composed of the parameter values of 86 individuals of both genders by combining the CIPIC and the Listen Project HRIR datasets. The method of computation involves some preprocessing of the anthropometric dataset. This includes removing distinct outliers from the data, and replacing them by the mean of the remaining data of that parameter. The 86 individuals which have been taken from the aforementioned two datasets for this study include humans of both sexes and from different ages. Hence an intrinsic clustering was noticed initially between individuals of the same gender and age groups; where all parameters showed a positive correlation to the ear size ($d_5 \times d_6$). To remove this bias, and to investigate only the variations of the ear shape, each ear is fitted to a mean size window. The square root of the ratio between an individual ear size to the mean ear size of the dataset is taken as the normalization factor and the values of the anthropometric landmarks of each individual are then divided by this factor. The processed data is then further normalized so that the difference in the ranges of the different anthropometric parameters would not affect the computation and only the inherent variability in the data is focused upon.

Therefore, the previously mentioned five independent anthropometric parameters namely d_1,

d_2, d_3, d_4, θ_1 are clustered using an adaptive classifier which does not take the prospective number of classes, or class centers as input. In this work a Genetic Algorithm based automatic clustering technique is used. The detail of the algorithm is available in (Yongguo et al. 2011). Hence, the detailed description of the automatic clustering algorithm is not given here for brevity.

The results obtained by this automatic classifier are corroborated by comparing them to the clustering done by the standard fuzzy c-means classifier available in MATLAB™.

4 RESULTS

As discussed before, the results are computed over 86 individual human subjects, the data being compiled from the CIPIC and the Listen Project HRIR datasets. The intrinsic pinna variability has been studied through the clustering of the five independent variables taken from the total anthropometric dataset. The clustering done via the aforementioned adaptive classifier yields 7 clusters.

To corroborate this finding the same dataset is clustered using the standard fuzzy c-means clustering where the number of classes is input as 7 in accordance to the result obtained above. The corresponding observation is listed in the following Table 2 where the cluster centers obtained from fuzzy c-means clustering are listed.

Table 1. Cluster determination.

Class number		Anthropometric features				
		d_1 (cm)	d_2 (cm)	d_3 (cm)	d_4 (cm)	θ_1 (deg)
1	Mean	1.83	0.66	1.27	1.65	24.98
	Standard Deviation	0.26	0.06	0.02	0.41	7.57
2	Mean	1.76	0.71	1.41	1.70	18.27
	Standard Deviation	0.19	0.09	0.04	0.25	8.20
3	Mean	1.76	0.76	1.52	1.71	17.88
	Standard Deviation	0.13	0.11	0.01	0.18	7.17
4	Mean	1.89	0.75	1.62	1.79	20.65
	Standard Deviation	0.18	0.10	0.03	0.23	9.25
5	Mean	1.88	0.66	1.74	1.51	21.50
	Standard Deviation	0.17	0.13	0.03	0.27	8.15
6	Mean	1.73	0.75	1.86	1.61	27.81
	Standard Deviation	0.11	0.08	0.02	0.18	6.73
7	Mean	1.88	0.74	2.02	1.53	16.63
	Standard Deviation	0.13	0.15	0.01	0.25	11.28

Table 2. Cluster centers using fuzzy c-means algorithm.

Class number	d_1 (cm)	d_2 (cm)	d_3 (cm)	d_4 (cm)	θ_1 (deg)
1	1.99	0.64	1.76	1.28	27.03
2	1.83	0.81	1.54	1.68	23.85
3	1.79	0.72	1.47	1.44	17.10
4	1.80	0.76	1.74	1.74	21.20
5	1.83	0.78	1.79	1.51	20.83
6	1.73	0.73	1.50	1.88	16.64
7	1.87	0.73	1.62	1.95	19.65

The individualization of the HRTFs will be based upon this finite number of clusters. These clusters will give rise to 7 different HRTF templates.

5 CONCLUSION

In the reported paper, a clustering of the human pinna shapes based upon the major independent anatomical landmarks, which interfere with the way sound is received and guided into the eardrum, is provided; giving the specific anatomical dimensions of the different classes in which the pinna shapes are clustered. This work will lead the way into a coarse individualization of the HRTF for each class of human pinna shape depending upon the significant anthropometric parameters of the pinna, as identified by the present study. Accordingly, 7 different templates of the HRTF will be made based upon the 7 different pinna shapes obtained through the study. Any unknown HRTF will be classified into one of these different HRTF templates thus giving a coarse individualization. Fine tuning of the closest model is required to have an effective match of the HRTF for a particular subject. Hence, a parameter tuning algorithm will be developed for this purpose to achieve finer levels of individualization. After finalizing the tuning method, the developed scheme will be implemented in an integrated form in general purpose microcontroller/microprocessor to fit into 3D assisted hearing modules which may include PIC microcontroller or ARM Processor or FPGA based systems.

REFERENCES

Algazi, R. V., C. Avendano, & O. R. Duda (2001). Estimation of a spherical head model from anthropometry. *J. Audio Eng. Soc.* 49, 472–479.

Algazi, R. V., O. R. Duda, M. D. Thompson & C. Avendano (2001). The CIPIC hrtf database. *IEEE Workshop on Applications of Signal Processing to Audio and Acoustics,* October.

Avanzini, F. & P. Crosato (2006). Integrating physically-based sound models in a multimodal rendering architecture. *Comp. Anim. Virtual Worlds.* 17, 411–419.

Bandyopadhyay, S. & U. Maulik (2002). An evolutionary technique based on K-means algorithm for optimal clustering in R. *Inf. Sci.* 146, 221–237.

Batteau, W. D. (2014). The role of pinna in human localization. *Proceedings of the Royal Society of London. Series B, Biological Sciences.*

Brown, P. C. (1996). *Modeling the Elevation Characteristics of the Head-Related Impulse Response.* San Jose State Univ., San Jose, CA.

Brown, P. C. & O. R. Duda (1998). A structural model for binaural sound synthesis. *IEEE Trans. Speech Audio Process.* 6, 476–488.

Cheng, I. C. & H. G. Wakefield (2001). Introduction to Head-Related Transfer Functions (HRTFs): Representations of HRTFs in time, frequency, and space. *J. Audio Eng. Soc.,* 49, no. 4, 231–249.

Davis, U. C. (2001). The CIPIC HRTF Database files, released 1.1.

Durant, C. E. & H. G. Wakefield (2002). Efficient model fitting using a genetic algorithm: Pole-zero approximations of HRTFs. *IEEE Trans Speech Audio Process.* 10, 18–27.

Geronazzo, M. & S. Spagnol (2010). Estimation and modelling of pinna-related transfer functions. *Proc. of the 13th International Conference on Digital Audio Effects (DAFx-10),* 6–10.

Geronazzo, M. S. Spagnol & F. Avanzini (2013). Mixed structural modelling of head-related transfer functions for customized binaural audio delivery. *IEEE Trans. DSP,* 1–8.

Gierlich, W. H. (1992). The application of binaural technology, *Appl Acoust.* 36, 219–243.

Grijalva, F., L. Martini, S. Goldenstein & D. Florencio (2014). Anthropometric-based customization of head-related transfer functions using isomap in the horizontal plane. *IEEE International Conference on Acoustic Speech and Signal Processing (ICASSP),* 4493–4497.

Iwaya, Y., M. Toyoda, M. Otani & Y. Suzuki (2012). Evaluation of realism of dynamic sound space using a virtual auditory display. *IEEE Conference Proceedings,* 561–566.

Ju, Y., Y. Park & S. Lee (2010). Calibration of measured head-related transfer function database. *International Conference on Control, Automation and Systems (ICROS).*

Kahana, Y. & A. P. Nelson (2007). Boundary element simulations of the transfer function of human heads and baffled pinnae using accurate geometric models. *J. Sound Vibr.* 300, 552–579.

Katz, B. (2001). Boundary element method calculation of individual head-related transfer function. I. Rigid model calculation, *J. Acoust. Soc. Amer.* 110, 2440–2448.

Kimberly, J. Fink & L. Ray (2014). Individualization of head related transfer functions using principal component analysis. *Elsevier. Applied Acoustics.* 87, 162–173.

Kumar, A. & W. Chenye (2012). Automated human identification using ear imaging. *Elsevier. Pattern Recognition.* 45, 956(13).

LISTEN HRTF DATABASE (2003).

Mokhtari, P., H. Takemoto, R. Nishimura & H. Kato (2007). Comparison of simulated and measured HRTFs: FDTD simulation using MRI head data. *Proc. 123rd AES Conv.* 7240, 1–12.

Mokhtari, P., P. Takemoto, R. Nishimura & H. Kato (2011). Pinna sensitivity patterns reveal reflecting and diffracting surfaces that generate the first spectral notch in the front median plane, in *Proc. IEEE Int. Conf. Acoust., Speech, Signal Process. (ICASSP'11)*, 2408–2411.

Moller, H. (1992). Fundamentals of binaural technology. *Appl Acoust.* 36, 171–218.

Morimoto, M. (2001). The contribution of two ears to the perception of vertical angle in sagittal planes. *J. Acoust. Soc. Amer.* 109, 1596–1603.

Nishimura, R., P. Mokhtari, H. Takemoto & H. Kato (2011). An Attempt to Calibrate Headphones for Reproduction of Sound Pressure at the Eardrum. *IEEE Trans on Audio, Speech and Language Processing.* 19, 2137–2145.

Paquier, M. & V. Koehl (2015). Discriminability of the placement of supra-aural and circumaural headphones. *Int. Journal of Applied Acoustics, Elsevier.* 93, 130–139.

Pralong, D. & S. Carlile (1999). The role of individualized headphone calibration for the generation of high fidelity virtual auditory space. *J Acoust Soc Am.* 100, 3785–3793.

Qu, T., Z. Xiao, M. Gong, Y. Huang, X. Li & X. Wu (2009). Distance-dependent head-related transfer functions measured with high spatial resolution using a spark gap. *IEEE Trans. Audio, Speech, Lang. Process.* 17, 1124–1132.

Spagnol, S., M. Geronazzo & F. Avanzini (2013). On the Relation Between Pinna Reflection Patterns and Head-Related Transfer Function Features. *IEEE Trans on Audio, Speech and Language Processing.* 21, 508–519.

Verma, K., J. Bhawana & V. Kumar (2014). Morphological variation of ear for individual identification in forensic cases: A study of an Indian population. *Research Journal of forensic sciences.* 2, 1–8.

Xie, B. (2013). This is a chapter 7. In Ning Xiang (2nd Ed.), *Head-Related Transfer Fucntion and Virtual Auditory Display: Customization of Individualized HRTFS*, (pp. 233). Rensselaer Polytechnic Institute.

Yongguo, L., W. Xindong & Y. Shen (2011). Automatic clustering using genetic algorithms. *Applied Mathematics and Computation, Elsevier.* 218, 1267–1279.

Zahorik, P., P. Bangayan, V. Sundareswaran, K. Wang & C. Tam (2006). Perceptual recalibration in human sound localization: learning to remediate front-back reversals. *J Acoust Soc Am.* 120, 343–359.

Zotkin, D. Y. N., J. Hwang, R. Duraiswami & L. S. Davis (2003). HRTF personalization using anthropometric measurements. *IEEE Workshop on Applications of Signal Proccrvnp to Audio and Acoustics*, 157–160.

Computer, Communication and Electrical Technology – Guha, Chakraborty & Dutta (Eds)
© 2017 Taylor & Francis Group, ISBN 978-1-138-03157-9

Preterm birth prediction using electrohysterography with local binary patterns

Finky Francis
Department of Electronics and Communication, MES College of Engineering, Kuttippuram, Kerala, India

M. Bedeeuzzaman
Department of Applied Electronics and Instrumentation, MES College of Engineering, Kuttippuram, Kerala, India

Thasneem Fathima
Department of Electrical and Electronics, MES College of Engineering, Kuttippuram, Kerala, India

ABSTRACT: Many techniques have been raised for improving the survival of preterm infants. However, premature delivery remains as a common problem for the new born deaths. The complications due to preterm deliveries may even lead to significant health problems of the new born babies and also many other economical problems. There exists a strong evidence that the analysis of uterine electrical signals could provide an easy path towards preterm birth detection and even prediction. Exploring this idea, the paper focus on the Electrohysterography (EHG) signal processing and efficient machine learning algorithms. Features based on the one dimensional Local Binary Patterns (1D-LBP), a powerful texture classification feature was proposed here. This study uses an open dataset, Term-Preterm Electro Hystero Gram (TPEHG) database, containing 300 delivery records (38 preterm and 262 term) for data acquisition. This paper provides a simple, but efficient way of approach for predicting premature deliveries. An accuracy of about 60% is evaluated as the performance of classifying term and preterm records.

1 INTRODUCTION

Studies by UNICEF (2013) reported that India stands first in the number of neonatal deaths from preterm complications. Preterm deliveries remain as a common problem in the field of obstetrics. It increases the rate of infant morbidity and mortality. By the World Health Organisation (WHO), Preterm birth, also known as premature birth or delivery, is described as the delivery of babies who are born, alive, before 37 completed weeks of gestation. Then, the other one, Term births are the live delivery of babies after 37 weeks, and before 42 weeks (Hussain et al., 2015). Preterm deliveries will cause long term health problems.

Preterm delivery can occur due to various reasons. From studies given by Hussain et al. (2015), one-third of the preterm deliveries are due to medically indicated or induced labour. This type of labour is strictly undertaken for the well being of mother or baby. Another third occurs due to the rupture of membranes, prior to labour (PPROM). Last part cases are due to the spontaneous contractions. Along with this, studies done by Fele-Zorz et al. (2008) and Hussain et al. (2015) show that several

health and lifestyle factors of the mother like short cervix, uterine abnormalities, infections, any invasive surgery, underweight or obesity, long working hours, alcohol and drug use, diabetes, hypertension and folic acid deficiency will also lead to preterm labour. Preterm birth has an increased risk of death and health effects. These include impairments to hearing, vision, and the lungs, the cardiovascular system and non communicable diseases.

Many efforts have been made to improve the survival of preterm infants. Predicting preterm delivery before it occurs led a great interest because of its necessity in nowadays. Perhaps one of the most efficient approaches in the prediction of preterm births is the analysis of electrical activity of uterine muscles. This results the study named Electrohysterography (EHG) (Hussain et al., 2015). Electrohysterogram monitors the uterine contractions and is equivalent to uterine Electromyogram (EMG) (Fele-Zorz et al., 2008). EHG signal results from the propagation of electrical activity of uterine muscles. This provides the potential difference between the electrode leads. The use of the EHG signal processing and suitable machine learning algorithms can make an adventurous result in predicting preterm delivery and diagnosing preterm labor.

This paper focus on using EHG signals to detect the onset of preterm births. The work was done using an open dataset named Term-Preterm ElectroHysteroGram (TPEHG) database described in Fele-Zorz et al. (2008), containing 300 records on pregnant subjects. This study was done with the term and preterm EHG records of 22.29 weeks of gestation from the dataset. Features based on one dimensional Local Binary Patterns (1D-LBP) were then proposed. Statistical features like variance, Median Absolute Deviation (MAD), Interquartile Range (IQR) and their combinations have been extracted from the obtained LBP feature vector. The final concatenated feature vector of term and preterm EHG data was the input to the linear classifiers. This approach was quite different from the previous studies and also provide a good result in the field of EHG signal processing and the preterm birth prediction.

The structure of our paper is as follows. Section 2 discusses related studies in this field. Section 3 describes the experimental methodology. Section 4 presents the results and Section 5 narrates the conclusions and future directions.

2 RELATED STUDIES

The electrical activity of the uterus had been known from late 1930s (Bozler, 1938). But techniques for recording the EHG signal have been appeared from only last 20 years. Different approaches have been made for recording the signal using two, four and even sixteen electrodes.

KNN algorithm has been used by Diab et al. (2007) to classify term and preterm records. Two electrodes were used and parameters were calculated based on Autoregressive (AR) model. Wavelet decomposition converts the signal into frequency domain. 5 vectors of variance from each detail was extracted as features. Classification was done over two groups, G1 and G2. G1 corresponds to the signals of women recorded at same weeks of gestation but, delivered at different weeks of gestation. G2 corresponds to the signals of women delivered at same weeks of gestation but, recorded at different weeks of gestation. Classification error for G1 and G2 is 2.4% and 8.3% respectively. A modification to this was done by Moslem et al., (2011) extracting features from the power density spectrum of the signal. Here Artificial Neural Networks (ANN) classification had been used. Groups G1, G2 were considered for preterm and term labour contractions and a classification error of 3.33% was obtained. Again, Moslem et al. (2011) proposed a method in which

Support Vector Machines (SVM) had been used to classify term and preterm deliveries. Sixteen electrodes were used. Power of the EMG signal and median frequency are the extracted features. An accuracy of 83.34% is obtained.

Baghamoradi et al. (2011) used TPEHG database and proposed a technique of classifying EHG data by cepstral analysis. Cepstrum techniques are suited for the analysis of the non linear signals. Four electrodes were used for the recording and the extracted features were cepstral coefficients, cepstral cofficients selected using sequential feature selection and sample entropy. Classification is done individually by Multilayer Perceptron (MLP) method. Accuracy of 72.73% is got by sequential feature selection method.

A comparison of various classification methods had been discussed to classify term and preterm deliveries. In all these works, TPEHG dataset was used for data acquisition and features like peak frequency, median frequency, root mean square and sample entropy were extracted. Fergus et al. (2013) found out the performance of density based classifiers like Linear Discriminant Classifier (LDC), Quadratic Discriminant Classifier (QDC), Uncorrelated normal Density Classifier (UDC) and linear and polynomial based classifiers like Polynomial Classifier (POLYC), Logistic Classifier (LOGLC) and non linear based K Nearest Neighbour Classifier (KNNC), decision Tree Classifier (TREEC), Parzen Classifier (PARZENC) and Support Vector Classifier (SVC). POLYC, LOGLC, KNNC and the TREEC classifiers perform very well. The best classifier is the POLYC with 97% sensitivity, 90% specificity, 95% Area Under Curve (AUC) value and a global mean error of 8%. Artificial Neural Networks (ANN) have been extensively used by Idowu (2014) to classify between term and preterm deliveries. Six advanced artificial neural network classifiers have been evaluated. This includes the Back Propagation trained feed-forward neural Network Classifier (BPXNC), Levenberg-Marquardt trained feed-forward neural Network Classifier (LMNC), automatic Neural Network Classifier (NEURC), Radial Basis function neural Network Classifier (RBNC), Random Neural Network Classifier (RNNC) and Perceptron Linear Classifier (PERLC). Peak frequency, median frequency, root mean square, sample entropy were the extracted features. LMNC has dramatically improved with a value of sensitivity 96%, specificity 91% and Area Under Curve (AUC) 95.46%. Dynamic neural network architecture inspired by the Immune Algorithm (DSIA) has been used by Hussain et al. (2015) to predict the preterm deliveries in pregnant women. A comparison of the MLP, TREEC, SVC, DSIA and Self-Organized multilayer Network inspired by

Immune Algorithm (SONIA) has been discussed. Both DSIA and SONIA have the highest values of sensitivity with 91.23% and SONIA gives highest specificity with 94.51%. An accuracy of 92.77% and 89.81% was obtained with SONIA and DSIA respectively.

From the above discussed studies, EHG could be considered as the strong predictor of term and preterm deliveries. Many of the studies have been used frequency related parameters for the classification. So an enhanced study on other non linear parameters and simpler classification techniques can be introduced as a new approach.

3 METHODOLOGY

True labour begins within 1 day for term deliveries. In the case of preterm deliveries, it may begin within 7 to 10 days. There is a dramatic change in EHG activity, from non-labour to labour. Therefore, it is expected that, classifying the records, into preterm and term, is a challenging work.

The EHG data used for the study was shown in the Figures 1 and 2. Figure 1 shows the term data filtered at 0.3–3 Hz, recorded at 22.29 weeks of gestation and delivered at 40.29 weeks of gestation. Figure 2 shows the preterm data filtered at 0.3–3 Hz, recorded at 22.29 weeks of gestation and delivered at 32.86 weeks of gestation. Term data samples are arranged in a random fashion. But a regular pattern of EHG bursts can be seen in the preterm EHG data and it gives a symptom of labour.

Once EHG signal was acquired from the abdomen of pregnant woman, several steps are to be followed for the further analysis and studies. EHG data processing mainly consists of three steps: Preprocessing, Feature Extraction and Classifica-

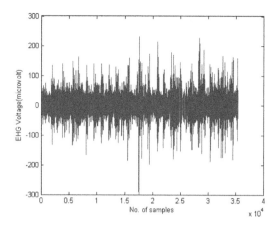

Figure 2. Preterm EHG data.

Figure 3. Steps of EHG data processing.

tion. These are depicted in the Figure 3. Preprocessing includes the techniques done before processing EHG signals. Feature extraction includes the step of extracting features from preprocessed data. This feature vector inputs to the classifier and then classifies the data.

3.1 Data acquisition

The data was collected from a publicly available database called Term-Preterm ElectroHysteroGram DataBase (TPEHG DB) from Physionet. This includes uterine EMG obtained from 1997 to 2005 at the University Medical Centre Ljubljana, Department of Obstetrics and Gynecology (Fele-Zorz et al. 2008). This study uses the term and preterm data of 22.29 weeks of gestation. Each record is composed of three channels, recorded from 4 electrodes. The first electrode (E1) was placed 3.5 cm to the left and 3.5 cm above the navel. The second electrode (E2) was placed 3.5 cm to the right and 3.5 cm above the navel. The third electrode (E3) was placed 3.5 cm to the right and 3.5 cm below the navel and the fourth electrode (E4) was placed 3.5 cm to the left and 3.5 cm below the navel. The difference between potentials of electrode leads were recorded and it gives 3 channels. Difference between first and second, second and third, third and fourth electrodes produce the first, second and third channel respectively. The individual records were 30 minutes in duration. Each signal has been sampled at 20 samples per second per channel with 16-bit resolution over a range of 2.5 millivolts.

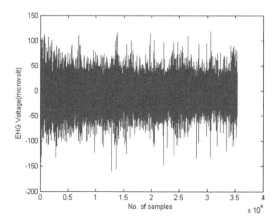

Figure 1. Term EHG data.

Each signal was digitally filtered using 3 different digital butterworth filters with a double-pass filtering scheme. The band-pass cut-off frequencies were from 0.08–4 Hz, 0.3–3 Hz and 0.3–4 Hz. Here 0.3–3 Hz filtered term and preterm EHG signals, recorded at 22.29 weeks of gestation were acquired for the study.

3.2 Signal processing and feature extraction

After acquiring, entire data was divided into a set of frames, each with a duration of 1 second. For each frame, a Local Binary Pattern (LBP) feature vector was then obtained. After that, from the obtained LBP feature vector of each frame, a set of statistical features like, variance, Median Absolute Deviation (MAD), Interquartile Range (IQR) and their combinations were extracted. Then the individual feature vectors of all frames, obtained after extracting statistical features are concatenated and formed a final feature vector of the given record. This was the required feature vector, which is to be given as an input to the classifier. The idea about the local binary patterns and the steps required for creating the LBP feature vector was discussed in the next section.

3.3 One dimensional Local Binary Patterns

Local Binary Pattern (LBP) is a simple, efficient feature commonly used in images for texture classification. It is used to capture intensity of local neighbourhood of a pixel by thresholding and considers the result as a binary number (Sunil et al., 2015). LBP texture operator has become a popular approach in various applications due to its discriminative power and computational simplicity. The LBP feature vector was first described by Ojala et al. (1996) as a non parametric, grey scale texture invariant analysis model, which summarizes the local spatial structure of an image. It was based on the 3×3 local neighbourhood as shown in the Figure 4. For each pixel P in a cell, it was compared with each of its 8 neighbours P_i, $0 \leq i \leq 7$. Then the pixels are followed along clockwise or counter clockwise direction. When the center pixel's value is lesser than the neighbour's value, replace it with 1, Otherwise, 0. This gives an 8-digit binary number and convert it to decimal for convenience. Each pixel in an image creates a single LBP code. Then a histogram is computed based on the occurrences of LBP codes from all pixels of the image and treated as a feature vector (Iakovidis et al., 2008). This same procedure is also applicable to signals. One dimensional local binary patterns arise there. Computation of one dimensional local binary patterns are described by Navin et al. (2010) as shown in the Figure 5.

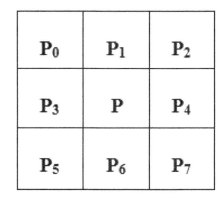

Figure 4. 3×3 local neighbourhood.

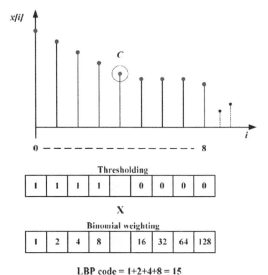

Figure 5. Computation of 1D LBP.

3.4 Classification

With the extracted features, classification is to be done. In the previous studies, complex classification methods are chosen for the term and preterm discrimination. Apart from this, simple linear classifiers are taken for this study. Linear Discriminant Classifier (LDC) and Support Vector Machine Classifier (SVMC) are used in this work. After the classification process, a comparison is done with the performance of both classifiers. Performance evaluation of the classifiers are done using the metrics - Sensitivity, Specificity and Accuracy. This is based on the values of True Negative (TN), True Positive (TP), False Negative (FN) and False Positive (FP).

The flow chart of the algorithm was shown in the Figure 6. The 0.3–3 Hz term and preterm EHG

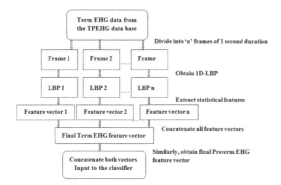

Figure 6. Flow chart of the algorithm.

Table 1. Linear discriminant classifier performance.

Features	Sensitivity (%)	Specificity (%)	Accuracy (%)
Variance	58.61	62.00	60.23
MAD	48.65	71.41	60.02
IQR	50.36	68.85	59.89
Variance-MAD	58.60	62.73	60.67
Variance-IQR	59.74	62.8	61.30
MAD-IQR	53.48	66.71	60.1

Table 2. Support vector machine classifier performance.

Features	Sensitivity (%)	Specificity (%)	Accuracy (%)
Variance	61.45	61.30	61.38
MAD	51.49	69.98	60.74
IQR	52.99	67.42	60.31
Variance-MAD	61.45	61.31	61.38
Variance-IQR	62.59	61.45	62.02
MAD-IQR	56.33	65.29	60.81

data of 22.29 weeks of gestation was undergone with the steps as shown in the figure. After extracting statistical features from the obtained LBP feature vector, a final feature vector was obtained for the term and preterm EHG data. Concatenated term and preterm final feature vectors was then input to the LDC and SVM classifiers.

4 RESULTS AND DISCUSSIONS

With the use of 0.3–3 Hz filtered data in all channels, simulation results of the statistical features were obtained from the LBP feature histogram. Performance of the LDC and SVMC is evaluated by the above described metrics. Performance evaluated was tabulated in the Tables 1 and 2.

From the Tables 1 and 2, a summary of the performance of the linear discriminant classifier and support vector machine classifier towards the discrimination of term and preterm data was obtained. Both the classifiers gave almost 60% accuracy. Sensitivity gives the rate at which a classifier can detect the abnormality accurately, when the condition is present. Then the specificity gives the rate at which a classifier can detect the normal condition accurately, when the condition is present. SVC provide a slightly greater sensitivity than LDC. But, at the same time, specificity of the SVC is lower than that of the LDC performance.

5 CONCLUSIONS

Medical information systems play a vital role in the biomedical domain. This has led to the various preprocessing algorithms and feature extraction techniques. Electrohysterography led as a strong predictor to classify term and preterm records. The 0.3–3 Hz filtered TPEHG data is used for the classification between term and preterm records. After framing, one dimensional local binary patterns were found out from each frame. With the help of 1D-LBP, computational simplicity was improved and a better classification model was developed. Variance, median absolute deviation, interquartile range, combination of variance and median absolute deviation, combination of variance and inter quartile range, combination of median absolute deviation and interquartile range were extracted from each LBP feature vector. Performance of two classifiers were measured in terms of the metrics like accuracy, specificity and sensitivity. Accuracy of the both classifiers is about 60%. But the SVM classifier provide a slightly better sensitivity and accuracy.

With the help of simple, less complex features and algorithm, the obtained accuracy is good. A better way would be to develop for improving the performance of the classifiers using the same dataset. This will be the focus of future research, alongside a more extensive investigation into simple, efficient features.

REFERENCES

Baghamoradi, S., M. Naji, & H. Aryadoost (2011). Evaluation of cepstral analysis of EHG signals to prediction of preterm labor. *18th Iranian Conference on Biomedical Engineering,* 81–83.

Diab, M., C. Marque, & M. Khalil (2007). Classification for uterine EMG signals: Comparison between AR model and statistical classification method. *International Journal of Computational Cognition 5(1),* 8–14.

Fele-Zorz, G., G. Kavsek, Z. Novak-Antolic, & F. Jager (Sep. 2008). A comparison of various linear and nonlinear signal processing techniques to separate uterine EMG records of term and pre-term delivery groups. *Med. Biol. Eng. Comput. 46*, 911–22.

Fergus, P., P. Cheung, A. Hussain, D. Al-Jumeily, C. Dobbin, & S. Iram (Jan. 2013). Prediction of preterm deliveries from EHG signals using machine learning. *PLos One 8(10)*, e77154.

Hussain, A., P. Fergus, H. Al-Askar, D. Al-Jumeily, & F. Jager (2015). Dynamic neural network architecture inspired by the immune algorithm to predict preterm deliveries in pregnant women. *Neurocomputing. 151*, 963–974.

Idowu-Olatunji, I., P. Fergus, A. Hussain, C. Dobbins, & H. Al-Askar (2014). Advance artificial neural network classification techniques using EHG for detecting preterm births. *Complex, Intelligent and Software Intensive Systems (CISIS), 2014 Eighth International Conference on IEEE*, 95–100.

Moslem, B., M. Diab, M. Khalil, & C. Marque (2011). Classification of uterine EMG signals by using supervised competitive learning. *IEEE Workshop on Signal Processing Systems*, 267–272.

Moslem, B., M. Khalil, M. Diab, A. Chkeir, & C. Marque (Dec. 2011). A multisensor data fusion approach for improving the classification accuracy of uterine EMG signals. *Proceedings of the 18th IEEE International Conference on Electronics, Circuits and Systems (ICECS) 2011*, 93–96.

Navin, C. & S. John, J. (Aug. 23–27, 2010). Local binary patterns for 1-d signal procesing. *18th European signal processing conference (EUSIPCO-2010)*, 95–99.

Ojala, T., M. Pietikinen, & T. Maenpaa (1996). Multiresolution gray-scale and rotation invariant texture classification with local binary patterns. *IEEE Transactions Pattern Annnal Machine Intelligence 24*, 971–987.

UNICEF, WHO, W. Bank, & D. UN-DESA (2013). Population levels and trends in child mortality 2013. *Estimates Developed by the UN Interagency Group for Child Mortality Estimation*, 1–34.

Computer, Communication and Electrical Technology – Guha, Chakraborty & Dutta (Eds)
© 2017 Taylor & Francis Group, ISBN 978-1-138-03157-9

Arrhythmia classification by nonlinear kernel-based ECG signal modeling

S. Ghorai & D. Ghosh
Department of AEIE, Heritage Institute of Technology, Kolkata, India

ABSTRACT: A new kernel-based modeling technique of Electrocardiogram (ECG) signal is presented in this work to discriminate normal and arrhythmia beats. ECG signals are characterized by time series modeling using Fast Regularized Kernel Function Approximation (FRKFA) technique. The characteristics parameters of the nonlinear regression models are considered as the features of ECG beats. The ability of these features to discriminate normal and arrhythmia beats are verified using Support Vector Machine (SVM) classifier in a global beat classification approach. The results are compared with the existing linear Autoregressive (AR) signal modeling technique. The test results on large data sets show that the performance of kernel-based modeling technique achieves an accuracy as high as 95.43% and this shows an improvement in performance by more than 11% to that of linear AR modeling technique to discriminate normal and arrhythmia ECG beats.

1 INTRODUCTION

Electrocardiogram (ECG) is a noninvasive technique primarily used as the diagnostic tool for cardiovascular diseases (Adamec & Adamec 2008). A cleaned ECG signal provides necessary information about the electrophysiology of the heart diseases and ischemic changes that may occur. Continuous observation and detection of abnormal ECG signals in clinics or intensive care units is tricky due presence of large number of patients and human fatigue. In addition to a simple ECG test, examination of a longer recording using a portable Holter monitor is usually used to determine abnormality in heart rhythm of a patient. Testing of such long record by human being is a time consuming process and an automated ECG beat testing system may assists the cardiologist in diagnostics.

A number of methods have been presented in the past two decades for detection of arrhythmia by means of computer aided diagnostic. Most of the proposed techniques rely on QRS peak detection, beat segmentation and then extraction of features from the beats for classification (Chazal et al. 2004, Chen and Yu 2007, Inan et al. 2006, Lagerholm et al. 2000, Osowski et al. 2004). Researchers have reported performances of different types of features, such as ECG morphology (Chazal et al. 2004), heart beat intervals (Chazal et al. 2004, Chen and Yu 2007, Inan et al. 2006), combination of both morphology and time intervals (Chazal et al. 2004), combination of higher order statistics

and time interval features (Ebrahimzadeh et al. 2014), Hermite function (Lagerholm et al. 2000), transformed features (Chen and Yu 2007) and combination of complex wavelet coefficients with AC power, kurtosis, skewness and timing information features (Thomas et al. 2015), etc. Application of several classifiers such as self-organizing map (Lagerholm et al. 2000), neural networks (Inan et al. 2006, Chen and Yu 2007), Support Vector Machines (SVM) (Melgani and Bazi 2008, Osowski et al. 2004), extreme learning machine based interactive ensemble learning approach (Rahhal et al. 2015) etc., are found in literature for the purpose of beat classification.

In the recent past, there are some attempts to predict beats using model coefficients (Andreão et al. 2006, Ge et al. 2002, Ouelli et al. 2012). Andreão et al. (2006) recently used HMM to segment different waveforms of ECG signal and detected Premature Ventricular Contraction (PVC) beat by an unsupervised classification approach using multichannel fusion strategy. They achieved 87.2% of Sensitivity (SE) and 85.6% of Positive Predictivity (PP) on 59 recordings collected from QT database (Laguna 1997). Edla et al. (2014) proposed classification of normal and arrhythmia beats by modeling ECG beats using multiharmonic component with an adaptive parameter estimation technique and Bayesian Maximum-Likelihood (ML) classifier. Auto-regressive (AR) modeling has been utilized to classify various cardiac arrhythmias (Ge et al. 2002, Ouelli et al. 2012). Ge et al. (2002) used

AR model to characterize Normal Sinus Rhythm (NSR) including Atrial Premature Contraction (APC), Premature Ventricular Contraction (PVC), Supra Ventricular Tachycardia (SVT), Ventricular Tachycardia (VT) and Ventricular Fibrillation (VF). They used AR coefficients as features and reported an accuracy of 93.2% to 100% using a Generalized Linear Model (GLM) based algorithm in a number of stages performing experiments on 866 beats from MIT-BIH database (Goldberger et al. 2000). Ouelli et al. (2012) used second order AR model individually to model P, QRS and T segments of each beat to classify six different types of beats in their work similar to Ge et al. (2002). They obtained classification accuracy of 96.7% to 100% by using a Quadratic Discriminant Function (QDF) based algorithm on 900 beats collected from 10 recordings of MIT-BIH database Goldberger et al. 2000. But performances of most of the methods are limited by results on small number of samples using simple classifiers. As a consequence, these algorithms fail to produce good result on large data sets. Further, most of the methods reported good classification results on patient specific data or local data. Usually these techniques provide poor performance on global beat classification approach.

In this work, we have presented a new technique of arrhythmia beat classification by employing nonlinear kernel-based modeling of the ECG beats. In order to model the beats, we have used Fast Regularized Kernel Function Approximation (FRKFA) technique proposed by Ghorai et al. (2008) in the recent past. The high dimensional model coefficients are used to characterize the normal and arrhythmia beats. Performance of kernel-based modeling technique is compared with that of linear Autoregressive (AR) modeling (Kobayshi, Mark, & Turin 2012) by using a Support Vector Machine (SVM) classifier (Cristianini & Taylor 2000). Experimental results indicate the superiority of the proposed technique over linear AR modeling technique in a local beat classification approach.

2 AR AND FRKFA MODELING

2.1 *Linear Autoregressive (AR) model*

Autoregressive analysis models the ECG signal as the output (Ge et al. 2002, Ouelli et al. 2012, Kobayshi et al. 2012) of a linear system driven by white noise of zero mean and unknown variance. AR models have the form

$$v[k] = \sum_{i=1}^{p} a_i \, v[k-i] + n[k]; k = 1,2,3,\ldots, \qquad (1)$$

where $v[k]$ is the ECG time series, $n[k]$ is zero mean white noise, a_i's are the AR coefficients, and p is the AR order. A critical issue in AR modeling is the AR order used to model a signal. An appropriate AR model order is crucial to model the signal with high accuracy for its reconstruction.

2.2 *Nonlinear modeling by FRKFA*

Fast Regularized Kernel Function Approximation (FRKFA) (Ghorai et al. 2008) is a kernel-based regression technique which may be implemented in a space lower than the number of training patterns present in the data set. The basic aim of FRKFA formulation was to reduce the computational complexity of approximating a nonlinear kernel function compared to Support Vector Regression (SVR) (Vapnik et al. 1997, Smola and Schölkopf 1998) both in terms of speed and memory usage by storing thin rectangular kernel matrix. Its objective is to fit a linear hyperplane

$$f(x) = k(x^T, B^T)\theta + b \qquad (2)$$

in the feature space, which corresponds to a nonlinear function in the input space. The regression function $f(x)$ given by (2) is obtained by solving following optimization problem:

$$\min_{(\theta, b)} \frac{C}{2} \left\| \begin{bmatrix} \theta \\ b \end{bmatrix} \right\|^2 + \frac{1}{2} \left\| K(A, B^T)\theta + eb - Y \right\|^2. \qquad (3)$$

Here, $A \in \Re^{m \times n}$ is the training data with m samples and each containing n features, $B \in \Re^{d \times n}$ is the basis matrix formed by randomly picking up d training samples (a small fraction ~1 to 15% of A) from training data if requires for large number of training samples, $\theta \in \Re^d$ and $b \in \Re$ are normal vector to the hyperplane in d-dimensional space and bias term, respectively, $x \in \Re^n$ is the input feature vector in n-dimensional space, e is the vector of ones, Y is the target vector, $k(.,.)$ is the kernel function that determines dot product of two vectors in feature space and K is a kernel matrix. The solution of (3) can be obtained by solving a system of linear equation in $d+1$ dimensional space. If θ and b are known, then expression (2) is used to predict new data. In this work, we have used θ and b as the features of an ECG beat.

3 PROPOSED ECG BEAT CLASSIFICATION METHOD

In this work, we have proposed characterization of normal and arrhythmia beats by nonlinear kernel-based regression technique using FRKFA. At the same time, we have also compared the performance of kernel-based FRKFA model to that of linear AR models. We have used one step prediction model for both AR and FRKFA methods. To accomplish this, training sample and target pair $\{(x_i, y_i)\}_{i=1}^{l}$ are formed for each of the segmented ECG beats with the ith training sample $x_i = [y_{i-1}, y_{i-2}, ..., y_{i-p}]$, formed by talking p number of past signal values, where p is the model order. For an AR model of pth order, the feature vector consists of p coefficients. In order to develop a pth order FRKFA model of an ECG beat, having N number of samples, a moving window of length p is used. The past p samples within the window are used to predict the present sample. The window is shifted by one sample. So, there will be $(N - p)$ number of input feature vector—target pair in a training set to train a FRKFA model. Since FRKFA employs a kernel

trick, these p input features are mapped to a feature space of dimension $(N - p)$. So, the regression model has $(N - p)$ coefficients and a bias term. Thus, for a FRKFA model of ECG beat these $(N - p + 1)$ coefficients are used as features for classification using a SVM classifier. Fig. 1 shows prediction ability of one normal ECG beat by 4th order AR model and 16th order FRKFA model, respectively. Both the methods can predict a signal accurately, but some spikes arises at the ends of the signal in case of AR modeling which is not observed in case of FRKFA modeling.

4 EXPERIMENTAL RESULTS

4.1 Data set

We have used data from MIT–BIH database (Goldberger et al. 2000) for experimental verification of two types of modeling techniques. This database contains 48 ECG records; each contains two-channel ECG signals for 30 min duration selected from 24-hr recordings of 47 individuals which are collected from 25 men aged 32 to 89 years, and 22 women aged 23 to 89 years. In this database, the ECG signals are digitized at a sampling rate of 360 Hz. Twenty-three of the recordings (numbered in the range of 100–124) are in-tended to serve as a representative sample of routine clinical recordings and 25 recordings (numbered in the range of 200–234) contain complex ventricular, junctional, and supraventricular arrhythmias.

In this work, normal and arrhythmia beats are collected from the 1st channel of ECG signals within 1–10 minutes duration of each signal. For beat segmentation, we have used the annotation of the database. Each signal is preprocessed with a 4th order bandpass FIR filter with pass band frequencies 0.1 and 100 Hz, respectively, to eliminate noise and artifacts. The length of each beat is considered as 301 number of samples with 100 samples in the left of R-peak and 200 samples in the right of R-peak, such that morphology of all the significant waveforms are captured within this. The total no. of normal and arrhythmia beats and their distribution in training and testing set for the experiment is shown in Table 1.

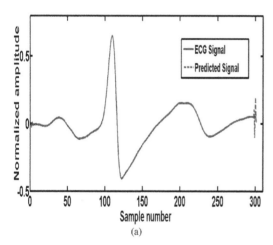

(a)

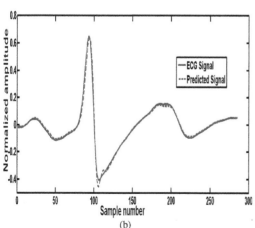

(b)

Figure 1.   A normal ECG beat modeled by (a) 4th order AR model and (b) 16th order FRKFA model.

Table 1.   Description of data set.

Beat type	No. of beats	No. of beats in	
		Training set	Testing set
Normal	24731	1500	23231
Arrhythmia	12408	1500	10908
Total beats	37139	3000	34139

4.2 Experimental setup

All the modeling techniques are implemented using MATLAB programming language. To find the model coefficients of FRKFA method, it has been implemented in MATLAB using a Gaussian kernel of the form $k(x_i, x_j) = \exp(-\gamma)\|x_i - x_j\|^2$. We have evaluated performance of the model parameters to discriminate normal and arrhythmia beats by binary SVM classifier (Cristianini & Taylor 2000). For SVM classifier implementation, we have used LIBSVM Toolbox (Chang & Lin 2001). The optimal value of the regularization parameter C and Gaussian kernel parameter γ for SVM are selected from the set of values $[2^i | i = -5, -4, \ldots, 18]$. We have performed the experiment 50 times by randomly dividing the total beats into training and testing sets. The optimal parameter set is selected by the performance of a classifier on a tuning set consists of 20% of training data set (Mitchell 1997). Finally, the performance of the trained classifier has been evaluated by using all the training samples with the selected optimal parameters. Each training feature is normalized in the range [0 1]. The testing features are also normalized in the range [0 1] by using the minimum and maximum values of each training feature. The final accuracy and confusion matrix presented here are the average of results of 50 runs of the experiment. The experiment is performed by observing the performance of classification accuracy with the variation of model order of AR coefficients. But, for FRKFA model we have chosen a window of length of 16 to model normal and arrhythmia beats.

4.3 Results

Table 2 shows the performance of SVM classifier on AR and FRKFA model parameter for normal and arrhythmia beat classification. We have observed the effect of model order on classification accuracy for AR model. For FRKFA model we have chosen a window of length 16. For each model, we have provided average % classification accuracy, standard deviation and selected optimal regularization and kernel parameters. It is observed from the results that 4th order AR model parameters offer the highest classification accuracy of 84.25%. On the other hand, FRKFA model parameters offer an accuracy of 95.43% on the same binary beat classification problem. It is also noticed from the Table 2 that as the AR model order increases its performance decreases.

4.4 Analysis of performance

We have compared the performance of AR and FRKFA models by evaluating sensitivity and specificity of the results obtained in Table 2. These parameters are defined as follows:

$$\text{Sensitivity} = \frac{TP}{TP + FN} \quad (4)$$

$$\text{Specificity} = \frac{TN}{TN + FP}. \quad (5)$$

where TP, TN, FP and FN stands for the words true positive, true negative, false positive and false negative, respectively. Thus, sensitivity is related to the fraction of positive events correctly detected while the specificity measures the fraction of negatives which are correctly detected. The confusion matrices of the performances of AR and FRKFA models are shown in Tables 3 and 4, respectively. Sensitivity and specificity of both AR and FRKFA models are shown in the Table 5. It shows that both sensitivity and specificity of FRKFA model are above 95% and simultaneously both of these values are improved approximately by 11%

Table 2. Performance of AR and FRKFA models.

Types of models	Order of models/ window size	% Average accuracy ± std. dev.	Selected kernel parameters
AR	4	84.25 ± 0.694	$C = 2^{14}, \gamma = 2^0$
	5	82.26 ± 0.552	$C = 2^{15}, \gamma = 2^2$
	6	82.88 ± 0.445	$C = 2^{15}, \gamma = 2^3$
	7	79.84 ± 0.416	$C = 2^{15}, \gamma = 2^2$
	8	79.12 ± 0.586	$C = 2^{16}, \gamma = 2^3$
	9	77.95 ± 0.815	$C = 2^{16}, \gamma = 2^3$
FRKFA	16	95.43 ± 0.372	$C = 2^{14}, \gamma = 2^{-1}$

Table 3. Confusion matrix obtained by 4th order AR model parameters.

True value	Predicted value	
	Normal	Arrhythmia
Normal	19511	3720
Arrhythmia	1652	9256

Table 4. Confusion matrix obtained FRKFA model parameters.

True value	Predicted value	
	Normal	Arrhythmia
Normal	22154	1077
Arrhythmia	483	10425

Table 5. Sensitivity and Specificity of AR and FRKFA models.

Model type	% Sensitivity	% Specificity
AR	83.99	84.86
FRKFA	95.36	95.57

Table 6. Statistics of 4th order AR model coefficients.

| Type of beats | Mean and std. dev. of the coefficients a_i | | | |
	a_1	a_2	a_3	a_4
Normal	−3.6492	5.2098	−3.4495	0.8941
	0.064	0.144	0.112	0.034
Arrhythmia	−3.6170	5.1426	−3.4151	0.8932
	0.078	0.168	0.121	0.035

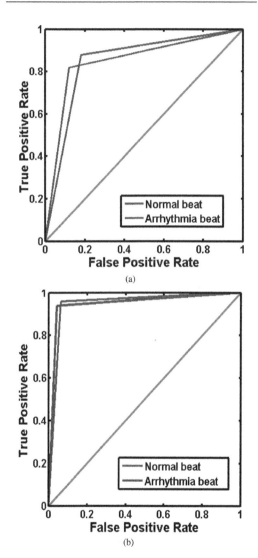

Figure 2. ROC curves of results obtained with (a) AR (b) FRKFA coefficients.

for the classification of normal and arrhythmia beats by FRKFA model parameters. The experimental results show that performance of kernel-based FRKFA model parameters improves approximately by an accuracy of 11 than that of linear AR models. The poor performance of the AR parameters is analyzed using their statistics. Table 6 shows the means and standard deviations of the coefficients of 4th order AR model for normal and arrhythmia beats. This shows that model coefficients or characteristic features of both the classes of ECG beats are scattered and overlapped with each other. This fact limits the classification performance of AR coefficients by a classifier. FRKFA models, on the other hand, maps a pattern in high dimensional feature space equivalent to the number of features in the training set. For example, FRKFA model with window length 16 maps the pattern in 285-dimensional space as the training set has 301−16 = 285 instances. As a result, FRKFA model uses 285 coefficients and a bias term totaling 286 features to characterize normal and arrhythmia beats. This large number of coefficients capturing the characteristics of the beats in a better way and hence performing better compared to linear AR models.

5 CONCLUSIONS

In this work, we have set up a new kernel-based modeling technique for normal and arrhythmia beats of ECG signal and their classification using the model parameters. The concept of kernel-based modeling technique is a new one and has not been previously employed in ECG beat prediction. FRKFA model parameters provide an average accuracy of 95.43% on 34139 beats of two different categories in a global beat classification framework. Performance of this model is much higher compared to the performance obtained by linear AR model parameters using SVM classifier. Thus, the high performance of kernel-based FRKFA model is of great significance in view of early detection of arrhythmia by a computer aided diagnosis system.

than that of the sensitivity and specificity of AR model. The ROC curves for both 4th order AR model and FRKFA model are shown in Fig. 2. It is observed that ROC curve of FRKFA model is far better than that of AR model. This shows a clear picture of superiority of the proposed method

This classification technique may be used to study performance of multiclass beat classification using full 30 minutes duration of each signal. Further, it will be worth to study the effect of variation of classification accuracy with the variation of window length of FRKFA model for binary and multiclass beat classification. Other kernel based regression techniques as well as classifiers may be employed to test the performance on these features. In addition, this type of kernel-based modeling techniques may be used for missing beat prediction of ECG signal.

REFERENCES

Adamec, J. & R. Adamec (2008). *ECG Holter: Guide to Electrocardiographic Interpretation.* Springer.

Andreão, R. V., B. Dorizzi, & J. Boudy (2006). Ecg signal analysis through hidden markov models. *IEEE Trans. Biomed. Eng. 53*(8), 1541–1549.

Chang, C. C. & C. J. Lin (2001). Libsvm: A library for support vector machines. Technical report. [Online]. Available: http://www.csie.ntu.edu.tw/cjlin/libsvm.

Chazal, P. D., M. ÒDwyer, & R. Reilly (Jul, 2004). Automatic classification of heartbeats using ecg morphology and heartbeat interval features. *IEEE Trans. Biomed. Eng. 51*(7), 1196–1206.

Chen, Y.-H. & S.-N. Yu (2007). Electrocardiogram beat classification based on wavelet transformation and probabilistic neural network. *Pattern Recognit. Lett. 28*(12), 1142–1150.

Cristianini, N. & J. S. Taylor (2000). *An Introduction to Support Vector Machines.* Cambridge, Mass.

Ebrahimzadeh, A., B. Shakiba, & A. Khazaee (2014). Detection of electrocardiogram signals using an efficient method. *Applied Soft Computing 22*, 108–117.

Edla, S., N. Kovvali, & A. Papandreou-Suppappola (2014). Electrocardiogram signal modeling with adaptive parameter estimation using sequential bayesian methods. *IEEE Trans. Signal Processing 62*(10), 2667–2679.

Ge, D., N. Srinivasan, & S. M. Krishnan (2002). Cardiac arrhythmia classification using autoregressive modeling. *Biomedical Engineering OnLine.* [Online]. http://www.biomedicalengineering-online.com/content/1/1/5.

Ghorai, S., A. Mukherjee, & P. K. Dutta (2008). Fast regularized kernel function approximation. In *IEEE Region Ten Conference, TENCON 2008*, Hyderabad, India.

Goldberger, A. L., L. A. N. Amaral, L. Glass, J. M. Hausdorff, P. C. Ivanov, R. G. Mark, J. E. Mietus, G. B. Moody, C.-K. Peng, & H. E. Stanley (2000). Physiobank, physiotoolkit, and physionet: Components of a new research resource for complex physiologic signals. *Circulation Electron. 101*(23), e215–e220.

Inan, O. T., L. Giovangrandi, & G. T. A. Kovacs (2006). Robust neural-network based classification of premature ventricular contractions using wavelet transform and timing interval features. *IEEE Trans. Biomed. Eng. 53*(12), 2507–2515.

Kobayshi, H., B. L. Mark, & W. Turin (2012). *Probability, Random Process and Statistical Analysis,* Cambridge, UK.

Lagerholm, M., C. Peterson, G. Braccini, L. Edenbrandt, & L. Sornmo (2000). Clustering ecg complexes using hermite functions and self-organizing maps. *IEEE Trans. Biomed. Eng. 47*(7), 838–848.

Laguna, P. (1997). A database for evaluation of algorithms for measurement of QT and other waveform intervals in the ECG. In *Proc. Computers in Cardiology*, pp. 673–676.

Melgani, F. & Y. Bazi (2008). Classification of electrocardiogram signals with support vector machines and particle swarm optimization. *IEEE Trans. Inf. Tech. Biomed. Eng. 12*(5), 667–677.

Mitchell, T. M. (1997). *Machine Learning.* Singapore: The Mc-GRaw Hill Companies.

Osowski, S., L. T. Hoai, & T. Markiewicz (2004). Support vector machine-based expert system for reliable heartbeat recognition. *IEEE Trans. Biomed. Eng. 51*(4), 582–589.

Ouelli, A., B. ElhadadiL, H. Aissaoui, & B. Bouikhalene (2012). AR modeling for automatic cardiac arrhythmia diagnosis using QDF based algorithm. *International Journal of Advanced Research in Computer Science and Software Engineering 2*(5), 493–499.

Rahhal, M. M. A., Y. Bazi, N. Alajlan, S. Malek, H. Al-Hichri, F. Melgani, & M. A. A. Zuair (2015). Classification of AAMI heartbeat classes with an interactive ELM ensemble learning approach. *Biomedical Signal Processing and Control 19*, 56–67.

Smola, A. J. & B. Schölkopf (1998). A tutorial on support vector regression. Technical report, Royal Holloway College, Neuro COLT Tech. Rep. TR-1998-030, London, UK.

Thomas, M., M. K. Das, & S. Ari (2015). Automatic ECG arrhythmia classification using dual tree complex wavelet based features. *International Journal of Electronics and Communications 69*, 715–721.

Vapnik, V., S. Golowich, & A. J. Smola (1997). Support vector method for function approximation, regression estimation, and signal processing. In *Proc. NIPS, Camb., MA: MIT Press*, Volume 9.

Computer, Communication and Electrical Technology – Guha, Chakraborty & Dutta (Eds)
© 2017 Taylor & Francis Group, ISBN 978-1-138-03157-9

Characterization of obstructive lung diseases from the respiration signal

S. Sarkar & S. Pal
Department of Applied Physics, University of Calcutta, Kolkata, India

S. Bhattacherjee
School of Mechatronics and Robotics, IIEST, Shibpur, India

P. Bhattacharyya
Institute of Pulmocare and Research, New Town, Kolkata, India

ABSTRACT: Obstructive Lung Disease (OLD) is a major respiratory disease with airflow limitation due to an obstruction in the airway. Lung function gradually decreases in patients with OLD if the disease remains untreated. Early detection and continuous treatment with a change in lifestyle might improve the quality of life and also prevent the possibility of disease progression. Conventionally, OLD is detected through a cumbersome procedure using a spirometer. In this work, a new method is proposed in which the respiration signal from normal subjects and OLD patients were analyzed and morphological changes were studied. Features from the respiration signal were extracted and used for classification using the two-class Support Vector Machine (SVM) classifier. The present study showed up to 100% accuracy with a specified set of features for 22 subjects as discussed in the result section.

1 INTRODUCTION

Obstructive Lung Disease (OLD) is a type of respiratory disease characterized by airflow obstruction, dyspnea, and airway inflammation (Antoniu, 2010). Asthma and Chronic Obstructive Pulmonary Disease (COPD) are two major obstructive lung diseases triggered by exposure to smoking, indoor and outdoor pollution, allergens, etc. (Antoniu, 2010; Buist, 2003). OLD is becoming a major cause of illness with its increasing morbidity and mortality rate worldwide (Lazovic, Zlatkovic, Mazic, Stajic, & Delic, 2013; Renwick & Conolly, 1996). It often remains untreated until it becomes severe, causing a significant burden to society (Lazovic et al., 2013). OLD also results in an increased risk of cardiovascular diseases (van Schayck et al., 2003; Siegler, 1977). For these reasons, the necessity of developing a non-obstructive and non-invasive technique for early detection and diagnosis of this disease has witnessed a rapid surge of interest in recent years.

Among all the techniques available for COPD and asthma detection, spirometry is the most widely used pulmonary function test in the diagnosis of airway obstruction and routine follow-up of pulmonary diseases (Finkelstein, Cha, & Scharf, 2007; Vandevoorde, Verbanck, Schuermans, Kartounian, & Vincken, 2005). In this disease, airway constriction occurs during expiration, and it is defined by a decreased forced expiratory volume in the first second to forced vital capacity ratio (FEV1/FVC) (Ferguson, Enright, Buist, & Higgins, 2000; Sahin, Ubeyli, Ilbay, Sahin, & Yasar, 2010). But apart from its vast use, it often performs poorly because of its high patient-effort dependency, which makes the test uncomfortable for most of the patients. Spirometry is also unsuitable for non-ambulatory and ICU patients and very young or old patients. The dependence on patient effort for spirometric measurement leads the result to be a function of some non-measurable parameters, which must be eliminated for the standardization of the detection of respiratory diseases. Apart from this, methods like lung imaging technique by Milne and King (2014), volatile organic compound analysis from exhaled breath by Berkel et al. (2009), CO_2 concentration analysis using capnography (Mieloszyk et al., 2014), exhaled breath analysis using an electronic nose (Velasquez, Duran, Gualdron, Rodriguez, & Manjarres, 2009; Valera, Togores, & Cosio, 2012), and a nasal obstruction measurement technique (Manjunatha, Mahapatra, Prakash, & Rajanna, 2015) have also been developed. But these methods are costly and time-consuming, and most of them are still in the development stage in research laboratories.

The measurement of respiration signal is crucial in health monitoring (Cretikos et al., 2008; Fekr, Radecka, & Zilic, 2015), especially in the case of

OLDs to diagnose the current condition and severity. In this study, the respiration signal acquired from a thoracic belt was studied in both normal subjects and patients with OLD. The technique used in this study is completely non-obstructive, suitable for all age groups, and can be used even in critical conditions. The features extracted from the respiration signals were used to demonstrate the characterization of these two groups.

2 MATERIALS

2.1 *Study population*

In this study, ten healthy individual subjects with no history of pulmonary or cardiac disease or even tobacco use were selected as a normal subject group. Twelve patients previously diagnosed with OLD by pulmonary function testing constituted the disease group. Patients under the disease group were either smokers or had a history of long-term asthma or heavy indoor or outdoor pollution exposure. All studies were carried out in the biomedical laboratory of Department of Applied Physics, University of Calcutta and Institute of Pulmocare & Research. The participants gave their informed consent to take part in this study. Each of them underwent a medical history checkup along with physical examination by an expert investigator prior before being accepted for the study. The respiration signal of each subject was collected while the subject was resting in supine position. The breathing protocol was maintained in a normal and relaxed way. The details of the 22 subjects are given in Table 1.

2.2 *Data acquisition*

A data acquisition system MP-45, designed by Biopac Systems Inc., was used to collect study data (Pflanzer & McMullen, 2014). For the measurement of the respiration signal, the respiratory effort transducer (SS5 LB) was tied across the upper chest of the subject. The thoracic belt measured the volume change in the thoracic cavity and

lungs during inspiration and expiration. The transducer was connected to the MP-45 unit via cables. Programs written in the MATLAB platform were used for data processing and feature extraction purposes. Each real-time data was recorded for a 300 second time duration with the signals being sampled at a frequency of 1000 Hz.

3 METHOD

3.1 *Pre-processing*

During the pre-processing of raw and noisy respiration signals, the precision of the respiratory frequency should be considered for extracting the original respiration signal without losing information. The frequency band of the respiration signal was generally <1 Hz (Chan, Ferdosi, & Narasimhan, 2013; Park, Noh, Park, & Yoon, 2008). A second-order IIR Butterworth filter in the frequency range of 0.2–0.7 Hz was used to eliminate unwanted noise, preserving necessary information. The amplitude of the filtered respiration signal was normalized before feature extraction. Figure 1 illustrates the measured respiration signal along with the signal after filtering.

3.2 *Feature extraction*

Respiration is a vital parameter in the clinical and diagnostic field (Sarkar, Bhattacherjee, & Pal, 2015). Abnormal respiration and sudden change in the respiratory rate indicate a major physiological imbalance. In OLD, one of the main symptoms is the variable expiratory airflow, i.e., difficulty in exhalation due to bronchoconstriction, airway wall thickening, and excessive mucus production (Buist, 2006; FitzGerald, 2015). These physiological changes in lungs are also reflected in respiration leading to the extraction of features from the respiration signal in this study.

3.2.1 *Average time ratio*
Following the normalization of the respiration signal, the location of the starting and ending points and the peak of a respiration cycle were detected

Table 1. Details of the normal and disease group.

Database	No. of subjects	Age range	Smoker	Chronic asthma/ pollution exposure	Non-smoker
Normal group	10 (5 male, 5 female)	21–63	–	–	10
Disease group	12 (10 male, 2 female)	23–67	9	3	–

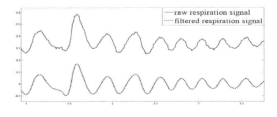

Figure 1. Original respiration signal (solid line) and respiration signal after filtering (dotted line).

using a peak-detection algorithm. The inspiration time (T_I) was derived by calculating the time interval between the starting point and the peak and the expiration time (T_E) was derived from the time difference between the peak and the ending point of a cycle. The time ratio (F_1) was calculated using the following equation:

$$F_1 = \frac{T_E}{T_I} \qquad (1)$$

The process was repeated five times and the average of all time ratios was produced for each subject.

3.2.2 Average area ratio

To evaluate the area ratio, five cycles from a previously normalized respiration signal were selected randomly. The maximum and minimum points were detected from each cycle and the area under the curve against the baseline for both inspiration and expiration of that cycle were calculated to derive the inspiration area (A_I) and expiration area (A_E), respectively. The area ratio (F_2) was calculated using equation 2 and the whole process was repeated five times to derive the average area ratio.

$$F_2 = \frac{A_E}{A_I} \qquad (2)$$

3.2.3 Average respiration rate

Respiration rate is actually the number of inspiration-expiration cycles, i.e., the number of respiration cycles per unit of time. To estimate the respiration rate from the detected peaks, the time interval between two successive maxima were computed and the instantaneous breathing rates (F_3) were calculated using equation 3:

$$F_3 = \frac{t_{(n+1)} - t_n}{60} \qquad (3)$$

where F_3 = respiratory rate, $t_{(n+1)}$ = time period for the occurrence of the (n+1)th respiratory peak, and t_n = time period for the occurrence of the nth respiratory peak.

The respiration rate of each individual thus obtained for every consecutive cycle for a specified time window length of 60 seconds and the average was calculated.

3.3 Classification

In this study, to discriminate the respiration signal from the extracted features for both the normal and the disease group, a binary Support Vector Machine (SVM) classifier was implemented. The SVM is a classification tool that maps the input features into a higher-dimensional feature space through some non-linear mapping, and a decision surface is constructed over there. The formulation of the SVM classifier (Christianini & Taylor, 2000; Melgani & Bazi, 2008) is briefed below. In this study, the training data are considered to be of two class as

$$\lambda = [(i_1, k_1), \ldots, (i_p, k_p)], i \in \Gamma^N, k \in (\pm 1) \qquad (4)$$

Here, Γ = the radius of a hyper-sphere enclosing all the data points and N = number of parameters.

For the features cannot be separated linearly, the kernel function $F(i_k, i) = \langle \phi(i_k), \phi(i) \rangle$ can be used and the nonlinear classifier is given by

$$g(i) = \text{sgn}[\langle u^*, i \rangle + c] = \mathbf{sgn}\left[\sum_{m=1}^{p} k_i \alpha^* F(i_m, i) + c^*\right] \qquad (5)$$

where α = Lagrange multiplier.

The test performance of the classifier was determined by calculating the specificity, sensitivity, and accuracy, where specificity is the number of true negative decisions/total number of the actual negative subjects, sensitivity is the number of true positive decisions/total number of actual positive subjects, and accuracy is the number of correct decisions/total number of subjects.

4 RESULT AND DISCUSSION

The respiration signals of 22 subjects acquired from the data acquisition device from both the normal and the OLD groups were analyzed. The filtered respiratory signals of the subjects from both groups showed some visual differences between them, as shown in Figure 2.

Three features were extracted for all 22 subjects in this study. A partial table of the extracted features for subjects from both groups are shown in Table 2.

Figure 3 illustrates the output of the SVM classifier using the specified features for respiration with two feature sets at a time.

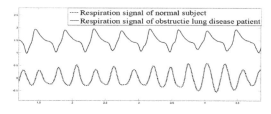

Figure 2.   Respiration signal of OLD patient (solid line) and normal subject (dotted line).

Table 2. Partial list of extracted features for both groups.

Subject	Average area ratio	Average time ratio	Average respiration rate
Normal* 1	0.84	0.89	16
Normal 2	1.06	1.01	21
Normal 3	1.07	1.02	20
Normal 4	0.78	0.83	17
Normal 5	1.06	1.08	21
Disease# 1	1.17	1.26	22
Disease 2	3.10	2.38	12
Disease 3	1.66	1.66	22
Disease 4	1.38	1.78	14
Disease 5	2.53	2.32	13

*Normal denotes normal subjec
#Disease denotes patients with OLD

Table 3. Classification performance of different features.

Feature	Specificity	Sensitivity	Accuracy
Area ratio vs. respiration rate	90%	91.67%	91%
Area ratio vs. time ratio	100%	100%	100%
Time ratio vs. respiration rate	90%	91.67%	91%

on the three spirometric parameters for classifying normal, restrictive, and obstructive patterns. In this work, the differentiation from normal to obstructive has been done purely based on the extracted features from the ECG and EDR signals. The spirometric method for monitoring patients with respiratory diseases is effort-specific, whereas the proposed method is patient-effort-independent, exerting no physical or psychological stress in the patients. The current study shows a classification accuracy of 91%, 100%, and 91% in the case of area ratio vs. respiration rate, area ratio vs. time ratio, and time ratio vs. respiration rate, respectively.

5 CONCLUSION

Respiration is a very important physical property and, if properly used, can be an excellent source of physiological information. Conventional and some newly developed methods of OLD detection face various drawbacks. In this situation, a non-obstructive method based on the respiration signal is proposed in this study. This method acquired only the respiration signal using a thoracic belt and extracted various features from the signal. The purpose of this study was to inquire the characteristic differences between the normal and the disease groups based on the extracted features. The classification between different features for the normal group and the disease group indicates that the present study may prove effective for the detection and monitoring of OLD. This study can be extended to the classification of specific obstructive diseases with more number of patient data and healthy control groups.

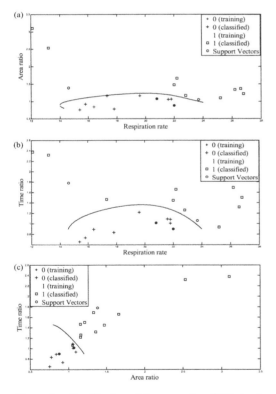

Figure 3. Classification of features using SVM classifier for (a) area ratio vs. respiration rate, (b) time ratio vs. respiration rate, and (c) time ratio vs. area ratio.

Table 3 elaborates the test performance of the classifier. It shows excellent accuracy with the subjects considered in this study.

A previous study by Sahin et al. (2010) showed a total of 97.32% accuracy using multiclass SVM

REFERENCES

Antoniu, S.A. (2010). Descriptors of dyspnea in obstructive lung diseases. *Multidisciplinary Respirator Medicine*, 5(3), 216–219. doi:10.1186/2049-6958-5-3-216.

Berkel, J.J.B.N.Van, Dallinga, J.W., Möller, G.M., Godschalk, R.W.L., Moonen, E.J., Wouters, E.F.M., &

Van Schooten. F.J. (2009). A profile of volatile organic compounds in breath discriminates COPD patients from controls. *Respiratory Medicine,* 104, 557–563. doi:10.1016/j.rmed.2009.10.018.

Buist, A.S. (2003). Similarities and differences between asthma and chronic obstructive pulmonary diseases: treatment and early outcomes. *European Respiratory Journal,* 39, 30–35. doi:10.1183/09031936.03.00404903.

Buist, A.S. (2006). Global Initiative for Chronic Obstructive Lung Disease. Global strategy for the diagnosis management, and prevention of chronic obstructive pulmonary disease. Retrieved from http://www.who.int/respiratory/copd/GOLD_WR_06.pdf.

Chan, A.M., Ferdosi, N., & Narasimhan, R. (2013, July). *Ambulatory Respiratory Rate Detection using ECG and a Triaxial Accelerometer.* Paper presented at the Annual International Conference of the IEEE EMBS, Osaka, Japan.

Christianini N. & Taylor J.S. 2000, *An Introduction to Support Vector Machines and Other Kernel-based Learning Methods,* Cambridge University Press, Cambridge.

Cretikos, M.A., Bellomo, R., Hillman, K., Chen, J., Finfer, S., & Flabouris, A. (2008). Respiratory rate: the neglected vital sign. *MJA,* 188(11), 657–659. Retrieved from https://www.mja.com.au/journal/2008/188/11/respiratory-rate-neglected-vital-sign

Fekr, A.R., Radecka, K., & Zilic, Z. (2015). Design and Evolution of an Intelligent Remote Tidal Volume Variability Monitoring System in E-Health Application. *IEEE Journal of Biomedical and Health Informatics,* 19(5), 1532–1548. doi:10.1109/JBHI.2015.2445783.

Ferguson, G.T., Enright, P.L., Buist, A.S., & Higgins, M.W. (2000). Office spirometry for lung health assessment in adults: a consensus statement from the National Lung Health Education Program. *Chest,* 117, 1146–1161. doi:10.1378/chest.117.4.1146.

Finkelstein, J., Cha, E., & Scharf, S.M. (2007). Chronic obstructive pulmonary disease as an independent risk factor for cardiovascular morbidity. *International Journal of COPD,* 4, 337–349. Retrieved from http://www.ncbi.nlm.nih.gov/pmc/articles/PMC2754086/pdf/copd-4-337.pdf

FitzGerald, J.M. (2015). Global Initiative for Asthma-Pocket Guide for Asthma Management and Prevention. http://ginasthma.org/wp-content/uploads/2016/01/GINA_Pocket_2015.pdf

Lazovic, B., Zlatkovic M.S., Mazic, S., Stajic, Z., & Delic, M. (2013). Analysis of Electrocardiogram in Chronic Obstructive Pulmonary Disease patients. *Med. Pregl,* LXVI(3–4), 126–129. doi:10.2298/MPNS1304126 L.

Manjunatha, G.R., Mahapatra, D.R., Prakash, S., & Rajanna, K. (2015). Validation of polyvinylidene fluoride nasal sensor to assess nasal obstruction in comparison with subjective technique. *American Journal of Otolaryngology,* 36(2), 122–129. doi:10.1016/j.amjoto.2014.09.002.

Melgani, F., & Bazi Y. (2008). Classification of Electrocardiogram Signals with Support Vector machines

and Particle Swarm Optimization. *IEEE Transactions on Information Technology in Biomedicine,* 12(5), 667–677. doi: 10.1109/TITB.2008.923147.

Mieloszyk, R.J., Verghese, G.C., Deitch, K., Cooney, B., Khalid, A., Mirre-González, M.A., Heldt, T., & Krauss, B.S. (2014). Automated Quantitative Anlysis of Capnogram Shape for COPD-Normal and COPD-CHF Classification. *IEEE Transactions on Biomedical Engineering,* 61(12), 2882–2890. doi: 10.1109/TBME.2014.2332954.

Milne, S., & King, G.G. (2014). Advanced imaging in COPD: insights into pulmonary pathophysiology. *Journal of Thoracic Disease,* 6(11), 1570–1585. doi:10.3978/j.issn.2072–1439.2014.11.30.

Park, S.B., Noh, Y.S., Park, S.J., & Yoon, H.R. (2008). An improved algorithm for respiration signal extraction from electrocardiogram measured by conductive textile electrodes using instantaneous frequency estimation. *Med Bio Eng Comput.,* 46, 147–158. doi:10.1007/s11517-007-0302-y.

Pflanzer, R., & McMullen, W. (2014). This is a chapter. Biopac Student Lab Laboratory Manual: *Respiratory Cycle I Procedure* (P1-P9). Goleta, CA: BIOPAC Systems, Inc.

Renwick, D.S., & Conolly, M.J. (1996). Prevalence and treatment of chronic airways obstruction in adults over the age of 45. *Thorax,* 21, 164–168. doi:10.1136/thx.51.2.164.

Sahin, D., Ubeyli, E.D., Ilbay, G., Sahin, M., & Yasar, A.B. (2010). Diagnosis of Airway Obstruction or Restrictive Spirometric Patterns by Multiclass Support Vector Machine. *J. Med. Syst.,* 3, 967–973. doi:10.1007/s10916-009-9312-7.

Sarkar, S., Bhattacherjee, S. & Pal, S. (2015, September). *Extraction of Respiration Signal from ECG for Respiratory Rate Estimation.* Michael Faraday IET International Summit: MFIIS-2015, Kolkata, India.

Siegler, D. (1977). Reversible electrocardiographic changes in severe acute asthma. *Thorax,* 32, 328–332. doi:10.1136/thx.32.3.328.

Valera, J.L., Togores, B., & Cosio, B.G. (2012). Use of the Electronic Nose for Diagnosing Respiratory Diseases. *Arch Bronconeumol,* 48(6), 187–188. doi:10.1016/j.arbres.2011.08.004.

Vandevoorde, J., Verbanck, S., Schuermans, D., Kartounian, J. & Vincken, W. (2005). FEV1/FVC6 and FEV6 as an alternative for FEV1/FVC and FVC in the spirometric detection of airway obstruction and restriction. *Chest,* 127(5), 1560–1564. doi:10.1378/chest.127.5.1560.

van Schayck, O.C.P., D'Urzo, A., Invernizzi, G., Roman, M., Stallberg, B., & Urbina, C. (2003). Early detection of chronic obstructive pulmonary disease (COPD): the role of spirometry as a diagnostic tool in primary care. *Primary Care Respiratory Journal,* 12(3), 90–93. doi:10.1038/pcrj.2003.54.

Velasquez, A., Duran, C.M., Gualdron, O., Rodriguez, J.C., & Manjarres, L. (2009). Electronic Nose To Detect Patients with COPD From Exhaled Breath. *AIP Conference Proceedings,* 452–454.

Computer, Communication and Electrical Technology – Guha, Chakraborty & Dutta (Eds)
© 2017 Taylor & Francis Group, ISBN 978-1-138-03157-9

Studies on a formidable dot and globule related feature extraction technique for detection of melanoma from dermoscopic images

S. Chatterjee, D. Dey & S. Munshi

Department of Electrical Engineering, Jadavpur University, Kolkata, West Bengal, India

ABSTRACT: Among the wide variety of skin abnormalities, malignant melanoma is the deadliest and causes a vast majority of deaths. Existence of dark dots/globules is a feature demarcating the melanocytic skin lesions from the others. In this reported research work, a methodical approach for the identification and segmentation of dark dots/globules from the dermoscopic images using mathematical morphology has been proposed and implemented. Different morphological gradient operations have been performed with varying size of the Structuring Element (SE) to identify the small circular structures present in the skin lesion area. In the segmentation stage the threshold value has been selected according to the range of intensity values present in those dot/globule regions. Subsequently, a number of dot/globule related features have been extracted for the classification of the dermoscopic images. Furthermore, it has been shown that the dot/globule related features have a significant effect on the classification of the melanoma, and has the potential for scoring even 100% sensitivity, provided it is used in conjunction with appropriately chosen classification algorithms.

1 INTRODUCTION

Non-invasive and non-contact imaging techniques have a great influence in the field of dermatology for the identification of various skin diseases. Dermoscopy or dermatoscopy is a gold standard imaging tool for the identification of melanocytic skin lesions with a high degree of diagnostic sensitivity (Serup et al. 2006). Experts use this tool to examine different morphological patterns including pigment network, irregular streaks, blue white veils, dots and globules (Marks 1995). Identification of this complex structures and consequently providing a proper diagnosis based on visual evaluation only, is an extremely challenging task for the expert radiologists and dermatologists. According to the existing literature, computer based image analysis techniques have been used for image segmentation, and border detection along with the identification of differential structures. Sadeghi et al. (2013), proposed a methodical approach for the identification of streaks and also the classification of dermoscopy images based on the presence and irregular nature of the streaks in the images. Bank of directional filters and a connected component analysis technique has been used for the detection of pigment network by Barata et al. (2010). The identification of dots and globules present in an image and improvement of classification accuracy have been described by Maglogiannis et al. (2015). Some state-of-the-art techniques for dermoscopic image segmentation and border detection have

been discussed (Ma et al. 2016, Sadri et al. 2013). Mathematical morphology has been considered as an efficient tool for the extraction of various shape-features from an image. An outline of the application of angiographic image processing based on grey-level hit-or-miss transform has been given by Naegel et al. (2007). A detailed and up-to-date survey of adaptive mathematical morphology using adaptive Structuring Element (SE) has been discussed by Ćurić et al. (2014). Bai et al. (2015) proposed a different approach of grey level hit-or-miss transform using multi structuring element of varying width and direction, for the extraction of linear features of different images. Rastgoo et al. (2015) extracted different local as well as global shape, texture and color features for the automatic classification of melanoma and dysplastic nevi using different classification methods. A computational image analysis system for the estimation of melanoma thickness and the classification of images using machine learning algorithm has been explained by Sáez et al. (2016).

For the categorization of melanoma, identification of dots and globules has been considered as an important step by the expert dermatologists and clinicians. Dots having small structures and globules with some circular or oval shaped area of black, brown or grey color, may be observed in center area or may be spread throughout the entire lesion.

In the research work that is being reported here, mathematical morphology has been used for the identification of dots and globules in different skin lesions. For the classification of malignant melanoma, the classification performance has been improved by the inclusion of the dot and globules related features. This paper has been organized as follows. A brief introduction of mathematical morphology has been given in section 2. A detailed outline of the proposed research work has been provided in section 3, followed by results and discussions in section 4. Finally the paper has been concluded in section 5.

2 MATHEMATICAL MORPHOLOGY

Morphological image processing has become a gold standard imaging toolbox and now a days, has been utilized in a wide range of industrial applications, biomedical image processing, pattern recognition, robot vision, etc. The foundation of mathematical morphology is the set theory (Soille 2004). Mathematical morphology deals with different geometrical structures of various shape and size. The aim of the different morphological operators is to extract relevant structures present in that image by probing the image with another set of structures small in size and known shape, referred to as SE. Choice of the SE is based on the prior knowledge of the relevant shapes and sizes of the structures present in that image. Erosion and dilation are the fundamental operations of mathematical morphology.

The erosion of a set F (an image) with a SE K can been defined as,

$$\left(\mathbb{E}(F)\right)(f) = \min_{k \in K} F(f + k) \tag{1}$$

The dual operation of erosion is dilation. The dilation of a set F (an image) with a SE K can been defined as,

$$\left(\mathbb{D}(F)\right)(f) = \max_{k \in K} F(f + k) \tag{2}$$

Erosion has an effect of shrinking the input image, eliminating small objects. Dilation has an expanding effect, filling in small structures compared to the SE in an image.

Besides these two primary operations, there are two secondary operations namely opening and closing, which play an important role in morphological image processing. Opening of an image F with an SE K can be defined as a combination of erosion and dilation,

$$\mathbb{O}(F) = \mathbb{D}(\mathbb{E}(F)) \tag{3}$$

Similarly closing of an image F with a SE K can be defined as,

$$\mathbb{C}(F) = \mathbb{E}(\mathbb{D}(F)) \tag{4}$$

3 PROPOSED METHODOLOGY

The proposed technique as shown graphically in Figure 1 has been discussed in following sub sections.

3.1 *Preprocessing and segmentation*

Most of the dermoscopic images are contaminated by noise due to external effects such as uneven illumination, and also have hair artifacts. The noises have been removed by using median filter and morphological filters have been applied for the identification of hair artifacts. After the removal of the artifacts, the region of interest, i.e. the skin lesion area has been segmented using different morphological operations. The entire preprocessing and segmentation method has been discussed (Chatterjee et al. 2015). Example of an image preprocessing and segmentation has been shown in Figure 2.

3.2 *Dots and globules detection*

As discussed in the earlier section, the presence of dots and globules has been considered as one of the demarcating feature for the identification of melanocytic skin lesions. This paper focuses mainly on the identification of small dot and globule-like structures present on the skin lesion area, from the dermoscopic images. To exclude the dots/globules present outside the skin lesion area, the original grey scale image has been masked with the segmented binary image. The algorithm for identification of circular structures consists of two main steps—the identification of the circular structures and the dot segmentation. In this proposed algorithm three variables, namely the threshold sensitivity (value between 0 and 1), and the minimum and the maximum radii of the circular structures have been considered.

Algorithm: Dots/globules identification and segmentation.

Input: I ← *Original grey scale image, masked with segmented image.*
   T_s ← *Threshold sensitivity*
   R_{min} ← *Radius of the minimum object to enhance.*
   R_{max} ← *Radius of the maximum object to enhance.*
for x = R_{min} : 2 : R_{max}
   SE ← *Circular_Kernel(x)*
   I_c ← *closing(I,SE);*
   I_r ← *Reconstruction (I$_c$,I,4);*
   I_d ← *I$_r$–I;* % **Identification of dots/globules** %

338

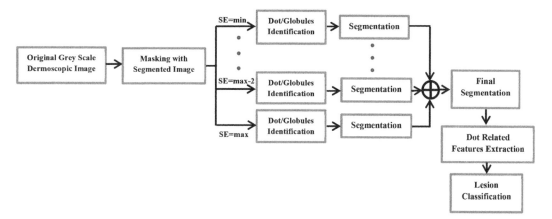

Figure 1.   Block diagram representation of dots/globules segmentation methodology.

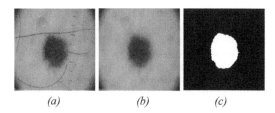

Figure 2.   (a) Original gray scale image; (b) preprocessed image; (c) segmented image.

$$
\begin{aligned}
I_{dmin} &\leftarrow min(min(I_d)); \\
I_{dmax} &\leftarrow max(max(I_d)); \\
T_h &\leftarrow T_s * (I_{dmax} - I_{dmin}) + I_{dmin}; \\
I_s &\leftarrow I_d >= T_h ; \%\ \textbf{Segmented image}\ \%
\end{aligned}
$$
end;

For the identification of the circular structures, morphological operations have been performed with a circular kernel as a SE of varying size. The size of the SE has been varied from the earlier specified minimum radius of the circular structure to the specified maximum radius, with an increment of two pixel distance. Morphological closing operation has been performed with each of the SE, followed by an image reconstruction using 4 connected neighbors. Morphological gradient operation has been performed by subtracting the original grey scale image from the reconstructed image, to identify the small circular structures with the pre-specified minimum and maximum radii.

After identification of the circular structures, they have been segmented out from the original dermoscopic images using operation. Firstly, the minimum and maximum intensity values present in those circular regions have been determined. According to the predefined threshold sensitivity, a percentage of the intensity range present in that regions have been selected as the threshold value.

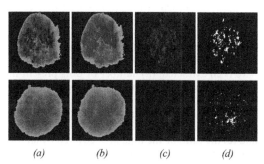

Figure 3.   (a) Original gray scale image, masked with the segmented image; (b) image after closing the circular regions, (c) identified dots/globules, (d) segmented dots/globules.

Finally, the segmented dots and globules have been replaced in the original grey scale image. Examples of the identified and segmented dots/globules for melanoma and non-melanoma cases have been shown in Figure 3.

3.3   Feature extraction

In the feature extraction stage, three different features have been extracted from each of the dermoscopic images, namely shape features, texture features and dot/globules related features.

3.3.1   Shape
Shape related features such as area, perimeter, equivalent diameter, eccentricity, solidity, rectangularity, aspect ratio and elongation, have been extracted from each of the segmented dermoscopic images.

3.3.2   Texture
The variations of the pixel grey intensities throughout the dermoscopic image have been assessed

339

in terms of some statistical measures obtained from Grey Level Co-occurrence Matrix (GLCM) (Haralick et al. 1973). The GLCM matrix characterizes a grey image by determining the probabilities of occurrence of a pair of pixels with specified values and spatial relationship. In this work GLCM matrices have been constructed by considering nine pixel distances with four different directions as {0°, 45°, 90° and 135°}. For the normalization of the GLCM matrices three normalization factors-8, 16 and 32 have been considered. From each of the GLCM matrices, fourteen statistical features namely, correlation, energy, homogeneity, contrast, autocorrelation, dissimilarity, entropy, and maximum probability, sum-square variance, sum average, sum variance, sum entropy, difference variance, and difference entropy have been extracted. Thus, in all, forty two texture features have been extracted from each of the dermoscopic images.

3.3.3 Dots/Globules related features

After the identification of dots/globules in the dermoscopic images, following dot-related features have been extracted: (a) number of dots present in the lesion area, (b) total number of pixels in dots, i.e. sum of the area covered by individual dots/globules, (c) mean area covered by the dots/globules, (d) variance of the individual dots/globules' area, and (e) the ratio of the total area covered by all the dots/globules to the total skin lesion area.

4 RESULTS AND DISCUSSION

4.1 Dataset description

To carry out this work, 66 dermoscopic images have been considered. Among these 66 dermoscopic images, 25 images have been collected from openly accessible Dermoscopy Atlas website (http://www.dermoscopyatlas.com) and remaining 41 images have been collected from International Dermoscopy Society website (http://www.dermoscopy-ids.org).

4.2 Experiment

In this work the preprocessing, noise removal and exclusion of hair artifacts and the skin-lesion segmentation have been done using morphological operations as explained (Chatterjee et al. 2015). For the identification and segmentation of the dot/globules, the original grey scale image, masked with the segmented binary image has been considered as an input image. In this algorithm, the minimum and the maximum radius of the dots has been considered as 4 pixels and 16 pixels respectively. So, as already explained, the size of the circular kernel (SE) has been varied from 4 pixels distance to 16 pixels distance with an increment of

2 pixels distances. The threshold sensitivity for the segmentation of dots/globules has been considered as 0.2, i.e. 20% of the difference between the minimum and maximum intensity values within that particular region has been considered as the threshold value.

After extraction of shape, texture and dot/globule related features, the melanoma images have been classified using k-Nearest Neighbor (KNN) classifier, decision tree and Support Vector Machine (SVM) classifier, (Vapnik 1995) with different types of kernel functions.

4.3 Performance of the proposed scheme

For the classification of melanoma, two combined feature sets have been considered, as shape {S} and texture {T} feature combination and shape, texture and dot/globule related features {D} combination. The features have been utilized by a decision tree classifier, a KNN classifier with k = 2, and SVM classifier with three different types of kernel functions, namely linear kernel, Radial Basis Function (RBF) kernel and polynomial kernel have been used. The classification performance has been evaluated by computing classification sensitivity from the resultant confusion matrix of classification using the equations described below with the following conventions.

TP (True Positive)   = Positive samples classified as positive.
TN (True Negative) = Negative samples classified as negative.
FP (False Positive)   = Negative samples classified as positive.
FN (False Negative) = Positive samples classified as negative.

$$Sensitivity = \frac{TP}{TP + FN};$$

The classification sensitivities of the classifiers deployed are tabulated in Table 1. The entries of Table 1 reveal that the inclusion of dot/globule related features with the shape and texture features combination, improve the classification sensitivity remarkably. The performances of the same classifiers have also been evaluated considering only the dot/globule related features {D}. On comparing the sensitivity values in Tables 1 and 2, it is amply clear that the {D} features have an overwhelming effect on sensitivities of all classifiers, with the exception of the SVM linear kernel classifier, and in effect, swamp the contributions from the {S} + {T} combine. It is worth noting that the KNN classifier and SVM classifiers with RBF and polynomial kernels classify the melanoma images with 100% sensitivity.

Table 1. Classification performance using combined features set.

Classifier	Features	Classification sensitivity (%)
Decision Tree	$\{S\} + \{T\}$	87.5
	$\{S\} + \{T\} + \{D\}$	**93.75**
kNN, k = 2	$\{S\} + \{T\}$	93.75
	$\{S\} + \{T\} + \{D\}$	**100**
SVM linear kernel	$\{S\} + \{T\}$	56.25
	$\{S\} + \{T\} + \{D\}$	**75**
SVM RBF kernel	$\{S\} + \{T\}$	100
	$\{S\} + \{T\} + \{D\}$	**100**
SVM polynomial kernel	$\{S\} + \{T\}$	93.75
	$\{S\} + \{T\} + \{D\}$	**100**

Table 2. Classification performance using only {D} features set.

Classifier	Classification sensitivity (%)
Decision Tree	93.75
kNN, k = 2	**100**
SVM, linear kernel	68.75
SVM, RBF kernel	**100**
SVM, polynomial kernel	**100**

5 CONCLUSION

In this reported work a methodical approach has been proposed for the identification and segmentation of dots/globules present in the skin lesion area using mathematical morphology. Here, the dots/globules have been identified so accurately by excluding irrelevant dot/globule features, that 100% sensitivity for the identification of melanoma has been achieved by most of the classifiers using individual dots related features and also using the combined features of shape, texture and dot features. It has also been shown that the inclusion of the shape and texture features with the dots/globules related features have no significant effect on the classification sensitivity of majority of the classifiers considered.

ACKNOWLEDGEMENT

This work is supported by the Department of Electronics and IT, Govt. of India through 'Visvesvaraya PhD scheme' awarded to Jadavpur University, India. The authors would also like to thank International Dermoscopy Society and Dermoscopy Atlas, Australia and New Zealand for kindly allowing the authors to access the dermoscopy database for this work.

REFERENCES

Bai, X., T. Wang, & F. Zhou (2015), "Linear feature detection based on the multi-scale, multi-structuring element, gray-level hit-or-miss transform", *Comp. Elect. Engg.,* vol. 46, pp. 487–499.

Barata, C., J. S. Marques, & J. Rozeira (2010), "A System for the Detection of Pigment Network in Dermoscopy Images Using Directional Filters", *IEEE Trans. Biomed. Engg.,* vol. 59, no. 10.

Chatterjee, S., D. Dey, & S. Munshi (2015), "Mathematical Morphology aided Shape, Texture and Color Feature Extraction from Skin Lesion for Identification of Malignant Melanoma", in Proc. IEEE CATCON 2015, pp. 200–203.

Ćurić, V., A. Landström, M. J. Thurley & C. L. L. Hendriks (2014), "Adaptive mathematical morphology—A survey of the field", *Patt. Recog. Let.,* vol. 47, pp. 18–28.

Dermoscopy Atlas, www.deroscopyatlas.com.

Haralick, M. R., K. Shanmugam & I. Dinstein (1973), "Textural features for image classification," *IEEE Trans. Syst., Man, Cybern.,* vol. 3, no. 6, pp. 610–621.

International Dermoscopy Society, www.dermoscopy-ids.org.

Ma, Z. & J. M. R. S. Tavares (2016), "A Novel Approach to Segment Skin Lesions in Dermoscopic Images Based on a Deformable Model", *IEEE J. Biomed. Health Info.,* vol. 20, no. 2, pp. 615–623.

Maglogiannis, I. & K. K. Delibasis (2015), "Enhancing classification accuracy utilizing globules and dots features in digital dermoscopy", *Comp. Method. Prog. Biomedic,* vol. 118, pp. 124–133.

Marks, R. (1995), "An overview of skin cancers: Incidence and causation," *Cancer Suppl.,* vol. 75, no. S2, pp. 607–612.

Naegel, B., N. Passat, & C. Ronse (2007), "Grey-level hit-or-miss transforms—Part I: Unified theory", *Patt. Recog.,* vol. 40, pp. 635–647.

Rastgoo, M., R. Garcia, O. Morel, & F. Marzani (2015), "Automatic differentiation of melanoma from dysplastic nevi", *Comput. Med. Imag. Graph,* vol. 43, pp. 44–52.

Sadeghi, M., Lee, T. K., McLean, D., Lui, H., & M. S. Atkins (2013), "Detection and Analysis of Irregular Streaks in Dermoscopy Images of Skin Lesions", *IEEE Trans. Med. Img.,* vol. 32, no. 5, pp. 849–861.

Sadri, A. R., M. Zekri, S. Sadri, N. Gheissari, M. Mokhtari, & F. Kolahdouzan (2013), "Segmentation of Dermoscopy Images Using Wavelet Networks", *IEEE Trans. Biomed. Engg.,* vol. 60, no. 4, pp. 1134–1141.

Sáez, A., J. Sánchez-Monedero, A. P. Gutiérrez, & C. Hervás-Martínez (2016), "Machine Learning Methods for Binary and Multiclass Classification of Melanoma Thickness from Dermoscopic Images", *IEEE Trans. Med. Img.,* vol. 35, no. 4, pp. 1036–1045.

Serup, J., G. B. E. Jemec, & G. L. Grove (2006), "Handbook of Non-Invasive Methods and the Skin," Second Edition, CRC Press.

Soille, P. (2004), "Morphological Image Analysis Principles and Applications", 2nd Edition, Springer.

Vapnik, V. (1995), "The Nature of Statistical Learning Theory", New York: Springer-Verlag.

Computer, Communication and Electrical Technology – Guha, Chakraborty & Dutta (Eds)
© 2017 Taylor & Francis Group, ISBN 978-1-138-03157-9

A method for automatic detection and classification of lobar ischaemic stroke from brain CT images

Aparna Datta
Department of Master of Computer Application, Meghnad Saha Institute of Technology, Nazirabad, Kolkata, West Bengal, India

Ashis Datta
Department of Neurology, Institute of Neurosciences Kolkata, Kolkata, West Bengal, India

ABSTRACT: This paper presents a method for the automatic detection and location of lobar ischaemic stroke lesion from brain CT images. The proposed method consists of three main steps: image enhancement, detection of abnormal slices and location of ischaemic stroke. Image enhancement based on fuzzy logic and histogram features were used for the detection of abnormal slices. Subsequently, CT images were divided into 10 parts. The relative density of each part was compared with that of the other parts. The part with the maximum relative density value indicated the location of ischaemic stroke. The results were computed from the brain CT images of ischaemic stroke patients. The objective of this paper was to propose a method to assist clinicians or neurologists for the early detection of ischaemic stroke and its proper management, thus reducing mortality and morbidity.

1 INTRODUCTION

Ischaemic stroke is the most common type of stroke accounting nearly 85% of all stroke cases, as clearly explained by Rosamond W et al. (2008). It occurs due to the occlusion of blood vessels supplying the brain. As neurons are solely dependent on glucose and oxygen carried through the blood for their survival, in the absence of blood supply, they gradually stop functioning, ultimately leading to neuronal death. However, if early intervention (thrombolysis) is provided in an attempt to revascularise ischaemic penumbra, the functional recovery becomes much better. Thus, the initial few hours (four and half) are very crucial because beyond this time frame, revascularisation does not benefit the patient. CT remains the imaging of choice in the emergency evaluation of acute stroke patients. Although stroke is very common in India, there are only a few neuroradiologists. Therefore, automated detection of ischaemic stroke will be helpful for clinicians or neurologists for the early identification of ischaemic stroke and decision regarding thrombolysis, rather than waiting for a neuroradiologist to interpret the CT images and thus losing the critical time frame for thrombolysis. Ischaemic stroke appears as a dark region (hypodense area) in contrast to its surroundings. Hypodensity gradually increases over time until it reaches that of the cerebrospinal fluid.

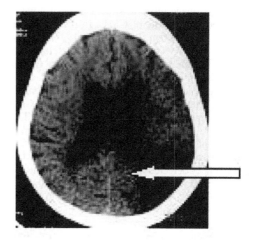

Figure 1. A CT image showing the ischaemic area that appears as a dark region.

In this paper, we present a method for the detection of ischaemic stroke from a given brain CT image. The proposed method is based on the observation that stroke changes the natural contralateral symmetry of a CT slice. Accordingly, we characterise stroke as a distortion between the two halves of the brain in terms of tissue density and texture distribution. Our approach will be able to

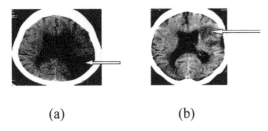

(a)   (b)

Figure 2.   CT images showing the different locations of ischaemic stroke lesions.

distinguish frontal, parietal and occipital ischaemic stroke, and thus will be helpful to locate ischaemic stroke lesions.

2   RELATED WORK

Histogram features are used for the classification of haemorrhagic and ischaemic stroke, whereas wavelet-based features are used for the classification of ischaemic infarct and normal slices, as described by Chawla M et al. (2009). Knowledge-based approaches have been proposed by Cosic D. and Loncaric S (1997). A content-based 3D neuro-radiologic image retrieval system was developed at the Robotics Institute of CMU. In the preliminary results of this system, Liu Y et al. (1997) proposed a symmetry detection approach to detect brain lesions. The method checks the symmetry for each slice. If the midline of the brain shifts, a haemorrhage is considered to be present. In addition, it is assumed that in pathological brain slices, the symmetric property of normal brain slices is lost. Thus, in any brain slice, if the abnormal region appears only at one side, it is considered as a haemorrhage. A similar method was also adopted by Hara T et al. (2007). They detected the midline based on the skull contour. Midline detection was also adopted by Chan T (2007). A number of approaches for the segmentation of head or brain images have been presented over the past few years: pattern recognition technique by Bezdek J. C. et al. (1993), rule-based system by Raya S. P. (1990), and knowledge-based approach by Li C. et al. (1993). An unsupervised clustering and backtracking tree search algorithm was used by Loncaric S et al. (1996). Dhawan A et al. (1990) used the intelligent system for the segmentation of medical images. Research on brain image segmentation using rule-based expert systems has been presented in papers by Cosic D et al. (1997), Li C et al. (1993) and Sonka M et al. (1996). Fuzzy set theory has been used to extract the meaningful regions from medical images by Menhardt W (1988). Early automated identification of ischaemic stroke can minimise the mortality and morbidity of stroke victims, which has been truly emphasised in the paper by Abreu T (2002). However, the problem of classification of lobar ischaemic stroke from a given brain CT image has not been addressed properly. This study will primarily focus on the early detection of ischaemic stroke and its location using an automated image analysis. This will be helpful for the early institution of thrombolysis without unduly waiting for a neuroradiologist to interpret the CT images, thus reducing the residual neurodeficit.

3   DETECTION AND CLASSIFICATION METHOD

The proposed algorithm has the following steps. In the first step, the given CT slice taken at the level of the third ventricle is enhanced and denoised. The next step is to divide the CT image into left and right parts. Histogram features are used for the detection of ischaemic stroke. Finally, the classification of different types of lobar ischaemic strokes is made.

3.1   *Noise reduction*

Noise removal is performed using Wiener filtering to remove the graininess from the image.

3.2   *Contrast enhancement*

The technique used here is based on the modification of pixel values in the fuzzy property domain. Fuzzy image enhancement is based on grey-level mapping into a fuzzy plane, using a membership transformation function. The aim is to generate an image of higher contrast than the original image by giving a larger weight to the grey levels that are closer to the mean grey level of the image than to those that are farther from the mean. An image I of size $M \times N$ and L grey levels can be considered as an array of fuzzy singletons, each having a value of membership denoting its degree of brightness relative to some brightness levels.

Algorithm 1: *To enhance the image using the fuzzy technique*
     The algorithm starts with the initialisation of the image parameters; size, minimum, mid and maximum grey level. The fuzzy rule-based approach is a powerful and universal method used for many tasks in image processing.
     The algorithm is described as follows:

Step 1 Convert the image data into fuzzy domain data.

a) If grey_ value between zero and min
     Then *fdata=0*;
b) Else if grey_ value between min and mid
     Then *fdata = (1/(mid-min) * min + (1/mid- min)* data:*

c) If grey _ value between mid and max
  Then *fdata = (1/(max-mid))* mid + (1/(max-mid))*data;*
d) If grey_ level between max and 255
  Then *fdata=1;*

Step 2 Membership modifications
a) If grey _ value ben 0 and min
  Then *fdata= 0;*
b) If grey _ value between min and mid.
  i) If fdata between 0 and 0.5
    Then *fdata=2*(fdata) ^2;*
  ii) Else if fdata between 0.5 and 1
    Then *fdata= 1-2*(1-fdata) ^2;*
c) If grey _ value between mid and max
  i) If fdata between 0 and 0.5
    Then *fdata=2*(fdata) ^2;*
  ii) If fdata between 0.5 and 1
    Then *fdata= 1- 2* (1-fdata) ^2;*
  iii) If grey_ value between max and 255
    Then *fdata = 1;*

Step 3 Deffuzification
a) If grey _ value between 0 and min
  Then enhanced_ data = grey _ value
b) If grey_ value between min and mid
  Then *enhanced_ data=-(mid – min) * fdata+ min*
c) If grey _ value between mid and max
  Then *enhanced_data= (max –mid) * fdata+mid;*

Step 4 Displaying the enhanced image
i) Show the original image.
ii) Show the enhanced image

3.3 *Detection of abnormal slices using histogram features*

We modelled our approach based on the procedure followed by the neurologist who detects abnormality by examining the dissimilarity between the left and right hemispheres of the brain. Therefore, in our approach, we relate the appearance changes in the two hemispheres to the changes in the overall shapes of their respective histograms. The shape of the histogram is equal for both sides in the case of a normal CT image. In the case of ischaemic stroke, the shape of the histogram is markedly different between the affected side and the normal side. This is because chronic ischaemic affects the lower part and acute ischaemic affects the middle part of the greyscale. Histogram-based comparison is therefore used to identify these two cases.

3.4 *Finding different locations of the CT image*

Algorithm 2: *The algorithm for counting the number of pixels in each region after image enhancement of the input CT image*

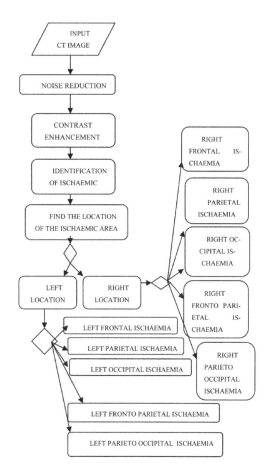

Figure 3.   Block diagram of detection and classification of the images.

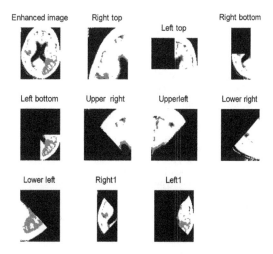

Figure 4.   CT images showing the division of the image into different parts, which is useful for finding the location of ischaemic stroke.

345

Let X[m, n] be part of image matrix of m × n size

Count pixel = 0

Count pixel = Count pixel + 1 for all X [i,j] < T where T is the threshold value, and i ranges from 1 to m and j ranges from 1 to n

3.5 Location of ischaemic stroke

The image obtained after image enhancement is divided into two equal parts (left and right). The relative density of the left part is compared with the right part of the CT. The part with the maximum relative density value indicates whether the left or the right part is affected.

3.5.1 Location of ischaemic stroke on the right side
The image obtained after the above method is divided into five parts. The relative density of each part is compared with the other parts. The part with the maximum relative density value indicates the exact location of ischaemic stroke.

Algorithm 3: *Label the output image according to the location of ischaemic stroke*

If ischaemic location is Upper_Right
Then the type is Right Frontal Intra Cerebral ischaemic
elseif ischaemic location is Lower_Right
    Then the type is Right Occipital ischaemic
elseif ischaemic location is Right
    Then the type is Right Parietal ischaemic
elseif ischaemic location is Right_top
Then the type is Right Fronto Parietal ischaemic
elseif ischaemic location is Right_bottom
 Then the type is Right Parieto-Occipital ischaemic

3.5.2 Location of ischaemic stroke on the left side
The left part of the image is divided into five parts. The relative density of each part is compared with the other parts. The part with the maximum relative density value indicates the exact location of ischaemic stroke.

Algorithm 4: *Label the output image according to the location of ischaemic stroke*

If ischaemic location is Upper_Left
 Then the type is left Frontal Intra Cerebral ischaemic
elseif ischaemic location is Lower_Left
        Then the type is Left Occipital ischaemic
elseif ischaemic location is Left
        Then the type is Left Parietal ischaemic
elseif ischaemic location is Left_top
        Then the type is   Left Fronto Parietal ischaemic
elseif ischaemic location is Left_top
        Then the type is   Left Parieto-Occipital ischaemic

4 RESULTS AND DISCUSSIONS

The dataset consists of volume CT data of 13 patients who attended the emergency department of a neurology superspeciality hospital. All patients had their CT scan performed using the same machine [Siemens (32 slices)]. Of the 13 patients, four had a normal CT image. Therefore,

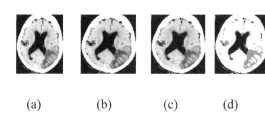

(a)   (b)   (c)   (d)

Figure 5. (a) The input CT image, (b) the grey image, (c) the filtered image obtained by the Wiener filtering method and (d) the enhanced image.

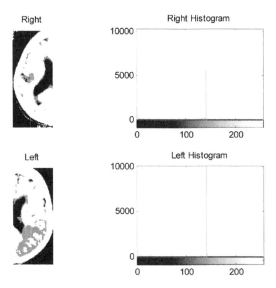

Figure 6. Division of the image into the left and right parts and their histograms that are useful for the detection of the ischaemic region.

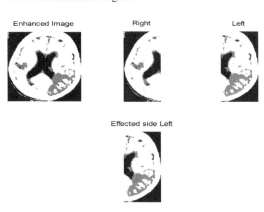

Figure 7. Division of the image into the left and right parts; left part of the image is shown to be affected.

346

we analysed the CT images of the remaining nine patients in this study. As this was a pilot study, we did not analyse the performance of the method.

From Table 1 it can be seen that the number of pixels on the left part of the image is greater than that on the right part of the image. This means the left part of the image is the affected region.

Table 2 provides the number of pixels at different locations of the left part of the image. The number of pixels at the left bottom is maximum among the other parts. This means the left bottom of the image is the affected region.

Table 1. Pixel count 1.

Part of the image	Number of pixels
Right	5479
Left	14285

Table 2. Pixel count 2.

Part of the image	Number of pixels
Upper left	1427
Lower left	5234
Left1	7683
Left top	2753
Left bottom	11565

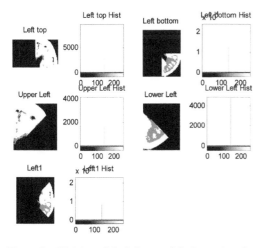

Figure 8. Division of the left part of the image into five parts and corresponding histograms that are useful for finding the location of ischaemic stroke.

Input Image   Left Parieto Occipital Ischaemia

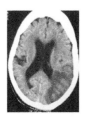 

Figure 9. CT image showing that ischaemia is in the left parieto-occipital region.

5 CONCLUSION

In this paper, we have proposed an algorithm based on contralateral symmetry to detect stroke-affected slices in a given brain CT image. The key features of our algorithm are: ability to detect ischaemic stroke and its location. This approach will be helpful in building a stroke analysis system that can detect and classify ischaemic stroke at an early stage, thereby facilitating early intervention (thrombolysis) without waiting for a neuroradiologist to interpret the CT scan, thus reducing mortality and morbidity. The contralateral symmetry condition that we have used fails when ischaemic stroke occurs simultaneously in both hemispheres, but such cases are very rare in practical situation and in the condition of very early ischaemic stroke when CT changes are not apparent. It is expected that this system can benefit patient care especially in emergency situation when timely management decision needs to be made by acute care physicians.

REFERENCES

Abreu, T. (2002). Early CT signs in acute ischaemic stroke. *Medicina Interna.* Vol. 9, N. 1.
Bezdek, J. C., L. O. Hall, & L. P. Clarke (1993). "Review of MR image segmentation techniques using pattern recognition," *Medical Physics. 20*, 1033–1048.
Chan, T. (2007). "Computer aided detection of small acute intracranial hemorrhage on computer tomography of brain", *Computerized Medical Imaging and Graphics. 07(4–5)*, 285–298.
Chawla, M., S. Sharma, J. Sivaswamy, L. T. Kishore (2009). "A method for automatic detection and classification of stroke from brain CT images", *in 31st annual international conference of the IEEE EMBS*, USA.
Cosic, D. & S. Loncaric (1997). "Computer system for quantitative analysis of ICH from CT head images," in *19th Annual International Conference of the IEEE.*
Ćosić, D. & S. Lončarić (1997). "Rule-Based Labeling of CT Head Image", In *Proceedings of the 6th European Conf. of AI in Medicine Europe (AIME97),* Grenoble.
Dhawan, A. & S. Juvvadi (1990). "Knowledge-Based Analysis and Understanding of Medical Images", *Computer Methods and Programs in Biomedicine*, vol. 33, 221–239.
Hara, T., N. Matoba, X. Zhou, S. Yokoi, H. Aizawa, H. Fujita, K. Sakashita & T. Matsuoka (2007). "Automated Detection of Extradural and Subdural Hematoma for Content-enhanced CT Images in Emergency Medical Care", In *Proceeding of SPIE.*

Li, C., D. B. Goldgof & L. O. Hall (1993). "Knowledge-Based Classification and Tissue Labeling of MR Images of Human Brain", *IEEE Transactions on Medical Imaging*, *12*, 740–750.

Liu, Y., W. E. Rothfus & T. Kanade (1997). Content-based 3D Neuroradiologic Image Retrieval: Preliminary Results, IEEE *Content-based video and image retrieval workshop associated with CVPR97*.

Liu, Y., N. A. Lazar, W. E. Rothfus, F. Dellaert, A. Moore, J. Schneider & T. Kanade (2004). "Semantic-based Biomedical Image Indexing and Retrieval", *Trends and Advances in Content-Based Image and Video Retrieval*, Shapiro, Kriege, and Veltkamp, ed.

Loncaric, S., D. Cosic & A. P. Dhawan (1996). "Segmentation of CT head images", In *Proceedings of the 10th International Symposium Computer Assisted Radiology*. Elsevier, Amsterdam, pp. 1012.

Loncaric, S., D. Cosic & A. P. Dhawan (1996). "Hierarchical segmentation of CT images", In *Proceedings of the 18th Annual Int'l Conference of the IEEE EMBS, IEEE*.

Menhardt, W. (1988). Image analysis using iconic fuzzy sets, In *Proceedings of European Conference on Artificial Intelligence*, pp. 612–674.

Raya, S. P. (1990). "Low-level segmentation of 3-D magnetic resonance brain images A rule based system", *IEEE Transactions on Medical Imaging 9*, 327–337.

Rosamond, W., K. Flegal & K. Furie (2008). "Heart disease and Stroke Statistics—2008 update: a report from American Heart Association Statistics Committee and Stroke Statistics subcommittee." *Circulation, 117*, 25–146.

Sonka, M., S. K. Tadikonda & S. M. Collins (1996). "Knowledge-Based Interpretation of MR Brain Images", *IEEE Transactionson Medical Imaging, 15*, 443–452.

A new modified adaptive neuro fuzzy inference system-based MPPT controller for the enhanced performance of an asynchronous motor drive

B. Pakkiraiah & G. Durga Sukumar

Department of Electrical and Electronics Engineering, Vignan's Foundation for Science Technology and Research University, Guntur, Andhra Pradesh, India

ABSTRACT: Solar energy is one of the most important renewable energy sources. On an average, the sun shines in India for about 6 h per day and for about 9 months in a year. To generate electricity from the sun, the Solar Photovoltaic (SPV) modules are used. SPV comes in various power outputs to meet the load requirements. The power from a solar photovoltaic module is maximized in special instances to increase the efficiency of the PV system. The Adaptive Neuro Fuzzy Inference System (ANFIS) based Maximum Power Point Tracking (MPPT) controller is used to track the maximum power. DC-DC boost converter and space vector modulation based inverters are used to provide the required supply to the load. The proposed ANFIS-based MPPT improves the system efficiency even at abnormal weather conditions. Here, a lot of reduction in torque and current ripple contents is obtained with the help of ANFIS-based MPPT for an asynchronous motor drive. Also, better performance of an asynchronous motor drive is analyzed with the comparison of a conventional and proposed MPPT controller using Matlab-simulation results. Practical validations are also carried out and tabulated.

1 INTRODUCTION

With the earth's natural resources decreasing day by day, to meet the increase in the power demand, the power sector is looking at alternate energy resources. Due to usage of renewable energy sources, the carbon content in the atmosphere can be reduced to overcome the issue of global warming. Out of various renewable sources, the solar photovoltaic (PV) system is most used due to its simple structure. Literature so far has discussed the various structures of the PV panel system and their suitability for certain locations (Saadi, A. et al. 2003, Adel et al. 2008, Roberto et al. 2008, Pakkiraiah, B. et al. 2016). The efficiency of the PV system can be increased by using power electronic devices along with Maximum Power Point (MPP) controllers.

Due to the nonlinearities of PV systems, artificial intelligence paradigms such as artificial neural networks (Abdessamia, E. et al. 2012), fuzzy logic control (Theodoros, L. et al. 2006), and particle swarm optimization (Hu, Y. et al. 2011) could be effectively employed to enhance system performance. ANFIS is an intelligent regime comprising an adaptive network performing the function of a Sugeno-type fuzzy model (Jang, J. S. et al. 1997). Optimization algorithms were employed through adaptive networks to make the system performance similar, with minimal error to a targeted training data set. ANFIS combines the optimization strength of adaptive networks with the ability of fuzzy systems to handle difficult situations and process uncertain data. Such attributes enabled ANFIS to find many immediate engineering applications including, but not limited to, decision making, problem solving, pattern recognition, nonlinear mapping, system modeling, and adaptive control (Jang, J. S. 1993, Mellit, A. et al. 2006).

In the area of solar energy research, ANFIS was successfully employed to extract the MPP of PV modules (Iqbali, A. et al. 2010). Most of the research centered on the principle of adjusting the voltage of the solar PV module by changing the duty ratio of DC-DC chopper circuits. The duty ratio is controlled for a given solar irradiation and cell temperature condition by a closed-loop scheme (Farhat, M. et al. 2012, Haitham, A. et al. 2013). The PV module energy conversion efficiency lies between 12–20%. The energy conversion loss depends on the PV system and also on the loads that are connected. This can be overcome using an MPPT with DC-DC converter to get the required load voltage at the MPP voltage (Qiang et al. 2011, Adly, M. et al. 2012, Pakkiraiah, B. et al. 2015). An MPPT controller with inverter is connected to the asynchronous motor drive with a space vector modulation technique to get better performance with the

PV system. Various strategies are used for selecting the order of vectors with zero vectors to reduce the harmonic content and the switching losses (Aleenejad, M. et al. 2012, Joshi, J. J. et al. 2013).

The space vector modulation diagram of an inverter is composed of a number of sub hexagons. The sector identification can be done by determining the triangle, which encloses the tip of the reference space vector diagram with the formation of six regions (Mbarushimana, A. et al. 2011, Sreeja, C. et al. 2011, Pakkiraiah, B. et al. 2016). To overcome the distortions in the output voltage and currents of an inverter, the single-phase SVM-based cascaded H-Bridge multilevel inverter is used for the PV system to improve the quality of power even under abnormal weather conditions. A better torque ripple and performance is obtained with the help of genetic algorithm-particle swarm optimization based indirect vector control for the optimal torque control of an induction motor drive (Dong 2007). A comparison of neuro fuzzy based space vector modulation with neural network and conventional based system has been presented (Durga Sukumar et al. 2014).

The advantage of this proposed ANFIS-based MPPT is to control the MPP even under abnormal weather conditions, compared to other conventional algorithms. Section 2 discusses the mathematical modeling of a PV array. Section 3 explains the proposed MPPT algorithm. Section 4 discusses the mathematical modeling of an asynchronous motor drive. Section 5 states a brief note on the proposed space vector modulation technique. Section 6 discusses the use of the proposed MPPT along with a DC-DC converter to boost up the PV output and to feed the asynchronous motor drive. Section 7 discusses Matlab-simulation results with the comparison of both MPPT techniques. Section 8 consists of the concluding remarks.

2 MATHEMATICAL MODELING OF PV ARRAY

Photovoltaic technology converts solar energy directly into electricity through solar cells. A photovoltaic generator, also known as a photovoltaic array, is the total system consisting of all PV modules connected in series or in parallel with each other. A solar cell constitutes the basic unit of a PV generator. This, in turn, is the main component of a solar generator. For a photo voltaic system, power generated is calculated as $P = V \cdot I$, where P represents power in watts (W), V the voltage in volts (V), and I the current in Ampere (A). Figure 1 shows a model of a solar cell equivalent circuit.

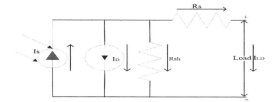

Figure 1. Model of a solar cell equivalent circuit.

Table 1. The calculated i_s values for different irradiance and different temperatures.

Irradiance in (W/m²)	Temperature (in °C)				
	20°C	30°C	40°C	50°C	60°C
1000	8.661	8.678	8.695	8.712	8.729
800	6.929	6.942	6.956	6.97	6.983
500	4.330	4.339	4.347	4.356	4.364
250	2.165	2.169	2.173	2.178	2.182
100	0.866	0.867	0.869	0.871	0.872
50	0.433	0.433	0.434	0.435	0.436
10	0.086	0.086	0.086	0.087	0.087

$$I_{LD} = I_S - I_D - I_{sh} \tag{1}$$

where I_S refers to the sunlight current and I_D is the diode current. From the Shockley equation, the diode current can be expressed as

$$I_D = I_0 \left[exp \left(\frac{q(V_{LD} + I_{LD}R_s)}{\Upsilon KT_s} \right) - 1 \right] \tag{2}$$

where I_0 refers to the actual reverse saturation current, q the electron charge, and Υ, K the shape factor and Boltzmann constant. Substituting (2) in (1), we get

$$I_{LD} = I_S - I_0 \left[exp \left(\frac{q(V_{LD} + I_{LD}R_s)}{\Upsilon KT_s} \right) - 1 \right] \tag{3}$$

where $\gamma = Z. N_A. N_{SC}$, Z is the completion factor, N_A is the number of modules in an array, and N_{SC} is the number of series connected cells.

$$I_S = \frac{G}{G_R} \left(I_{SCR} + \mu_{SCR} (T_S - T_{SCR}) \right) \tag{4}$$

In the equations given above, I_{scr} refers to the short circuit current at the reference condition; G and G_R are the irradiance at the actual and at reference condition, respectively; μ_{SCR} the manufacturer supplied temperature coefficient of short-circuit current; and T_S, T_{SCR} the solar cell temperatures at actual and reference conditions, respectively.

$$\frac{T_S}{T_{SCR}}$$

$$I_O = I \tag{5}$$

I_{OR} refers to the reverse saturation current at the reference condition. The module photo current I_S of the photovoltaic module depends linearly on the solar irradiation and is also influenced by the temperature according to equation (4). The value of the module short-circuit current I_{SCR} is taken from the data sheet of the reference model.

3 PROPOSED MAXIMUM POWER POINT TRACKING ALGORITHM

Maximum Power Point Tracking (MPPT) is a technique that ties the inverters and solar chargers to get the maximum possible power from one or more photovoltaic devices, typically solar panels. The purpose of the MPPT system is to sample the output of the solar cells and apply the proper resistance (load) to obtain maximum power for any given environmental conditions. MPPT devices are typically integrated into an <u>electric power converter</u> system that provides voltage or current conversion, filtering, and regulation for driving various loads, including power grids, batteries, or motors. MPPT is a control technique mainly used to extract the maximum capable power of the PV modules with the respective solar irradiance and temperatures at a particular instant of time by the MPPT controller.

The appearance of multi-peak output curves of partial shading in PV arrays is common, where the development of an algorithm for accurately tracking the true MPPs of the complex and nonlinear output curves is crucial. A typical solar panel converts only 30–40% of the incident solar irradiation into electrical energy.

MPPT is used to increase the efficiency of the panel. The requirement to get the maximum power is that the source resistance should be equal to that of the load resistance. In the source side, a boost converter is connected to a solar panel in order to enhance the output voltage to use in different applications like motor loads by changing the duty cycle of the boost converter. Most of the existing MPPT algorithms suffer from the drawbacks of slow tracking, wrong tracking, and oscillations during rapidly changing weather conditions. Due to this, the utilization efficiency is reduced.

To overcome these drawbacks, an ANFIS-based MPPT control technique is introduced in this paper. Here, it improves the performance of the system and efficiency much better than any other conventional methods. The boost converter and inverter are used to provide maximum output voltage to the load. The proposed ANFIS-based MPPT structure is shown in the Figure 2.

3.1 *Practical outputs of a conventional MPPT controller with variable irradiance and constant temperature*

When the irradiance varies to 100, 250, 500, 800, and 1000 W/m², it is observed that the PV current and voltage will increase with irradiance levels. Due to this, the net PV array power also gets increased. These characteristics are observed in Figure 3.

3.2 *Practical outputs of a conventional MPPT controller with variable temperature and constant irradiance*

When the temperature varies to 20°C, 30°C, 40°C, 50°C and 60°C, it increases the PV current marginally with a drastic decrease in PV array voltage. Due to this, the net PV array output power reduces. These characteristics are presented in Figure 4.

3.3 *Practical outputs of the proposed MPPT controller with variable temperature and variable irradiance*

When both the temperature and irradiance are variable, it increases the PV module current and decreases

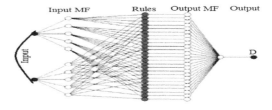

Figure 2. ANFIS-based MPPT.

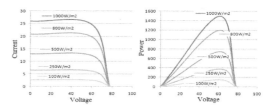

Figure 3. Practical I-V & P-V characteristics with variable irradiance and constant temperature.

351

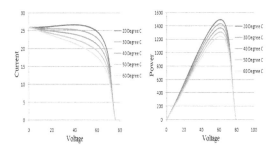

Figure 4. Practical I-V & P-V characteristics with variable temperature and constant irradiance.

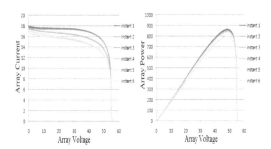

Figure 5. Practical I-V & P-V characteristics of PV array with variable temperature and variable irradiance.

the voltage till the temperature rises and vice versa. Also, it increases the array current and slightly increases the voltage till the irradiance rises and vice versa. These results are illustrated in Figure 5.

4 MATHEMATICAL MODELING OF ASYNCHRONOUS MOTOR DRIVE

The mathematical modeling of a three-phase, squirrel-cage asynchronous motor drive can be described with a stationary reference frame as

$$V_{qS} = (R_S + pL_S)I_{qS} + PL_M I_{qR} \tag{6}$$

$$V_{dS} = (R_S + pL_S)I_{dS} + PL_M I_{dR} \tag{7}$$

$$0 = pL_M I_{qS} - \omega_R L_M I_{dS} + (R_R + pL_R)i_{qR} - \omega_R L_R i_{dR} \tag{8}$$

$$0 = \omega_R L_M i_{qS} + pL_M i_{dS} + \omega_R L_R i_{qR} + (R_R + pL_R)i_{dR} \tag{9}$$

where $\omega_R = \frac{d\theta}{dt}$ and $p = \frac{d}{dt}$. Suffixes S and R represent the stator and rotor, respectively. V_{dS} and V_{qS} are d-q axis stator voltages. i_{dS}, i_{qS} and i_{dR}, i_{qR} are d-q axis stator currents and rotor currents, respectively. R_S and R_R are the stator and rotor resistances per phase, respectively. L_S, L_R are the self-inductances of the stator and rotor, respectively, and L_M is the mutual induct-

ance. The stator and rotor flux linkages can be expressed as

$$\lambda_{qS} = L_S i_{qS} + L_M i_{qR} \tag{10}$$

$$\lambda_{dS} = L_S i_{dS} + L_M i_{dR} \tag{11}$$

$$\lambda_{qR} = L_R i_{qR} + L_M i_{qS} \tag{12}$$

$$\lambda_{dR} = L_R i_{dR} + L_M i_{dS} \tag{13}$$

From the equations (6)–(9), the squirrel-cage asynchronous motor can described by following equations in stator reference frame as

$$\begin{bmatrix} V_{qS} \\ V_{dS} \\ 0 \\ 0 \end{bmatrix} = \begin{bmatrix} R_S + pL_S & 0 & pL_M & 0 \\ 0 & R_S + pL_S & 0 & pL_M \\ pL_M & -\omega_R L_R & R_R + pL_R & -\omega_R L_R \\ \omega_R L_M & pL_M & \omega_R L_R & R_R + pL_R \end{bmatrix} \begin{bmatrix} i_{qS} \\ i_{dS} \\ i_{qR} \\ i_{dR} \end{bmatrix} \tag{14}$$

The electromagnetic torque T_e of the induction motor is given by

$$T_e = \frac{3}{2} \left(\frac{p}{2} \right) (\lambda_{qR} i_{dR} - \lambda_{dR} i_{qR}) \tag{15}$$

From the dynamic model of the asynchronous machine, the rotor flux is aligned along with the d-axis then the q-axis rotor flux $\lambda_{qR} = 0$. So, from the equations (12) and (15) described in the previous section and putting $\lambda_{qR} = 0$, the electromagnetic torque of the motor in the vector control can be expressed as

$$T_e = \frac{3}{2} \left(\frac{p}{2} \right) \frac{L_M}{L_R} (\lambda_{dR} i_{qS}) \tag{16}$$

If the rotor flux linkage λ_{dR} is not disturbed, the torque can be independently controlled by adjusting the stator q-component current i_{qS}. As the rotor flux is aligned on the d-axis, this leads to $\lambda_{qR} = 0$ and $\lambda_{dR} = \lambda_R$. Then,

$$\omega_{sl} = \frac{L_M R_R}{\lambda_R L_R} i_{qS} \tag{17}$$

5 THE PROPOSED SVM TECHNIQUE FOR A TWO-LEVEL INVERTER

In this, the space vector modulation algorithm for the two-level inverter is introduced for which the solar panels are connected to provide the dc supply. The SVM basic principle and switching sequence is given in order to get symmetrical algorithm pulses and voltage balancing. This scheme is used to control the output voltage of the two-level inverter with the ANFIS-based MPPT controller. In the SVM algorithm, the d-axis and q-axis voltages are converted into three-phase instantaneous reference voltages. The imaginary switching time periods are proportional to the instantaneous values of the reference phase voltages, which are defined as

$$T_{U1} = \left(\frac{T_S}{V_{DC}}\right)V_{U1}, T_{V1} = \left(\frac{T_S}{V_{DC}}\right)V_{V1}, T_{W1} = \left(\frac{T_S}{V_{DC}}\right)V_{W1}$$

$$(18)$$

where T_S and V_{DS} are the sampling interval time and dc link voltage, respectively. Here, the sampling frequency is twice the carrier frequency. Then, the maximum (MAXI), middle (MID), and minimum (MINI) imaginary switching times can be in each sampling interval by using the following equations (19)–(21):

$$T_{MAXI} = MAXI\left(T_{U1}, T_{V1}, T_{W1}\right) \quad (19)$$

$$T_{MINI} = MINI\left(T_{U1}, T_{V1}, T_{W1}\right) \quad (20)$$

$$T_{MID} = MID\left(T_{U1}, T_{V1}, T_{W1}\right) \quad (21)$$

The active voltage vector switching times T_1 and T_2 are calculated as

$$T_1 = T_{MAXI} - T_{MID} \text{ and } T_2 = T_{MID} - T_{MINI} \quad (22)$$

The zero voltage vectors switching time is calculated as

$$T_Z = T_s - T_1 - T_2 \quad (23)$$

The zero state time will be shared between two zero states as T_0 for V_0 and T_7 for V_7, respectively, and can be expressed as

$$T_0 = K_0 T_Z \quad (24)$$

$$T_7 = \left(1 - K_0\right)T_Z \quad (25)$$

6 THE PROPOSED MPPT SYSTEM WITH DC-DC CONVERTER, INVERTER, AND ASM DRIVE

The system given below represents the proposed system structure with a DC-DC converter. In this,

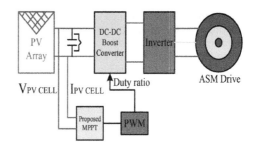

Figure 6. The proposed MPPT system with a DC-DC converter and asynchronous motor drive.

the PV array contains 6 PV modules with 250 watts each; these modules are connected in series and in parallel to yield a better output voltage and current. The proposed ANFIS-based MPPT algorithm extracts the maximum power from the solar PV array. This is a new technique when compared to the other conventional methods.

These individual case results are presented in sections 3 and 7. The proposed system structure with the asynchronous motor drive is presented in Figure 6.

The point of operation of the PV array is adjusted by varying the duty cycle. The DC-DC converter boosts the PV array voltage and also increases the maximum utilization of the PV array by operating at MPP. The boost converter increases the array output voltage up to 400 Volts with the help of the SVM-based inverter.

The minimum inductor value (L_{MIN}) is calculated from Eq. (26) to ensure the continuous inductor current:

$$L_{MIN} = \left(\frac{V_0\left(1 - D\right)^2 \times D}{2}\right) \times f_S \times I_{AVG} \quad (26)$$

where V_0 is DC output voltage, D is duty ratio, f_S is switching frequency of the converter, and I_{AVG} is the average output current. The minimum capacitance value (C_{MIN}) can be calculated using Eq. (27):

$$C_{MIN} = \frac{V_0 \times D}{R \times \Delta V_0 \times f_S} \quad (27)$$

The switching frequency selection is chosen based on the switching losses, cost of switch, and converter efficiency.

7 RESULTS AND DISCUSSION

The proposed model has been developed with Matlab/Simulink. The input to the module is temperature and solar irradiance. At Standard Test Conditions (STC) containing 60 cells to produce 250-watt power, six such modules are connected in order to form solar PV array. From the simulation results, we got the array-generated open circuit voltage as 75.96 volts with a short-circuit current of about 26.01 amps and the maximum power obtained at MPP of 1500 watts. These results are shown in Figures 7 and 8.

7.1 Practical set up for measuring and calculation of the practical PV array output power at different instants

The PV array output power is measured from morning 8 am to evening 5:30 pm at different instants with different temperatures. The respective values are tabulated in Table 2 with the help of the experimental setup diagrams as shown in Figure 9.

7.2 Simulation results of an asynchronous motor drive with inverter

Simulation results are obtained with the reference speed of 1400 RPM and switching frequency of 5 KHz. The performance of motor parameters such as stator phase currents, torque, and speed are analyzed in Figures 11–16.

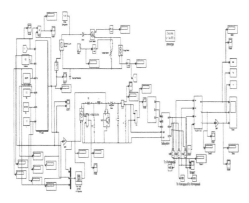

Figure 7. Matlab-Simulink diagram with the PV array, proposed MPPT system, DC-DC converter, inverter, and asynchronous motor drive.

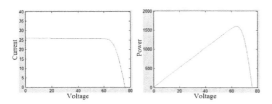

Figure 8. I-V & P-V characteristics obtained from PV array.

Here, the motor drive is fed with 400 volts supply with the help of a boost converter and inverter. The output voltages of the inverter are shown in Figure 10.

7.3 Simulation results of an asynchronous motor drive at starting

For the asynchronous motor drive, the maximum current and the ripple content in the torque is reduced in the beginning to reach the early steady state. With the proposed MPPT, the maximum torque, stator phase current, and speed are obtained as 12.38 N-m, 3.396 amps, and 1400 RPM, respectively. It is observed that the ripple content in the torque is reduced to 0.22 as compared to the conventional method. Due to this, a better speed response is obtained. These results are presented in Figures 11–13.

Table 2. Practical PV array output power.

Time (am/pm)	Temperature (°C)	Array power (Watts)
08.00	30.75	237.51
09.06	33.35	679.07
10.00	38.25	821.76
11.00	39.55	919.32
12.00	40.50	1050.99
12.30	40.68	1157.56
13.30	41.35	1108.88
14.30	40.08	919.49
15.30	38.42	638.32
16.30	37.32	244.77
17.30	37.6	46.32

Figure 9. PV array experimental set up diagrams.

7.4 Simulation results of an asynchronous motor drive in steady-state condition

The steady-state responses of the stator phase currents, torque, and speed with the conventional and proposed MPPT are observed in Figures 14–16. Here, the torque ripple with the proposed MPPT is reduced a lot, i.e., it is observed that the torque ripple with the conventional and proposed MPPT are 0.35 and 0.08, respectively. The better speed response is obtained with the proposed MPPT controller.

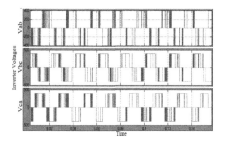

Figure 10.   Inverter output voltages.

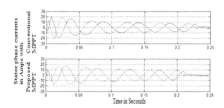

Figure 11.   Stator phase current responses with a conventional and proposed MPPT controller at starting.

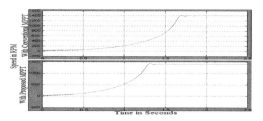

Figure 12.   Speed responses with a conventional and proposed MPPT controller at starting.

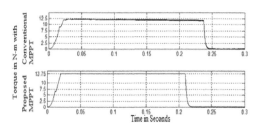

Figure 13.   Torque responses with a conventional and proposed MPPT controller at starting.

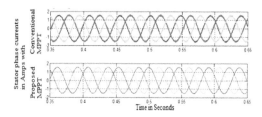

Figure 14.   Stator phase current responses with a conventional and proposed MPPT controller at steady state.

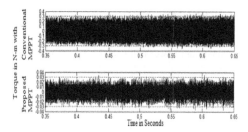

Figure 15.   Torque responses with a conventional and proposed MPPT controller at steady state.

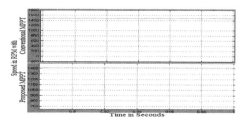

Figure 16.   Speed responses with a conventional and proposed MPPT controller at steady state.

8   CONCLUSION

The PV array model with the Adaptive Neuro Fuzzy Inference System (ANFIS) based MPPT controller is tested. From this, the performance of the asynchronous motor drive is analyzed by comparing both conventional and proposed ANFIS MPPT controller results. Also, the behavior of the proposed ANFIS MPPT is observed with practical validations during a partially cloudy day. The PV system with a DC-DC boost converter and space vector modulation based technique inverter enhances the system performance by improving the power quality even under abnormal weather conditions. The ripple contents in the torque and stator phase currents are reduced a lot with the proposed ANFIS-based MPPT controller. Here, the early steady-state response of the motor drive is reached along with attaining better speed response. Thus, the utilization and efficiency of the system is improved much with the proposed ANFIS-based MPPT controller.

ACKNOWLEDGEMENT

The solar-based research project was funded by SERB, Department of Science and Technology (DST), Government of India with vides SERB Order No: SERB/ET-069/2013.

REFERENCES

Abdessamia, E., M. Dhafer, & M. Abdelkader (2012). A maximum power point tracking method based on artificial neural network for a PV system. *International Journal of Advances in Engineering and Technology. 5(1)*, 130–140.

Adel Mellit, Soteris A., & Kalogirou (2008). Artificial Intelligence Techniques for Photovoltaic Applications. *A Review Progress in Energy and Combustion Science. 34*, 574–632.

Adly, M., M. Ibrahim, & H. El Sherif (2012, March). *Comparative study of improved energy generation maximization techniques for photovoltaic systems.* 2012 IEEE Asia-Pacific Power and Energy Engineering Conference (APPEEC), Shanghai, 1–5.

Aleenejad, M., H. Iman-Eini, & S. Farhangi (2012, May). *A minimum loss switching method using space vector modulation for cascaded H-bridge multilevel inverter.* IEEE 20th Iranian Conference on Electrical Engineering (ICEE), Tehran, 546–551.

Dong Hwa Kim (2007). GA-PSO based vector control of indirect three phase induction motor. *Elsevier Science Direct Applied Soft Computing. 7(2)*, 601–611.

Durga Sukumar, Jayachandranath Jitendranath, & Suman Saranu (2014). Three-level Inverter-fed Induction Motor Drive Performance Improvement with Neuro-fuzzy Space Vector Modulation. *Electrical Power Components and Systems-Taylor & Francis. 42(15)*, 1633–1646.

Farhat, M., & S. Lassaad (2012, March). *Advanced ANFIS-MPPT control algorithm for sunshine photovoltaic pumping systems.* 2012 First International Conference on Renewable Energies and Vehicular Technology, Hammamet, 167–172.

Haitham, A., I. Atif, A. Moin, Z. Fang, Yuan Li, & B. Ge (2013). Quasi-Z-source inverter-based photovoltaic generation system with maximum power tracking control using ANFIS. *IEEE Transactions on Sustainable Energy. 4(1)*, 11–20.

Hu, Y., J. Liu, & B. Liu (2012, January). *A MPPT control method of PV system based on fuzzy logic and particle swarm optimization.* Second IEEE International Conference on Intelligent System Design and Engineering Application (ISDEA), Sanya, Hainan, 73–75.

Iqbali, A., H. Abu-Rub, & Sk. Ahmed (2010, December). *Adaptive neuro-fuzzy inference system based maximum power point tracking of a solar PV module.* IEEE International Energy Conference and Exhibition (EnergyCon), Manama, 51–56.

Jang, J. S. (1993). ANFIS adaptive-network-based fuzzy inference system. *IEEE Transactions on Systems, Man, and Cybernetics. 23(3)*, 665–685.

Jang, J. S., C. T. Sun, & E. Mizutani (1997). Neuro-Fuzzy and Soft Computing—A Computational Approach to Learning and Machine Intelligence. *IEEE Transactions on Automatic Control. 42(10)*, 1482–1484.

Joshi, J. J., P. Karthick, & R. S. Kumar (2013, April). *A solar panel connected multilevel inverter with SVM using fuzzy logic controller.* IEEE International Conference on Energy Efficient Technologies for Sustainability (ICEETS), Nagercoil, 1201–1206.

Mbarushimana, A., & Xin Ai (2011, July). *Real time digital simulation of PWM converter control for grid integration of renewable energy with enhanced power quality.* IEEE 4th International Conference on Electric Utility Deregulation and Restructing and Power Technologies (DRPT), Weihai, Shandong, 712–718.

Mellit, A., & S. A. Kalogirou (2006, November). *Neuro-fuzzy based modeling for photovoltaic power supply system.* IEEE First International Power and Energy Conference, Putra Jaya, 88–93.

Pakkiraiah, B., & G. Durga Sukumar (2015, November). *A New Modified MPPT Controller for Solar Photovoltaic System.* 2015 IEEE International Conference on Research in Computational Intelligence and Communication Networks (ICRCICN), Kolkata, West Bengal, India, 294–299.

Pakkiraiah, B., & G. Durga Sukumar (2016). A New Modified MPPT Controller for Improved Performance of an Asynchronous Motor Drive under Variable Irradiance and Variable Temperature. *International Journal of Computers and Applications-Taylor & Francis.* 1–14.

Pakkiraiah, B., & G. Durga Sukumar (2016). Research Survey on Various MPPT Performance Issues to Improve the Solar PV System Efficiency. *Journal of Solar Energy. 2016(2016)*, 1–20.

Qiang Mei, Mingwei Shan, Liying Liu, & M. Josep Guerrero (2011). A novel improved variable step-size incremental-resistance MPPT method for PV systems. *IEEE Transactions on Industrial Electronics. 58(6)*, 2427–2434.

Roberto Faranda, & Sonia Leva (2008). Energy Comparison of MPPT techniques for PV systems. *WSEAS Transactions on Power Systems. 3(6)*, 446–455.

Saadi, A., & A. Moussi (2003). Neural Network use in the MPPT of photovoltaic pumping system. *Rev. Energy. Ren.: ICPWE.* 39–45.

Sreeja, C., & S. Arun (2011, December). *A novel control algorithm for three phase multilevel inverter using SVM.* IEEE PES Innovative Smart Grid Technologies-India (ISGT India), Kollam, Kerala, 262–267.

Theodoros, L., Yiannis, S., & Athanassios, D. (2006). New maximum power point tracker for PV arrays using fuzzy controller in close cooperation with fuzzy cognitive networks. *IEEE Transactions on Energy Conversion. 21(3)*, 793–803.

Computer, Communication and Electrical Technology – Guha, Chakraborty & Dutta (Eds)
© 2017 Taylor & Francis Group, ISBN 978-1-138-03157-9

Study of nonlinear phenomena in a free-running current controlled Ćuk converter

P. Chaudhuri & S. Parui
*Department of Electrical Engineering, Indian Institute of Engineering Science and Technology, Shibpur,
Howrah, West Bengal, India*

ABSTRACT: The nonlinear phenomena have been explored in a free-running current controlled Ćuk Converter. At first, the study has been conducted with regulated dc power supply input. As in most of the cases, the supply to the converter is from a rectified dc source, our study is then extended to rectified dc input and the changes in the nonlinear phenomena have also been explored when the input to the converter is a rectified dc source instead of a dc regulated power supply.

1 INTRODUCTION

A detailed exploration of power electronic systems deals with the study of nonlinear dynamics. It has already been explored in different published literatures that such power electronic converters are prone to nonlinear phenomena like bifurcation, chaos, sub-harmonics etc. In (Iu, Lai, & Tse 2000) the nonlinear dynamics of Ćuk converter is studied. It is observed that the system loses stability via Hopf bifurcation as stable spiral develops into an unstable spiral in the locality of the equilibrium point. Further cycle-by-cycle computer simulations done, by varying the circuital control parameters, to see the system developing into limit cycle as it loses stability, and further develops into quasi-periodic and chaotic orbits. The occurrence of bifurcation and chaotic behaviour in dc-dc autonomous converters is reported first in this paper. In (Tse, lai, & Iu 1998) as well, it is shown that the system loses its stability via Hopf bifurcation. Observations revealed that at small values of k (control parameter), the trajectory spirals into a fixed period-1 orbit, with further increase of k, period-1 becomes unstable leading to outward spiralling of trajectory and settling into limit cycle. For larger k, a Poincaré section indicates quasi periodic orbit and with further increase in k chaos is observed. In (Daho, Giaonris, Zahawi, Picker, & Banerjee 2008), Filippovs method is employed to investigate the stability of an autonomous Ćuk converter with hysteresis current controller. Here it is seen that the converter loses stability via Neimark Bifurcation. Non-linear dynamic behaviour in a Zero Average Dynamics (ZAD) control is investigated here in (Deivasundari, Uma, & Ashita 2013). Moment matching technique is implemented to obtain reduced order model, for computing ZAD control parameters. Here it is

shown, even for small change in control parameters, the system exhibits period-doubling bifurcation. It is also shown that the onset of chaos can be delayed by including a time delay component in ZAD control strategy. In (Fuad, de Koning, & van der Woude 2004), the authors have used multi-frequency averaging as a generalisation of state space averaging method to analyse different stability aspects of open loop as well as closed loop converter. In (Iu & Tse 2000), two Ćuk converters, connected in a well-known drive response configuration, operating under free-running current-mode control is considered, in order to study the synchronization property of a chaotically operated system. Here it is first mathematically shown that the Conditional Lyapunov Exponents (CLEs) of the coupled system under study are negative, and therefore proven that synchronization of such systems are possible. This paper for the first time has highlighted the synchronization phenomenon in power electronic converters. In all the above mentioned papers, dynamics of the system is explored, while being fed from regulated dc supply. But in reality, most of the time the converters are fed from rectifiers instead of regulated power supply and the input voltage will contain ripple if the input is from a rectifier. So, there will be changes in behaviour when the supply is from a rectified dc voltage source. Hence this paper deals with the modelling of an autonomous current controlled Ćuk converter and then making a comparative study of dynamic behaviour of the system, when fed with either of the supplies.

2 SYSTEM DESCRIPTION

System under study consists of a Ćuk converter being operated by a free-running hysteretic current

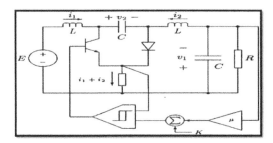

Figure 1. Schematic diagram of Ćuk converter under Hystereis controller.

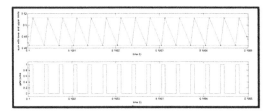

Figure 2. A sample plot of sum of the inductor currents, i.e, the switch current and the gate pulse of the switch.

mode control (Figure 1). Turning on and off of the switch is done in a hysteretic fashion, based on the values of sum of inductor currents, $I_{sum} = (i_1 + i_2)$, falling below and above a certain preset hysteretic band (Figure 2). The governing control equation of the hysteresis controller is given by $(i_1 + i_2) = g(\upsilon_1)$, where, i_1 and i_2 are the inductor currents respectively, υ_1 is the output voltage, $g(.)$ is the control function (Tse, lai, & Iu 1998). A simple proportional control takes the form $\Delta(i_1 + i_2) = -\mu\Delta\upsilon_1$, μ being the gain factor. The following equivalent form of the above equation, assuming regulated output is given by, $(i_1 + i_2) = k - \mu\Delta\upsilon_1$, where k and μ are control parameters.

3 SIMULATION RESULTS

The modelling of the system done with the following values of the parameters: $E = 30$ V, $L = 0.01$ H, $C = 100$ μ F, $R = 25$ Ω, $k = 0.4$. Trajectory plotting of the same, using output voltage (υ_1), voltage across input capacitor (υ) and inductor current through L_1 (i_{L1}) is done for different sets of parameter values. When the input is from a dc regulated power supply, $E = V_{dc}$ and if the input is from a rectifier, the input voltage will contain a dc component along with a ripple component of the voltage, $E = V_{dc} + V_m sin(n\omega t + \theta)$, where V_m is the peak value of the ripple component, n is the order

of harmonic present in the output of the rectifier, ω is the angular frequency of the ac supply, θ is the angle between the instant of initial switching of the converter and zero crossing of the ripple voltage. For our model, it has been assumed that the input is fed from a single phase diode rectifier with maximum 10% ripple peak of 100 Hz. So $n = 2$, $V_m = 0.1$ V_{dc}, $\omega = 314$ rad/s, and θ has been taken as zero. Now, analysis of the system behaviour is done using both a regulated dc supply and rectifier input respectively and changes in the nature of the trajectory is marked.

3.1 For k = 1

Keeping all the other controlling and converter parameters fixed and varying value of k, system trajectories are plotted. With regulated dc supply, a stable limit cycle is obtained. We see from Figures 3(a),(b) that almost steady dc components of individual i_{L1} and i_{L1} obtained. Proper switching is exhibited with I_{sum} remaining within the hysteresis band throughout. The same study is done with a rectified dc supply shows that an additional ripple content of 10% is there in the input voltage. $V_{dc} = 30$, $V_m = 10\%$ of 30 V, i.e. 3 V, value chosen for $n = 2$, i.e. it is assumed that an additional ripple peak of 3 V of 100 Hz is fed from a single phase diode rectifier input. The trajectory in Figure 4 shows the occurrence of a limit cycle in this case even though having a different structural pattern. Reflection of switching frequency in the structure, being superimposed on the ripple frequency is observed in the structure of limit cycle. From Figure 5(a) we see ripple content of the input is reflected in the output quantities. Second order harmonic, ripple frequency of 100 Hz superimposed on the waveform,

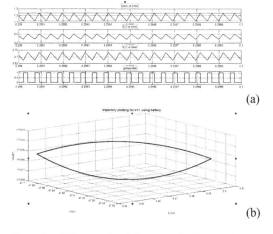

(a)

(b)

Figure 3. Using regulated dc source, $k = 1$ (a) trajectory plotting for $k = 1$, (b) time plot of I_{sum}, i_{L1}, i_{L2}.

358

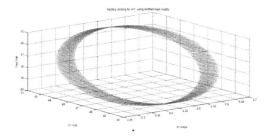

Figure 4. Trajectory plotting for $k = 1$ using rectified dc source.

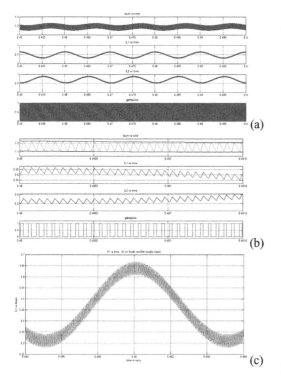

Figure 5. Using rectified dc voltage source for $k = 1$, time plots for (a) I_{sum}, i_{L1}, i_{L2}, (b) I_{sum}, i_{L1}, i_{L2} using extended scale, (c) i_{L1}.

which is reflected in the output waveforms, of the individual components as well as in the band envelope. The hysteresis band, formed by the upper and lower limits respectively is also remaining bounded within the envelope consisting of 100 Hz frequency. Figure 5(b) is plotted, by enlarging the time shows the switching frequency of I_{sum}, along with i_{L1} and i_{L2} individually respectively. Figure 5(c) is used to show one half cycle (corresponding to 100 Hz) of the above waveform.

3.2 For $k = 4$

We see from Figure 6(a) that a limit cycle is obtained with dc regulated supply, the structure depicting the occurrence of oscillatory current, in repetitive manner. Figure 6(b) depicts I_{sum} and the individual inductor currents respectively. In (Parui & Basak 2014) it is pointed that due to chosen circuit parameters in a non-autonomous current controlled Ćuk converter, i_{L1} and i_{L2} be oscillatory. In this case, a current component of high frequency (at switching frequency) is superimposed on a sinusoidal current component of comparatively lower frequency decided by the circuit L and C values. Figure 7 shows the path (shown in dotted lines) traversed by the oscillatory current (Parui & Basak 2014). The frequency of this oscillatory current, is given by

$$f_{osc} = 1/(2\pi\sqrt{L_{eq}.C_{eq}}) \tag{1}$$

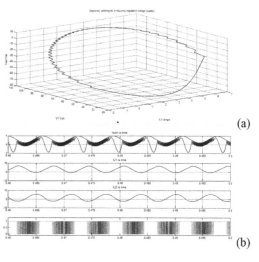

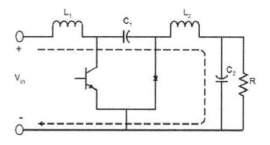

Figure 6. (a) Trajectory plot for $k = 4$ using regulated voltage source, (b) Time plot of I_{sum}, i_{L1}, i_{L2} for $k = 4$ (Regulated dc voltage source).

Figure 7. Path of LC oscillatory current.

where, $L_{eq} = L_1 + L_2$ (2)

$$C_{eq} = C_1 . C_2 / (C_1 + C_2) \qquad (3)$$

The above mentioned oscillatory component of current is termed as LC oscillation current. In (Wong, Wu, & Tse 2008), a slow scale oscillation has been reported in Ćuk converter, but no quantitative information and reason for the onset of such oscillation are available regarding the oscillation frequency. As shown in Figure 7, we see that with the chosen values for capacitor and inductor respectively, an LC oscillatory current is generated over here as well in free running current controlled Ćuk converter. I_{sum} eventually leaves the band envelope in a periodic manner. Theoretically frequency of LC oscillatory current should be (from (1)) = 160 Hz. Frequency obtained from time plot = 166.67 Hz. The calculated frequency of the oscillatory current is approximately equal to the frequency as obtained from the time plot of the individual waveforms.

Figure 8 shows the trajectory with rectified dc source as input. Here we see that multiple non-overlapping loops with similar structure is obtained. Because of the existence of the input ripple, there is a disruption in the repetitive occurrence of a single structure, giving rise to a complex dynamics in the state system. The reason for such complexity can be understood from the time plots of inductor currents and switch current, I_{sum}.

Figure 9(a) consists of time plot of I_{sum} and the individual currents respectively, when a rectified dc voltage source is used, for the value of $k = 4$. Here we see that I_{sum} fails to remain within the band for a continuous period of time in a similar manner as found with dc regulated power supply. But, here the input ripple is changing the oscillation frequency which is not same as obtained with mathematical expression given in (1). Now oscillation frequency is found to be 150 Hz which is neither 100 Hz, nor 160 Hz. Figure 9(b) shows the blow up

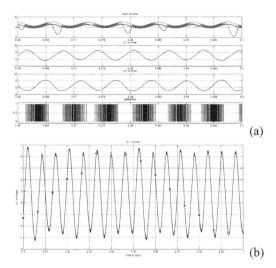

Figure 9. Using rectified dc voltage source for $k = 4$ (a) Time plot for I_{sum}, i_{L1}, i_{L2}, (b) Time plot for i_{L1}.

of the time plot of i_{L1}, for $k = 4$, using rectified dc voltage source. Here we see that the faster switching dynamics of i_{L1} is aperiodic and it is riding on the LC oscillation of much lower frequency. In a non-autonomous system, there is a fixed switching frequency, so we can refer the occurrence of quasi-periodicity or phase locking due to the interaction of switching frequency and LC oscillation frequency. But here as it is an autonomous system, the switching frequency in not fixed. So we can not define it as a quasi-periodic orbit, but it is giving rise to a complex trajectory in the state space similar to a quasi-periodic attractor in a non-autonomous system.

4 CONCLUSION

The nonlinear phenomena have been observed with regulated dc input as well as with rectified dc input. A stable limit cycle is observed for lower values of k with regulated dc input. With the input from a single phase rectifier, there is a 2nd order harmonic component (because of 100 Hz input ripple) in the current waveforms as well as in the band envelope. Oscillatory behaviour is observed for values of k above 4. LC oscillation is been generated with dc regulated voltage supply, compelling the switch current to come out of the hysteresis band. Whereas, when rectified DC voltage source is been used, the input voltage ripple affects the LC oscillation and a third frequency (neither LC oscillation frequency nor 100 Hz ripple) is reflected at the output.

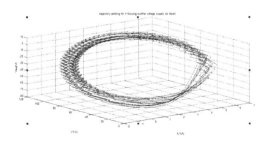

Figure 8. Trajectory plot for $k = 4$ using rectified dc voltage source.

REFERENCES

Daho, I., D. Giaonris, B. Zahawi, V. Picker, & S. Banerjee (2008). Stability analysis and bifurcation control of hysteresis current controlled Cuk converter using filippov's method. In *4th IET Conference on Power Electronics, Machines and Drives (PEMD)*.

Deivasundari, P., G. Uma, & S. Ashita (2013). Chaotic dynamics of a zero average dynamics controlled DC-DC Cuk converter. *IET Power Electronics 7*(2), 289–298.

Fuad, Y., W. L. de Koning, & J.W. van der Woude (2004). On the stability of the pulse width-modulated Cúk converter. *IEEE Transactions on Circuits & Systems-II 51*(8), 412–420.

Iu, H. H. C. & C. K. Tse (2000). A study of synchronization in chaotic autonomous Cúk DC/DC converters.

IEEE Transactions on Circuits & Systems–I 47(6), 913–918.

Iu, H. H. C., Y. M. Lai, & C. K. Tse (2000). Hopf bifurcation and chaos in a free-running current-controlled Cúk switching regulator. *IEEE Transactions on Circuits & Systems–I 47*(4), 448–457.

Parui, S. & B. Basak (2014). Inherent oscillation in Cúk converter and its effect on bifurcation phenomena. In *IEEE TENCON*.

Tse, C. K., Y. lai, & H. H. C. Iu (1998). Hopf bifurcation and chaos in a hysteretic current-controlled Cúk regulator. In *IEEE Power Electronics Specialists' Conference*.

Wong, S. C., X. Wu, & C. K. Tse (2008). Sustained slow-scale oscillation in higher order current-mode controlled converter. *IEEE Transactions on Circuits and Systems-II 55*(5), 489–493.

Computer, Communication and Electrical Technology – Guha, Chakraborty & Dutta (Eds)
© 2017 Taylor & Francis Group, ISBN 978-1-138-03157-9

Study of the power fluctuation of the DC motor in chaotic and non-chaotic drive systems

Mriganka Roy, Partha Roy & Samar Bhattacharya
Department of Electrical Engineering, Jadavpur University, Kolkata, India

ABSTRACT: Chaos in a fixed-frequency DC chopper-fed PMDC drive system is identified with variations of its parameters. The dynamical system is described by the state space representation in the continuous conduction mode. The system shows different chaotic behaviors at different values of switching frequency (f_s). The chaotic and non-chaotic region is separated by the boundary, which is obtained at error amplifier gain (g) against switching frequency (f_s) and a mathematical relationship is built. A special observation shows that the power wave fluctuations of the selected PMDC motor become less when the system is in chaotic zone. The study is helpful in selecting the range and combination of parameters to run the motor as per the requirement.

1 INTRODUCTION

Chaos is often mixed up with disorder or even random, but it is disordered and its random-like behavior is governed by a rule. Chaos has some typical features like nonlinearity determinism, dependence on initial conditions, and aperiodicity. The study of chaos theory and chaotic systems is called chaology. The mathematical definition of chaos was introduced by Tien-Yien Li and James A. Yorke in 1975. The electrical motor and the speed and current transducers of the power converter circuit and control electronics are referred to as an electrical drive. Power electronics switches are used for both the speed and torque control of electrical motor. The PWM switching action makes the entire electrical drive system time-varying and nonlinear. Chaos was identified in power electronics in the late 1980s. The occurrence of chaos in a simple buck converter by using iterated nonlinear mappings was analyzed by Hamill, Deane, and Jefferies in 1992, which is considered as the milestone of investigations of chaos in power electronics. Chaos in electric drive systems was first identified in induction drive systems in 1989. In 1997, the chaotic behavior in a simple DC drive system was reported. It could be conducted in voltage mode-controlled operation and current mode-controlled operation.

In voltage mode control, the nonlinear dynamics and chaotic behavior of industrial drive systems have been investigated both numerically and analytically (Chen et al. 1997, Chau et al. 2010, Chakrabarty et al. 2015). It is operated in a fixed-frequency continuous conduction mode.

PWM full-bridge converter system exhibits rich period-doubling route to chaos at forward and reverse rotating conditions, respectively (Tang et al. 2006, Okafor et al. 2010). Chaotic behavior and the coexistence of multiple stable attractors were investigated in DC series motor with different switching element (Chakrabarty et al. 2013). A current mode buck-type DC chopper-fed PMDC drive is considered for case study (Chau et al. 1997, Basak et al. 2009, Chakrabarty et al. 2013). A separately excited DC motor drive is considered (Chen et al. 2000). The nonlinear dynamics of both voltage mode- and current mode-controlled operations has been reported in DC drive systems during the continuous conduction mode of operation (Dai et al. 2011, Roy et al. 2014). In 2000, the research was extended to the stabilization of chaos in DC drive systems by using a time delay feedback control.

In this paper, the chaotic behavior in simple PMDC drive in continuous conduction mode is identified. A 12 V DC chopper-fed PMDC drive system is considered in this work without ignoring switching effect. The effect of switching frequency (f_s) on the dynamic behavior of the PMDC motor is observed and a mathematical relationship is built. It is seen that the power wave fluctuations of the DC motor are less when the drive system is in chaotic condition.

The rest of this paper is organized as follows. In section 2, we describe and present the dynamic model of simple PMDC motor drive system. In section 3, we show the results with variation of switching frequency (f_s) and observe the power wave fluctuations in chaotic and non-chaotic condition. This paper ends with section 4 presenting the conclusion and future scopes of research of our works.

2 SYSTEM DESCRIPTION AND DYNAMIC MODEL

The schematic diagram and the equivalent circuit model of voltage mode-controlled PMDC drive system are shown in Fig. 1. The DC drive is operated by a DC voltage and MOSFET is used as a switch in the chopper.

The speed of the motor $\omega(t)$ is sensed by the speed sensor and compared with the reference speed (ω_{ref}) at the speed error amplifier gain, and the proportional controller signal can be expressed as:

$$V_c = g(\omega(t) - \omega_{ref}) \quad (1)$$

The ramp generator voltage can be expressed as:

$$V_r = v_l + \frac{(v_u - v_l)}{T} t \quad (2)$$

where v_u and v_l are the upper and lower voltages of the ramp signal, respectively, and T is the time period. Then, both V_r and V_c are fed to the comparator, whose output signal goes to power switch (S). The switch S will be in "ON" and the diode D is "OFF" until exceeds, otherwise S is in "OFF" and the diode is "ON". Thus, the system is operated in two stages.

State 1:   When $V_c \geq V_r$. Switch (S) is in OFF, hence Diode (D) becomes ON:

$$\begin{bmatrix} \dfrac{d\omega(t)}{dt} \\ \dfrac{di(t)}{dt} \end{bmatrix} = \begin{bmatrix} \dfrac{-B}{J} & \dfrac{K_T}{J} \\ \dfrac{-K_E}{L} & \dfrac{-R}{L} \end{bmatrix} \begin{bmatrix} \omega(t) \\ i(t) \end{bmatrix} + \begin{bmatrix} \dfrac{T_L}{J} \\ 0 \end{bmatrix} \quad (3)$$

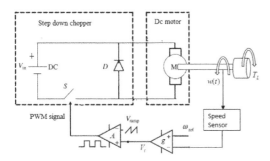

Figure 1.   Block diagram of DC drive.

State 2:   When $V_c < V_r$. Switch (S) is in ON, hence Diode (D) becomes OFF

$$\begin{bmatrix} \dfrac{dw(t)}{dt} \\ \dfrac{di(t)}{dt} \end{bmatrix} = \begin{bmatrix} \dfrac{-B}{J} & \dfrac{K_T}{J} \\ \dfrac{-K_E}{L} & \dfrac{-R}{L} \end{bmatrix} \begin{bmatrix} \omega(t) \\ i(t) \end{bmatrix} + \begin{bmatrix} \dfrac{T_L}{J} \\ \dfrac{V_{in}}{L} \end{bmatrix} \quad (4)$$

where i is armature current, R is armature resistance, L is armature inductance, DC supply voltage, back-EMF constant, K_T is torque constant, B is viscous damping, J is load inertia, and T_L is load torque. Therefore, the system equation can be written in the following sequence of state equations:

$$\dot{X}(t) = A_1 X(t) + B_1 E \quad \text{for } V_c \geq V_r \quad (5)$$

$$\dot{X}(t) = A_2 X(t) + B_2 E \quad \text{for } V_c < V_r \quad (6)$$

These system equations can be rewritten as:

$$\dot{X}(t) = A_K X(t) + B_K E \quad (\text{where } \mathbf{K} = 1,2) \quad (7)$$

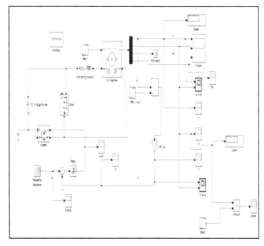

Figure 2.   Simulation model.

Table 1.   Parameter Values of the Motor.

Motor voltage constant K_E	0.05 (volts-s/rad)
Motor torque constant K_T	0.05 (Nm/amp)
Motor resistance R	0.5 (ohms)
Motor inductance L	1.5 (mH)
Motor inertia J	0.00025 (Nm/rad/s^2)
Damping coefficient B	0.0001(Nm/rad/s)
Reference speed ω_{ref}	100 rad/s

This system equation given by (7) is a time-varying state equation. Thus, this drive system is a second-order non-autonomous dynamical system. We simulate this model using Table 1 [8].

3 SIMULATION RESULTS

In this part, we have observed the effect of switching frequency (f_s) on the dynamic behavior of the PMDC motor and the power wave fluctuations of the DC motor in chaotic and non-chaotic conditions. MAT-LAB SIMULINK is used to carry out the study at a reference speed of 100 rad/s (Table 1) [8].

The chaotic behavior is identified on the selected PMDC motor drive at g = 1.2 and keeping Load Torque (T_L) = 0.3 Nm and switching frequency (f_s) = 100 Hz. Then, the behavior of that particular drives is changed while switching frequency (f_s) is varied, but gain (g) and Load Torque (T_L) are kept constant.

It is evident from Figure 3 that when switching frequency (f_s) is varied, the dynamic behavior the DC drive is changed. Chaotic behavior is observed in the switching frequency (f_s) range of 100 to 200 Hz (Figure 1). Further variation of switching frequency (f_s), that is, from 250 to 350 Hz (Figure 4) and 450 Hz (Figure 5H), leads to period doubling. The nominal orbit or Period-1 is observed at switching frequencies (f_s) of 400 Hz (Figure 5G) and 500 Hz (Figure 5I).

The chaotic boundary is calculated for different switching frequencies (f_s) according to Table 2.

The chaotic and non-chaotic region is separated by the boundary in Figure 6.

Figure 6 shows that if the error amplifier gain (g) is y and switching frequency (f_s) is x, then they are related to each other in a simple relationship given below:

Condition-1:
If $y \geq 4e^{-14}x^6 - 6e^{-11}x^5 + 4e^{-08}x^4 - 2e^{-05}x^3 + 0.0029x^2 - 0.2689x + 10.967$

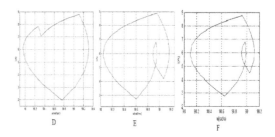

Figure 4. Phase portrait of motor speed and armature current at different switching frequencies (f_s) (D) f_s = 250 Hz, (E) f_s = 300 Hz, (F) f_s = 350 Hz.

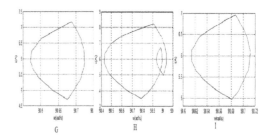

Figure 5. Phase portrait of motor speed and armature current at different switching frequencies (f_s) (G) f_s = 400 Hz, (H) f_s = 450 Hz, (I) f_s = 500 Hz.

Table 2. Occurrence of chaos at different switching frequencies (f_s) and error amplifier gain (g).

Switching frequency (f_s)	Error amplifier gain (g)	Switching frequency (f_s)	Error amplifier gain (g)
100 Hz	1.2	350 Hz	2.1
150 Hz	1.2	400 Hz	2.8
200 Hz	1.2	450 Hz	3.2
250 Hz	1.5	500 Hz	4.1
300 Hz	1.6		

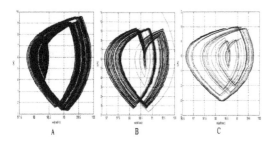

Figure 3. Phase portrait of motor speed and armature current at different switching frequencies (f_s) (A) f_s = 100 Hz, (B) f_s = 150 Hz, (C) f_s = 200 Hz.

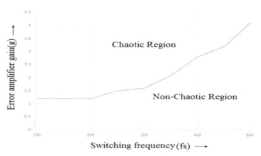

Figure 6. Error amplifier gain (g) versus switching frequency (f_s) to identify chaotic and non-chaotic regions.

365

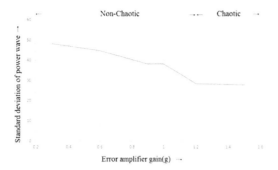

Figure 7. Standard deviation of power wave in chaotic and non-chaotic conditions ($f_s = 100$ Hz and $T_L = 0.30$ Nm).

Case-1: at switching frequency (f_s)=100 Hz

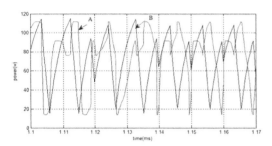

Figure 8. (A) Chaotic and (B) non-chaotic power waveforms.

Case-2: at switching frequency (f_s) =150 Hz

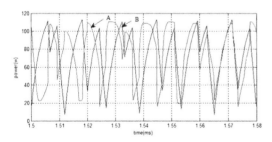

Figure 9. (A) Chaotic and (B) non-chaotic power waveforms.

Then, the system turns out to be a Chaotic one.
Condition-2:
If $y < 4e^{-14}x^6 - 6e^{-11}x^5 + 4e^{-08}x^4 - 2e^{-05}x^3 + 0.0029x^2 - 0.2689x + 10.967$
Then, the system becomes non-Chaotic.
Two mathematical tools, average value and stadrad deviation, are used to observe the power

Case-3: at switching frequency (f_s)=200 Hz

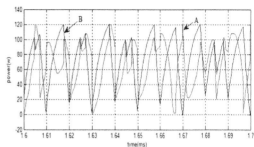

Figure 10. (A) Chaotic and (B) non-chaotic power waveforms.

Case-4: at switching frequency (f_s)=250 Hz

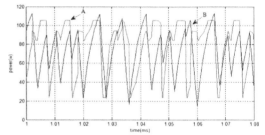

Figure 11. (A) Non-chaotic and (B) chaotic speed waveforms.

Table 3. Average power and standard deviation of power wave.

Case	Gain (g)	Condition	Average power (W)	Standard deviation
1	1	Non-Chaotic	73.80809	30.73599
	1.2	Chaotic	71.55887	29.5689
2	1	Non-Chaotic	73.64423	32.31095
	1.2	Chaotic	71.69843	30.15689
3	1	Non-Chaotic	73.56	31.32491
	1.2	Chaotic	68.35	29.2568
4	1.2	Non-Chaotic	78.77	28.1131
	1.5	Chaotic	72.952	25.9128

wave fluctuation of the selected PMDC motor at chaotic and non-chaotic conditons.

Four different cases studies are conducted at different switching frequencies (f_s) to observe the power wave fluctuations of the PMDC motor at Chaotic and Non-Chaotic conditons.

Table 3 shows that the average power and standard deviation of the power wave become less when the system is in Chaotic zone.

4 CONCLUSION

In this paper, the effect of switching frequency (f_s) on the dynamic behavior of electrical drive systems is presented. It is reported that the dynamic behavior of the PMDC drives is changed with the switching frequency (f_s). The chaotic and non-chaotic region is separated by the boundary obtained at error amplifier gain (g) against switching frequency (f_s). A special observation shows that the average value and standard deviation of the power wave fluctuations become less when the system is in chaotic zone. This phenomenon is investigated at different switching frequencies (f_s) to justify the realization. This realization can make chaos a worthy phenomenon in a PMDC Motor Drives system.

REFERENCES

Basak, B., & S. Parui, (2009) "*Bifurcation and Chaos in Current Mode Controlled DC Drives In Continuous and Discontinuous Conduction Mode of Operation,*" IEEE Transactions on Power Electronics and Drives, pp. 4244–4247.

Chakrabarty, K., UrmilaKar, (2015) "*Stabilization of unstable periodic orbits in DC drives*", IEEE Transactions On ICEEICT.

Chakrabarty, K., U. Kar & SusmitaKundu, (2013) "*Control of Chaos in Current Controlled DC Drives*" J CIRCUIT SYST COMP 22, 1350035.

Chau, K.T., J.H. Chen, & C.C. Chan, (1997) "*Dynamic bifurcation in dc drives,*" In Proceedings of IEEE Power Electronics Specialists Conference, St. Louis, USA, pp. 1330–1336.

Chau, K.T., J.H. Chen, C.C. Chan, J.K.H. Pong, & D.T.W. Chan, (1997) "*Chaotic behavior in a simple dc drive,*" In Proceedings of IEEE Power Electronics and Drive Systems, Singapore, pp. 473–479.

Chen, J.H., K.T. Chau & C.C. Chan, (2010) "*Chaos in voltage-mode controlled DC drive systems,*" Taylor & Francis Transaction on International Journal of Electronics, vol. 86, no. 7, pp. 857–874.

Chen, J.H. & K.T. Chau, (2000) "*Analysis of Chaos in Current-Mode-Controlled DC Drive Systems*", IEEE Transactions On Industrial Electronics, vol. 47, No. 1.

Dai, D., X. Maa, B. Zhang & Chi K. Tse, (2011), "*Hopf bifurcation and chaos from torus breakdown in voltage-mode controlled DC drive systems*", ELSEVIER, Chaos, Solitons & Fractals, vol. 44.

Kundu, S., U. Kar, & K. Chakrabarty, (2013), "*Co-existence of multiple attractors in the PWM controlled DC drives,*" In the European Physical Journal Special Topics, pp. 699–709.

Okafor, N., B. Zahawi & D. Giaouris & S. Banerjee, (2010) "*Chaos, Coexisting Attractors, and Fractal Basin Boundaries in DC Drives with Full-Bridge Converter*", IEEE Transactions on Power Electronics and Drives, pp. 129–132.

Roy, P., S. Ray & S. Bhattacharya, (2014) "*Control of Chaos in Brushless DC Motor: Design of Adaptive Controller following Back-stepping Method,*" International Conference on Control, Instrumentation, Energy & Communication (CIEC), IEEE, Kolkata, pp. 91–95.

Tang, T., M. Yang, H. Li & D. Shen, (2006.) "*A New Discovery and Analysis on Chaos and Bifurcation in DC Motor Drive System with Full-bridge Converter*", IEEE Transactions on Power Electronics and Drives.

Author index

Agarwal, A. 105
Alam, S. 3
Anwer Askari, S.S. 189

Bandyopadhyay, R. 267
Bandyopadhyay, S. 313
Bandyopadhyay, S.K. 19
Banerjee, A. 33
Banik, B.G. 19
Basak, S. 125
Basu, A. 9
Bedeeuzzaman, M. 319
Begum, S. 43
Bera, J.N. 199
Bera, S. 273, 279
Bera, S.P. 43
Bera, T.K. 261, 273, 279
Bezboruah, T. 75
Bhattacharya, S. 245
Bhattacharya, S. 363
Bhattacharyya, A. 163
Bhattacharyya, N. 267
Bhattacharyya, P. 331
Bhattacherjee, S. 331
Bhowmik, S. 15
Bora, A. 75
Bose, S. 115
Brahma, N. 111

Chakraborty, A. 91
Chakraborty, B. 255, 261,
  273, 279
Chakraborty, D. 43
Chakraborty, M. 245
Chakraborty, S. 139, 209
Chatterjee, A. 9, 49, 209
Chatterjee, B. 203
Chatterjee, S. 337
Chattopadhyay, A.K. 97
Chattopadhyay, S. 195
Chaudhuri, P. 357
Chaudhuri, S. 313
Choudhury, H. 111
Chowdhury, A. 261, 279
Chowdhury, T. 255

Dalai, S. 203
Das, A. 313

Das, B. 175
Das, B.K. 27
Das, K. 239
Das, M. 33, 39
Das, M.K. 159, 185, 189
Das, S. 249, 255
Dasgupta, D. 91
Datta, A. 343
Datta, A. 343
Datta, L. 85
Datta, S. 9
Deb, K. 131
Deb, P. 245
Dey, D. 313, 337
Deyasi, A. 175, 181
Dhar, S. 3
Dubey, H. 221, 227
Durga Sukumar, G. 295,
  349
Dutta, H.S. 239, 255
Dutta, R. 181
Dutta, S. 39

Fathima, T. 319
Francis, F. 319

Ghorai, S. 67, 267, 325
Ghosal, A. 139
Ghosh, B. 153
Ghosh, D. 325
Ghosh, R. 203
Ghosh, S. 61
Ghosh, S. 125
Ghosh, S. 181
Ghoshal, D. 261, 279
Gope, S. 289
Goswami, A.K. 289
Goswami, S.K. 209
Guha Thakurta, P.K. 105
Gupta, S. 55

Halder, T. 307
Halsana, S. 33
Haque, N. 203
Hazra, S. 153
Hoque, MD.T. 307
Hore, A. 255
Hori, N. 15

Karmakar, R. 9
Karwa, V.K. 125
Khaliluzzaman, Md. 131
Kumar Chy, D. 131
Kumar, A. 203
Kumar, R. 163

Maity, S. 139
Majumder, K. 97
Mandal, D. 233
Mandal, S. 55
Manna, S. 121
Medhi, S. 75
Mishra, D. 215
Mohan, K. 215
Mohanty, A. 215, 221,
  227
Mohanty, S.P. 215
Mondal, A. 9
Mondal, A.K. 121
Mondal, R.K. 81
Mondal, S.K. 49
Mukhopadhyay, S. 153
Munshi, S. 337

Nag, A. 61, 97
Nagaraju, J. 273
Nandi, A. 143
Nandi, E. 81
Nandi, S. 143
Nandi, S.K. 149
Naresh Kumar, M.L. 261
Nasipuri, M. 15
Nath, N. 233
Nayak, S. 249

Pakkiraiah, B. 295, 349
Pal, A. 61
Pal, S. 199, 331
Panigrahi, B.K. 221, 227
Pareek, P. 185
Parui, S. 357
Pathak, N.N. 143
Pati, R. 61
Patra, S.N. 153
Paul, A. 139
Poddar, M.K. 19
Prasad, T.J. 169

Ranjan, R. 159
Ravi, N. 169
Ray, D.K. 195
Ray, P. 81
Ray, P.K. 215, 221, 227
Rout, P.K. 221, 227
Roy, A. 39
Roy, A. 245
Roy, M. 363
Roy, P. 27, 363
Roy, P. 139
Roy, S. 91
Roy, S. 163

Sadhu, S. 61
Saha, P. 3

Saha, P. 267
Sahoo, A. 33
Sahoo, P.K. 233
Santra, M. 3
Sarcar, P. 121
Sarddar, D. 81
Sarkar, M. 143
Sarkar, P.P. 153
Sarkar, R. 15, 43
Sarkar, S. 9
Sarkar, S. 331
Sarkar, T.S. 249, 255
Sen, P. 81
Sen, S. 15
Sengupta, S. 195, 199
Seth, A. 245

Sharma, B. 111
Sharma, K.D. 195
Shaw, V. 181
Singha Roy, S. 91
Sinha, A.K. 307
Supraja, B. 169
Sur, A. 115

Thakur, S. 3
Tiwari, P.K. 289
Tribedi, R. 115
Tudu, B. 49, 267

Verma, D. 105

Printed and bound by CPI Group (UK) Ltd, Croydon, CR0 4YY

18/10/2024

01776219-0006